BAYESIAN
BIOSTATISTICS

STATISTICS: Textbooks and Monographs

A Series Edited by

D. B. Owen, Founding Editor, 1972–1991

W. R. Schucany, Coordinating Editor
Department of Statistics
Southern Methodist University
Dallas, Texas

R. G. Cornell, Associate Editor
for Biostatistics
University of Michigan

W. J. Kennedy, Associate Editor
for Statistical Computing
Iowa State University

A. M. Kshirsagar, Associate Editor
for Multivariate Analysis and
Experimental Design
University of Michigan

E. G. Schilling, Associate Editor
for Statistical Quality Control
Rochester Institute of Technology

BAYESIAN BIOSTATISTICS

DONALD A. BERRY
*Institute of Statistics and Decision Sciences
and Comprehensive Cancer Center
Duke University
Durham, North Carolina*

DALENE K. STANGL
*Institute of Statistics and Decision Sciences
and Terry Sanford Institute of Public Policy
Duke University
Durham, North Carolina*

CRC Press
Taylor & Francis Group
Boca Raton London New York

CRC Press is an imprint of the
Taylor & Francis Group, an **informa** business
A TAYLOR & FRANCIS BOOK

CRC Press
Taylor & Francis Group
6000 Broken Sound Parkway NW, Suite 300
Boca Raton, FL 33487-2742

First issued in paperback 2019

ISBN-13: 978-0-8247-9334-0 (hbk)
ISBN-13: 978-0-367-40139-9 (pbk)

Library of Congress Cataloging-in-Publication Data

Bayesian biostatistics / edited by Donald A. Berry, Dalene K. Stangl.
   p. cm. — (Statistics, textbooks and monographs ; 151)
   Includes index.
   ISBN 0-8247-9334-X (hardcover : alk. paper)
   1. Medicine—Research—Statistical methods. 2. Bayesian
statistical decision theory. 3. Biometry. I. Berry, Donald A.
II. Stangl, Dalene K. II. Series.
R853.S7B38 1996
610'.72—dc20
                                                      95-51158
                                                      CIP

Visit the Taylor & Francis Web site at
http://www.taylorandfrancis.com

and the CRC Press Web site at
http://www.crcpress.com

Preface

Bayesian methods are useful in the design and analysis of health research. The goals of this book are to make this utility clear and to demonstrate Bayesian methods by presenting applications of the Bayesian approach to problems in the health sciences.

We write for applied statisticians, students of statistics and biostatistics, and others who use statistical methods in their professional life. Our objective is to teach, and we teach mostly by example. A typical chapter begins with real data sets and real scientific or medical questions. The authors then guide the reader through the development of methodology while relating to those examples.

Some chapters demonstrate the individual components of Bayesian analysis: how to quantify belief into a probability distribution (prior elicitation), how to choose a mathematical model (model specification), how to design studies that maximize the information gained, how to weigh ethical and resource constraints (design), how to update beliefs (posterior calculations), how to check models (model criticism and sensitivity analysis), and how best to convey empirical evidence to nonstatisticians (graphical display and predictive calculations). Other chapters employ several of these components in carrying out actual data analysis and solving decision problems. The main focus in each chapter is the substantive medical question.

The level of chapters ranges from elementary to advanced. Each chapter starts from first principles and proceeds to state-of-the-art techniques upon which there are many open research questions suitable for graduate projects and dissertations. We try to address controversies when they exist and make appropriate yet practical recommendations. These recommendations are derived with the notion that statisticians must be able to persuade nonstatisticians as to the correctness of their conclusions. To this end we accent pictorial presentations that are backed up by the appropriate mathematical analyses.

As a result of the many advances in computing, Bayesian methods have become easier to implement, and the scope and complexity of Bayesian applications have greatly increased. Many of the chapters implement modern developments in computation. When appropriate, the authors provide computer code for their analyses through StatLib, a public access library of computer software stored and maintained by the Department of Statistics at Carnegie Mellon University, Pittsburgh.

This book is primarily a reference. However, it is ideal as a supplemental text for graduate statistics and biostatistics courses. An undergraduate course in statistical theory and methods will provide the necessary background for most of the chapters. By selecting various chapter combinations, an instructor can use the book in a variety of courses. Each chapter could serve as a basis for a student project. In such a project, the student can present the analysis, think through the pros and cons of the methods, investigate other evidence that bears on the medical questions, and suggest improvements or further steps that could be carried out.

Donald A. Berry
Dalene K. Stangl

Contents

Contributors

Keith Abrams Department of Epidemiology and Public Health, University of Leicester, Leicester, England

James Albert Department of Mathematics and Statistics, Bowling Green State University, Bowling Green, Ohio

Mark Ancukiewicz Center for Health Policy, Research and Education, Duke University, Durham, North Carolina

Deborah Ashby Department of Public Health and Department of Statistics and Computational Mathematics, University of Liverpool, Liverpool, England

Donald A. Berry Institute of Statistics and Decision Sciences and Comprehensive Cancer Center, Duke University, Durham, North Carolina

Kathryn Chaloner Department of Applied Statistics, University of Minnesota, St. Paul, Minnesota

Siddhartha Chib Olin School of Business, Washington University, St. Louis, Missouri

Cindy L. Christiansen Department of Ambulatory Care and Prevention, Harvard Medical School, and Harvard Pilgrim Health Care, Boston, Massachusetts

Merlise Clyde Institute of Statistics and Decision Sciences, Duke University, Durham, North Carolina

Petros Dellaportas Department of Statistics, Athens University of Economics, Athens, Greece

Dennis O. Dixon Division of AIDS, National Institute of Allergy and Infectious Diseases, National Institutes of Health, Bethesda, Maryland

Roxane du Berger Division of Clinical Epidemiology, Department of Medicine, Montreal General Hospital, Montreal, Quebec, Canada

William DuMouchel Medical Informatics Department, Columbia University, New York, New York

Laurence S. Freedman National Cancer Institute, National Institutes of Health, Bethesda, Maryland

Boris Freidlin The Emmes Corporation, Potomac, Maryland

Stephen George Comprehensive Cancer Center, Division of Biometry, Duke University Medical School, Durham, North Carolina

Irwin Guttman Statistics Department, University of Toronto, Toronto, Ontario, Canada

Vic Hasselblad Center for Health Policy Research and Education, Duke University, Durham, North Carolina

Joan Houghton Cancer Research Campaign Clinical Trials Centre, King's College School of Medicine and Dentistry, London, England

Jane L. Hutton Department of Public Health and Department of Statistics and Computational Mathematics, University of Liverpool, Liverpool, England

Annie M. Jarabek Environmental Criteria and Correction Office, U.S. Environmental Protection Agency, Research Triangle Park, North Carolina

Lawrence Joseph Division of Clinical Epidemiology, Department of Medicine, Montreal General Hospital, and Department of Epidemiology and Biostatistics, McGill University, Montreal, Quebec, Canada

Joseph B. Kadane Department of Statistics, Carnegie Mellon University, Pittsburgh, Pennsylvania

Dennis Kinney McLean Hospital, Belmont, Massachusetts

Roger J. Lewis UCLA School of Medicine, Los Angeles, and Department of Emergency Medicine, Harbor–UCLA Medical Center, Torrance, California

Jane C. Lindsey Department of Biostatistics, Harvard School of Public Health, Boston, Massachusetts

Roseann M. Lyle Health Promotion Section, Purdue University, West Lafayette, Indiana

David Matchar Center for Health Policy, Research, and Education, Duke University, Durham, North Carolina

Carl N. Morris Department of Statistics, Harvard University, Cambridge, Massachusetts

Peter Müller Institute of Statistics and Decision Sciences, Duke University, Durham, North Carolina

Mahesh K. B. Parmar Medical Research Council Cancer Trials Office, Cambridge, England

Giovanni Parmigiani Institute of Statistics and Decision Sciences, Duke University, Durham, North Carolina

Amy Racine-Poon Biometrics Division, Ciba-Geigy, Basel, Switzerland

Jiang Qian Abbott Laboratories, Abbott Park, Illinois

Michael Racz Department of Mathematics and Statistics, State University of New York at Albany, Albany, New York

Adrian E. Raftery Department of Statistics, University of Washington, Seattle, Washington

Sylvia Richardson INSERM, Villejuif, France

Di Riley Cancer Research Campaign Clinical Trials Centre, King's College School of Medicine and Dentistry, London, England

Gary L. Rosner Duke University Medical Center, Durham, North Carolina

Louis M. Ryan Department of Biostatistics, Harvard School of Public Health, and Division of Biostatistics, Dana Farber Cancer Institute, Boston, Massachusetts

J. Sedransk Department of Statistics, Case Western Reserve University, Cleveland, Ohio

Richard Simon Biometrics Research Branch, National Cancer Institute, National Institutes of Health, Bethesda, Maryland

Mitchell J. Small Department of Civil and Environmental Engineering and Department of Engineering and Public Policy, Carnegie Mellon University, Pittsburgh, Pennsylvania

Adrian F. M. Smith   Department of Mathematics, Imperial College, London, England

Teresa C. Smith   Medical Research Council Biostatistics Unit, Institute of Public Health, Cambridge, England

David J. Spiegelhalter   Medical Research Council Biostatistics Unit, Institute of Public Health, Cambridge, England

Dalene K. Stangl   Institute of Statistics and Decision Sciences and Terry Sanford Institute of Public Policy, Duke University, Durham, North Carolina

David A. Stephens   Department of Mathematics, Imperial College, London, England

Jon Wakefield   Department of Mathematics, Imperial College, London, England

Christine Waternaux   Division of Biostatistics, Columbia University, New York, New York

Russell D. Wolfinger   SAS Institute, Inc., Cary, North Carolina

David B. Wolfson   Department of Mathematics and Statistics, McGill University, Montreal, Quebec, Canada

Lara J. Wolfson   Department of Statistics and Actuarial Science, University of Waterloo, Waterloo, Ontario, Canada

I
GENERAL OVERVIEW

1
Bayesian Methods in Health-Related Research

Donald A. Berry and Dalene K. Stangl Duke University, Durham, North Carolina

ABSTRACT

This book introduces the Bayesian approach to health-related research. The first chapter provides an overview of Bayesian methods, presenting key definitions and the basic framework that will be needed to understand remaining chapters. Key components of a Bayesian analysis include prior, posterior, and predictive distributions, likelihood functions, and utility functions. Each of these components is defined and discussed in this chapter. Because the overriding goal of all health research is to guide decision making, most discussion is embedded in a decision theoretic context. Differences between Bayesian and frequentist paradigms are highlighted, with special attention given to the difference in their approach to scientific reasoning and decision making.

1 INTRODUCTION

By far the most common statistical approach to designing and analyzing medical research is what is known as the frequentist or classical method. Frequentist methods pervade the statistical analyses in medical journals. The purpose of this chapter is to describe a Bayesian approach as an alternative or perhaps as a supplement. The two approaches focus so differently that they can be viewed as distinct paradigms. Yet, because both use laws of probability and empirical evidence, the distinction is poorly understood by nonstatisticians. The distinction is further blurred when frequentists and Bayesians act and think alike—which they are wont to do, despite "anti" rhetoric coming from both sides. Both approaches have good characteristics. The most important advantages of the Bayesian approach are an attitude more consistent and helpful in the accumulation of knowledge and an emphasis on prediction and decision making. Other advantages include more direct interpretations, use of all available information, and more flexible inferences. An example will help show important differences between the two approaches.

Suppose a clinical trial has been carried out to compare an experimental therapy and a standard therapy. A frequentist summary of a clinical trial is often reported as a p-value: the probability of results more extreme than those observed in the trial given the null hypothesis, H_0, that the two therapies have identical effectiveness. In symbols: p-value = $P(\text{DATA} \mid H_0)$; DATA refers to the totality of data more extreme than that observed. (The reason for including "more extreme" data is that the probability of any particular observation is usually tiny, regardless of the hypothesis assumed.) Another type of frequentist summary is the probability of results more extreme than those observed in the trial under a particular alternative hypothesis, H_A. For example, H_A might be that the new therapy increases survival by 20%. These probabilities depend in part on the results of the trial at hand and so are descriptors of the trial results. However, which data constitute the set considered more extreme depends on the way the trial was conducted. For example, Berry (1987) and Berger and Berry (1988a) show how various stopping rules result in different sets of data being considered more extreme.

A Bayesian presents the probability that the experimental and standard therapies are equally effective given the results of the trial at hand. In symbols: $P(H_0 \mid \text{data})$, where data refers to the data observed. Sometimes a Bayesian may find the posterior probability that the experimental therapy is at least 20% more effective than the standard, again given the trial's results. These are called *posterior probabilities* in that they apply *after* the trial. Yet another Bayesian conclusion is the probability distribution for the outcome of interest for the next patient. Probabilities that refer to future observations on individual patients or on finite sets of patients (again given the trial's results) are called *predictive probabilities*. So Bayesians use probabilities of hypotheses and probabilities of future observations given the study results, whereas frequentists use probabilities of sets of results given hypotheses.

The Bayesian approach is so natural that users often interpret frequentist measures as though they were Bayesian probabilities. For example, if the p-value for a therapy comparison is small, people tend to conclude that there probably is a difference in therapies. It is difficult to give a p-value any other interpretation! However, the p-value does not say how different the treatments are. With a large sample, a very small p-value can arise even when there is very little difference between treatments. On the other hand, with a small sample size a large p-value can occur even when there is a large difference between treatments.

The difference in the two paradigms is also apparent in the design of clinical trials. Consider choosing a sample size for the test of a new treatment. The usual frequentist approach is to select a "clinically" meaningful improvement of experimental over standard therapy, say 30%, and a statistical power, say 80%. Both choices are arbitrary, chosen at least in part to give a sample size that is acceptable to investigators and sponsors. In many, or even most, cases the actual sample size attained does not equal the sample size in the design. Reasonable frequentists would proceed as though the actual sample size was the designed size, but the corresponding sample space and resulting p-values and confidence intervals are incorrect. In Bayesian analysis, the choice of a particular sample size is unnecessary. Decisions to stop or continue the trial can be made at any time. The Bayesian paradigm allows similar flexibility in other aspects of the protocol as well.

1.1 Bayes' Theorem

Bayes' theorem is the basic tool of Bayesian analysis. It provides the means by which one learns from data. Given a prior state of knowledge or belief, it tells how to update beliefs based upon observations. To update belief about a hypothesis, H, Bayes' theorem is used to calculate the posterior probability of the hypothesis as follows:

$$P(H \mid \text{data}) = \frac{P(\text{data} \mid H)P(H)}{P(\text{data})}$$

We can also consider the posterior probability of any parameter θ. For example, θ could be the difference in effect between an experimental and standard therapy, such as the difference in median survival or the odds ratio of survival; Bayes' theorem says that the posterior probability or density of any θ is

$$p(\theta \mid \text{data}) = \frac{P(\text{data} \mid \theta)p(\theta)}{P(\text{data})}$$

The factor $P(\text{data} \mid H)$ or $P(\text{data} \mid \theta)$ is the *likelihood function* evaluated at H or θ. So Bayes' theorem relates the conditional density $p(\theta \mid \text{data})$

of a parameter θ with its unconditional density $p(\theta)$. Since the latter depends on information present *before* the experiment, it is a *prior* probability. Think of $1/P$(data) as a factor that makes the total probability equal to 1 when adding over all possible θ's; that is, the denominator P(data) is the sum (or integral) of the numerator over all θ's. It is often referred to as the normalizing constant. Rewriting Bayes' theorem:

Posterior probability $\propto$ likelihood $\times$ prior probability

where $\propto$ means proportional to. Again, Bayes' theorem provides a formalism for learning: The prior represents what was thought before seeing the data, the likelihood represents the data now available, and the posterior represents what is thought given both prior information and the data just seen. Continual and instantaneous updating can occur as today's posterior becomes tomorrow's prior.

1.2 Prior Distributions

In a Bayesian analysis, prior information about a parameter θ is assessed as a probability distribution on θ. This distribution depends on the assessor and so is *subjective*. Since posterior probabilities depend on prior probabilities, they too are subjective. A subjective probability can be calculated any time a person has an opinion. When a person's opinion about a parameter includes a broad range of parameter possibilities, perhaps even the whole real line, and all possible values are thought to be roughly equally probable, a diffuse prior distribution is chosen. In some cases a prior may be chosen that is so diffuse that it does not integrate to 1. This is called an *improper* prior. In other cases, opinions may allow for only a few values of the parameter. The most extreme case would be a prior that places all its mass on one value. A prior that is uniform on the whole real line and the prior that is a point mass on a single value represent two possible extreme prior distributions. In the first case, when the prior is a uniform distribution on the whole real line, the prior will have little influence on the posterior distribution and the posterior and likelihood will be proportional. In the latter case, when the prior is a point mass at a single value, the likelihood will not influence the posterior and the posterior will be identical to the prior. Priors that fall between these two extremes are useful for representing states of belief neither completely uncertain nor completely certain.

One type of prior distribution that makes it simple to assess influence is the *conjugate* prior. A family of distributions is called a conjugate family of prior distributions, for samples from a particular distribution, if the posterior distribution at each stage of sampling is of the same family as the prior regardless of the values observed in the sample. For example, the family of beta distributions is conjugate for samples from a binomial distribution, and the family of gamma distributions is conjugate for exponential samples. These two conjugate families are explained further in Sections 2.3 and 2.4 of this chapter. For further examples and an explanation of the construction of conjugate priors, see DeGroot (1970) and Berger (1985). It is simple to assess the influence of conjugate prior distributions because multiplying the likelihood by the prior has the same effect as increasing the sample size. This, too, is explained in more detail later in this chapter.

The prior distribution of a parameter θ includes all information available before the current trial. Interpreting this information in the context of the current trial and assessing prior probabilities is subjective. This is so even if some of the prior information itself arose from clinical trials. The patients in earlier trials may not be exchangeable with those in the current trial. So prior information has to be partially discounted (subjectively) when applied to the current setting.

1.3 Decisions and Utilities

An important difference between Bayesian and frequentist philosophies is that the Bayesian approach is prediction and decision oriented. Clinical trial results alter one's state of knowledge concerning various therapies. Measures of lead contamination in soil alter one's state of knowledge concerning health risk. Epidemiological studies of disease prevalence alter one's state of knowledge concerning risk factors. But knowledge without application has no impact. How should future patients be treated? Should parties deemed responsible for a toxic contamination be forced to clean it up? Should the public be made aware of discovered risk factors of a disease? The answers depend on one's knowledge and on the consequences of the various decisions. In a Bayesian approach, consequences are evaluated explicitly by associating a *utility* with each. Also, each consequence has a predictive probability. So the *utility of a decision* can be found by averaging utilities of consequences with respect to these predictive probabilities. Examples are presented later in this chapter and throughout this book. Although frequentist methods for decision theory exist, Bayesian methods are conceptually simpler and more directly interpretable.

1.4 Bayesian Versus Frequentist Approaches

The principal objection to the Bayesian approach is that it is subjective. Different people can draw different conclusions from the same results.

This criticism is specious for several reasons. First, frequentist methods too are subjective. Anyone who has done statistical data analysis is well aware of the many subjective decisions that go into any analysis. Subjectivity in model choice is not reserved just for the prior. Choices of likelihood, variables, tests, and critical regions all implicitly entail subjective choices. Second, although it is comforting when two analysts give the same answer, subjectivity leading to diversity is quite appropriate. Differences of opinion are the norm in health science and in science generally, so an approach that explicitly recognizes differences openly and honestly is realistic, forthright, and welcome.

Bayesian and frequentist probabilities are inverses of each other in the sense that the roles of the arguments and the conditions (data and hypotheses) are exchanged. This is an important distinction, but it is not the most important. A more important difference is that a Bayesian conclusion depends on previous evidence, whereas the frequentist conclusion is restricted to the data at hand. Bayesians assess evidence other than that in the current study and incorporate it into their analysis through a subjective assessment of prior probabilities. This makes the posterior probabilities relevant scientifically and medically, but it makes the Bayesian approach more difficult to use because it requires investigators to incorporate prior information formally and consumers to practice an extra step of vigilance in digesting published results. Both consequences are desirable.

Much of the Bayesian statistical literature does not consider utilities or exploit the decision analytic aspects of the Bayesian approach. In the clinical trials literature, this can be seen in Spiegelhalter, Freedman, and Parmar (Ch. 2) and many of their references. A *fully Bayesian* approach can be distinguished from partial Bayesian approaches, without meaning to imply that less than fully Bayesian is less than good. A fully Bayesian approach is decision theoretic and posterior probabilities are based on all available evidence, including that separate from the trial at hand. There are at least two ways to be less than fully Bayesian. First, one can calculate posterior distributions as data summaries without incorporating them into a decision analysis. Second, one can calculate posterior distributions using canonical prior distributions rather than prior distributions based on the available evidence. Bayesian approaches that are missing both of these characteristics are similar to standard frequentist approaches that focus on data summary. But there are differences. An important difference is flexibility: Accumulating data can be used to update Bayesian measures, independent of the design of the research. Frequentist measures are tied to the design, and interim analyses must be planned for frequentist measures to have meaning. Its flexibility makes the Bayesian approach ideal for analyzing data in health-related research.

The principal advantage of the Bayesian approach is its attitude toward accumulating evidence and updating belief. Bayesian analysis is not data analysis per se, because its conclusions use evidence beyond any particular data set. The Bayesian approach attempts to bring possibly different types of evidence to bear on questions of importance—questions such as whether a substance is toxic, whether a therapy is beneficial, or whether a characteristic or behavior is a risk factor for disease.

2 ASSESSING PROBABILITIES

Subjective probability is based on degrees of belief. Consider an event whose occurrence is uncertain. How strongly an individual believes that this event will occur (or has occurred) depends on the individual. The purpose of this section is to introduce methods for assessing probabilities as degrees of belief. Early works on the general principles for quantifying subjective probabilities and specific practical methods are presented by Savage (1971) Hampton *et al.* (1973), Smith (1965), and Tversky (1974). This section provides a brief introduction to the elicitation of probabilities. Later in this book, Chaloner (Chapter 4) reviews the literature on elicitation of personal beliefs. Chapters by Kadane and Wolfson (Chapter 5) and by Chaloner provide elicitation examples. Smith *et al.* (Chapter 15) show an example of assessing the sensitivity of results to the selection of a prior distribution.

Anyone can assess his or her probabilities. For the purposes of designing research, the most important subjects are the investigators and other experts whose beliefs can be elicited. When results of research are published, the reader too becomes an appropriate subject.

2.1 Calibrating for Probability Assessment

A basic requirement for assessing an individual's probabilities is the existence of a calibration scale. One must be able to imagine experiments in which the outcomes are exchangeable, in the following sense. Suppose the person gets to choose among a set of outcomes and will receive a prize if the outcome chosen occurs. Outcomes in the set are *exchangeable* if the person is indifferent among the various outcomes and would strictly prefer any one outcome over all others if the reward on it were increased by an arbitrarily small amount. (Statements about preferences are always somewhat delicate because other people set the ground rules. The assessor should be able to imagine that there's no chicanery afoot.)

An experiment is a *calibration experiment* for someone if all outcomes of the experiment are exchangeable for that person. There are many candidates. But whether a particular experiment serves to calibrate depends on the assessor. What's required is that the assessor be indifferent and not that the probabilities be equal in a universal sense. A convenient example is selecting a chip from a bowl that contains chips of the same size and shape.

Consider a specific setting. You would like to know your probability that for a particular population, the average drop in diastolic blood pressure on a certain drug is less than 10 mm Hg—call this event A. Consider a (calibration) bowl with one green and one red chip. We offer you the choice of getting a prize if a chip selected from the bowl is green or getting the same prize if A is true. If you choose to select from the bowl and the chip is red, or if you choose A and it turns out that A is false, then you receive nothing. Suppose you choose A. Then we take this to mean that your $P(A)$ is at least $\frac{1}{2}$. Now consider a (calibration) bowl with three green chips and one red chip. Again you get to choose between the prize if a chip selected from the bowl is green or the prize if A is true. If you now prefer the chip, then, taken together, your two answers mean that $\frac{1}{2} \leq P(A) \leq \frac{3}{4}$. Proceeding in this way, each time halving the interval by doubling the total number of chips in the bowl, will give $P(A)$ sufficiently accurately.

There are several problems with this approach. The most obvious is that the bet cannot be settled. How can it be determined whether A is true? Luckily, in most practical circumstances the assessor is motivated to take the assessment procedure seriously by imagining rewards in the comparison of events with calibration experiments. Another problem is that the assessor soon faces very difficult decisions. By the time the bowl contains 16 chips most assessors have a hard time deciding between green and A. Again luckily, a high degree of accuracy in specifying $P(A)$ is seldom required.

2.2 Probability Distributions

Most problems require a probability distribution and not a single probability. Suppose the task is to elicit prior belief about the efficacy of a blood pressure treatment. The above process can be carried out for various events of the form: average decrease in diastolic blood pressure on the drug is *less* than X mm Hg. The prior distribution of decrease in blood

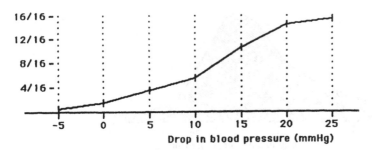

Figure 1 Subjective distribution function of change in blood pressure. Vertical lines show range indicated by the assessor. The curve drawn goes through the middle of the range at each point and so approximates the assessor's opinion.

pressure can be determined within a specified accuracy by varying X. For example, Figure 1 shows the result of hypothetically carrying out this process for seven different changes in blood pressure, $X = -5, 0, 5, 10, 15, 20, 25$. The assessor in question has probabilities 0 to 1/16, 1/16 to 2/16, 3/16 to 4/16, 5/16 to 6/16, 10/16 to 11/16, 14/16 to 15/16, and 15/16 to 1 for each level of X, respectively. The corresponding density function is shown in Figure 2, and a smoothed version is shown in Figure 3. The version in Figure 3 is more plausible and cosmetically appealing, but it has no computational advantage.

Extreme values are of special importance in assessing probabilities. For example, suppose the drug is given to 20 subjects and the diastolic blood pressure of all 20 subjects *increases* by more than 10 mm Hg (that

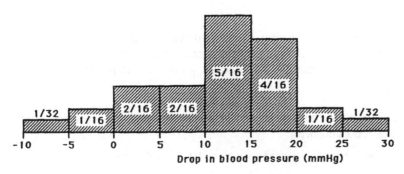

Figure 2 Subjective density function of drop in blood pressure estimated from Figure 1.

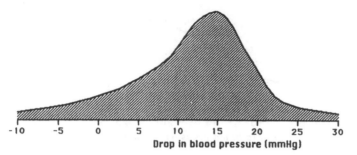

Figure 3   Smoothed version of density function from Figure 2.

is, they decrease by less than − 10). Because the prior probability for the density in Figure 2 gives all its probability to the right of − 10, the posterior density would give all its probability to the region just to the right of − 10. This may not be appropriate and is likely due to a somewhat cavalier assessment. We recommend being open-minded to the extent that you give some probability to all events, even those that seem implausible.

2.3   Beta Densities for Proportions

Many questions raised in health-related research involve two possible outcomes, such as success/failure, improved/not improved, response/non-response, or life/death. The parameter of interest is the proportion p of successes in the population. Suppose in a sample of n patients there are s successes and $f = n - s$ failures. Assuming independence of the observations conditional on p, the likelihood of p is a binomial distribution and is proportional to

$$p^s(1 - p)^f$$

Bayes' theorem says to multiply the prior density by this likelihood. There is a convenient way to calculate the posterior distribution when the prior density is in the beta family. The beta(α, β) density is proportional to

$$p^{\alpha - 1}(1 - p)^{\beta - 1}$$

By Bayes' theorem the posterior density is proportional to the product of the likelihood and prior:

$$p^s(1 - p)^f p^{\alpha - 1}(1 - p)^{\beta - 1} = p^{\alpha + s - 1}(1 - p)^{\beta + f - 1}$$

This product is proportional to a beta($\alpha + s$, $\beta + f$) density. The prior parameters have been updated by replacing α with $\alpha + s$ and β with

$\beta + f$. As discussed in Section 1.2, when observations have a binomial distribution, beta prior distributions for p are said to be *conjugate*, because the posterior distribution of p is also in the beta family. We see here the ease of interpreting the influence of the prior, because multiplying the likelihood by the prior has the same effect as increasing the sample size.

Choosing a prior density in the beta(α, β) family requires specification of α and β. One way of specifying α and β is presented here. Two assessments are required. First, the assessor may specify the probability of a success on the first trial as the mean of the beta density—call it m—which has the simple form

$$m = \frac{\alpha}{\alpha + \beta}$$

For the second requirement, imagine that the first trial is a success. Then the probability of success on the second trial is

$$m^+ = \frac{\alpha + 1}{\alpha + \beta + 1}$$

Solving simultaneously gives

$$\alpha = \frac{m(1 - m^+)}{m^+ - m} \quad \text{and} \quad \beta = \frac{(1 - m)(1 - m^+)}{m^+ - m}$$

Consider an example. The most important prognostic factor in early breast cancer is the number of axillary lymph nodes testing positive during a pathology review. The number of lymph nodes dissected during surgery varies. (This number and the nodes dissected may depend on clinical characteristics, but we will assume that the sampling is random and therefore that nodes selected are no different a priori from those not selected.) Suppose the probability that any particular node is positive is 3%, and so $m = 0.03$. However, if the first one sampled tests positive, then the probability that the next is also positive increases dramatically to 20%: $m^+ = 0.20$. Therefore

$$\alpha = \frac{0.03(1 - 0.2)}{0.2 - 0.03} = 0.14 \quad \text{and} \quad \beta = \frac{(1 - 0.03)(1 - 0.2)}{0.2 - 0.03} = 4.56$$

So the prior density of p is beta(0.14, 4.56). This prior density has a mean of 0.03 and a standard deviation of 0.07 and is shown in Figure 4.

An alternative elicitation method would be to assess the mean, $\alpha/(\alpha + \beta)$, and sample size, $\alpha + \beta$, and then back-solve for α and β. Many alternatives exist, but none has been shown to be the easiest to implement or to provide the most accurate assessments.

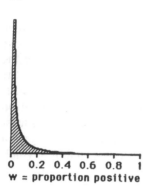

0   0.2 0.4 0.6 0.8   1
w = proportion positive

Figure 4 Beta(0.14, 4.56) density for proportion of positive axillary lymph nodes. The mean is 0.03, which is also the probability of a positive lymph node when not conditioned on p.

Further examples of Bayesian methods for dichotomous outcomes are in Ashby and Hutton (Chapter 3), Albert and Chib (Chapter 22), and Racz and Sedransk (Chapter 25). Ashby and Hutton present examples of the analysis of dichotomous outcomes in epidemiological studies, such as the analysis of a case-control study on occupational exposure and malignant lymphoma. Albert and Chib present a Bayesian random-effects probit model for binary repeated measures data. The model is applied in the context of a clinical trial to determine the effect of an analgesic on pain relief in primary dysmenorrhea. Racz and Sedransk demonstrate Bayesian predictive inference for a finite population proportion given data from a two-stage cluster sample. They apply their methods to a survey on the quality of care that cancer patients receive.

2.4 Exponential Densities for Event Times

Health research is also often interested in outcomes that are event times. Two examples are survival time and time between transitions of disease stages. The simplest model for event times is the exponential distribution. The parameter of interest is the hazard rate, θ, of event occurrences. Again assuming a sample of n patients with event times $t_i, i = 1, \ldots, n$, and no censored observations, the likelihood is

$$\theta^n e^{-\theta \sum t_i}$$

A conjugate prior when observations are exponentially distributed is the gamma distribution. A gamma prior has two parameters, a shape parameter, α, and a scale parameter, β:

$$\frac{\beta^{\alpha}}{\Gamma(\alpha)} \, \theta^{\alpha-1} e^{-\beta\theta}$$

Updating the prior, the posterior is proportional to

$$\frac{\beta^{n+\alpha}}{\Gamma(\alpha)} \, \theta^{n+\alpha-1} e^{-\theta(\beta + \Sigma \, t_i)}$$

Because a conjugate prior has been used, the posterior is a gamma distribution with updated parameters $\alpha + n$ and $\beta + \Sigma \, t_i$.

Choosing a prior density in the gamma(α, β) family requires specification of α and β. Because the mean of a gamma is α/β and the variance is α/β^2, one way of specifying α and β is to specify the mean and variance and then back-solve for α and β. Wolfson (1995) presents an alternative method. She suggests eliciting the quantiles (at least two) of the prior predictive distribution which is a shifted Pareto distribution. Then she relies on the idea that α can be interpreted as the prior sample size to reconcile incoherencies.

An example of Bayesian methods using exponentially distributed data with censored observations is presented in Stangl (Chapter 16). Other examples of event-time analyses using proportional hazards are in Abrams *et al.* (Chapter 20) and Simon *et al.* (Chapter 21). Qian *et al.* (Chapter 6) show an example using a Weibull distribution. Abrams *et al.* consider the problem of making inferences about the efficacy of tamoxifen and cyclophosphamide in the treatment of patients with early-stage breast cancer. Simon *et al.* are concerned with estimating the heterogeneity of treatment effects among subsets of patients. Qian *et al.* consider the problem of choosing a stopping time for clinical trials. The example presented is a clinical trial comparing radiotherapy to a combination of radiotherapy and chemotherapy for patients with stage III non–small cell lung cancer.

2.5 Checking for Consistency

Assessors should make consistency checks of their prior probabilities. Various checks are possible. For the beta-binomial case one can propose probabilities of intervals and compare these probabilities with areas under the beta density. If they disagree, then some adjustment in α and β may be in order. In addition, consistency may be checked in the beta case by specifying the probability of success on the second trial assuming that the first trial results in a *failure*. This probability is

$$m^- = \frac{\alpha}{\alpha + \beta + 1}$$

The m^- found by using this expression with α and β as found from the formulas for m and m^+ can be compared with the assessed value of m^-. If the two disagree, then α and β should be adjusted until they do agree.

Another consistency check in the beta case involves $\alpha + \beta$. The larger $\alpha + \beta$ is, the greater the influence of the prior on the posterior. The updating rule for beta densities says that, when s successes and f failures are observed, the new density is beta($\alpha + s$, $\beta + f$). So the new sum of the parameters is the old sum plus n, the sample size. This gives an interpretation for $\alpha + \beta$ as a prior sample size. Suppose the assessor had experienced a number of observations deemed to be roughly exchangeable with the current observations. Then the assessor might use these observations in setting α to be the number of prior successes and β to be the number of prior failures. So $\alpha + \beta$ is a measure of information in the prior distribution that compares directly with the number of observations in the experiment. This might provide a good primary means of assessing α and β, or at least $\alpha + \beta$. However, people can't remember their prior observations very well. Also, prior observations should seldom if ever be regarded as exchangeable with current observations. A possible solution is to discount prior observations as compared with current observations, taking α and β smaller than they would be otherwise. This reflects greater open-mindedness and may therefore be reasonable.

3 PREDICTIVE PROBABILITIES

Assuming probability distributions for parameters allows calculation of probabilities of responses for *future* observations. Consider a patient with a particular set of characteristics. How will that patient respond to therapy A? The patient's response is unknown. Like all unknowns in the Bayesian approach, it has a probability distribution. Because it refers to a future observation, it is called a predictive distribution. Predictive probabilities are useful for communicating results and making decisions such as selecting a patient's dosage or determining whether a clinical trial should continue. How to use predictive distributions is introduced below, but further examples can be found in chapters by Racine-Poon and Wakefield (Chapter 13), Qian *et al.* (Chapter 6), and Kadane and Wolfson (Chapter 5). Racine-Poon and Wakefield use hierarchical models to fit pharmacokinetic/pharmacodynamic models. They demonstrate that predictive probabilities can be used to predict the required dose for an individual as well as for model validation and diagnostic purposes. Qian *et al.* demonstrate the utility of predictive distributions for determining whether a clinical trial should continue. Kadane and Wolfson use predictive distributions for prior probability assessments.

3.1 Number of Positive Lymph Nodes

To continue the example of Section 2.3, suppose a surgeon removes three axillary lymph nodes from a woman with breast cancer and none tests positive. Another doctor suggests to the surgeon that had more nodes been removed, perhaps some might have been positive. Given what we now know, should the surgeon have removed more? Should the patient have additional surgery? These are complicated questions that require addressing the purpose and utility of nodal dissection—would therapy be different if the patient was node positive, and how beneficial would the different therapy be? We do not hope to do justice to this issue here. But we can address the probability that none of the next 10 nodes removed would test positive.

Let p be the proportion of the axillary lymph nodes that would test positive. As in the previous section, suppose p has a beta(0.14, 4.56) prior density. Since the surgeon found no positive nodes in the sample of three, the posterior density of p is then beta(0.14, 7.56). Given the current values of the beta parameters, $\alpha = 0.14$ and $\beta = 7.56$, the predictive probability that the next (i.e., fourth) node selected would be negative is $\beta/(\alpha + \beta)$ = 7.56/7.70. Given that the fourth is negative, the predictive probability that the fifth would also be negative is $(\beta + 1)/(\alpha + \beta + 1) = 8.56/8.70$, and so on. So the predictive probability that every one of the next 10 is negative is

$$\frac{\beta}{\alpha + \beta}\frac{\beta + 1}{\alpha + \beta + 1}\cdots\frac{\beta + 9}{\alpha + \beta + 9} = \frac{7.56}{7.70}\frac{8.56}{8.70}\cdots\frac{16.56}{16.70}$$

$$= 0.982 \times 0.984 \times \cdots \times 0.992 = 0.88$$

(The individual factors in this sequence show how the probability of "negative" test results increases as additional negative test results accrue.) So this patient is very likely to continue to be regarded as node negative, even if an additional 10 nodes are tested. Whether an 88% chance is small enough to recommend more surgery is open to question, but the point is that this predictive calculation is an appropriate consideration in such a decision.

3.2   Calculations for Design

Predictive probabilities help in choosing from among possible research designs. When calculated during the course of an ongoing project, they aid in deciding whether to alter the project's design—for example, in deciding whether to stop the project. Consider a phase II clinical trial for evaluating a particular new agent in the treatment of breast cancer that is newly diagnosed as metastatic. (We have a particular agent in mind.) Now let p be the rate of response (complete plus partial; i.e., at least 50% of the symptoms have disappeared) in this population. Numerous treatment agents exist. Although their response rates are imperfectly known, it is safe to say that some are as large as 30% but that no currently available agents have a response rate much greater than 30%. We have assessed probabilities for p as described in Section 2.3 and find that they are well approximated by the beta(2, 4) density, shown in Figure 5. The mean of this density is $2/(2 + 4) = 1/3$, which is the prior probability of response on the first patient. This is rather large compared with existing agents, but previous experience with this agent in patients with other types of solid tumors indicates that it is promising.

    Suppose the trial is ongoing and 10 patients have been treated, with only 1 of the 10 responding. In deciding whether to continue this trial and include 10 more patients, say, it would be important to know the predictive probabilities for the number of future responders. Using the result of the previous section, the updated distribution of p is the beta(2 + 1, 4 + 9) = beta(3, 13) density. The predictive probability of the number of successes, k, in the next 10 trials is easy to find.

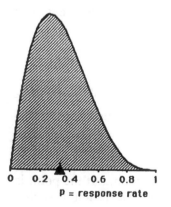

Figure 5   Beta(2, 4) prior density for response rate p of an experimental therapy for metastatic breast cancer. The mean of 1/3 is shown as a triangle.

$$P(k \mid \text{data}) = \int_0^1 P(k \mid \text{data}, p) f(p \mid \text{data}) \, dp$$

$$= \int_0^1 \binom{10}{k} p^k (1 - p)^{10-k} \frac{\Gamma(\alpha + \beta)}{\Gamma(\alpha)\Gamma(\beta)} p^{\alpha - 1}(1 - p)^{\beta - 1} \, dp$$

$$= \binom{10}{k} \frac{\Gamma(\alpha + \beta)}{\Gamma(\alpha)\Gamma(\beta)} \frac{\Gamma(\alpha + k)\Gamma(\beta + 10 - k)}{\Gamma(\alpha + \beta + 10)}$$

$$= \frac{10!}{k!(10 - k)!} \frac{15!}{2!12!} \frac{(2 + k)!(22 - k)!}{25!}$$

The values of $P(k \mid \text{data})$ for all values of k are shown in Table 1. Also shown in Table 1 are the observed response rates after 20 patients, $(k + 1)/20$. With the addition of 10 patients, k of whom are responders, the posterior density of p will be beta$(k + 3, 23 - k)$. The mean of this density is $(k + 3)/26$, which is also shown in Table 1. This is another estimate of the response rate, one that includes the prior information as well as the results of the trial. Hence, they are shrunk toward the prior mean of $2/(2 + 4) = 1/3$.

The question of whether such a trial should continue is addressed in Section 4.1. The point here is that the predictive probabilities of these various response rates can be calculated and are relevant for the decision problem. Qian *et al.* (Chapter 6) show another example of how predictive distributions can be used to determine stopping rules for phase III clinical

Table 1 Predictive Probabilities of the Number of Responses in Next 10 Patients Given One Response in the First 10

k	$P(k \mid \text{data})$	Total no. of responses	$\dfrac{k + 1}{20}$	$\dfrac{k + 3}{26}$
0	.198	1	0.05	0.12
1	.270	2	0.10	0.15
2	.231	3	0.15	0.19
3	.154	4	0.20	0.23
4	.085	5	0.25	0.27
5	.040	6	0.30	0.31
6	.016	7	0.35	0.35
7	.005	8	0.40	0.38
8	.001	9	0.45	0.42
9	.000	10	0.50	0.46
10	.000	11	0.55	0.50

trials. Finding predictive distributions is relatively straightforward. Using them is more involved. This is the subject of the next section.

4 DECISION PROBLEMS

In this section we indicate how to address two types of decision problems in a Bayesian fashion: deciding whether to stop a clinical trial and allocating patients to therapy to achieve an overall measure of successful health care delivery. For other examples, see Berry (1991). A decision approach for these problems is an alternative to choosing a significance level, an alternative hypothesis, and a power level.

Whether to stop a clinical trial depends on the available information (from the trial and otherwise), given as the current probability distribution of any unknowns. The utility of stopping can be evaluated by weighing the utilities of the various consequences of the status quo by their probabilities. Continuing the trial entails the additional randomness provided by the future observations. The utility of continuing can be evaluated by weighing the utilities of the various consequences of future observations by their predictive probabilities. Allocating patients to therapy on the basis of currently available information is a similar problem, one in which predictive probabilities again play a pivotal role.

We give two examples of stopping clinical trials. The first is a continuation of the phase II trial example of Section 3.2. The second is a randomized vaccine study. Two chapters later in this book further demonstrate the utility of interim analysis for stopping clinical trials. Lewis (Chapter 10) examines data from a clinical trial evaluating the prophylactic effect of phenytoin on early post-traumatic seizures in children. Qian *et al.* (Chapter 6) examine data from a clinical trial comparing radiotherapy alone to radiotherapy plus chemotherapy in the treatment of non–small cell lung cancer. Further reading can be found in Lewis and Berry (1994). They provide a classical evaluation of Bayesian group-sequential clinical trial designs.

4.1 Stopping a Phase II Clinical Trial

Deciding whether and when to stop a clinical trial is a common problem and includes stopping a trial before it starts! In this section we consider the phase II trial of Section 3.2, making assumptions about utilities. Recall that 1 patient responded among the first 10 patients in the trial. Only two possibilities are considered: stopping immediately or adding another 10 patients to the trial. Consider utilities measured in terms of effective treatment of breast cancer patients. Because there are therapies with a 30% response rates and this agent has an estimated $3/16 = 19\%$ response rate, continuing the trial may not be in the best interest of the patients in the trial.

If the trial stops, then assume this agent will not be investigated further. What would be the resulting impact for the women who have or will have metastatic breast cancer? This is not easy to evaluate but we will delay the assessment problem by taking it to be 0 and considering other eventualities relative to it. So continuing is appropriate if continuing has utility greater than zero.

Now suppose the trial continues to 20 patients. The predictive probabilities for the 11 possible outcomes in the second 10 patients are given in Table 1. A utility must be specified for each and is taken in terms of (our estimates of) the increment in number of responses over the next several years (as compared with the currently available therapies) effected by having a trial involving 20 patients with that number of responses. (There is almost certainly a positive relationship between response and survival, but the exact relationship is not clear.) To assess utilities requires addressing several issues. For patients in the trial, how serious is a delay in alternative treatment should it turn out that the experimental agent is not very effective? What other therapies are available and how effective are they? What are the possibilities that other, perhaps more effective experimental agents will be developed? In what time frame? If the experimental agent turns out to be promising, what other trials will be necessary before the agent becomes commonly used? As a function of the results from this trial and from other trials, how extensively will the agent be used?

Some assessments are shown in Table 2. Such assessments should be made by a team of oncologists, pharmacologists, and other experts. The tabled utilities permit demonstration of the method. The first number listed under "incremental utility" refers to the expected difference in responses among the next 10 patients in the current trial if they receive this agent as opposed to some other therapy with a response probability of .3, that is, $k - (.3 \times 10)$. The second number, the one following the " + ," corresponds to patients who present after the trial. This is zero when the response rate is sufficiently low ($\leq 25\%$) that, in our estimation, the agent would not be pursued and so would have no incremental utility. If the response rate is sufficiently high ($>25\%$), then later patients are more likely to receive the agent (depending on the results of further clinical trials, an uncertainty included in the assessment), and hence the benefit will be greater. For example, an agent with an estimated response rate of $50\% \, [(2 + 1 + 10)/(2 + 4 + 10 + 10)]$ would be very widely used—per-

Table 2 Predictive Probabilities from Table 1 with the Utilities of the Various Possible Number of Responses, k, in the Next 10 Patients

k	Mean rate	Predictive probability	Incremental utility	Product
0	0.12	.198	−3 + 0	−0.59
1	0.15	.270	−2 + 0	−0.54
2	0.19	.231	−1 + 0	−0.23
3	0.23	.154	0 + 0	0
4	0.27	.085	1 + 10	0.94
5	0.31	.040	2 + 20	0.87
6	0.35	.016	3 + 60	0.98
7	0.38	.005	4 + 150	0.77
8	0.42	.001	5 + 500	0.63
9	0.46	2E−4	6 + 2000	0.44
10	0.50	2E−5	7 + 5000	0.10
Sums:		1.0000		3.37

haps on 25,000 patients—resulting in the incremental utility of 5000 (30% would have responded on another agent and so an estimated excess of 20% would respond on this agent). Again, the utility is relative to not having this agent available for use.

The (expected) incremental utility of continuing is the average of the fourth column of Table 3, with respect to the predictive probabilities in the third column. The result is 3.37, shown as the sum of products in the fifth column. Units are numbers of responses, and so this is not a dramatic improvement. But the sum is positive, so continuing is appropriate.

The sensitivity of the decision (although not its utility) can be judged by varying the utilities in Table 2 and also considering prior distributions for p other than the beta(2, 4). Such considerations show that the decision to continue the trial is quite robust. The sum of the negative products in the fifth column of Table 2 is −1.36; the sum of the positives is 4.73. The overall sum would be positive even if the utilities were greatly reduced—with all numbers greater than 10 set equal to 10, say. Only an assessment as extreme as ignoring the patients who present after the trial would make the sum negative. (Because of the evident trade-off, the assessment team should include a medical ethicist.) Also, changing the prior distribution (and hence the predictive probabilities) has little effect on the final result. Prior distributions with more variability (more open-minded) give more probability to smaller values of k but also to larger values of

Table 3 Bimonthly Results of the Trial and Expected Numbers of Future Cases[a]

	Cumulative cases		Expected number of future cases if	
Month	Vaccine	Placebo	Stop	Continue
0	0	0	608	439[a]
2	0	2	807	505[a]
4	0	4	954	498[a]
6	0	6	1042	442[a]
8	0	7	997	366[a]
10	0	7	864	299[a]
12	0	8	496	246[a]
14	0	10	238	200[a]
16	1	12	364	313[a]
18	1	13	297	269[a]
20	1	15	242	234[a]
22	1	18	200[a]	201
24	1	21	172[a]	176
26	1	22	153[a]	158

[a] These numbers indicate the better decision.

k; because of the asymmetry in the utilities, the net effect is positive. Prior distributions with less variability give less probability to smaller and larger values of k; again, the effect is positive because an agent with a response rate near 30% has positive utility. The only type of prior distribution that results in a negative utility for continuing is one that has a smaller mean and not a very large variance.

4.2 Sequential Randomized Vaccine Efficacy Trial

Berry *et al.* (1992, 1994) consider a vaccine trial of *Hæmophilus influenzæ* type b (HIB). The vaccine was designed to be effective in infants, and the trial led to the licensure of the vaccine for children as young as 2 months of age. Because American Indian and Eskimo children are at very high risk, the trial was conducted on Navajo reservation children between 2 and 18 months of age. All subjects were vaccinated, with children in the control group receiving placebo innoculations.

The authors take the goal of the trial to be minimizing the number of HIB cases among Navajo children over a horizon of 20 years. They assess prior information concerning the rates of HIB and model HIB occurrence among infants as a Poisson process. [Qian (1994) generalizes this assumption and models the population as a mixture of two groups: a group in which HIB occurrence is Weibull and a group that is not susceptible.] They consider historical information about the regulatory process and assess probabilities of licensure and time required for licensure as a function of the available data. They also assess the possibility that competing vaccines will become available and the timing of such. For specific assumptions, refer to Berry *et al.* (1994).

Accumulating information is evaluated without penalty. At the beginning of the trial and bimonthly thereafter, a decision is made whether to continue the trial. The preferred action is the one with the smaller expected number of future cases—determined by dynamic programming and exploiting predictive probabilities. (Further details of this approach are presented in Section 4.3.) Figure 6 is taken from Berry *et al.* (1994) and

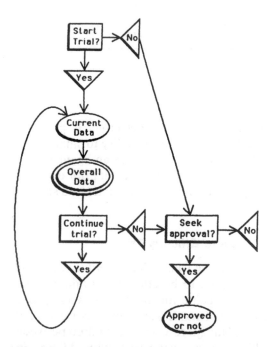

Figure 6   Schema of the decision process. The trial is stopped once the predetermined maximum number *N* of months is reached. It is also stopped if the expected number of cases over the subject horizon is greater for continuing than for stopping.

shows the decision schema. The predetermined maximum length of the trial is N months. At each decision time prior to the Nth month, the trial must be continued or stopped. If it is stopped, a decision is made to seek licensure or not. If licensure is sought, the vaccine may or may not be approved.

Table 3 gives the data, the numbers of cases over time in the two treatment groups. Approximately 450 Navajo infants are born each month on the reservations under observation. Not all infants were randomized, but they are all at risk for HIB. We assume they would all receive an approved vaccine. The average accrual rate was 105 children per treatment group per month, so about 240 Navajo infants per month do not participate in the trial. The number of current subjects in each treatment group increased linearly until month 16, when the number remained constant at $16 \times 105 = 1680$.

The last two columns of Table 3 show the results of the dynamic programming. These columns give the expected numbers of future cases of HIB when stopping and when continuing. "Continuing" assumes that subsequent decisions to stop or continue are optimal. The smaller number is marked with a small a. In particular, starting the trial is optimal because 439 is smaller than 608. (For comparison, the expected number of future cases using a fixed, rather than sequential, design is 553).

4.3 Adaptive Allocation in Clinical Trials

A standard way to compare two therapies is to randomize patients equally to them. For any fixed sample size this procedure gives maximal information about the difference in their effectiveness. This in turn will help in treating patients who present once the trial's results become known. In designing a trial using a decision approach, one can explicitly consider effective treatment of patients, those in the trial and those not.

Consider a trial in which information accrues relatively quickly concerning each patient's response. (Adaptive allocation has no benefit when there are long delays in deciding whether a patient is a responder.) Suppose the objective is to maximize the number of patients who respond, over some patient horizon N. Also, suppose the number of patients in the trial is n. The patients in the trial can be allocated to either therapy, depending on the accumulating results, and the $N - n$ patients outside the trial will receive the therapy that performs better during the trial. How should the patients in the trial be allocated? To demonstrate the method of finding the optimal procedure, we will carry out an example using $n = 7$ and $N = 100$. A small value of n might be considered in a phase II study, but our reason for choosing this small an n is to be able to draw

figures showing the method in a manageable space. In general, increasing
n increases the expected number of responses (except that in the example
below there is no benefit in increasing an odd n by 1, and so $n = 7$ gives
the same expected number of responses as does $n = 8$).

Label one therapy as A and the other as B. Take the two population
proportions of responses to be w_A and w_B. Take them to be independent
and both having a uniform prior distribution on the interval (0, 1), which
is beta(1, 1). Dynamic programming proceeds from the last step of the
decision process, which in this case is at the end of the trial. But it requires
that all possibilities be considered. Suppose n_A is the number of patients
assigned to A and n_B is the number assigned to B. At the end of the trial,
$n = n_A + n_B$. One possibility is $n_A = 5$ and $n_B = 2$. Let s_A be the number
of responses among the n_A patients assigned to A and s_B the number of
responses among the n_B patients assigned to B; s_A takes values 0, 1, . . . ,

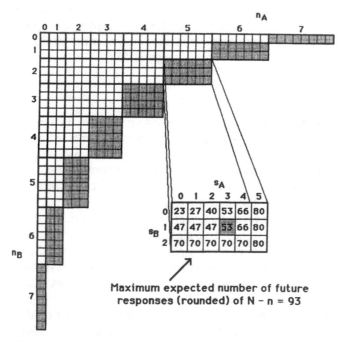

Figure 7 Beginning the dynamic program, for each of the cells with $n_A + n_B$
$= 7$ (cells that are shaded), the expected number of future responses is calculated
and entered into this tableau. These entries are shown for cells with $n_A = 5$ and
$n_B = 2$.

n_A and s_B takes values $0, 1, \ldots, n_B$. For $n_A = 5$ and $n_B = 2$, the 18 possible combinations are magnified in Figure 7.

Consider possibility $n_A = 5$, $n_B = 2$, $s_A = 3$, and $s_B = 1$, which is shown shaded among the magnified cells in Figure 7. The updated mean of w_A is $(3 + 1)/(5 + 1 + 1) = \frac{4}{7}$ and that of w_B is $(1 + 1)/(2 + 1 + 1) = \frac{2}{4}$. Since $\frac{4}{7} > \frac{2}{4}$, the remaining $N - n = 93$ patients are assigned to A, with an expected future number of successes of $93 \times \frac{4}{7} = 53.14$. Values for cells with $n_A + n_B = 7$ are calculated similarly. Entering these numbers into the tableau initializes the dynamic program.

Now consider the cells for which $n_A + n_B = 6$. The solidly shaded cell in Figure 8 shows the one with $n_A = 4$, $n_B = 2$, $s_A = 3$, and $s_B = 1$.

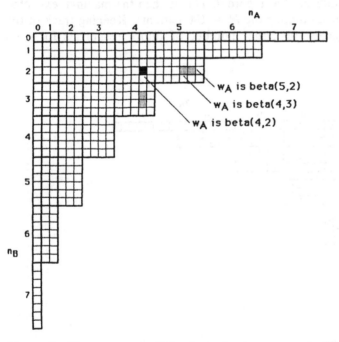

Figure 8 The next steps of the dynamic program are to fill in the tableau for cells with $n_A + n_B = 6$. The solid cell in the figure has $n_A = 4$, $n_B = 2$, $s_A = 3$ and $s_B = 1$. Using treatment A moves the process to one of the two shaded cells to the right, and using treatment B moves down to one of the two shaded cells. The distribution of w_A for this cell is beta $(4, 2)$ and that of w_B is beta$(2, 2)$. If treatment A is used, then the distribution of w_A changes, to either beta$(5, 2)$ [with probability $4/(4 + 2) = \frac{2}{3}$] or beta$(4, 3)$ (with probability $\frac{1}{3}$), and the distribution of w_B stays the same.

For this cell w_a is beta(4, 2) with mean $4/(4 + 2) = \frac{2}{3}$. The expected number of future responses from this cell is calculated under the two possibilities: use A and use B. If A, then the process moves to one of the two cells shaded in the $n_A = 5$, $n_B = 2$ block of cells. The process moves to the $s_A = 4$, $s_B = 1$ cell with probability that the patient given A is a responder and to the $s_A = 3$, $s_B = 1$ cell with probability that the patient given A is not a responder. Referring back to Figure 7, the maximal expected number of future responses is 66.43 (with probability $\frac{2}{3}$) and 53.14 (with probability $\frac{1}{3}$). So the expected number of future responses when using A is $(1 + 66.43) \times \frac{2}{3} + 53.14 \times \frac{1}{3} = 62.67$. This is compared with the expected number of future responses when using B and the larger number is entered in the tableau. After the cells with $n_A + n_B = 6$ come those with $n_A + n_B = 5$, etc., until reaching $n_A = n_B = 0$. The entry for this cell is 63.05, as indicated in Figure 9. This is then the maximal expected number of responses among the $N = 100$ patients. Keeping track of the treatment that gives the larger number of future responses for each

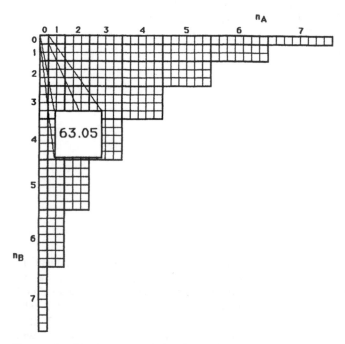

Figure 9 The last step of the dynamic program gives the expected number of responses over the $N = 100$ patients, namely 63.05.

cell provides the optimal allocation procedure. This is shown in Figure 10.

For comparison, letting $N = n = 100$ gives a maximal expected number of responses of 64.92. [When $N = n$ the decision problem is a classical bandit; see Berry and Fristedt (1985).] For a randomized trial with $n = 7$ with three patients assigned to A and four to B, or vice versa, the expected number of responses is 62.4.

Berry and Eick (1994) call the optimal Bayesian procedure for uniform prior distributions found as above the "robust Bayes" procedure. They compare it with various adaptive procedures and with a balanced randomized controlled trial with sample size n. No procedure can perform better than the robust Bayes procedure on the average. But they compare procedures for fixed w_A and w_B.

For example, suppose $N = 10,000$ and $n = 100$. Figure 11 shows the expected number of responses lost as compared with robust Bayes, for two particular procedures. One is RCT, the randomized controlled trial,

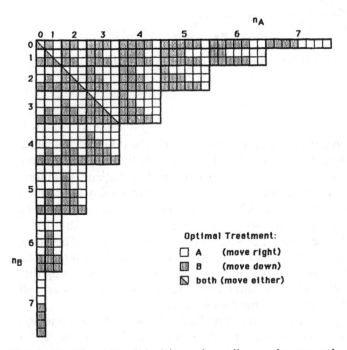

Figure 10 The optimal decisions, depending on the currently available data, as given by n_A, n_B, s_A, and s_B.

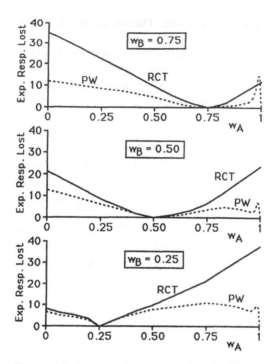

Figure 11 Expected successes lost by RCT and PW as compared with robust Bayes.

in which patients are assigned equally to A and B. The other is PW, which is the play-the-winner procedure, one of several adaptive procedures considered by Berry and Eick (1994). Under PW, the first patient is randomized and then the same treatment is used after a response (play the winner) and the other treatment is used after a nonresponse (switch on a loser). For all three procedures, following the experimental phase (the first n patients), the better performing treatment is used exclusively. The figure considers three values of w_B and considers w_A varying from 0 to 1. An interesting aspect of the curves is that they stay positive, indicating that robust Bayes is better than both alternatives for given w_A and w_B as well as on the average, hence the name robust Bayes.

More generally, RCT loses to robust Bayes for all n, N, w_A, and w_B; but it fares relatively well when the patient horizon N is very large. So from a decision analytic perspective, an RCT is a reasonable choice when a disease or condition is at least moderately common.

A decision approach is an alternative to choosing a significance level, an alternative hypothesis, and a power level. For those interested in power, it may be difficult to evaluate analytically for an adaptive design. However, it is rather easy to evaluate power for any design using simulation.

Several chapters in this book specifically focus on decision making. They are by Parmigiani *et al.* (Chapter 7), and Wolfson *et al.* (Chapter 9). Parmigiani *et al.* provide a general overview and practical illustration of the role of Bayesian methods in studies to develop clinical practice guidelines. These studies require assimilating data from multiple sources to come up with recommendations for diagnosis and treatment of specific disorders. Their goals are clearly decision oriented and require the strengths of the Bayesian paradigm. Wolfson *et al.* present results of a statistical analysis for policy-making decisions not in the health clinic but rather in the environment. They consider two decision-making examples in which conclusive evidence exists that environmental contamination leads to clear health risks. The specific decision problem they address is, who should remediate such environmental contamination? Related but not directly incorporating a decision model are chapters by Hasselblad and Jarabek (Chapter 8) and Lindsey and Ryan (Chapter 19). Hasselblad and Jarabek demonstrate the Bayesian paradigm for risk characterization, that is, estimating dosages of toxic substances yielding an adverse risk. These risk characterizations are used as the basis for risk management decisions. Lindsey and Ryan use Bayesian methods to estimate dose effect of potentially carcinogenic compounds on the tumor incidence rate in rodents.

5 DESIGN OF EXPERIMENTS

Statisticians have had a substantial impact on medical research, but most of the impact has been in the area of data analysis. Although experimental design is clearly regarded as an important component of the discipline, some basic principles of design are largely ignored in health-related research. In this section an introduction to the Bayesian approach to designing experiments is presented. Because predictive distributions, stopping rules, and subject allocation play a key role, there is considerable overlap with Sections 3 and 4 of this chapter.

Most simply, the core of Bayesian design of experiments is deriving the value of particular design strategies by averaging the costs and benefits of the consequences of each strategy over the probability distribution of the consequences. Two important reasons for taking a Bayesian approach are that it allows for calculating predictive probabilities or probabilities of as yet unobserved data based on currently available information and it does not require that all aspects of the design, including the stopping rule, be specified. The importance of these advantages will be highlighted by considering a drug development program.

The purpose of a drug development program is to obtain information that will convince the regulatory agency that the drug should be marketed and that will determine whether marketing the drug will be profitable. At any time during a drug development program, the program can take various directions. In the Bayesian approach, accumulating data can be analyzed to change the course of a clinical trial or of the entire drug development program. In such analyses, the worth of available strategies is evaluated using costs and benefits of the various consequences. We will consider monetary costs and income for the phamaceutical company trying to maximize profit.

Consider the simple setting of a matched-pair experiment, where the patients within each pair are regarded as exchangeable—one is assigned randomly to an experimental drug and the other serves as a control. The parameter θ is the probability that the patient receiving the experimental drug has a better response than the control patient. More complicated scenarios in which there are multiple efficacy and safety parameters—including survival—can be handled in a straightforward way, but with an increase in computational complexity.

The Bayesian approach requires assessing the available information concerning θ and quantifying it as a prior probability distribution. For simplicity, assume this has been done and suppose θ has a uniform prior distribution. Suppose also that, for considerations of safety, the minimum number of pairs required for marketing approval is 34. For a fixed sample of 34, a minimum of 24 successes is necessary to achieve statistical significance at the .05 level. Take this 24 out of 34 as a requirement for approval. Although, fortunately, this is not the decision rule used by regulatory agencies, they use *some* decision rule, and the following analysis can be adapted to other rules or to the case in which the rule is unknown.

The Bayesian decision theoretic approach to design requires specification of costs and benefits. For convenience and without losing generality, take the monetary unit to be the clinical trial costs of a single pair of patients. We require an assessment of expected profit as a function of θ should the drug be marketed. If the drug has no effect, then $\theta = \frac{1}{2}$. The drug is better than the control if $\theta > \frac{1}{2}$, and it is worse if $\theta < \frac{1}{2}$. Suppose company marketers assess the eventual profit (not counting clinical development costs) to be 0 if the drug is not approved for marketing and proportional to $\theta - \frac{1}{2}$ if it is approved. To be specific, take the proportionality constant to be 200. Other constants and other profit structures can be handled easily. Berry and Ho (1988) carry out an example.

The simplest design is to admit 34 pairs and follow them until they respond. The cost of such a trial is 34. Before data are collected, the predictive distribution of the number of successes is uniform on 0, 1, 2, 3, . . . , 34, and the expected profit is

$$\sum_{s=0}^{23} \frac{1}{35}(0) + 200 \sum_{s=24}^{34} \frac{1}{35} E(\theta - .5 \mid s \text{ successes})$$

The mean of θ given s successes and $34 - s$ failures is expressed as $E(\theta \mid s \text{ successes}) = (s + 1)/36$. Substituting the expectation into the equation gives a value of 21.0, so the corresponding net profit is $21.0 - 34 = -13.0$. So this fixed-sample design results in an expected net loss: the monetary gains expected from the information in the trial are not sufficient to warrant the expenditures.

Another possible design is to analyze the available data at one intermediate point and continue the trial and further development of the drug only if these interim data are sufficiently promising. For example, Figure 12 provides the predictive probabilities for the next 17 pairs of patients given 13 successes in the first 17 pairs. The probability of at least 24 successes in the total of 34 pairs is 80%. Weighing the expected profit by the predictive probabilities shows that continuing the trial from this point has expected net worth of $42.0 - 17 = 25.0$, a value greater than 0.

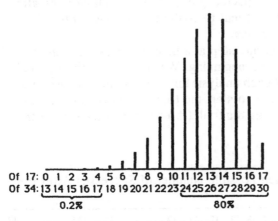

Figure 12 Predictive probabilities for successes in the next 17 pairs of patients given 13 successes in the first 17 pairs.

Similarly, it is easy to calculate the maximal expected profit (counting the sampling costs for the second set of 17 pairs but not for the first set) for each possible number of successes after the first 17 pairs. These values are shown in Table 4.

The expected profit (averaged with respect to the predictive distribution of the number of successes, which is uniform on the 18 possibilities) is 14.4, as opposed to 21 with the fixed sample design. But the expenses have been halved. Now the expected net profit is $14.4 - 17 = -2.6$. This is better than under the fixed-sample design but is still negative.

The one-look design incorporates the ability to stop the trial at its halfway point when interim data suggest that the drug may not be effective. This ability provides a substantial improvement over the fixed-sample-size design, but the improvement is not enough to make the trial worthwhile.

Further improvement can be attained by allowing more than one interim look. Consider the extreme: continual updating of the current probability distribution of θ and continual assessment of the question of continuation. Determining appropriate stopping times is a typical problem in dynamic programming (or backward induction), of which the preceding analysis of the one-look design was a simple special case.

To use dynamic programming for continual analysis, find the maximal expected profit for all possible final data (ignoring sunken costs of the trial). Then "back up" one step to the penultimate analysis time—to $n = 33$ in our case. For each possible piece of data after 33 pairs, calculate the predictive distribution of the next observation given the present data and evaluate the expected profit from continuing (include -1 for the incremental sampling cost). If the expected profit from continuing is greater than that from stopping, it is optimal to continue *if we ever have these data*, and we would stop otherwise.

Backing up in this fashion one step at a time, we pass through each analysis time and end up at the beginning. We then know the optimal decision (stop or continue) at each analysis time and for all possible data. In particular, we know whether to start the trial. We also learn the expected net profit from conducting the trial. (If other characteristics of the

Table 4 Expected Profit for Each Possible Number of Successes After the First 17 Pairs

Successes	0	1	2	3	4	5	6	7	8	9	10	11	12	13	14	15	16	17
Expected profit	0	0	0	0	0	0	0	0	0	0	0	0	8.9	25.0	39.2	51.1	61.9	72.5

optimal strategy besides the mean net profit are of interest—for example, the entire distribution of profit or the power function of the resulting hypothesis test—then these can be found easily once the optimal strategy is known.)

Consider the various possible data after 34 pairs. The decision problem is then trivial: If the number of successes is less than 24, then the drug cannot be marketed and the expected profit is 0 (ignoring sunken costs). And for $s \geq 24$ successes among the 34 pairs, the expected profit is

$$200E(\theta - .5 \mid s \text{ successes}) = 200 \frac{s - 17}{36}$$

These profits are shown in Table 5 and the first row of Table 6.

Now suppose 33 pairs are available, with one more observation possible. And suppose 23 of the 33 are successes. If the 34th pair results in a success, then the expected profit is 38.9. If it is a failure, then the expected profit is 0. If there are fewer than 23 successes among the first 33 pairs, 24 successes among 34 pairs are not possible and so it is obviously best to not take the last observation.

Again consider 23 successes among the first 33. If another pair is observed, then marketing will be possible if it is a success (resulting in a profit of 38.9) and not if it is a failure (profit = 0). Given 23 successes among the first 33 pairs, the probability of a success on the next pair is 24/35. So, after subtracting 1 for the clinical trial costs of this next pair, the expected net profit of continuing is

$$\tfrac{24}{35}(38.9) + \tfrac{11}{35}(0) = 25.7$$

This is greater than 0 and so continuing is better than stopping at $(s, n) = (23, 33)$. The other possible numbers of successes after 33 pairs are considered in the same way, yielding the second row in Table 6. To calculate the second row of Table 6 (after 33 pairs are observed) required only the top row of the table and the corresponding predictive probabilities. The third row of Table 6 (after 32 pairs are observed) requires only the second row and the predictive probabilities for moving from a point on the third row to the various points on the second row. Moreover, the

Table 5 Expected Profits for 24 or More Successes in 34 Pairs

Successes	24	25	26	27	28	29	30	31	32	33	34
Expected profit	38.9	44.4	50.0	55.6	61.1	66.7	72.2	77.8	83.3	88.9	94.4

Table 6   Expected Net Profits for Continuous Updating Procedure[a]

NUMBER OF SUCCESSES (columns 0–34); rows are NUMBER OF PAIRS REMAINING

Pairs \ Successes	0	1	2	3	4	5	6	7	8	9	10	11	12	13	14	15	16	17	18	19	20	21	22	23	24	25	26	27	28	29	30	31	32	33	34
34	0	0	0	0	0	0	0	0	0	0	0	0	0	0	0	0	0	0	0	0	0	0	0	0	39	45	50	56	62	67	73	78	84	89	95
33	0	0	0	0	0	0	0	0	0	0	0	0	0	0	0	0	0	0	0	0	0	0	0	26	42	48	54	59	65	71	77	82	88	94	
32	0	0	0	0	0	0	0	0	0	0	0	0	0	0	0	0	0	0	0	0	0	0	17	37	46	51	57	63	69	75	81	87	93		
31	0	0	0	0	0	0	0	0	0	0	0	0	0	0	0	0	0	0	0	0	0	10	30	42	49	55	61	67	73	79	85	91			
30	0	0	0	0	0	0	0	0	0	0	0	0	0	0	0	0	0	0	0	0	6	23	38	46	53	59	65	71	78	84	90				
29	0	0	0	0	0	0	0	0	0	0	0	0	0	0	0	0	0	0	0	3	16	32	43	50	57	63	70	76	83	89					
28	0	0	0	0	0	0	0	0	0	0	0	0	0	0	0	0	0	0	1	11	26	39	48	54	61	68	74	81	88						
27	0	0	0	0	0	0	0	0	0	0	0	0	0	0	0	0	0	0	6	21	35	45	52	59	66	73	80	87							
26	0	0	0	0	0	0	0	0	0	0	0	0	0	0	0	0	0	3	15	30	41	49	57	64	71	78	85								
25	0	0	0	0	0	0	0	0	0	0	0	0	0	0	0	0	1	10	24	37	47	54	62	69	77	84									
24	0	0	0	0	0	0	0	0	0	0	0	0	0	0	0	0	6	19	33	43	52	60	67	75	83										
23	0	0	0	0	0	0	0	0	0	0	0	0	0	0	0	3	14	28	40	49	57	65	73	81											
22	0	0	0	0	0	0	0	0	0	0	0	0	0	0	1	9	23	36	46	55	63	72	80												
21	0	0	0	0	0	0	0	0	0	0	0	0	0	0	5	18	31	43	53	61	70	79													
20	0	0	0	0	0	0	0	0	0	0	0	0	0	3	13	27	39	50	59	68	77														
19	0	0	0	0	0	0	0	0	0	0	0	0	1	8	22	35	47	57	66	76															
18	0	0	0	0	0	0	0	0	0	0	0	0	5	17	31	44	54	64	74																
17	0	0	0	0	0	0	0	0	0	0	0	2	12	26	40	52	62	73																	
16	0	0	0	0	0	0	0	0	0	0	1	8	21	36	49	60	71																		
15	0	0	0	0	0	0	0	0	0	0	4	16	31	45	58	70																			
14	0	0	0	0	0	0	0	0	0	2	11	27	42	55	68																				
13	0	0	0	0	0	0	0	0	0	7	22	38	53	66																					
12	0	0	0	0	0	0	0	0	4	16	33	49	64																						
11	0	0	0	0	0	0	0	2	12	28	46	62																							
10	0	0	0	0	0	0	0	7	23	42	60																								
9	0	0	0	0	0	0	4	18	38	57																									
8	0	0	0	0	0	1	13	33	54																										
7	0	0	0	0	0	8	27	51																											
6	0	0	0	0	4	21	47																												
5	0	0	0	2	15	42																													
4	0	0	0	10	37																														
3	0	0	5	30																															
2	0	2	23																																
1	0	15																																	
0	7																																		

[a] When an entry is 0, stopping the trial is best should those data arise. (To eliminate ambiguities regarding stopping or not, all positive entries have been rounded up.)

algorithm for generating the second row from the first is the same as that for generating the third from the second. This simple algorithm can be incorporated into a computer program.

The bottom line in Table 6 shows that this design is a substantial improvement over the previous designs, and now the expected net profit is positive: 6.22 (which rounds up to the 7 shown in the last entry of the table). The zeroes in the table indicate the interim results for which stopping is better than continuing. So the complete design is a bit complicated but is not difficult to describe.

The entry 0 in the next to the bottom row under 0 successes means that stopping is in order after but a single pair if that pair results in a failure. This sensitivity to the data has at least two sources. First, as we saw in analyzing the previous two designs, the prospects for profit with this drug are so tenuous that it is not even clear that beginning the trial is appropriate. Second, the uniform prior distribution reflects little prior information, with the result that a single observation can have a substantial effect.

Regarding the second point, the entries near the top of Table 6 are more nearly similar to neighboring entries than are entries near the bottom of the table. This is because information obtained early has a more dramatic effect when updating the distribution of θ. For example, the variance of θ drops by $\frac{1}{3}$ with the first observation [from $\frac{1}{12}$ for the beta(1, 1) distribution to $\frac{1}{18}$ for both beta(2, 1) and beta(1, 2) distributions]. On the other hand, the average drop in variance resulting from one observation is less than 5% for distributions near the top of the table. (Certain observations can increase the variance of θ, but on average it decreases.)

Average sample sizes can be calculated for each of the designs just presented. The sample size for the fixed-sample design is, of course, 34. For the one-look design the average sample size is only

$$\tfrac{6}{18}(34) + \tfrac{12}{18}(17) = 22.7$$

The average sample size for the design with continual analysis is smaller yet: 10.5.

It will not usually be appropriate to require exactly 34 pairs for marketing approval. The design of Table 6 sometimes takes too many observations and sometimes it takes too few. We can modify the sample size restriction as follows. Suppose marketing approval requires a minimum of 34 patients and $P(\theta > \tfrac{1}{2}) \geq .99$. Dynamic programming applies to find the best procedure in the same way as previously described, but now each step along the way of the backward process includes a stopping criterion. Results of such an analysis are not presented here, but interested readers are referred to Berry (1991b).

Such developments are not extensively used in making actual pharmaceutical or other decisions. They should be, at least to serve as guides.

If the actual decision differs from the one found in a formal decision analysis, the decision maker should be able to identify the aspects of the analysis (probabilities or utilities) that are inappropriate. Should it happen that the modified assumptions do not change the implied decision, then the decision maker would have to face the possibility that the actual decision is a bad one. In any case, the decision analytic process highlights the important aspects of the problem and leads to more enlightened decisions.

We mentioned that much of the Bayesian literature uses posterior probabilities to play the role usually played by p-values but does not explicitly address decision issues. Examples in the clinical trials literature include Spiegelhalter et al. (Chapter 2), Carlin et al. (1994), and Rosner and Berry (1994). The latter paper describes the design of a multiple-arm clinical trial in which an arm is dropped if the probability that other arms are better becomes sufficiently large. The motivating example is a trial in patients with metastatic breast cancer comparing various infusion schedules and doses of taxol. Stangl (1995) formally addresses the use of predictive distributions within a decision theoretic context in a multicenter clinical trial. The chapter by Clyde et al. (Chapter 11) more explicitly addresses decision issues. The authors incorporate historical data via hierarchical models to come up with sampling designs for setting heart defibrillators. Both the energy level to be used and the number of observations to be taken are design parameters. The high cost and sequential nature of sampling make Bayesian methods attractive for such a problem.

6 MODELING HISTORICAL INFORMATION

Science is about synthesizing all information available with the goal of advancing knowledge. However, many researchers take the view that statistical methods apply only to the data from an experiment at hand. Researchers must place historical information in the context of the current information. As we have discussed, the usual way to incorporate historical information is subjectively, through the prior distribution. Eddy et al. (1992), Berry and Hardwick (1993), and Lin (1993) introduce methods for discounting historical data through modifications of the likelihood function of the historical data. Such discounting is in part subjective. And it depends partly on the historical data and partly on the current data—especially on the comparability of the two.

The "Discussion" sections of medical publications usually place the current study in the context of previous and concurrent studies, but the arguments are seldom statistical. In effect, the reader is left to synthesize the available information or wait for a published meta-analysis. Statistical analyses can and should address larger scientific questions. They should not be restricted to the current experiment. In this section we review a method proposed in Berry and Hardwick (1993) for analyzing clinical trial data that explicitly models the other available information.

Patients in historical studies may be different from those in the current study. Some ways in which they are different are known—time is an example. But many ways are not known. And the effects of any differences may not be understood. A reader usually has information from sources other than the study at hand and discounts it (subjectively) based on his or her information. The approach to be presented here discounts information formally and has at least two advantages over an ad hoc approach. First, it does not rely on a reader's memory of previous studies. Second, it allows explicit use of covariates. With such an approach, readers have a common basis for judging treatment differences. Of course, some readers have information beyond that considered in any synthesis of historical data. But readers' conclusions drawn from a comprehensive report of a study will be less variable than from a statistical analysis restricted to the study data.

A concern in comparing results across studies is that the treatment effect may be confounded with study. It will then be confounded as well with any covariates that vary with study. In addition, treatment assignment may be confounded with study. In the extreme case, experimental therapy E is used exclusively in the current trial, whereas only historical data are available concerning control therapy C. Treatment comparisons under complete or nearly complete confounding are problematic. Some statisticians view such comparisons as fundamentally flawed and so dismiss historical controls.

Using historical controls requires some confidence that the controls may have responded in a way similar to the experimental patients had they been in a randomized trial and happened to have been assigned the therapy actually assigned. Also, the use of historical controls requires out-of-the-ordinary data analytic methods. At the same time this approach must be convincing to the medical community, so it should be intuitive and easy to describe.

There are two extreme attitudes regarding historical observations: (1) They are exchangeable with the current observations (except possibly for therapy administered), and (2) they are unreliable and should be ignored. The first is always wrong, but it may provide an approximation or bound.

Attitude (2) may be appropriate in some circumstances. One possibility is when ample concurrent controls are available. But even then, historical controls contain information, and the analyst should address any differences between concurrent and historical controls.

Consider an experimental unit. In a clinical setting this may be a single patient or a set of patients. Suppose θ is a parameter or set of parameters of interest. $L(\theta)$ is the likelihood function evaluated at θ for a unit in the current trial. For example, if responses are dichotomous then $L(\theta) = \theta$ if success and $L(\theta) = 1 - \theta$ if failure.

The likelihood function may not be $L(\theta)$ for a historical unit or for another unit not part of the study of interest. If the unit is regarded as exchangeable with those in the current trial, then $L(\theta)$ applies. So assuming exchangeability weighs historical and current units equally. When the current population of patients is of interest, historical units should be weighed less. A way of modeling lack of exchangeability of historical units is to allow that their responses may have been different had they been in the current study. We consider mixtures of the two extremes mentioned above: historical and concurrent units are (1) exchangeable and (2) completely unrelated.

Suppose the probability that a historical unit is exchangeable with a current unit (on the same treatment) is α. And suppose the alternative likelihood corresponding to not being exchangeable is $L^*(\theta)$. For example, should the response of the historical unit contain no information about θ—historical and current responses are unconditionally independent—then $L^*(\theta)$ = constant. (In what may be viewed as a limitation of the model to be developed below, if L^* is constant, conclusions depend on the constant assumed.) This probability associated with $L^*(\theta)$ is $1 - \alpha$. Interpreting "probability" loosely (since these two likelihoods have separate and arbitrary scale parameters), take the likelihood for a historical unit to be

$$L(\theta, \alpha) = \alpha L(\theta) + (1 - \alpha)L^*(\theta)$$

where $0 \leq \alpha \leq 1$. Such modifications in the likelihood have been considered by Eddy *et al.* (1992, Chapter 15), but there are many alternatives to this model. For example, L and L^* might have interpretations similar to those above but combine multiplicatively rather than additively, as in Berry (1991a).

An important question to ask now is how one should choose α and L^*. These are unknown parameters, but they are related. One value of α might be appropriate for one functional form of L^*, whereas another value is appropriate for another. In the Bayesian approach, unknowns are random variables. So α and L^* have a joint prior probability distribution. All available information about the comparability of the various units can be used to formulate a prior distribution for (α, L^*). In the subjective view, this distribution depends on the person assessing it. Updating is via Bayes' theorem based on the available data (historical as well as concurrent). To keep calculations relatively simple, in our example below we will specify a particular form for L^* and consider various values for α. An appealing alternative is to let α be random and update its distribution based on the available data (Lin, 1993).

For any functional form L^*, the comparability parameter α depends on the unit. In particular, α can be a function of the unit's covariates. The basic unit in a clinical trial is the patient, and we can without loss consider individual patients. If historical patients in some subset are regarded as exchangeable, then each is assigned the same α. If a patient is in the current trial then $\alpha = 1$. If a historical patient seems nearly exchangeable with those in the current trial then α should be near 1. But if the historical patient is deemed to be quite different, then α should be small, with $\alpha = 0$ being the extreme case in which the patient is irrelevant for θ.

In 1982 Dr. Robert Bartlett and his co-workers at the University of Michigan initiated a clinical trial comparing extracorporeal membrane oxygenation (ECMO = E) with conventional therapy (CVT = C) for the treatment of severe respiratory failure in newborn infants (Bartlett et al., 1985). They cite prior evidence of ECMO's effectiveness in this population: 28 of 40 moribund newborns with birth weight ≥ 2 kg had survived on ECMO. And they claimed that their admission criteria included only infants who had "an 80% or greater chance of mortality despite optimal [conventional] therapy." However, they cited no data to support this claim for CVT.

ECMO is an extremely invasive procedure (Bartlett et al., 1985). It involves ligating the right carotid artery and (in 1982) leaving it permanently clamped. So ECMO would not be used unless the patient had a sufficiently bad prognosis without it. The investigators felt that a clinical trial comparing E with C was warranted. However, they were confident that ECMO was much better than CVT. They had ethical concerns about assigning patients to CVT should accumulating data suggest that ECMO was much better. They used a sequential treatment allocation scheme called "randomized play the winner." This scheme was symmetric in the two therapies: the current patient was more likely to be assigned to the better performing therapy.

Randomized-play-the-winner schemes aim to alleviate ethical concerns present in assigning fixed proportions of patients to the treatments. However, the probability of assigning a patient to either treatment is positive. A particular patient might be assigned to a therapy suggested by accumulating data to be inferior. So such schemes do not resolve ethical dilemmas. For a thorough discussion of ethical issues involving randomized trials of ECMO, see Truog (1992).

The first infant in the study was assigned to ECMO and survived. The second was assigned to CVT and died. The next 10 infants were assigned to ECMO and all 10 survived. The trial was then stopped, with 11 successes of the 11 infants on ECMO, and the only infant on CVT was a failure.

The analysis presented by the investigators and various subsequent analyses (including Ware and Epstein, 1985; Wei, 1988; Wei *et al.*, 1990; Begg 1990) were conditioned on the assignment scheme. (The following is a list of published P values: .00049, .001, .003, .009, .038, .045, .051, .083, .280, .500, .617, 1.000, and "undefined." A minor reason for such variety is that the study's design called for stopping the trial two patients before it was actually stopped, and some analysts ignore the last two patients.) Begg (1990) and many of his discussants feel that inferences from this study are weak or impossible. One reason is that frequentist measures are tied to a trial's design. Bayesian inferences are possible in this study because the stopping rule is not relevant in the Bayesian approach (Berger, 1985; Berger and Berry, 1988; Berger and Wolpert, 1984; Berry, 1987).

Ware and Epstein (1985) criticized the Bartlett *et al.* (1985) study. They claimed that the type I error rate for the study was "unacceptably high." (Their calculation gave $P = \frac{1}{2}$—assuming the therapies had identical effectiveness, E and C were equally likely to end up with the higher success proportion for the stopping rule used in the study.) They suggested that "further randomized clinical trials using concurrent controls . . . will be difficult but remain necessary." [The Epstein-Ware group at Harvard University, Brigham and Women's Hospital in Boston, and Massachusetts General Hospital conducted a randomized trial of ECMO. The results were reported and analyzed in Ware (1989).]

We agree that the study's conclusion is suspect. No evidence was provided to support the "80% mortality" claim. In our view it is important to examine the available data to evaluate the claim. If something like it is true, a randomized clinical trial seems inappropriate and perhaps unnecessary.

A way to view our goal is that we want to show how one might examine this claim. This requires assessing historical information about CVT and also about ECMO. We want to use the information available to Bartlett and his colleagues in 1985, when the last study patient responded. An obvious problem with our analysis is that it is not 1985. We now have more information concerning the relative efficacy of ECMO. Hindsight is perfect. So our example must be considered somewhat artificial. [However, our reaction (unpublished) upon seeing the Bartlett, *et al.* paper was to advocate precisely the analysis undertaken in the current paper before conducting a randomized trial.]

The measure of efficacy in the Bartlett study was survival. Although side effects are potentially serious, we restrict consideration to survival. Let $\theta = (\theta_E, \theta_c)$ where θ_j is the probability that an infant dies when assigned to treatment j.

The infants in the Bartlett study were not exchangeable. Some had better prognoses. Skeptics have suggested that the single CVT patient was the sickest of the 12 in the study (Begg, 1990; Paneth and Wallenstein, 1985). Although the calculation of prognosis considered in Berry and Hardwick (1993) suggests that it was not the sickest, the question is not resolved. But it is clear that patient covariates should play a role in analyzing data from any study.

We want a univariate measure of "sickness." A probability of death on either CVT or ECMO (or on any other therapy) is especially appealing. Berry and Hardwick (1993) use the probability of death on E based on disease type (respiratory distress syndrome, meconium aspiration syndrome, etc.) and neonatal pulmonary insufficiency index (NPII) as described in Wetmore *et al.* (1979). Wetmore *et al.* (1979) give data on 109 ECMO patients who weighed more than 1.5 kg. We estimated the relationship between NPII and death within patient groups for the Wetmore *et al.* data:

$x = P(\text{death} \mid \text{NPII, disease type})$

(We did not use the authors' estimate. It was apparently drawn by eye and greatly overestimates the probability of death.) We calculated x for each patient and exploit this estimate in our analysis. A by-product of our analysis is a revision of prognosis x for each therapy, ECMO and CVT, based on the available data.

The Bartlett *et al.* (1985) study admitted patients from October 1982 to April 1984. Our data set includes the 12 patients in this study. It also includes 14 newborn infants who were admitted to the University of Michigan Medical Center between February 1979 and June 1981 and who met the study admission criteria. The first 12 of these 14 were given CVT and 9 of the 12 died. The last two were given ECMO and both survived. Our data set also includes these 14 patients, but discounted by α, as described below.

We sought other available information about both CVT and ECMO. We found no further information (published or otherwise) about CVT. [A national database of the type recommended in Berry (1989a) would have been a boon to our endeavors.] Bartlett, *et al.* (1982) give an additional 34 ECMO patients with birth weight ≥ 2 kg. These were treated prior to 1981 at one of four hospitals. (The data were not presented by hospital, so we cannot judge hospital effect.) Twelve of these patients died. We included these 34 patients in our database, again discounted by α.

Figure 13 shows the full database of 60 patients: 13 on CVT and 47 on ECMO. Circles identify the patients in the clinical trial; all others are historical patients. This dot diagram shows the patients' $x = P(\text{death} \mid E)$

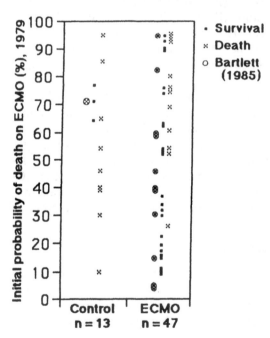

Figure 13 The full data set; one mark per patient. There are 47 ECMO and 13 CVT patients. Eleven ECMO patients and one CVT patient were treated between October 1982 and April 1984 at the University of Michigan as part of the Bartlett *et al.* (1985) study. Marks representing these 12 patients are shown in separate columns from the other 48 patients and are circled. Two ECMO and 12 CVT patients were treated between February 1979 and June 1981, also at the University of Michigan. The remaining 34 ECMO patients were treated prior to 1981 at four hospitals and were reported in Bartlett *et al.* (1982).

based on NPII, as described above, and their responses. The fit of x for ECMO patients is reasonably good, except for an evident shift in location. This shift reflects an improvement in ECMO for the Bartlett *et al.* (1982) patients over earlier patients (Wetmore *et al.*, 1979) for whom the estimate x is based. And there is a further improvement in the ECMO patients in Bartlett *et al.* (1985). Our method quantifies this improvement.

Consider a patient with prognosis x on ECMO as described above. We will use the logistic model as in Berry (1989c). The probability of death for a patient in the current clinical trial on therapy j is assumed to be

$$\theta_j(x) = \frac{x \exp(u_j)}{x \exp(u_j) + 1 - x}$$

for j = E and C. So each θ_j is transformed to u_j.

The same x is used whether j = E or C. It is not important that x is defined assuming ECMO rather than CVT or a third therapy. What matters is that x reflects the condition of the patient. Should it happen that both θ_j are constant in x then the analysis reduces to ignoring prognosis. However, if, for example, θ_E is increasing in x while θ_C is decreasing in x, then conclusions using the above model would be questionable.

Parameters u_E and u_C are unknown and so are random. Their distributions are assessed subjectively in the usual way Spiegelhalter *et al.* (Chapter 2) and Chaloner (Chapter 4). We assume u_E, u_C ~ iid $N(0, \sigma^2)$, with $\sigma^2 = 2$. [This value is effectively ∞ in the sense that values of $\sigma^2 > 2$ give essentially the same final distribution for (u_E, u_C).]

We assume $L^* = x$. So the probability of death for a historical patient on therapy j is

$$\alpha \frac{x \exp(u_j)}{x \exp(u_j) + 1 - x} + (1 - \alpha)x$$

for j = E and C. So the likelihood function of u_j is

$$L(u_j) = \prod_{\{i:i \in D_j\}} \left[\alpha_i \frac{x_i \exp(u_j)}{x_i \exp(u_j) + 1 - x_i} + (1 - \alpha_i)x_i \right]$$

$$\times \prod_{\{i:i \in S_j\}} \left[\alpha_i \frac{1 - x_i}{x_i \exp(u_j) + 1 - x_i} + (1 - \alpha_i)(1 - x_i) \right]$$

where α_i = 1 if patient i is a current patient and α_i = α if patient i is a historical patient. The first product is over the set D_j of patients i who received therapy j and died, and the second product is over the set S_j of patients i who received therapy j and survived. This is a two-sample problem; we assume the joint likelihood of u_E and u_C is the product $L(u_E)L(u_C)$.

The posterior distribution of (u_E, u_C) was calculated using Monte Carlo simulation and the rejection method. The posterior means of $\theta_E(x)$ and $\theta_C(x)$ are shown in Figures 14 through 16, assuming $\sigma^2 = 2$ and $\alpha = 0, \frac{1}{2}$, and 1. [Figure 17 shows the distribution of $\theta_E(x) - \theta_C(x)$ assuming $\alpha = \frac{1}{2}$, the same case as Figure 15.] Clearly, the value of α matters. One can choose whichever value of α is appealing or average over the three. We think $\alpha = \frac{1}{2}$ captures the historical data more appropriately than the other two. ECMO is apparently improving over time, so early data should perhaps not be counted fully. (This "learning curve" continues with ECMO. Results from individual centers show that the first 10 or so patients do not fare as well as later patients, leading some to suggest having only regional centers perform the operation. There may also be a time trend with CVT as the technique is improved and refined.)

Figure 18 shows the same curves as Figures 14 through 16, but ignoring the Bartlett *et al.* (1985) study data and counting the historical data

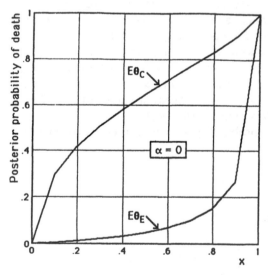

Figure 14 Historical patients are completely discounted ($\alpha = 0$) and so only the data from Bartlett *et al.* (1985) are used in this figure. The prior means of both $\theta_E(x)$ and $\theta_C(x)$ equal x. That the posterior mean of $\theta_C(x)$ is so close to x is expected, as there was only one CVT patient in the study. Moreover this patient's prognosis was quite bad ($x = 0.71$) and he died. On the other hand, the posterior mean of $\theta_E(x)$ is dramatically shifted from x, reflecting the fact that all 11 ECMO patients survived. (These patients had moderately poor prognoses, the average x being 0.43.)

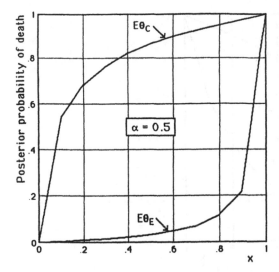

Figure 15   The 48 historical patients in Figure 1 are partially discounted ($\alpha = \frac{1}{2}$) compared with those reported in Bartlett *et al.* (1985). The posterior mean of $\theta_C(x)$ is shifted more than in Figure 14 because this figure includes an additional 12 patients (or rather, "half" of them). Nine of the 12 died, and the average value of x among the 12 was 0.55. The posterior mean of $\theta_E(x)$ is about the same as in Figure 14. It now includes 36 additional (half-)patients, but 12 of them died. Those who died tended to have worse prognoses.

fully ($\alpha = 1$). This study did indeed add information concerning ECMO. However, there was already good evidence that E was better than C.

An advantage of the method we describe is that it provides a means for updating initial prognosis x based on available data. One can calculate x as usual and then apply the appropriate curve from among those in Figures 14, 15, 16, and 18 (depending on treatment) to revise the calculated x.

The analysis just presented sets the stage for addressing an important decision problem. Given the information available in 1985, is it appropriate to design a randomized clinical trial (RCT) comparing ECMO with CVT? What is to be gained and what is to be lost? The information from an RCT might have value, but it has costs. The information has value only if it translates into more effective therapy of newborns suffering from respiratory failure. Costs are in terms of lives of patients in the trial. An RCT might help treat patients who present after the trial more effectively, but some patients in the trial may be treated poorly in the process.

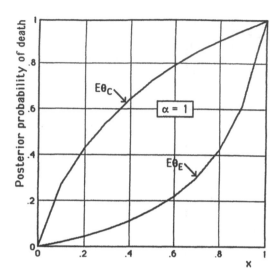

Figure 16 Now all 60 patients in Figure 13 count equally ($\alpha = 1$). The extra weight given to the historical patients (where the only deaths on ECMO occurred) increases the posterior mean of $\theta_E(x)$ compared with Figure 15.

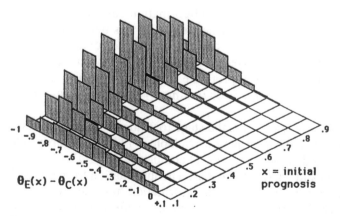

Figure 17 Posterior distribution of $\theta_E(x) - \theta_C(x)$ for $x = 0.1$ to 0.9 and for $\alpha = 1/2$ and $\sigma^2 = 2$. The means of $\theta_E(x)$ and $\theta_C(x)$ for this case are shown in Figure 15.

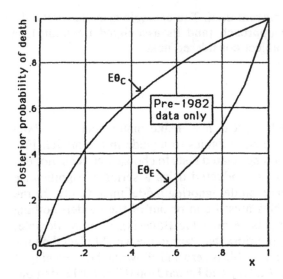

Figure 18 Now only the historical patients in Figure 13 count. We made no allowance for discounting the *current* patients in our model. We present this figure as a way of addressing the information available in 1982, before the Bartlett *et al.* (1985) study. Comparing Figures 14 and 16, say, shows that this study did indeed change the distribution of $\theta_E(x)$.

We have argued that the information available in 1985 concerning ECMO and CVT strongly suggests that ECMO was more effective. But reasonable people might reasonably disagree. If an RCT is not conducted, what would happen? Apparently, a substantial portion of the medical community had accepted ECMO and was using it (Toomasian *et al.*, 1988). The question is whether this use is appropriate. Weighing the possibility of saving lives after the trial as opposed to during the trial is a substantial question of medical ethics.

An important aspect of this decision problem is that therapies are evolving rapidly. An ECMO procedure that is standard at the start of a trial may be antiquated at the end; the same is true for CVT. So learning a lot about ECMO 1985 may solve no substantial clinical question about ECMO 1988. Thus, anything learned may be irrelevant. This dynamic aspect of both therapies argues strongly against an RCT and in favor of monitoring databases over time.

The example in this section shows the flexibility of the Bayesian approach. Historical data can be incorporated into an analysis even if they are not exchangeable with current data. The example also shows the dependence of the Bayesian approach (and its associated flexibility) on models. This leads to the subject considered next.

7 MODEL SELECTION

In building statistical models, researchers make many decisions about how to code variables and which variables to include in the model. Frequentists use likelihood ratios and related tests to choose between models. Typically, variable selection is conducted using a series of significance tests that condition on a single model ignoring model uncertainty. In the frequentist paradigm, variables must be in or out of the model; nothing in between is available. In this respect the Bayesian is more flexible, allowing the use of mixtures of priors that vary in the weight placed on the probability that the effect of a variable is zero and allowing the averaging of alternative plausible models. Raftery and Richardson (Chapter 12) demonstrate a Bayesian counterpart to frequentist significance tests that propagates the model uncertainty through to inference about quantities of interest. They describe a Bayesian modeling strategy using Bayes factors, the ratio of posterior to prior odds of competing models, to formalize the model selection process. They demonstrate their approach with an application that examines fat and alcohol consumption as risk factors for breast cancer.

8 HIERARCHICAL MODELS

Hierarchical models are ideally suited for, and are commonly used in, Bayesian analysis (Lindley and Smith, 1972; Berger, 1985). Indeed, most of Bayesian analysis can be thought of in this context. A typical problem in which a hierarchical model is appropriate is meta-analysis (DuMouchel, 1989; Eddy *et al.*, 1992). Each substudy incorporated in the meta-analysis is viewed as having unknown characteristics that set it apart from the others and hence as having its own distribution for responses. Associate a parameter with each distribution. Selecting a study means selecting a parameter—a random-effects model. If the parameter of the selected study were to be revealed, this would give direct information about the distribution of substudy distributions, and there is no hierarchy. But the substudy parameters are not known; we observe only a sample from each distribution. This makes a hierarchy necessary. To demonstrate, consider the following example taken from DuMouchel (1989). A meta-analysis was conducted to examine the effects of the antidepressant drug S-adenosyl-methionine (SAMe). Each of the nine study sites considered in the analysis had characteristics setting it apart from the others that affected the distribution of outcomes at that site. The number of patients at site j was n_j and the number of successes was x_j. These numbers are presented in Table 7.

Suppose that the patients at site j are exchangeable in the sense that all had the same probability p_j of success. The likelihood function of $(p_1, p_2, \ldots, p_9)$ is

$$\prod_{j=1}^{9} p_j^{x_j}(1 - p_j)^{n_j - x_j}$$

If we did a naive pooled analysis assuming all 150 patients were exchangeable, that is, all p_j are equal to a common p, the maximum likelihood estimate for p is about .7, and more than 95% of the area under the likelihood is between .6 and .8. This is a curious result because the observed success proportions in five of the nine sites is outside the range .6 to .8 (see Figure 19). Sampling variability accounts for some differences, but the variability in Table 7 is greater than would be expected from sampling alone. This suggests that the p_j are not equal.

Separate analyses of the nine sites are unsatisfactory as well. Presenting nine estimates or nine confidence intervals does not adequately assess the effect of the drug. This assumes we learn nothing about the effect of the drug at site 1 by observing the effect of the drug at site 2. This is not true, so by analyzing separately we have wasted information. In addition,

Table 7 Successes Observed with the Antidepressant Drug S-Adenosylmethionine

Site	x_i	n_i	$\hat{p}_j = x_j/n_j$
1	20	20	1.00
2	4	10	.40
3	11	16	.69
4	10	19	.53
5	5	14	.36
6	36	46	.78
7	9	10	.90
8	7	9	.78
9	4	6	.67
Totals	106	150	.71

Berry and Stangl

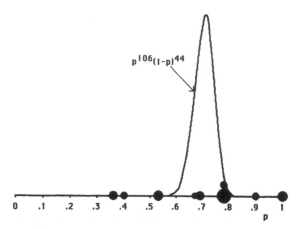

$p^{106}(1-p)^{44}$

| | | | | | | | | | | |
|0|.1|.2|.3|.4|.5|.6|.7|.8|.9|1|

p

Figure 19 Likelihood function of p for the data in Table 7 assuming the 150 patients in the nine studies are exchangeable. (The nine dots correspond to the observed proportions for the nine studies, with dot areas proportional to sample sizes.)

while we are interested in predicting the effect for the next patient at each of the nine sites represented, we are also interested in the effect for patients treated at sites not represented. So the question remains, how do we combine the data to best assess the effect of the drug at the sites represented as well as at those that were not?

The Bayesian hierarchical perspective is that each site's success proportion is selected from some population. To quantify information about the general population requires a probability distribution of subpopulation distributions.

Suppose $p_1, p_2, \ldots, p_9$ is a random sample from the population distribution F, which is itself random. Assume F is a beta distribution with parameters α and β, both of which are unknown. The variance of the beta distribution is

$$\frac{\alpha\beta}{(\alpha + \beta)^2(\alpha + \beta + 1)}$$

If $\alpha + \beta$ is large, the distribution of the p's is highly concentrated and consequently there is little heterogeneity in the treatment effect across sites.

As with all unknowns in the Bayesian approach, the user must assess a probability distribution for α and β; call it $\pi(\alpha, \beta)$. If the user has information suggesting that there is little variability in treatment effect across the sites, then the prior, $\pi(\alpha, \beta)$, should be concentrated on large values of α and β. Similarly, if information exists suggesting that there is a lot of variability in treatment effect across the sites, then the prior should be concentrated on small values of α and β.

Call $\pi^*(\alpha, \beta \mid x)$ the posterior probability distribution of α and β given x and n. From Bayes theorem,

$$\pi^*(\alpha, \beta \mid x) \propto f(x \mid \alpha, \beta)\pi(\alpha, \beta)$$

where

$$f(x \mid \alpha, \beta) = \int f(x \mid p, \alpha, \beta)f(p \mid \alpha, \beta)\, dp$$

$$= \int \binom{n}{x} p^x(1 - p)^{n-x} \frac{\Gamma(\alpha + \beta)}{\Gamma(\alpha)\Gamma(\beta)}\, p^{\alpha-1}(1 - p)^{\beta-1}\, dp$$

Completing the calculations, we get

$$\pi^*(\alpha, \beta \mid x) \propto \frac{\Gamma(\alpha)\Gamma(\beta)}{\Gamma(\alpha + \beta)} \frac{\Gamma(\alpha + \beta + n)}{\Gamma(\alpha + x)\Gamma(\beta + n - x)} \pi(\alpha, \beta)$$

Upon observing a sample $x_1 x_2, \ldots, x_9$, where x_j are binomial variables with parameters p_j and n_j and $p_1, p_2, \ldots, p_9$ is a random sample from F,

$$\pi^*(\alpha, \beta \mid x_1, \ldots, x_9) \propto \prod_{j=1}^{9} \frac{\Gamma(\alpha)\Gamma(\beta)}{\Gamma(\alpha + \beta)} \frac{\Gamma(\alpha + \beta + n_j)}{\Gamma(\alpha + x_j)\Gamma(\beta + n_j - x_j)} \pi(\alpha, \beta)$$

Next consider the response of an untreated patient from a represented site. Given the results in Table 7, the probability of success for the next patient at site j is

$$E(p_j \mid x_1, \ldots, x_9) = E\left(\frac{\alpha + x_j}{\alpha + \beta + n_j} \,\Big|\, x_1, \ldots, x_9\right)$$

This distribution is with respect to the posterior distribution of α and β. The probability of success for a patient at an unobserved site would be

$$E(p_{10} \mid x_1, \ldots, x_9) = E\left(\frac{\alpha}{\alpha + \beta} \,\Big|\, x_1, \ldots, x_9\right)$$

This is just the expected posterior mean. In Section 2.3 we discussed the selection of a prior for p in a simple beta-binomial model. Now we need to extend this to selecting the prior for the beta parameters α and β. This

Prior estimated density: beta (1,1)

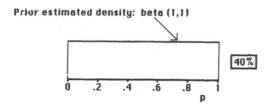

Figure 20 Assessor's prior estimate of population distribution of success proportions. This is a mixture of beta(α, β) distributions that happens to be itself a beta
density: $\alpha = \beta = 1$. The beta(1, 1) density has 40% of the prior probability.

prior can be discrete or continuous, but we will consider only the discrete
case in which α and β are integers between 1 and 10.

    Suppose an assessor's best estimate of the effectiveness of SAMe
over all sites is 50%. Moreover, the assessor's tentative opinion is that
the distribution F of success proportions over sites is that they are uniformly spread on the interval (0, 1). This corresponds to $\alpha = 1$ and $\beta = 1$
or the beta(1, 1) $= \pi(1, 1)$ distribution shown in Figure 20. Although this
is the assessor's best estimate, there is uncertainty and other F's are
plausible. Considering $\alpha + \beta = 3$ and $\alpha + \beta = 4$ gives the set of priors
in Figures 21 and 22. Note that the average of each set of distributions

Densities with $\alpha + \beta = 3$:

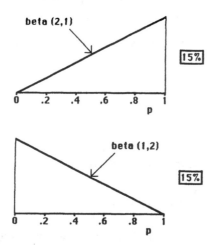

Figure 21 The assessor's prior probability of each of these two densities with
$\alpha + \beta = 3$ is about 15%. The average of the beta(2, 1) and beta(1, 2) densities
happens to equal the prior estimate, the beta(1, 1) density.

Densities with $\alpha + \beta = 4$:

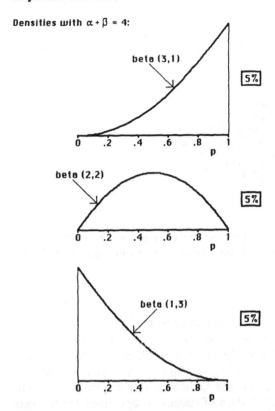

Figure 22 The assessor's prior probability of each of these three densities with $\alpha + \beta = 4$ is about 5%. The average of the beta(3, 1), beta(2, 2), and beta(1, 3) densities happens to equal the prior estimate, the beta(1, 1) density.

defined by $\alpha + \beta$ equal to a particular integer is the $\pi(1, 1)$ distribution. This is also true if we average across the sets of distributions specified. Similar sets can be derived for other integer values of $\alpha + \beta$. If the assessor gives prior probability of 0.40 for $\pi(1, 1)$, 0.15 for $\pi(1, 2)$ and $\pi(2, 1)$, and 0.05 for $\pi(3, 1)$, $\pi(2, 2)$, and $\pi(1, 3)$, etc., then the joint distribution $\pi(\alpha,\beta)$ implicit is the product of two independent geometric variables. This joint distribution is pictured in Figure 23. As an alternative, results under this model will be compared to the product of two independent uniforms on the integers between 1 and 10. This prior is shown in Figure 24. For this distribution all possible pairs of α and β are equally likely, so the posterior probabilities are proportional to the likelihood function on the lattice points where the prior probabilities are positive.

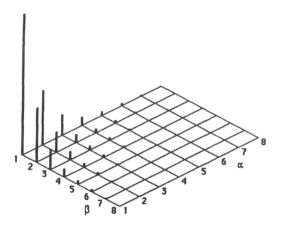

Figure 23 Independent geometric distributions on α and β. The points with the six largest probabilities correspond to the densities shown in Figures 20, 21, and 22.

Using the data from Table 6, the posterior probabilities of α and β can be calculated. Figures 25 and 26 show these probabilities assuming the geometric (Figure 23) and uniform (Figure 24) priors, respectively. Under the geometric prior $\pi(2, 1)$ has the highest posterior probability, and under the uniform $\pi(4, 2)$ has the highest posterior probability. The posterior estimates of the distribution of success proportions for the geometric and uniform priors are displayed in Figures 27 and 28. These distri-

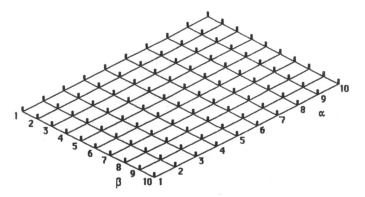

Figure 24 Uniform probabilities for beta parameters (α, β) for $\alpha = 1, \ldots, 10$ and $\beta = 1, \ldots, 10$.

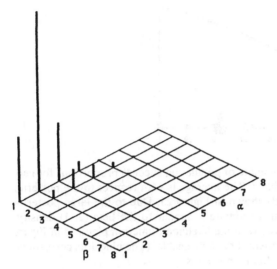

Figure 25 Posterior probabilities $\pi^*(\alpha, \beta \mid x_1, \ldots, x_9)$ assuming prior probabilities shown in Figure 23 and conditioning on the results of the nine studies shown in Table 7.

butions are a mixture of all $\pi(\alpha, \beta)$ in the lattice with weights equal to the posterior probability of $\pi(\alpha, \beta)$,

$$\pi^*(p \mid x_1, \ldots, x_9) = \sum_{\alpha, \beta} \pi(\alpha, \beta) P(\pi(\alpha, \beta \mid x_1, \ldots, x_9)$$

It is evident from the figures that the geometric prior is more heavily

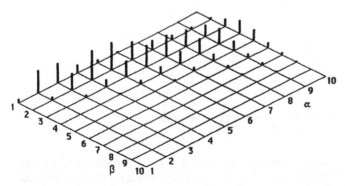

Figure 26 Posterior probabilities $\pi^*(\alpha, \beta \mid x_1, \ldots, x_9)$ calculated assuming the prior probabilities shown in Figure 24.

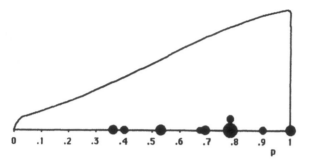

Figure 27 Posterior density estimate of population success proportions for geometric prior and the data in Table 7. The average is with respect to the posterior probability distribution of (α, β) shown in Figure 25. Due to the large posterior probability on $\alpha = 2$, $\beta = 1$, this estimate is similar to the beta (2, 1) density. This estimate should be compared with the likelihood function pictured in Figure 19, which assumes that all 150 patients are exchangable. (The nine dots correspond to the observed proportions for the nine studies.)

weighted toward site heterogeneity as the posterior density is less peaked and heavier tailed than under the uniform prior.

Table 8 presents the predictive probability or the probability of success for the next patient in each of the nine sites and for a patient at a tenth site. The predictive probability of success for the new site is the

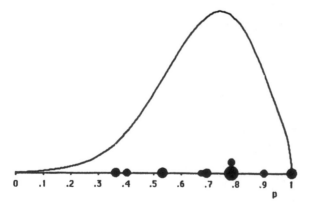

Figure 28 Posterior density estimate of population success proportions for uniform prior and the data of Table 7. This is the average density with respect to the posterior distribution of (α, β) shown in Figure 26. (As in Figure 27, the dots correspond to the observed proportions.)

Table 8   Successes Observed with SAMe and Predictive Probabilities of Success by Site

Site	x_i	n_i	$\hat{p}_j = x_j/n_j$	Predictive probability from Geometric	Uniform
1	20	20	1.00	.95	.90
2	4	10	0.40	.46	.53
3	11	16	0.69	.68	.69
4	10	19	0.53	.55	.57
5	5	14	0.36	.41	.48
6	36	46	0.78	.77	.77
7	9	10	0.90	.84	.80
8	7	9	0.78	.75	.73
9	4	6	0.67	.66	.68
Totals	106	150	0.71	.65	.68

overall mean and is shown as the column total. The table shows how the individual site probabilities are shrunk toward the overall mean. The shrinkage is less for the geometric prior because it gives more credence to site heterogeneity than does the uniform prior. Also, as is reasonable, shrinkage to the overall mean is greater for smaller sites. More interesting examples consider two treatments and model the relationship between them. Again, the Bayesian approach is to assume a hierarchical model.

This book contains numerous examples of hierarchical models. Du-Mouchel *et al.* (Chapter 17) use hierarchical linear models to explore the association between schizophrenia and extreme weather across 17 states. Racine-Poon and Wakefield (Chapter 13) use hierarchical models to fit pharmacokinetic-pharmacodynamic models. Smith *et al.* (Chapter 15) show how graphical models can be used to structure hierarchical random-effects models for meta-analysis. Albert and Chib (Chapter 22) present a random-effects model for binary repeated measures data. Wolfinger and Rosner (Chapter 14) use a mixed model with a repeated measures, normally distributed response variable to investigate the ability of hydralazine to reduce blood flow in tumors. Stangl (Chapter 16) and Simon *et al.* (Chapter 21) present hierarchical models for survival time data, and Christiansen and Morris (Chapter 18) provide a tutorial on iterative model checking for two-level Poisson hierarchical models and present an example for specifying and checking medical profile models for transplant hospitals.

9 TIME SERIES

Time series methods are not frequently seen in health-related journals, but if these methods were more widely taught in biostatistics programs it is likely they would be seen more often. Some examples of Bayesian time-series methods are presented by Smith and West (1983), Smith et al. (1983), Trimble et al. (1983), Gordon and Smith (1990), Gamerman (1991), and West (1992). The first three papers demonstrate Bayesian time-series modeling to monitor kidney transplant patients. Gamerman (1991) used Bayesian dynamic time-series models survival data with time-varying covariate effects and nonproportional hazards models.

10 SOFTWARE AVAILABILITY

Although much of the available software is not yet to the point of being user friendly and commercially marketed, software for Bayesian analysis is becoming widely available. An exhaustive description of all the products available is beyond the scope of this chapter, but several will be introduced. Software for many of the applications in this book are catalogued in Carnegie-Mellon's public domain statistical software library (StatLib). An index of all available subroutines in the library can be retrieved by sending the one-line e-mail message

   send index

to statlib@lib.stat.cmu.edu. Beyond this useful resource, several other programs are available that are useful to statisticians interested in trying their hand at Bayesian analysis. These include Proc Mixed, a set of SAS algorithms for mixed models; LISP-STAT, a general-purpose programming language for interactive experimental programming (Tierney, 1990; 1992); BATS, algorithms for Bayesian time series (Pole et al., 1994); BUGS, a program to perform Bayesian inference using Gibbs sampling (Thomas, 1992); Confidence Profile Method (CPM), a program for Bayesian meta-analysis; First Bayes, a program written by A. O'Hagan (http://www.maths.nott.ac.uk/personal/aoh) that is available free of charge to anyone interested in teaching or learning Bayesian statistics; and Bayesian Computation with Minitab (Albert, 1996), a set of Minitab macros to teach introductory Bayesian analysis. An abridged version of the latter is available on a 3.5-in. diskette accompanying Berry (1996).

   Other examples of the use of available software in Bayesian analysis were brought to our attention by Bernardo and Smith (1994). These include the following. Cook and Broemeling (1995), Ley and Steel (1992), Korsan (1992), and Albert (1990) demonstrate the use of Mathematica. Smith *et al.* (1985, 1987) and Racine-Poon *et al.* (1986) describe the use of Bayes Four. Lauritzen and Spiegelhalter (1988) use methods on which the commercial expert system builder Ergo is based. Cowell (1992) and Spiegelhalter and Cowell (1992) apply the probabilistic expert system shell BAIES. Wooff (1992) describes [B/D], an implementation of subjective analysis of beliefs, and Marriott and Naylor (1993) discuss the use of Minitab to teach Bayesian statistics.

11 OTHER TOPICS AND DISCUSSION

This chapter provides an introduction to Bayesian methods and decision making in health research. The utility of Bayesian models for the design and analysis of clinical trials is quite clear from the examples presented here as well as in many of the subsequent chapters of this book. As computing progresses Bayesian methods become easier to implement and the scope and complexity of applications increase as well. Examples of more complicated models that are being seen frequently in Bayesian analysis are mixture models and change-point models. Mixture models can be thought of as models in which subgroups of a population follow different distributions. Change-point models are models in which the distribution of observations changes at some point in time. Dellaportas *et al.* (Chapter 23) use mixture models to model subpopulations in their analysis of birth weight data. Joseph *et al.* (Chapter 24) use multipath change-point models to estimate the effects of calcium supplementation on blood pressure.

Many other chapters also demonstrate the ease with which Bayesian methods handle complicated models and show that most Bayesian statisticians working on applications use a blend of Bayesian and frequentist methods. We hope that readers learn from these chapters and are inspired to venture by blending Bayesian methods into their own work. For further reading and more detailed presentations of Bayesian methods, readers are referred to Barnett (1982), Berger (1985), Berry (1996), DeGroot (1970), Howson and Urbach (1989), Box and Tiao (1973), Press (1982), Lindley (1985), Bernardo and Smith (1994) and Gelman *et al.* (1995). To all those who venture, we wish you well.

ACKNOWLEDGMENTS

D. Berry's research was supported in part by the National Science Foundation under grant DMS 94-0422. Some of this chapter appears in other publications: Berry (1990, 1991a, 1991b, 1996) and Berry and Hardwick (1993).

REFERENCES

Albert, J. (1990). Algorithms for Bayesian computing using Mathematica. In *Computing Science and Statistics: Proceedings of the Symposium on the Interface*, eds. Page, C. and LePage, R., pp. 286–290. Springer, Berlin.

Albert, J. (1996). *Bayesian Computation Using Minitab*. Duxbury, Belmont, CA.

Barnett, V. (1982). *Comparative Statistical Inference*, 2nd ed. Wiley, New York.

Bartlett, R. H., Andrews, A. F., Toomasian, J. M. Haiduc, N. J., and Gazzaniga, A. B. (1982). Extracorporeal membrane oxygenation (ECMO) for newborn respiratory failure: 45 cases. *Surgery* 92: 425–433.

Bartlett, R. H., Roloff, D. W., Cornell, R. G., Andrews, A. F., Dillon, P. W., and Zwischenberger, J. B. (1985). Extracorporeal circulation in neonatal respiratory failure: A prospective randomized study. *Pediatrics* 4: 479–487.

Begg, C. (1990). On inferences from Wei's biased coin design for clinical trials (with discussion). *Biometrika* 77: 467–484.

Berger, J. O. (1985). *Statistical Decision Theory and Bayesian Analysis*, 2nd ed. Springer-Verlag, New York.

Berger, J. O. and Berry, D. A. (1988a). The relevance of stopping rules in statistical inference. In *Statistical Decision Theory and Related Topics IV* 1, pp. 29–72. Springer-Verlag, New York.

Berger, J. O. and Berry, D. A. (1988b). Statistical analysis and the illusion of objectivity. *American Scientist* 76: 159–165.

Berger, J. O. and Wolpert, R. L. (1984). *The Likelihood Principle*. Institute of Mathematical Statistics, Hayward, CA.

Bernardo, J. and Smith, A. F. M. (1994). *Bayesian Theory*. Wiley, West Sussex, England.

Berry, D. A. (1985). Interim analysis in clinical trials: classical versus Bayesian approaches. *Statistics in Medicine* 4: 521–526.

Berry, D. A. (1987). Interim analysis in clinical trials: The role of the likelihood principle. *The American Statistician* 41: 117–122.

Berry, D. A. (1988). Interim analysis in clinical research. *Cancer Investigation* 5: 469–477.

Berry, D. A. (1988a). Ethics and ECMO: Comment on a paper by Ware. *Statistical Science* 4: 306–310.

Berry, D. A. (1989b). Inferential aspects of adaptive allocation rules. *Proceedings of the Pharmaceutical Section of the American Statistical Association*, ASA, Washington, DC, pp. 1–8.

Berry, D. A. (1989c). Monitoring accumulating data in a clinical trial. *Biometrics* 45: 1197–1211.

Berry, D. A. (1990). Invited paper: A Bayesian approach to metaanalysis and multicenter trials. *Proceedings of the Pharmaceutical Section of the American Statistical Association*, ASA, Washington, DC, pp. 1–10.

Berry, D. A. (1991a). Bayesian methodology in phase III trials. *Drug Information Association Journal* 25: 345–368.

Berry, D. A. (1991b). Experimental design for drug development: A Bayesian approach. *Journal of Biopharmaceutical Statistics* 1: 81–101.

Berry, D. A. (1993). A case for Bayesianism in clinical trials (with discussion). *Statistics in Medicine* 12: 1377–1404.

Berry, D. A. (1995). Decision analysis and Bayesian methods. In *Recent Advances in Clinical Trial Design and Analysis*, ed. Thall, P. F., pp. 125–154. Kluwer Academic Publishers, Boston.

Berry, D. A. (1996). *Basic Statistics: A Bayesian Perspective*. Duxbury Press, Belmont, CA.

Berry, D. A. and Eick, S. G. (1994). Adaptive assignment versus balanced randomization in clinical trials: A decision analysis. *Statistics in Medicine* 14: 231–246.

Berry, D. A. and Fristedt, B. (1985). *Bandit Problems: Sequential Allocation of Experiments*. Chapman & Hall, London.

Berry, D. A. and Hardwick, J. (1993). Using historical controls in clinical trials: Application to ECMO. In *Statistical Decision Theory and Related Topics* V, eds. Berger, J. O. and Gupta, S., pp. 141–156. Springer-Verlag, New York.

Berry, D. A. and Ho, C.-H. (1988). One-sided sequential stopping boundaries for clinical trials: A decision-theoretic approach. *Biometrics* 44: 219–227.

Berry, D. A., Wolff, M. C., and Sack, D. (1992). Public health decision making: A sequential vaccine trial (with discussion). In *Bayesian Statistics*, eds. Bernardo, J. M., Berger, J. O., Dawid, A. P., and Smith, A. F. M., pp. 79–96. Oxford University Press, Oxford, UK.

Berry, D. A., Wolff, M. C., and Sack, D. (1994). Decision making during a phase III randomized controlled trial. *Controlled Clinical Trials* 15: 360–379.

Box, G. E. P. and Tiao, G. C. (1973). *Bayesian Inference in Statistical Analysis*. Wiley, New York.

Carlin, B. P., Chaloner, K. M., Louis, T. A., and Rhame, F. S. (1994). Elicitation, monitoring, and analysis for an AIDS clinical trial (with discussion). In *Case Studies in Bayesian Statistics*, eds. Gatsonis, C., Hodges, J., and Kass, R., in press.

Cook, P. and Broemeling, L. (1995). Bayesian statistics using Mathematica. *The American Statistician*, 49: 70–76.

Cowell, R. G. (1992). *BAIES*, a probabilistic expert system shell with qualitative and quantitative learning. In *Bayesian Statistics 4*, eds. Bernardo, J. M., Berger, J. O., Dawid, A. P., and Smith, A. F. M., pp. 595–600. Oxford, University Press, Oxford, UK.

DeGroot, M. H. (1970). *Optimal Statistical Decisions*. McGraw-Hill, New York.

DuMouchel, W. (1989). Bayesian metaanalysis. In *Statistical Methodology in the Pharmaceutical Sciences*, ed. Berry, D. A., pp. 509–529. Marcel Dekker, New York.

Eddy, D. M., Hasselblad, V., and Schachter, R. (1992). *Meta-analysis by the Confidence Profile Method: The Statistical Synthesis of Evidence*. Academic Press, New York.

Gamerman, D. (1991). Dynamic Bayesian models for survival data. *Applied Statistics* 40: 63–79.

Gatsonis, C., Hodges, J., Kass, R., and Singpurwalla, N. (1995). *Case Studies in Bayesian Statistics*, Springer-Verlag, New York.

Gaver, D. P., Draper, D., Goel, P. K., Greenhouse, J. B., Hedges, L. V., Morris, C. N., and Waternaux, C. (1991). On Combining Information: Statistical Issues and Opportunities for Research. Report to National Research Council, Washington, DC.

Gehan, E. A. and Freireich, E. J. (1974). Non-randomized controls in cancer clinical trials. *New England Journal of Medicine* 290: 198–203.

Gehan, E. A. and Freireich, E. J. (1981). Cancer clinical trials: A rational basis for use of historical controls. *Seminars in Oncology* 8: 430–436.

Gelman, A., Carlin, J. B., Stern, H., and Rubin, D. B. (1995). *Bayesian Data Analysis*, Chapman and Hall, London.

George, S. L., Li, C. C., Berry, D. A., and Green, M. R. (1994). Stopping a clinical trial early: Frequentist and Bayesian approaches applied to a CALGB trial in non-small cell lung cancer. *Statistics in Medicine* 13: 1313–1327.

Gordon, K. and Smith, A. F. M. (1990). Modeling and monitoring biomedical time series. *Journal of the American Statistical Association*, 85: 328–337.

Hampton, J. M., Moore, P. G., and Thomas, H. (1973). Subjective probability and its measurement. *Journal of the Royal Statistical Association Ser.* 136: 21–42.

Howson, C. and Urbach, P. (1989). *Scientific Reasoning*. Open Court, LaSalle, IL.

Janicak, P. G., Lipinski, J., Davis, J. M., Comaty, J. E., Waternaux, C., Cohen, B., Altman, E. and Sharma, R. P. (1988). S-Adenosyl-methionine (SAMe) in depression: A literature review and preliminary data report. *Alabama Journal of Medical Sciences* 25: 306–312.

Korsan, R. J. (1992). Decision analytica: An example of Bayesian inference and decision theory using Mathematica. In *Economic and Financial Modelling with Mathematica*, ed. HR Varian, H. R., pp. 407–458. Springer, Berlin.

Lauritzen, S. L. and Spiegelhalter, D. J. (1988). Local computations with probabilities on graphical structures and their application to expert systems (with discussion). *J. Roy. Statist. Soc. B* 50: 157–224.

Lewis, R. and Berry, D. A. (1994). Group sequential clinical trials: A classical evaluation of Bayesian decision-theoretic designs. *Journal of the American Statistical Association* 89: 1528–1534.

Ley, E. and Steel, M. F. J. (1992). Bayesian econometrics, conjugate analysis and rejection sampling. In *Economic and Financial Modelling with Mathematica*, ed. Varian, H. R., pp. 344–367. Springer, Berlin.

Li, C. C. (1994). Metaanalysis of Survival Data. Ph.D. dissertation, Duke University, Durham, NC.

Lin, Z. (1993). Statistical Methods for Combining Historical Controls with Clinical Trial Data. Ph.D. dissertation, Duke University, Durham, NC.

Lindley, D. V. (1985). *Making Decisions*, 2nd ed. Wiley, London.

Lindley, D. V. and Smith, A. F. M. (1972). Bayes estimates for the linear model (with discussion). *Journal of the Royal Statistical Association Ser. B* 34: 1–41.

Marriott, J. M. and Naylor, J. C. (1993). Teaching Bayes on *MINITAB*. *Applied Statistics* 42: 223–232.

Mosteller, F, Gilbert, J. P., and McPeek, B. (1983). Controversies in design and analysis of clinical trials. In *Clinical Trials*, eds. Shapiro, S. H. and Louis, T. A., pp. 13–64. Marcel Dekker, New York.

Paneth, N. and Wallenstein, S. (1985). Extracorporeal membrane oxygenation and conventional medical therapy in neonates with persistent pulmonary hypertension of the newborn: A prospective randomized study. *Pediatrics* 84: 957–963.

Peto, B., Pike, M. C., Armitage, P., Breslow, N. E., Cox, D. R., Howard, S. V., Mantel, N., McPherson, K., Peto, J., and Smith, P. G. (1976). Design and analysis of randomized clinical trials requiring prolonged observation of each patient: I. Introduction and design. *British Journal of Cancer* 34: 585–612.

Pole, A., West, M., and Harrison, P. J. (1994). *Applied Bayesian Forecasting and Time Series Analysis* (with computer software). Chapman & Hall, London.

Press, S. J. (1982). *Applied Multivariate Analysis: Using Bayesian and Frequentist Methods of Inference*, 2nd ed., Krieger, Malabar, FL.

Qian, J. (1994). A Bayesian Weibull Survival Model. Ph.D. dissertation, Duke University, Durham, NC.

Racine-Poon, A., Grieve, A. P., Fluheler, H., and Smith, A. F. M. (1986). Bayesian methods in practice: Experiences in the pharmaceutical industry (with discussion). *Applied Statistics* 35: 93–150.

Rosner, G. L. and Berry, D. A. (1994). A Bayesian group sequential design for a multiple arm randomized clinical trial. *Statistics in Medicine* 13: 381–394.

Savage, L. J. (1971). Elicitation of personal probabilities and expectations. *Journal of the American Statistical Association* 66: 783–801.

Smith, A. F. M. and West, M. (1983). Monitoring renal transplants: An application of the multi-process Kalman filter. *Biometrics* 39: 867–878.

Smith, A. F. M., West, M., Gordon, K., Knapp, M. S. and Trimble, I. (1983). Monitoring kidney transplant patients. *The Statistician* 32: 46–54.

Smith, A. F. M., Skene, A. M., Shaw, J. E. H., Naylor, J. C., and Dransfield, M. (1985). The implementation of the Bayesian paradigm. *Comm. Statist. Theory Methods* 14: 1079–1109.

Smith, A. F. M., Skene, A. M., Shaw, J. E. H., and Naylor, J. C. (1987). Progress with numerical and graphical methods for Bayesian statistics. *The Statistician* 36: 75–82.

Smith, C. A. B. (1965). Personal probability and statistical analysis. *Journal of the Royal Statistical Association Ser. A* 128: 469–499.

Souhami, R. L., Spiro, S. G., and Cullen, M. (1991). Chemotherapy and radiation therapy as compared with radiation therapy in stage III non–small cell cancer. *N. Engl. J. Med.* 324: 1136–1137.

Speigelhalter, D. J. and Cowell, R. G. (1992). Learning in probabilistic expert systems (with discussion). In *Bayesian Statistics 4*, eds. Bernardo, J. M., Berger, J. O., Dawid, A. P., Smith, A. P., pp 447–465. Oxford University Press, Oxford, UK.

Spiegelhalter, D. J. and Freedman, L. S. (1989). Bayesian approaches to clinical trials. In *Bayesian Statistics* 3, (eds. Bernardo, J. M., DeGroot, M. H., Lindley, D. V., Smith, A. F. M., pp. 243–259. Oxford University Press, Oxford, UK.

Stangl, D. (1995). Prediction and decision making using Bayesian hierarchical models. *Statistics in Medicine* 14(20): 2173–2190.

Toomasian, J. M., Snedecor, S. M., Cornell, R. G., Cilley, R. E., and Bartlett, R. H. (1988). National experience with extra corporeal membrane oxygenation for newborn respiratory failure: Data from 715 cases. *Transactions of the American Society of Artificial Internal Organs* 11: 140–147.

Trimble, I., West, M., Knapp, M. S., Pownall, R., and Smith, A. F. M. (1983). Detection of renal allograft rejection by computer. *British Medical Journal* 286: 1695–1699.

Truog, R. D. (1992). Randomized controlled trials: Lessons from ECMO. Unpublished manuscript, Multidisciplinary intensive care unit, Children's Hospital, Boston.

Tversky, A. (1974). Assessing uncertainty (with discussion). *Journal of the Royal Statistical Association Ser. B* 36: 148–159.

Ware, J. (1989). Investigating therapies of potentially great benefit: Application to ECMO, *Statistical Science* 4: 298–340.

Ware, J. and Epstein, M. (1985). Extracorporeal circulation in neonatal respiratory failure: A prospective randomized study (commentary). *Pediatrics* 78: 849–851.

Wei, L. J. (1988). Exact two-sample permutation tests based on the randomized play-the-winner rule. *Biometrika* 75: 603–606.

Wei, L. J., Smythe, R. T., Lin, D. Y., and Park, T. S. (1990). Statistical inference with data-dependent treatment allocation rules. *J. Amer. Statist. Assoc.* 85: 156–162.

West, M. (1992). Modelling time-varying hazards and covariate effects (with discussion). In *Survival Analysis: State of the Art*, eds. Klein, J. P. and Goel, K. Kluwer. Academic Publishers, Boston.

Wetmore, N., McEwen, D., O'Connor, M., Bartlett, R. H. (1979). Defining indications for artificial organ support in respiratory failure. *Transactions of the American Society of Artificial Internal Organs* 25: 459–461.

Wolfson, L. (1995). Elicitation of Priors for Statistical Models. Ph.D. dissertation, Carnegie Mellon University, Pittsburgh, PA.

Wooff, D. A. (1992). [B/D] works. *Bayesian Statistics 4*, eds. Bernardo, J. M., Berger, J. O., Dawid, A. P., and Smith, A. F. M., pp. 851–859. Oxford University Press, Oxford, UK.

2

Bayesian Approaches to Randomized Trials

David J. Spiegelhalter Medical Research Council Biostatistics Unit, Institute of Public Health, Cambridge, England
Laurence S. Freedman National Cancer Institute, National Institutes of Health, Bethesda, Maryland
Mahesh K. B. Parmar Medical Research Council Cancer Trials Office, Cambridge, England

ABSTRACT

Statistical issues in conducting randomized trials include the choice of a sample size, whether to stop a trial early, and the appropriate analysis and interpretation of the trial results. At each of these stages, evidence external to the trial is useful, but generally such evidence is introduced in an unstructured and informal manner. We argue that a Bayesian approach allows a formal basis for using external evidence and in addition provides a rational way for dealing with issues such as the ethics of randomization, trials to show treatment equivalence, the monitoring of accumulating data, and the prediction of the consequences of continuing a study. The motivation for using this methodology is practical rather than ideological.

Keywords: clinical trials, ethics, power, prediction, prior distribution, range of equivalence, shrinkage; stopping rules; subjective probabilities

1 INTRODUCTION

The accepted statistical techniques for the design, monitoring, and reporting of controlled clinical trials are based on the frequentist theory of hypothesis testing developed by Neyman and Pearson (see, for example, Pocock, 1983). The impact of the frequentist approach is reflected in the established use of type I and type II error for clinical trial design and p-values and confidence intervals for analysis and in the demand by both medical journals and regulatory authorities for reporting within this framework.

This chapter is about the use of an alternative, Bayesian, theory of statistical inference in clinical trials. Clinical investigation is an essentially dynamic process, in which any individual study takes place in a context of continuously increasing knowledge. New information emerges not only from a trial as it progresses but also from other studies that are relevant to the questions addressed by the trial. Furthermore, it is unlikely that a single trial will provide a definitive clinical conclusion, whether in industry in which drug registration is the goal or in publicly funded research in which influencing clinical practice is the main objective. It therefore seems vital to acknowledge the *context* in which a study takes place and to emphasize that the data from a single trial report *add* to available evidence, rather than form the basis for decision making in themselves. The Bayesian method is naturally dynamic, in that prior distributions of belief regarding treatment differences may be modified by new data in a formal way by using Bayes theorem.

Such an approach is standard in diagnostic testing (Ingelfinger *et al.*, 1987). The practical value of a diagnostic test cannot be decided on the basis of its sensitivity and specificity alone—to make decisions we need the prior probability of disease and the perceived threshold of belief that would make intervention worthwhile. We follow others (e.g., Peto *et al.*, 1976) in arguing that the same is true for clinical trials: the conclusions to be drawn from a classically positive or negative result depend on the degree of skepticism concerning the efficacy of the new treatment based on evidence external to the study and the level of improvement required before it is considered clinically superior.

Bayesian methods will not, however, be adopted in the absence of a clear demonstration of the practical advantages that accrue from their use. Moreover, their practical application requires that the statistician moves into statistical territories that may be unfamiliar, particularly the appropriate choice of a prior distribution. The aim of our chapter is therefore twofold: first, to highlight some advantages of Bayesian methods over current practice in clinical trials and, second, to familiarize statisticians with the concepts and practical methods involved in their application. In Section 2 we comment briefly on the classical hypothesis-testing approach to clinical trials, and in Section 3 we give an overview of the alternative Bayesian framework using the simplest formulation. In Sections 4 and 5

we discuss prior distributions, analysis, and monitoring, including suggestions for formal specification of priors representing skepticism and enthusiasm. Section 6 deals with possible roles for prediction methods with
sample size calculation as a special case, and Section 7 contains applications to real trials. In Section 8 we discuss issues arising in monitoring
and reporting of results. Some technical details are given in Appendixes
A and B.

The practical emphasis of this chapter means that Bayesian techniques
will not be claimed to be superior simply on the basis of foundational
arguments. We do not consider alternative trial designs such as adaptive
allocation rules, and we fully accept the need for randomization. However, we do question the standard basis for drawing inferences following
a classical randomized trial, and the next section introduces some of our
misgivings.

2 HYPOTHESIS TESTING AND CLINICAL TRIALS

2.1 Null Hypothesis and Ranges of Equivalence

Suppose that a trial is comparing two treatments, which we shall term the
"new" and the "old," and that the true treatment difference is summarized by a parameter δ, where large values of δ correspond to superiority
of the new treatment. Statistical analysis usually centers on a significance
test of a null hypothesis H_0: $\delta = 0$. However, in some circumstances the
treatments may be so unequal in their "costs," e.g., in their toxicity,
inconvenience, or monetary cost, that it is commonly accepted that the
more costly treatment will be required to achieve at least a certain margin
of benefit, δ_1, before it can be even considered: hence $\delta < \delta_1$ corresponds
to clinical inferiority of the new treatment (Schwartz *et al.*, 1980). Another
value, δ_s, may be postulated, where $\delta > \delta_s$ indicates clinical superiority
of the new treatment: δ_s is sometimes termed the "minimal clinically
worthwhile benefit." We call (δ_1, δ_s) the *range of equivalence* (Freedman
et al., 1984), such that if we were certain that δ lay within this interval
we would be unable to make a definitive choice of treatment. Clearly the
specification of δ_1 and δ_s is not straightforward, but the concept is becoming recognized as being both feasible and useful (Armitage, 1989; Fleming
and Watelet, 1989). We note that it is quite reasonable that the range of
equivalence will often change as a trial progresses.

As we discuss further in Section 5.1, conclusions may be drawn by
relating evidence on δ to the three intervals $\delta < \delta_1$, $\delta_1 < \delta < \delta_s$, and δ
$> \delta_s$. This formulation also includes trials that aim to demonstrate equiva-

lence of treatments (Dunnett and Gent, 1977; Makuch and Simon, 1982), where we may set the range of equivalence as $(-\delta_E, +\delta_E)$ for a specified δ_E.

2.2 Alternative Hypothesis

Current statistical practice in clinical trial design is to formulate a point alternative hypothesis H_1, and a sample size is chosen which guarantees, under H_1, an acceptably high statistical power (e.g., 90%) of rejecting H_0 by using a specified statistical test at a given significance level (e.g., .05). Usually H_0 corresponds to $\delta = 0$. Medical statisticians continue to use different prescriptions for specifying the value δ_A that should represent H_1 (Spiegelhalter and Freedman, 1986). The main variants are that δ_A should represent

1. The smallest clinically worthwhile difference (Lachin, 1981), or
2. A difference that the investigator thinks is "worth detecting" (Fleiss, 1981), or
3. A difference that is thought likely to occur (Halperin et al., 1982).

The first variant corresponds to taking δ_A as δ_S, whereas the second leaves the choice to the investigator, who may well choose δ_A to be unrealistically large to reduce the required sample size. Approach 3 differs in stressing the importance of *plausibility* as a concept for choosing H_1 (Peto et al., 1976). Suppose that we set δ_A as the *expectation* of the benefit of a new treatment. In practice, clinicians participating in trials often have such expectations close to the *demands* that they would make of the treatment before using it routinely—in our notation, δ_A will be near δ_s. This similarity provides a sound basis for randomization but often leads to confusion between expectations and demands.

In heart disease and cancer research, trials with much larger sample sizes than were previously common have now been conducted to detect relatively small but clinically worthwhile treatment differences, using the argument that only such relatively small differences are likely (Yusuf et al., 1984). Having accepted the role of the plausibility of a given improvement, it is a short step to an explicit expression of prior belief and to a Bayesian perspective.

2.3 Sequential Analysis

Interim analyses are becoming increasingly popular in both public sector and pharmaceutical industry trials, with an accompanying large body of statistical literature: see, for example, Jennison and Turnbull (1990) and

Whitehead (1992). Within the frequentist framework it is necessary to specify a stopping rule before the trial analysis begins, spending the overall type I error at the interim analyses through the course of the trial. The formal use of these methods raises many serious conceptual and practical difficulties (Cornfield, 1966; Berry, 1987; Freedman and Spiegelhalter, 1989), even if they are only considered to be stopping guidelines rather than strict rules.

3 BAYESIAN FRAMEWORK

The basic paradigm of Bayesian statistics is straightforward. Initial beliefs concerning a parameter of interest, which could be based on objective evidence or subjective judgment or a combination, are expressed as a prior distribution. Evidence from further data is summarized by a likelihood function for the parameter, and the normalized product of the prior and the likelihood forms the posterior distribution on the basis of which conclusions should be drawn (see, for example, Lee, 1987). In this chapter we illustrate each of these steps with practical examples, and later in Section 6 we emphasize additional properties related to prediction of future outcomes.

This inferential process is often extended to a theory for decision making by the introduction of utilities for certain outcomes. Although there have been many attempts to place clinical trials within such a decision theoretic framework, in our formulation we specifically do not include utility assessments. Our reason (Spiegelhalter and Freedman, 1988) is that when the decision is whether or not to discontinue the trial, coupled with whether or not to recommend one treatment in preference to the other, the consequences of any particular course of action are so uncertain that they make the meaningful specification of utilities rather speculative. Since we are dealing with a society with considerable freedom of choice of treatment, the implications of reporting a "conclusive" trial result cannot currently be accurately modeled.

3.1 Prior to Posterior Analysis

To emphasize conceptual rather than technical issues we deliberately present only the simplest analysis, which can be carried out without specialist software. More elaborate analyses are briefly discussed, although our formulation can cope with a wide variety of problems. Throughout this paper we use the notation that $p(\ )$ and $\phi(\ )$ are density functions.

Likelihood

We shall assume that our data after m observations can be summarized by a statistic x_m, whose distribution is

$$p(x_m) = \phi(x_m \mid \delta, \sigma^2/m) \tag{1}$$

where ϕ represents a Gaussian distribution with mean δ and variance σ^2/m, δ is the parameter of interest, and σ^2 is assumed known; in comparative trials δ is the true treatment difference and x_m is the sample difference.

This assumption of a normal likelihood covers many situations: if individual responses are assumed Gaussian with variance $\sigma^2/2$, δ is the true difference in mean response, and x_m is the difference in group sample means where m individuals are allocated to each treatment; in survival analysis with proportional hazards, if m is the total number of events observed, $x_m = 4L_m/m$ where L_m is the log-rank test statistic (observed − expected events in a treatment group), and δ is the log-hazard ratio, then x_m has approximately a distribution given by Eq. (1) with $\sigma^2 = 4$ (Tsiatis, 1981). For rare events, we have a similar approximation in which δ is the log-odds ratio, m is the number of events, x_m is the observed log-odds ratio and $\sigma^2 = 4$. For binomial responses with higher event rates, x_m is the difference in sample response rates and, strictly speaking, σ^2 depends on the unknown response rates, but in this case and in that of normal responses with unknown variance an estimate of σ^2 may be used in Eq. (1) which for sufficiently large m will be adequate. Rate ratios can also be handled within this framework (Pocock and Hughes, 1989). Whitehead (1992) exploited the use of normal likelihoods for efficient score statistics as the basis for sequential monitoring and showed the wide applicability of this approach.

Prior Distribution

We denote the prior distribution $p_0(\delta)$, indicating our belief having made zero observations in our trial. With a normal likelihood it is mathematically convenient, and often reasonably realistic, to make the assumption that $p_0(\delta)$ has the form

$$p_0(\delta) = \phi(\delta \mid \delta_0, \sigma^2/n_0) \tag{2}$$

where δ_0 is the prior mean. This prior is equivalent to a normalized likelihood arising from a (hypothetical) trial of n_0 patients with an observed value δ_0 of the treatment difference statistic. We use this normal assumption for the expressions shown below and in our examples; although it could be argued that a symmetric distribution will not adequately reflect opinion about a treatment benefit, it should be remembered that it is only the shape of the prior in the area supported by the likelihood that is important.

In general, to accommodate asymmetric distributions, we have found it convenient to use "mixture" priors with two normal components, each of which may have an upper or lower bound. Either of the components may have zero variance, formally corresponding to $n_0 = \infty$, thus making a lump of probability on δ_0. The resulting family is both flexible and mathematically tractable (see Appendix A). We note that Carlin *et al.* (1993) describe more elaborate analyses within a proportional hazards regression model.

Posterior Distribution

For the prior given in Eq. (2) and likelihood (1), we obtain by Bayes theorem a posterior distribution

$$p_m(\delta) \propto p(x_m \mid \delta)p_0(\delta) \tag{3}$$

$$= \phi\left(\delta \,\middle|\, \frac{n_0\delta_0 + mx_m}{n_0 + m}, \frac{\sigma^2}{n_0 + m}\right)$$

where the subscript indicates our belief after m observations. The posterior mean serves as a point estimate, while $100\gamma\%$ credible interval estimates take the form of regions I_γ such that $\int_{I_\gamma} p_m(\delta)\, d\delta = \gamma$. For Eq. (3), 95% estimates, for example, are formed simply from the posterior mean $\pm$ 1.96 posterior standard deviations.

3.2 Example 1: Medical Research Council Neutron Therapy Trial

The Trial

Errington *et al.* (1991) reported a trial of high-energy neutron therapy for treatment of pelvic cancers. An ad hoc independent data monitoring committee was set up by the Medical Research Council (MRC) in January 1990 to review the interim results after 151 patients with locally advanced cancers had received either neutron therapy (90 patients) or conventional radiotherapy (61 patients).

Range of Equivalence

Interviews were conducted in March 1988 with 10 selected clinicians and physicists, knowledgeable about neutron therapy, before the disclosure of any of the trial results. On average, respondents reported that they would require a change in 1-year survival from 50% (assumed for standard photon therapy) to 61.5% for neutron therapy to be recommended as routine treatment. Under a proportional hazards assumption such a change corresponds to a hazard ratio of log .50/log .615 = 1.426 against photon therapy. Using the log-hazard scale, we therefore take log 1.426 = 0.355 as our upper limit of the range of equivalence, δ_S, and $\delta_1 = 0$ as our lower limit: each is shown as a dotted line in Figure 1.

Prior Distributions

Prior information regarding the effect of neutron therapy was available from two sources. First, the consensus of those interviewed gave a 28% belief that $\delta < 0$ (neutrons worse than photons) and 26% belief that $\delta > \delta_S$ (neutrons preferable to photons), with the remaining 46% lying in the

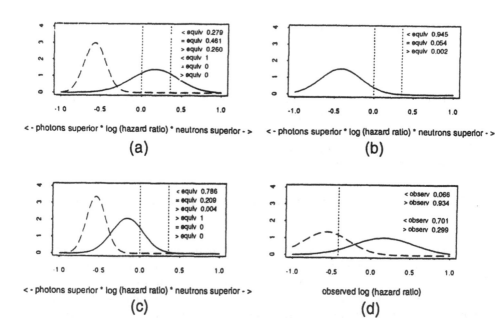

Figure 1 Neutron therapy prior, likelihood, and posterior distributions (the predictive distributions are discussed in Section 6.3) (———, prior based on clinical opinion; — — —, prior based on an overview of trials; the range of equivalence is indicated by the two vertical dotted lines; the probabilities of falling below, within, and above the range of equivalence are shown in the top right-hand corner, first for the clinical prior and then for the overview prior): (a) prior; (b) likelihood ($m = 59$, $x = -0.416$); (c) posterior; (d) predictive distribution.

range of equivalence. We have constructed, as an approximation, a normal prior distribution with mean log-hazard ratio $\delta_0 = 0.169$ and equivalent number of events $n_0 = 48$. We can think of this prior distribution as equivalent to having already observed 48 deaths, of which 22 were in the neutron therapy group and 26 in the photon therapy group. This prior is shown in Figure 1a (full curve) and represents a group clinical opinion.

The second source of prior information was a statistical analysis of the combined results of previous randomized trials, conducted using different dosage regimens, which gave an estimated odds ratio for 1-year mortality of 0.47 (95% interval 0.30–0.73) in favor of photon therapy. Relative to the assumed baseline survival of 50% under photons, this can be translated to an approximate normal prior, on the log-hazard-ratio scale, with $\delta_0 = -0.569$ and $n_0 = 138$. This prior distribution is particularly sceptical and, in view of the different regimens, might reasonably have been viewed with caution at the start of the trial.

Likelihood

The interim analysis, based on 59 observed deaths, showed an estimated hazard ratio of 0.66, with 95% interval (0.40, 1.10), comparing the death rate in the conventional radiotherapy group with that in the neutron therapy group. Thus we have an estimated log(hazard ratio) $= -0.416$ with standard error 0.260, and using the notation of Section 3.1.1 for log-hazard ratios we obtain a likelihood with $x_m = -0.416$, $\sigma = 2$, and $m = 59$. This likelihood is shown in Figure 1b.

Posterior Distributions

The probability that the effect of neutron therapy exceeds the minimal clinically worthwhile benefit δ_S is small (0.004) even under the more enthusiastic prior distribution. Moreover, the results of the current trial agree very closely with the combined results from previous trials. The data monitoring committee ratified the decision of the principal investigator to suspend entry of patients into the trial. The Bayesian analyses we have presented certainly support that decision.

4 PRIOR OPINION

4.1 What Type of Prior?

The term *prior* is used here to denote opinion based on evidence that is *external* to the trial: at the design stage this will be genuinely "prior," in the temporal sense, but as the trial continues it is quite feasible that this prior will change because, for example, of the publication of related studies. Such changes should preferably be carried out by translating other studies into appropriate likelihoods.

When reporting studies we should acknowledge that different individuals or groups hold different prior beliefs. There is therefore no reason to select one particular specification, and instead we may consider a *community* of priors (Kass and Greenhouse, 1989) covering the perspectives of a range of individuals. This may encompass a *reference* prior intended to add as little as possible to the data and a *clinical* prior expressing reasonable opinions held by individuals or derived from overviews (meta-analyses) of similar studies. However, it is also useful to develop "off-the-shelf" priors corresponding to a formal expression of *skeptical* and *enthusiastic* belief—these may be thought to provide reasonable bounds to the community of priors.

Reference Priors

Reference priors are supposed to represent minimal prior information, and for our simple normal likelihood they are obtained as the limit of the prior (2) as $n_0 \rightarrow 0$, leading to an improper uniform prior over the entire range of δ and an identification of the posterior distribution with the normalized likelihood.

In some ways this is the most unrealistic of all possible priors. It represents, for example, a belief that it is equally likely that the relative risk associated with neutron therapy (Section 3.2) is above or below 10. However, it could be argued that such a prior is the least subjective and as such plays a useful role as a baseline against which to compare other more plausible priors.

Clinical Priors

A clinical prior is intended to formalize opinion of well-informed specific individuals, often those taking part in the trial themselves. Deriving such a prior requires asking specific questions concerning a trial, and a variety of sources of evidence may be used as a basis for the opinion (see Sections 4.2 and 4.3).

Skeptical Priors

One step toward incorporation of knowledge into a prior is an attempt to formalize the belief that large treatment differences are unlikely. Such an opinion could be represented by a symmetric prior with mean $\delta_0 = 0$ and suitably spread to include a range of plausible treatment differences. Kass and Greenhouse (1989) explicitly considered several such priors in their reanalysis of the ECMO data and introduced the concept of a "cautious reasonable skeptic" who, if they held such a prior opinion, would consider it ethical to randomize patients, since they did not hold a strong preference in favor of either treatment. This correspondence between belief in clinical superiority and the ethics of randomization is discussed further in Section 5. Skeptical priors may be particularly appropriate for regulatory authorities who are considering new drug applications.

A natural form of scepticism is to consider that the trial designers were optimistic in their expectation regarding the new treatment. This leads to a particularly simple form for the prior. Suppose that the trial has been designed with size α and power $1 - \beta$ to detect an alternative hypothesis δ_A. Then we have the standard relationship

$$\sigma^2 \frac{(z_{1-\alpha/2} + z_{1-\beta})^2}{\delta_A^2} = n \tag{4}$$

between the proposed sample size n and δ_A, where $z_\beta = \Phi(\beta)$. We express scepticism concerning δ by having a prior distribution which is normal with mean 0 and such that $p(\delta > \delta_A)$ is a small value γ. Such a distribution is shown as a full curve in Figure 2 and has the property that

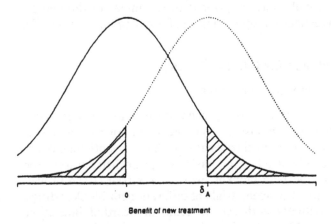

Benefit of new treatment

Figure 2 Skeptical (———) and enthusiastic ($\cdots$) priors for a trial with alternative hypothesis δ_A: the skeptics' probability that the true difference is greater than δ_A is γ (shown shaded), which is also the enthusiasts' probability that the true difference is less than 0.

$$\sigma \frac{z_{1-\gamma}}{\sqrt{n_0}} = \delta_A \tag{5}$$

Equating δ_A in Eqs. (4) and (5) gives

$$\frac{n_0}{n} = \left(\frac{z_{1-\gamma}}{z_{1-\alpha/2} + z_{1-\beta}} \right)^2$$

Reasonable values might be $\alpha = 0.05$, $\beta = 0.1$, and $\gamma = 0.05$, which gives $n_0/n = 0.257$. Thus a formal expression of skepticism is obtained by assuming a prior equivalent to already having observed a quarter of the trial with a zero treatment difference. Such a prior will obviously provide a considerable handicap to the data showing a positive difference; the interesting consequence of this particular handicap is explored in Section 8.1. Choosing n_0 in this manner only makes sense if the trial design is based on a realistic value for δ_A. If, as is often the case, δ_A is unrealistically large, then n_0 will no longer reflect the skepticism intended.

Enthusiastic Priors

To counterbalance the skeptical prior discussed above, it is natural to introduce an enthusiastic prior representing individuals who are reluctant to stop when results supporting the null hypothesis are observed. Such a prior may have mean δ_A and the same precision as the sceptical prior and hence provides a prior belief $p(\delta < 0) = \gamma$. This prior is shown in Figure 2. In practice, this formally derived prior is often similar to the clinical prior derived from investigators' opinions (see, for example, Section 4.3).

4.2 Sources of Evidence for Clinical Priors

Evidence from Other Randomized Trials

If the clinical prior distribution is to represent current knowledge accurately, we should base it, where possible, on objective information. When the results of several similar clinical trials are available, a statistical overview of those results can be used as the basis of a prior distribution. For example, the design of the beta-blocker heart attack trial (BHAT) was based on the results of five European trials of different beta-blocker drugs for the treatment of patients with recent acute myocardial infarction (BHAT Research Group, 1981). The investigators assumed that the drug propranolol would reduce mortality by 28% on the basis of these results. Although the researchers did not comment on the precision of this prior estimate, data from Yusuf et al. (1985) on these same five trials indicate that the standard error was approximately 10%. It is interesting that the final trial results showed a 28% reduction in mortality associated with propranolol.

Results from previous randomized trials should generally form the basis for a prior distribution but should not specify the distribution completely. As Kass and Greenhouse (1989) pointed out, a Bayesian who took this past evidence directly as his prior would be treating the historical and experimental subjects as exchangeable and essentially pooling the results. In the case of the BHAT, the evidence from previous studies might itself have been considered sufficiently strong to call into question the necessity for a further trial. However, it seems reasonable to combine previous data with some initial scepticism. Belief in heterogeneity of treatment effects across trials or patient subgroups, combined with reasonable scepticism, should suggest either shrinking the apparent treatment effect, expanding the variance or both. Random-effects models in meta-analysis might be an appropriate tool, in which case the prior distribution would correspond to the predictive distribution for the effect in a new trial (Carlin, 1992).

Evidence from Nonrandomized Studies

Often, however, no relevant randomized trial results are available, but nonrandomized studies may have been conducted. The difficulty of constructing a prior distribution from the results of such studies is related to the assessment of the possible biases that can exist in such studies (Byar *et al.*, 1976). Such difficulty may often lead quite reasonably to a prior distribution with a large variance.

Subjective Clinical Opinion

We have emphasized that even when evidence in the form of randomized studies is available the prior distribution should still be constructed by *using* that evidence but possibly applying a subjective adjustment. It may happen that there are no similar previous studies, in which case the distribution needs to be based on subjective judgment alone. One approach to eliciting opinion is to conduct individual interviews with clinicians who will participate in the trial.

We have reported this for several trials including the MRC trial of thiotepa for superficial bladder cancer (MRC Urological Working Party, 1985), and the European osteosarcoma intergroup trial of two adjuvant chemotherapy schedules for osteosarcoma (Spiegelhalter and Freedman, 1988). Two statisticians (Freedman and Spiegelhalter) interviewed individual clinicians in the manner described in Freedman and Spiegelhalter

(1983). Interactive graphics could enhance this elicitation procedure, by providing fast quantitative feedback and easy adjustment of distributional shape (Chaloner et al., 1993).

Individual interviewing is time consuming and it may be preferable to use telephone or postal elicitation. A form for postal elicitation of prior opinion has been designed and is being used in several trials currently being coordinated from the MRC Cancer Trials Office. Below we report results from this questionnaire: similar techniques were used by Gore in trials of artificial surfactant (Ten Centre Study Group, 1987) and neutron therapy (Errington et al., 1991).

4.3 Example 2: Continuous Hyperfractionated Accelerated Radiotherapy Study in Head and Neck Cancer

Our second example is a trial currently comparing standard therapy with continuous hyperfractionated accelerated radiotherapy, CHART (Parmar et al., 1993). In designing the CHART study, baseline 2-year disease-free survival under control therapy was estimated to be 45%. Nine individual prior distributions are summarized in Table 1, which presents the proportions of each distribution falling within intervals for treatment difference provided on the questionnaire.

As expected there are differences in the positions and shapes of the individual distributions. The opinions of clinicians 1 and 9 do not even intersect, whereas clinician 5 has a very confident (narrow) distribution, in contrast with clinician 6. How should we combine these individual distributions to arrive at a prior distribution for the group? Many methods have been proposed (Genest and Zidek, 1986). The two simplest methods are arithmetic and logarithmic pooling, corresponding to taking the arithmetic and (normalized) geometric mean, respectively, within each column of Table 1. The former takes the opinions as data and averages, whereas the latter takes the opinions as *representing* data and pools those implicit data. Our strong preference is for arithmetic pooling, to obtain an estimated opinion of a typical participating clinician.

Arithmetic pooling for the CHART study gives a median of 11% improvement, with a prior probability .08 that there is a deterioration (δ_0). Assuming a 45% 2-year disease-free survival rate under standard therapy and transforming to a log-hazard-ratio scale, this corresponds to $p_0(\delta)$ having a median of $\log(\log 0.45/\log 0.56) = 0.32$, with $p_0(\delta < 0) = .08$. Assuming a normal prior distribution with these characteristics gives parameters $\delta_0 = 0.32$ and $n_0 = 77$. This prior distribution is shown in Figure 3, superimposed on a histogram derived from the group distribution shown in Table 1—it can be seen that the normal approximation is adequate,

Table 1 Summary of Results of Questionnaires Completed by Nine Clinicians in the MRC CHART Trial for Head and Neck Cancer

Clinician	Range of equivalence	Prior distribution (absolute advantage of CHART over standard therapy in 2-year % disease-free survival)							
		−10 to −5	−5 to 0	0–5	5–10	10–15	15–20	20–25	25–30
1	5–10	10	30	50	10				
2	10–10	10	10	25	25	20	10		
3	5–10			40	40	15	5		
4	10–10			20	40	30	10		
5	10–15			20	20	60			
6	15–20	5	5	10	15	20	25	10	5
7	5–10				10	20	40	30	
8	20–20				10	20	40	20	10
9	10–15					10	50	30	10
Group mean	10–13	3	5	18	19	22	20	10	3

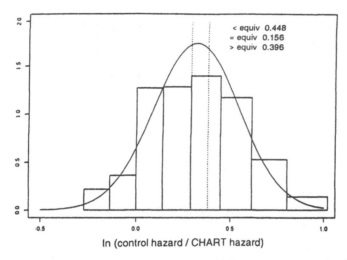

In (control hazard / CHART hazard)

Figure 3   Prior for log(control hazard/CHART hazard) derived from elicited subjective opinions.

and we have generally found this to be the case. The mean of the ranges
of equivalence that were elicited from the same nine clinicians (10%, 13%)
is shown transformed to (0.29, 0.38) on the δ-scale. The ethical basis for
the randomization is clear, since the prior probabilities of being on either
side of the range of equivalence are almost equal (0.45, 0.40). Such a
juxtaposition of prior opinion and range of equivalence formalizes a
grouped version of the "uncertainty principle" (see, for example, Byar
et al., 1990), under which it is considered ethical to randomize a patient
if the clinician has reasonable uncertainty about which is their most appropriate treatment.

   Section 6.2 contains an example of the use of this distribution in pretrial power calculations.

5   MONITORING OF TRIALS

5.1   Bayesian Monitoring Criteria

Ethical considerations place an obligation on organizers to take into account all sources of evidence when considering the continuation of a trial
(Pocock and Hughes, 1989), not purely the data from the trial itself. A
rational way to do this is to use the trial data to provide the likelihood

and have all external evidence concerning the effect of treatment on the main clinical outcome represented by a prior distribution. Their combination into a posterior distribution yields the intervals that can be contrasted with the range of equivalence, which combines evidence regarding the relative differences between the treatments in toxicity, cost, and inconvenience. In this way internal evidence and external evidence on the relative benefits and risks of the treatments are being combined and weighted quantitatively.

Our proposal for monitoring simply extends the preceding discussion on ethical randomization, the judgment being based on the posterior rather than the prior distribution. If clinicians are essentially certain that one treatment is clinically superior, then they should not randomize further patients; if they are almost certain that one treatment is not superior, and may be inferior, then they should carefully consider stopping randomization. This approach has already been illustrated by the neutron therapy example in Section 3.2 but is illustrated in generality in Figure 4.

In Figure 4 are shown six hypothetical credible intervals for the treatment difference (i.e., intervals containing a nominal percentage, say 95%, of the posterior distribution) in relation to the range of equivalence. Each interval corresponds to a different statement that may be made regarding

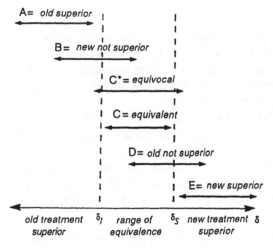

Figure 4 Possible situations at any point in a trial's progress, derived from superimposing an interval estimate (say 95%) on the range of equivalence (see the text for the relationship between situations A, B, C, C*, D, and E and monitoring trials).

the comparison of the treatments. If the interim analysis yields (credible) intervals which have the same form as A or E, then a recommendation to terminate the trial would generally be indicated (Armitage, 1989). Intervals of the form of B or D may also warrant termination, depending on the nature of the two treatments. Errington *et al.* (1991) described a trial that was stopped in situation B. Interval C* is of particular interest within equivalence studies. Meier (1975) has discussed similar procedures for monitoring trials.

From a posterior distribution it is straightforward to calculate the probability content of the three crucial areas: below, within, and above the range of equivalence. Using the interval B or D focuses attention on whether the area above or the area below the range of equivalence has become a sufficiently small value ϵ to consider stopping the trial. We are reluctant to express a firm opinion about what this critical size ϵ should be, since the choice should, in principle anyway, be made from decision theoretic considerations of expected utility. But we have already said that this is unrealistic, and so we are left with the conventional benchmarks such as 2.5% and 5%. However, the selected values can be informally varied according to the perceived importance of the conclusion; for trials of therapy likely to have wide implications for health care, we recommend a stringent criterion. Mehta and Cain (1984) have described essentially this procedure for use in phase II studies.

5.2 Which Prior to Use?

The preceding discussion has presupposed that a single posterior distribution is available. However, we have already argued that a community of prior distributions should be considered, and hence monitoring will give rise to a range of possible posterior distributions, possibly based on reference, clinical, skeptical, and enthusiastic priors.

Although tail areas based on a clinical prior may best reflect the opinions of the trial organizers and participants, they should keep in mind the need for a trial to provide convincing evidence to a spectrum of reasonable opinion. This fits with the recommendation within, for example, the AXIS study (United Kingdom Coordinating Committee on Cancer Research, 1989) that the data monitoring committee should only alert the steering committee if there is

> **both** (a) "proof beyond reasonable doubt" that for all, or for some, types of patient one particular treatment is clearly indicated . . . , **and** (b) evidence that might reasonably be expected to influence the patient management of many clinicians who are already aware of the results of other main studies.

Therefore we recommend that when considering stopping a trial on the basis of results that indicate an effect, either beneficial or harmful, of a new treatment, a *skeptical* prior should be examined; when confronted with data that indicate little or no effect of the new treatment an *enthusiastic* prior should be considered (see Section 7 for examples of this approach). In this way the trial will stop early only if sufficient evidence has been provided to counterbalance the prior opinions of someone who would doubt the observed results. In part, the prior is set up as representing an adversary who will need to be disillusioned by the data to stop further experimentation—for a formalization of this process see Lindley and Singpurwalla (1991).

6 PREDICTIONS

6.1 Making Predictions

A major strength of the Bayesian approach is the ease of making predictions concerning events of interest. Suppose that we have observed m observations and are interested in the possible consequences of continuing the trial for a further n observations. If we denote our future statistic by X_n, then it has predictive distribution

$$p_m(X_n) = \int p(X_n \mid \delta)p_m(\delta) \, d\delta$$

and is straightforward to show that given the posterior distribution (3) we obtain

$$p_m(X_n) = \phi \left\{ X_n \, \middle| \, \frac{n_0\delta_0 + mx_m}{n_0 + m}, \sigma^2 \left(\frac{1}{n_0 + m} + \frac{1}{n} \right) \right\} \tag{6}$$

As a special case, if we have not observed any data so far ($m = 0$) we have that

$$p_0(X_n) = \phi \left\{ X_n \, \middle| \, \delta_0, \sigma^2 \left(\frac{1}{n_0} + \frac{1}{n} \right) \right\} \tag{7}$$

The use of this expression in sample size determination is described in Section 6.2. Section 6.3 concerns a measure of conflict between the prior distribution and the likelihood, and interim predictions are discussed in Section 6.4.

6.2 Pretrial Predictions

Suppose that at the start of a trial ($m = 0$) we wish to assess the chance that the trial will arrive at a firm positive conclusion; i.e., the final interval will exclude a value δ_1 in favor of the new treatment. Here we assume that the target final interval will be based on the reference posterior, equivalent to the normalized likelihood. As outlined in Section 2.2, the standard recommended procedure in this context is to select an alternative hypothesis δ_A and to use the resulting "power" calculation as a basis for selecting a suitable sample size. We have argued that this alternative hypothesis should reflect realistic expectation, and hence it is a natural extension to acknowledge the uncertainty in the anticipated treatment difference expressed in the full prior distribution. Power over the plausible range of δ should be calculated, and a summary measure might be the expected power.

This can be derived as follows. The critical value of X_n is

$$X_n > \delta_1 - \frac{1}{\sqrt{n}} z_\epsilon \sigma = x^*$$

which from Eq. (7) will occur with probability

$$p_0(X_n > x^*) = \Phi \left\{ z_0(\delta_1) \sqrt{\left(\frac{n}{n_0 + n}\right)} + z_\epsilon \sqrt{\left(\frac{n_0}{n_0 + n}\right)} \right\} \qquad (8)$$

where $z_0(\delta_1) = (\delta_0 - \delta_1) \sqrt{n_0}/\sigma$ is the standardized distance of the prior mean from δ_1. As $n_0 \to \infty$ we obtain the classical power curve

$$\Phi \left\{ (\delta_0 - \delta_1) \frac{\sqrt{n}}{\sigma} + z_\epsilon \right\}$$

This use of the full prior distribution for power calculations leads to what has been termed predictive power (Spiegelhalter *et al.*, 1986), expected power (Brown *et al.*, 1987), or the strength (Crook and Good, 1982) of the study. Spiegelhalter and Freedman (1986) provided a detailed example of the predictive power of a study using subjective prior opinions from a group of urological surgeons to form the prior distribution, whereas Moussa (1989) discussed the use of predictive power with group sequential designs based on ranges of equivalence. Brown *et al.* (1987) have also suggested the use of prior information in power calculations but conditioning on $\delta > \delta_s$; i.e., the predictive probability of concluding the new treatment is superior, given that it truly is superior.

Example 2 (continued)

The CHART trial was originally designed with an alternative hypothesis of an absolute 15% improvement of CHART over the 45% 2-year disease-free survival estimated for the standard therapy, equivalent to $\delta_A = \log \cdot (\log 0.45/\log 0.60) = 0.45$. With a two-sided 5% test the trial had 90% power for the 218 events expected. Using the prior shown in Figure 3, the predictive probability that a final 95% interval will exclude 0 is 83%, 62% that it will be in favor of CHART and 21% that it will be in favor of control. However, the ranges of equivalence provided by the clinicians should warn us that simply proving "significantly better than no difference" may not be sufficient to change clinical practice. In fact, at the first interim analysis the data monitoring committee extended recruitment to expect 327 events. This decision was made independently of knowledge of interim results: using the same prior opinion the predictive probability of concluding in favor of CHART is now 68% and in favor of control 25%.

Spiegelhalter and Freedman (1988) noted that the predictive power (8) could be expressed as the power against a fixed alternative δ_u, where

$$\delta_u = \delta_{0.5} \left\{ 1 - \sqrt{\left(\frac{n_0}{n_0 + n} \right)} \right\} + \delta_0 \sqrt{\left(\frac{n_0}{n_0 + n} \right)}$$

a weighted average of the prior mean and the point of 50% power ($\delta_{0.5}$ is the value which, if observed, would just lead to rejection of the null hypothesis). Thus the predictive power will always lie between 0.5 and the power calculated at the prior mean. Predictive power is a useful addition to simple power calculations for a fixed alternative hypothesis and may be used to compare different sample sizes and designs for a proposed trial and as an aid to judge competing trials considering different treatments.

6.3 Checking Prior-Data Compatibility

We have suggested that, before a trial starts, there should be extremely careful consideration of what the treatment difference is likely to be and that this should be used as an important component for trial design. However, it may easily happen that these initial judgments are misguided and the data do not fit the stated expectations. Far from invalidating the Bayesian approach, such a conflict between prior and data only emphasizes the importance of pretrial elicitation of belief: having these opinions explicitly recorded will help a monitoring committee to focus on the difference between expected and actual results. It is useful to have a formal means of assessing the conflict between the prior expectations and observed data, although the precise action to be taken in the face of considerable conflict will depend on the circumstances.

Box (1980) suggested using the prior to derive a predictive distribution and then to calculate the chance of a result with lower predictive ordinate than that actually observed. In our context we would use the form of Eq. (7) but substituting m for n, to give a pretrial predictive distribution

$$p_0(X_m) = \phi \left\{ X_m \mid \delta_0, \sigma^2 \left(\frac{1}{n_0} + \frac{1}{m} \right) \right\} \tag{9}$$

Given observed x_m, Box's generalized significance test is given by

$$p_0\{p_0(X_m) \leq p_0(x_m)\} = 2 \min\{P_0(x_m), 1 - P_0(x_m)\}$$

since the predictive distribution is unimodal.

For example, in Figure 1 the predictive distributions $p_0(X_m)$ are drawn for the two priors, and the dotted line indicates the sample log-hazard ratio. Under the clinicians' prior there was only a $2 \times 0.07 = 0.14$ probability of observing such an unlikely statistic, whereas under the overview prior the observed statistic was quite unsurprising.

We deliberately do not specify the action to take on observing substantial conflict between prior and data. As with any global significance test, this must depend on individual circumstances.

6.4 Interim Predictions

A question frequently asked by investigators at interim analyses is: given the data so far, what is the chance of obtaining a "significant" result? Many researchers have considered an appropriate response under the heading of stochastic curtailment (Halperin et al., 1982; Spiegelhalter et al., 1986; Choi and Pepple, 1989; Ware et al., 1985), and this work is reviewed by Jennison and Turnbull (1990). Within a Bayesian framework the question arises whether the prior opinion is included in making the predictions and in the analysis (the fully Bayesian case), just in making the predictions (the mixed case), or ignored completely (the likelihood case). The appropriate formulas for each of these options are given in Appendix B. We emphasize that this analysis assumes that future summary statistics are drawn from exactly the same distribution as for those statistics observed so far. However, often the predictions are made over an extension of the follow-up of recruited patients, rather than for extending the numbers of new patients. Hence the adequacy of the predictions depends on a strong, and possibly unwarranted, assumption of proportional hazards.

An important issue concerns the use of such predictions. We follow Armitage (1989) in claiming that the decision on whether there is sufficient evidence to warrant stopping a study should be based on the current opinion summarized by the posterior distribution. Predictions concerning the consequences of continuing, although superficially attractive, give an undue weight to achieving significance and are of secondary importance.

7 EXAMPLES OF BAYESIAN ANALYSIS

Most published Bayesian analyses are reexaminations of trial results originally published after using conventional techniques. Examples include the AIMS trial of APSAC against a placebo (AIMS Trial Study Group, 1988) for which Pocock and Hughes (1989) showed the shrinkage of the observed treatment effect brought about by considering a plausible prior distribution; they also considered the consequences of a range of priors. Pocock and Spiegelhalter (1992) showed a similar use of shrinkage after an extreme result was reported in a trial of home thrombolytic treatment. A trial in colorectal cancer originally reported by Poon *et al.*, (1989) has been reanalyzed and discussed by Dixon and Simon (1992), Freedman and Spiegelhalter (1992), and Greenhouse (1992). Here we consider two further examples in detail.

7.1 Example 3: Levamisole and 5-Fluorouracil in Bowel Cancer

The Trial

Moertel *et al.* (1990) have reported results from a randomized clinical trial investigating the effect of the drug levamisole (LEV) alone or in combination with 5-fluorouracil (5-FU) for patients with resected cancer of the colon or rectum. Patients entering the trial were allocated one of the three treatments LEV, LEV + 5-FU, or control. The main outcome measure of treatment was the duration of survival. Moertel *et al.* (1990) planned their study with an alternative of $\delta_A = \log 1.35 = 0.30$, and to have 90% power for a one-sided test with 5% size required 380 deaths.

Range of Equivalence

In this context Fleming and Watelet (1989) suggested a range of equivalence, on a control-to-treatment relative hazard scale, of (1.0, 1.33), so that on a log-hazard-ratio scale $\delta_1 = 0$ (log 1.0) and $\delta_s = 0.29$ (log 1.33); that is, the control treatment would be clinically preferable provided that it had no excess mortality, whereas the new treatment, with its "inconvenience, toxicity and expense," would be clinically superior only if it reduced the hazard by at least 25% (1 − 1/1.33).

Prior Distributions

Since we are conducting a retrospective analysis of this trial, it is reasonable to investigate skeptical and enthusiastic priors; we note that in this study, as with many others, the range of equivalence (δ_1, δ_s) and the interval $(0, \delta_A)$ essentially coincide. We adopt a skeptical prior with mean 0 and probability 0.05 of exceeding $\delta_A = 0.30$, and assuming $\sigma = 2$ gives a prior with $\delta_0 = 0$ and $n_0 = 120$; this prior is quite reasonable in view of the implausibility of dramatic gains in cancer adjuvant therapy. The enthusiastic prior has the same precision and mean 0.30. These priors are shown in Figure 5a.

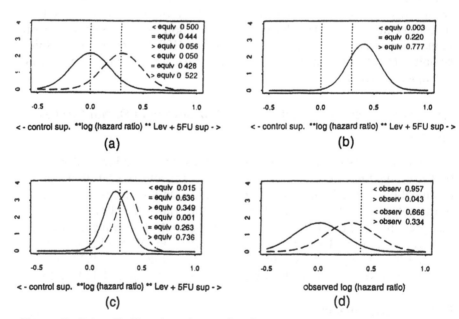

Figure 5 Prior, likelihood, and posterior distributions for LEV + 5-FU versus control (———, skeptical prior and posterior; − − −, enthusiastic prior and posterior; the probabilities of falling below, within, and above the range of equivalence are shown in the top right-hand corner, first for the skeptical prior and then for the enthusiastic prior): (a) prior; (b) likelihood ($m = 192; x = 0.4$); (c) posterior; (d) predictive distribution.

Likelihood

From Moertel *et al.* (1990), $x_m = 0.04$ and $m = 223$ for LEV versus control, and $x_m = 0.40$ and $m = 192$ for LEV + 5-FU versus control. The likelihoods for LEV + 5-FU versus control are shown in Figure 5b—the reference posterior (normalized likelihood) shows convincing evidence that $\delta > 0$ (prob = .003), but moderate evidence (prob = .78) that the treatment is clinically superior. The predictive distribution shows that there was only 4% chance of observing such a high result given the skeptical prior.

Posterior Distributions

The skeptical and enthusiastic posterior distributions for LEV + 5-FU versus control are shown in Figure 5c. Since the result is positive, we should concentrate on the skeptical prior to see whether the interim data are sufficient to convince someone holding this opinion. The skeptical posterior distribution has mean 0.25 with standard deviation 0.11, corresponding to an estimate of 1.28 and 95% interval (1.03, 1.60) for the hazard ratio. There is a small posterior probability (0.015) of the treatment difference being less than 0, i.e., of the control treatment being superior to LEV + 5-FU, whereas the posterior mean is within the range of equivalence. We might therefore argue that the result has not yet achieved "practical significance" to a reasonable skeptic and hence that the trial might have continued. This opinion is possibly supported by the fact that this treatment has not become accepted in Europe, where trials are still continuing.

Results are also available from an earlier study (Laurie *et al.*, 1989) that showed an apparent reduction in the death rate of patients receiving LEV + 5-FU, with an estimated log-hazard ratio of 0.14 with standard error 0.17. If we include these data in the analysis under the skeptical prior, our hazard ratio estimate is changed to 1.25 with 95% interval (1.04, 1.49). Including these data thus increases certainty that $\delta > 0$ (prob = .007) but decreases the certainty that the treatment is clinically superior, $p(\delta > 0.29 \mid \text{combined evidence}) = 0.22$, and so does not change the interpretation given above.

Interim Predictions

The investigators' original plan was to include approximately 188 further deaths in the comparison of control and LEV + 5-FU. Table 2 shows the predictive probability of different conclusions being drawn at the end of this period, calculated under the three assumptions

Table 2 Predictive Probability of the Final 99% Interval for the Hazard of Control Relative to LEV + 5-FU Having Different Positions Relative to the Range of Equivalence, After Observation of a Further 188 Events[a]

	Results for the following modes of prediction:		
Position of 99% interval	Bayesian	Mixed	Likelihood
A: old superior	0.000	0.000	0.000
B: new not superior	0.004	0.001	0.000
C: equivocal	0.407	0.256	0.091
D: old not superior	0.590	0.737	0.845
E: new superior	0.000	0.006	0.064

[a] For illustration of A, B, C, D, and E see Figure 3.

1. A fully Bayesian analysis using our skeptical prior
2. That the prior is used for predictions but excluded from the final analysis
3. That the predictions and analysis are based solely on the data (equivalent to using the reference prior throughout)

Ignoring the prior completely leads to a final interval that is very likely to exclude $\delta = 0$ (D or E), which might reinforce the decision to stop the trial. However, if the prior is used for prediction then the chance that a final 99% interval includes 0 rises to 0.26, whereas the chance that the skeptical posterior interval includes 0 is 0.41. Thus the eventual result of the study is not a foregone conclusion (we emphasize the strong dependence on the proportional hazards assumption for this analysis, as observing future events depends on extending the follow-up of patients already in the trial rather than new recruitment).

7.2 Example 4: Medical Research Council Misonidazole Trials

The Trial

In the summer of 1979 the MRC decided to investigate in three randomized trials the effect of the drug misonidazole (MISO) that was thought to enhance the effects of radiotherapy. The trials involved the treatment of three different cancers (brain, head and neck, and cervix cancer), for each of which X-ray therapy (XRT) was the main therapy. Each trial compared the outcomes of patients treated by XRT plus MISO with those treated by XRT alone. We focus here on the head and neck cancer trial. For this trial the primary end point was the duration of control of the primary tumor. Patients began to enter this trial in early 1979. The design required 292 events in two studies, giving 90% power to obtain a significant result at the two-sided 5% level under the alternative hypothesis $\delta_A = 0.38$.

Range of Equivalence

After about 1 year's experience it became apparent that MISO carried a higher risk of serious side effects, mostly peripheral neuropathies, than had been thought previously. Such neuropathy took its most severe form in complete numbness or cramp of the limbs and extremities. The problem occurred more frequently in the head and neck and the cervix cancer patients than in the brain cancer patients. Clinicians began to express concern and wished to know whether the early results indicated any benefits that would offset the increased toxicity due to MISO. In discussion with clinicians, minimum clinically worthwhile treatment differences were defined that could act as decision points in interim analyses. For head and neck cancer this was an increase in the 2-year primary control from 25% to 40%, a hazard ratio of 1.513; we may think of log $1.513 = 0.414$ as the upper limit δ_s of the range of equivalence on a log-hazard-ratio scale.

Prior Distributions

We consider three different prior distributions: a reference prior leading to an analysis based on the likelihood alone, a skeptical prior with mean 0 and implicit sample size equal to a quarter of the planned accrual in the trial ($n_0 = 73$) (i.e., a "handicap" of 0.25 in the language of Section 8.1), and a more enthusiastic prior with the same standard deviation as the skeptical prior but centered on the alternative hypothesis $\delta_A = 0.38$.

Likelihood

Interim analyses were conducted approximately once per year. The third such analysis was completed in September 1981 based on 108 events, which gave a hazard ratio estimate of 0.90 (95% interval 0.62–1.31), corresponding to a log-hazard-ratio estimate of -0.108 with standard error 0.193.

Posterior Distributions

Table 3 shows the consequences of assuming each type of prior opinion. The hazard ratio estimate for the skeptical posterior is 0.94 (95% interval

Table 3 Analysis of Head and Neck Trial of MISO Therapy with Different Prior Assumptions for δ = Log(control hazard/treatment hazard)[a]

Prior	Prior for δ mean	Prior standard deviation	Posterior mean	Posterior standard deviation	$p_m(\delta \geq \delta_S)$
Reference	0.0	∞	−0.108	0.193	.004
Skeptical	0.0	0.234	−0.064	0.149	.001
Enthusiastic	0.380	0.234	0.088	0.149	.016

[a] For the definitions of the various priors see the text.

0.70–1.26), whereas for the enthusiastic posterior we obtain 0.897 (95% interval 0.82–1.46). These analyses indicate that, even with the more enthusiastic prior opinion, the chances are very small that MISO leads to benefits that the clinicians considered to be sufficient to justify the risk of toxicity.

Following the September 1981 interim analysis the MRC decided to terminate this trial (and the other two trials). The analyses presented here support that decision, although we note that Bayesian methods were not used at the time. Final reports of these trials may be found in references MRC Working Party on Misonidazole in Gliomas (1982), MRC Working Party on Misonidazole in Head and Neck Cancer (1983), and MRC Working Party on Misonidazole in Cancer of the Cervix (1983).

8 PRACTICAL CONSIDERATIONS

The long-running debate between frequentist and Bayesian approaches to clinical trials (Cornfield, 1966; Berry, 1987; Whitehead, 1992) has largely been ideological, but in this chapter we are attempting to be pragmatic. Because the differences between the two approaches are greatest in the monitoring and reporting of clinical trial results, we concentrate on those issues here.

8.1 What About Type I Error (α-Levels)?

Frequentist sequential schemes aim to control the overall type I error α. An important consequence is conservatism, particularly at the early stages of a trial, where the degree of conservatism depends on the stopping rule chosen. Some have advocated that this choice should be determined by the plausible size of the treatment difference; for example, a Pocock-type rule (Pocock, 1977) may be suitable when large differences are more plausible, whereas when large differences are unlikely an O'Brien-Fleming–type rule (O'Brien and Fleming, 1979) may be appropriate (Armitage, 1985). Freedman and Spiegelhalter (1989) showed that explicit representation of such skepticism as a prior distribution centred on zero difference leads to conservative behavior similar to frequentist group sequential schemes. (We are assuming here that the range of equivalence is itself centered on 0 and is very narrow. If its center is nonzero, and the prior distribution has mean equal to this nonzero value, a simple transformation of the scale will result in a monitoring scheme with the same statistical properties.)

If we are very concerned about sampling properties of trial monitoring procedures it is possible (Grossman *et al.*, 1994) to consider the frequentist properties of a Bayesian monitoring procedure in which the trial is terminated at an analysis if a Bayesian interval excludes 0. Table 4 shows the handicaps, expressed as a fraction of the total sample size, which fix the

Table 4 Fraction of Study Size (the Handicap) Which is Assumed to Underlie a Prior Distribution with Mean 0 That Produces Bayesian Stopping Rules Based on Posterior $100(1 - \alpha)\%$ Intervals That Have Total Type I Error α

	Handicaps giving the following values of α:	
No. of analyses	$\alpha = 0.05$	$\alpha = 0.01$
1	0	0
2	0.16	0.11
3	0.22	0.15
4	0.25	0.17
5	0.27	0.18
6	0.29	0.20
7	0.30	0.21
8	0.32	0.22
9	0.33	0.22
10	0.33	0.23

type I error to be 0.05 (when monitoring with a 95% posterior interval) and 0.01 (when using a 99% interval). As an example, if there are five planned analyses and we wish to preserve the overall type I error at 0.05, we could use a 95% posterior interval assuming a prior equivalent to having already conducted a trial in which 27% of the currently planned total sample size had been entered and no treatment difference was observed. This particular degree of skeptical prior was derived by other means in Section 4.1.2. We note that the handicap is fairly insensitive to the exact number of looks, so this Bayesian monitoring scheme for a typical trial with three to five analyses will have type I error roughly corresponding to the associated critical tail area of the posterior distribution. Grossman *et al.* (1994) have shown that such Bayesian-frequentist rules have good properties in terms of average sample number.

8.2 Can We Let People Stop When They Want?

An extreme example of optional stopping is provided by the theoretical possibility of "sampling to a foregone conclusion," in which asymptotically we are guaranteed at some point to obtain a significant result even if the null hypothesis is true (McPherson, 1974). Cornfield (1966) argued that, if you are worried by this, it must reflect consideration of the null hypothesis as having a distinct probability of being true. It follows that we should put a lump of probability (however small) on the null hypothesis, and then the phenomenon will not occur.

Pocock and Hughes (1989) "feel that control of the overall type I error is a vital aid to restricting the flood of false positives in the medical literature." From a Bayesian perspective control of the type I error is not central to making valid inferences and we would not be particularly concerned with using a sequential scheme with a type I error that is not exactly controlled at a specified level. In fact, if it were not for reporting bias, we would see no objection to optional stopping. For example, if we were carrying out an overview of all randomized trials of a given question, and we knew the current results of all trials taking place, the type I error of any individual trial would be of little interest.

However, there are two pragmatic reasons why we might wish to employ monitoring schemes that have low type I error. First, reporting bias makes it more likely that trials with "significant" results are published in journals with high reputations and also more likely that "nonsignificant" trials remain unpublished (Easterbrook *et al.*, 1991). Hence a high type I error rate would lead to greater numbers of published false-positive results, some of which may remain uncountered by published negative reports. Second, a monitoring scheme that does not protect against early stopping is likely to lead to earlier publication of a false-positive result that will be countered only several years later by a negative trial that has gone its full course. Similar arguments hold for trials that are conducted in the pharmaceutical industry, where each trial is not necessarily published but is included in a submission to obtain a license for a new treatment. Thus, from a practical point of view, we are largely in agreement with Pocock and Hughes, although it should be emphasized that sequential analysis was not intended as a counterbalance to publication bias.

However, it must be understood that the two concerns above arise from a background in which many, perhaps most, new experimental treatments are found to be ineffective. It is this same context that leads us to recommend the use of skeptical prior distributions, which both reflect the background information and provide a conservative approach to early terminations of trials. In a research environment in which new treatments are usually found to be effective, the conservatism implicit in group sequential schemes and explicit in a skeptical prior distribution could be unattractive. Likewise, in exceptional circumstances, there may be such compelling evidence for a new treatment that the background skepticism is overcome and a less conservative approach to monitoring may reasonably be adopted. Such a situation may arise in a confirmatory trial, i.e., a trial that seeks to reproduce positive results in one or more previously conducted randomized studies.

8.3 Which Prior Should We Use and When?

An issue that is central to the application of Bayesian methods is the choice of prior distribution. Clearly the "stronger" (i.e., the more informative) the prior distribution, the more influence it has on the analysis. In particular, in monitoring trials, the prior distribution will have considerable influence in the early stages of a trial, when relatively few data are available. Rosenbaum and Rubin (1984) have also pointed out that with optional stopping the coverage properties of the final Bayesian intervals are sensitive to the chosen prior distribution.

These considerations lead us to recommend that a community of priors should be used in monitoring and reporting. These should include, for monitoring, a reference, skeptical, and enthusiastic prior. The skeptical prior should have mean 0 and precision discussed in Section 4.1.3; the effect will be to put a brake on early stopping in favor of a treatment difference. The enthusiastic prior might be an assessed clinical prior, or a prior from a meta-analysis or possibly a prior with mean δ_A, the alternative hypothesis, and precision the same as the skeptical prior. Such a prior tends to act against early stopping in favor of no difference or in favor of the control. Other options may be discussed when reporting the final results.

8.4   How Can Subjective Judgments Be Allowed Into Reporting Trial Results?

The methods that we have introduced are to aid interpretation of the trial results. Thus we suggest explicit separation of the results of a trial from their interpretation. In reporting, we suggest that the results are placed in their own traditional section, presenting means, standard errors, survival curves, etc., whereas the Bayesian methods would be included in an additional formal section on "interpretation." In this interpretation section it would be important to describe the source of any prior distributions used and to show the sensitivity of any conclusions to a community of prior distributions such as that discussed earlier. This section would include interval estimates based on posterior distributions and place particular emphasis on the posterior probabilities of the treatment difference lying below, within, and above the range of equivalence. This answers the crucial question: what is the chance, given evidence both internal and external to this trial, that a specific treatment is clinically superior?

8.5   How Can Bayesian Calculations Be Performed?

One of the main blocks against the use of Bayesian techniques has been the lack of suitable software. The calculations and plots for this chapter were all carried out by using a set of functions written in S (Becker *et al.*, 1990). This software is available free of charge from the first author on request. The software considerably reduces the time and effort required to perform Bayesian analyses, to the extent that there is only a slight addition to the time required for the standard analysis. It also provides the best means of presenting the results—graphically.

9   CONCLUSIONS

We have suggested a nonideological pragmatic set of tools to help in the design, monitoring, and reporting of clinical trials. In deciding whether to start a trial, how large to make it, whether to stop, and whether the results of the trial and other studies are together convincing to practicing clinicians, a team will generally make informal assessments of many factors, e.g., the likely variability, the plausible benefit, and what is sufficient evidence to render randomization inappropriate. All that we are proposing is that such assessments are made more *formal*, to clarify the issues and to provide a rational and explicit basis for both discussion of the issues and for decision making. In our experience such tools can only help in the communication between statisticians and their clinical colleagues.

ACKNOWLEDGMENTS

We thank the CHART steering committee for providing the data for example 2 and numerous referees for helpful comments.

APPENDIX A MORE GENERAL FAMILY OF PRIOR DISTRIBUTIONS

In Section 3 we introduced a simple Gaussian prior distribution and described how a flexible family of distributions may be obtained by allowing a two-component mixture of Gaussian, truncated Gaussian, and degenerate distributions (known as "lump" priors). Applications of such priors are contained in Sasahara *et al.* (1973) and Freedman and Spiegelhalter (1992). Here we provide the technical details for using this prior family for prior-to-posterior analysis, prior-likelihood comparison, and those predictive statements that are available in closed form, assuming a normal likelihood given in Eq. (1).

A.1 Lump Priors

Let $p_0^{\text{lump}}(\delta)$ denote a prior with a degenerate mass on $\delta = \delta_0$. Observing data x_m leads to a posterior distribution that is unchanged from the prior. The predictive ordinate of the data may be obtained by letting $n_0 \to \infty$ in expression (9), giving

$$p_0^{\text{lump}}(x_m) = \phi(x_m \mid \delta_0, \sigma^2/m)$$

A.2 Truncated Prior

We suppose a Gaussian density that has been truncated below at δ_L and above at δ_U. The prior density is denoted

$$p_0^{\text{trunc}}(\delta) = \phi(\delta \mid \delta_0, \sigma^2/n_0, \delta_L, \delta_U)$$

and is given by

$$p_0^{\text{trunc}}(\delta) = \begin{cases} \dfrac{p_0(\delta)}{\Phi\{z_0(\delta_L)\} - \Phi\{z_0(\delta_U)\}} & \delta_L < \delta < \delta_U \\ 0 & \text{otherwise} \end{cases}$$

where $p_0(\delta)$ is from expression (2) and $z_0(\delta) = (\delta_0 - \delta)\sqrt{n_0}/\sigma$ [see Eq. (8)]. Combined with likelihood (1) we obtain the posterior distribution

$$p_m^{\text{trunc}}(\delta) = \begin{cases} \dfrac{p_m(\delta)}{\Phi\{z_m(\delta_L)\} - \Phi\{z_m(\delta_U)\}} & \delta_L < \delta < \delta_U \\ 0 & \text{otherwise} \end{cases}$$

where $p_m(\delta)$ is from expression (3) and

$$z_m(\delta) = \left(\frac{n_0\delta_0 + mx_m}{n_0 + m} - \delta\right)\frac{\sqrt{(n_0 + m)}}{\sigma}$$

[see Eq. (11)].

The predictive ordinate of the data is given by

$$p_0^{\text{trunc}}(x_m) = p_0(x_m)\frac{\Phi\{z_m(\delta_L)\} - \Phi\{z_m(\delta_U)\}}{\Phi\{z_0(\delta_L)\} - \Phi\{z_0(\delta_U)\}}$$

where $p_0(x_m)$ is from expression (9). This predictive distribution does not appear to be integrable in closed form, and hence Box's test of prior-likelihood compatibility (Section 6.3) is not easily available, and also the chance of falling in critical regions cannot be explicitly calculated. Hence further predictive statements require numerical techniques for their computation.

A.3 Mixture Priors

We suppose that we have a prior that is a weighted mixture

$$p_0^{\text{mix}}(\delta) = w_0^A\, p_0^A(\delta) + w_0^B\, p_0^B(\delta)$$

where the components $p_0^A(\delta)$ and $p_0^B(\delta)$ may be standard, lump, or truncated normal densities and $w_0^A + w_0^B = 1$. The posterior distribution is a mixture of the component posterior densities but with revised weights w_m^A and w_m^B, so that

$$p_m^{\text{mix}}(\delta) = w_m^A p_m^A(\delta) + w_m^B p_m^B(\delta)$$

where the weights are obtained from the odds version of Bayes theorem

$$\frac{w_m^A}{w_m^B} = \frac{p_0^A(x_m)}{p_0^B(x_m)} \frac{w_0^A}{w_0^B}$$

where $w_m^A + w_m^B = 1$, and the predictive ordinates for the components $p_0^A(x_m)$ and $p_0^B(x_m)$ are obtained from expressions given previously. The distribution function of this posterior distribution is trivially obtained.

We also have the predictive ordinate

$$p_0^{mix}(x_m) = w_0^A \, p_0^A(x_m) + w_0^B \, p_0^B(x_m)$$

However, Box's procedure requires the distribution function of the predictive distribution and hence is unavailable in closed form if at least one of the components is a truncated normal. After having observed data, predictive statements concerning future critical events may only be explicitly calculated if the analysis is not to include the prior or there is no truncation in the prior.

APPENDIX B   INTERIM PREDICTIONS

By obtaining predictive distributions we can assess the chances that future data fall in critical regions such as A, B, D, or E (Figure 4). In our examples below, we consider the probability that the old treatment is clinically inferior (situation D), but it should be clear how to derive the formulas for other criteria of interest. We consider the fully Bayesian, mixed, and likelihood approaches.

B.1   Fully Bayesian Approach: Prior Used in Predictions and Analysis

For the lower extreme of the range of equivalence δ_1, we are interested in values of (X_n, x_m) that lead to $P_{m+n}(\delta_1) < \epsilon$, where P_{m+n} is the distribution function corresponding to the future posterior density $p_{m+n}(\delta) = p(\delta \mid X_n, x_m)$. For a simple normal prior we have that

$$p_{m+n}(\delta) = \phi \left(\frac{n_0 \delta_0 + m x_m + n X_n}{n_0 + m + n}, \frac{\sigma^2}{n_0 + m + n} \right)$$

and so this critical event will occur if

$$X_n > \frac{n_0 + m + n}{n} \delta_1 - \frac{\sqrt{(n_0 + m + n)}}{n} z_\epsilon \sigma - \frac{n_0 \delta_0 + m x_m}{n} \quad (10)$$

A similar expression for the critical value that leads to $P_{m+n}(\delta_S) > 1 - \epsilon$ can be obtained.

Using the predictive distribution (6) we obtain that the predictive probability that the future posterior tail area $P_{m+n}(\delta_1)$ will be less than ϵ is

$$p_m\{P_{m+n}(\delta_1) < \epsilon\} = \Phi\left\{z_m(\delta_1)\sqrt{\frac{(n_0 + m + n)}{n}} + z_\epsilon\sqrt{\frac{(n_0 + m)}{n}}\right\}$$

(11)

where

$$z_m(\delta_1) = \left(\frac{n_0\delta_0 + mx_m}{n_0 + m} - \delta_1\right)\frac{\sqrt{(n_0 + m)}}{\sigma}$$

is the statistic from which the current tail area is derived.

This analysis also yields useful expressions for those unwilling to incorporate prior distributions into their predictions or analyses. If we take $n_0 = 0$, the improper reference prior that yields a posterior distribution that is identical with the normalized likelihood, we obtain from Eq. (6) a predictive distribution

$$p_m^{\text{ref}}(X_n) = \phi\left\{X_n \mid x_m, \sigma^2\left(\frac{1}{m} + \frac{1}{n}\right)\right\}$$

(12)

B.2   Mixed Approach: Prior Used for Predictions but Not Analysis

If we wish not to include prior opinion in reporting, then the final inference will be based on a reference posterior distribution denoted $p_{m+n}^{\text{ref}}(\delta)$, and hence future critical regions are derived by setting $n_0 = 0$ in inequality (10). Using the predictive distribution (6) gives predictive probability

$$p_m\{P_{m+n}^{\text{ref}}(\delta_1) < \epsilon\} = \Phi\left[z_0(\delta_1)\sqrt{\left\{\frac{n_0 n}{(n_0 + m)(n_0 + m + n)}\right\}}\right.$$
$$\left. + z_m^{\text{ref}}(\delta_1)\sqrt{\left\{\frac{m(n_0 + m + n)}{n(n_0 + m)}\right\}} + z_\epsilon\sqrt{\left\{\frac{(m + n)(n_0 + m)}{n(n_0 + m + n)}\right\}}\right]$$

(13)

where

$$z_m^{\text{ref}}(\delta_1) = \frac{(x_m - \delta_1)\sqrt{m}}{\sigma}$$

is the current standardized statistic using the reference prior. A particular example of Eq. (13) occurs when we have no data so far ($m = 0$) from which we obtain expression (8) used in predictive power assessments. If a classical stochastic curtailment procedure is desired (although it is not recommended by us), predictions conditional on a particular value δ' are obtained by placing a lump prior on δ', in which case letting $n_0 \to \infty$ in Eq. (13) gives

$$\Phi \left\{ (\delta' - \delta_1) \frac{\sqrt{n}}{\sigma} + z_m^{\mathrm{ref}}(\delta_1) \sqrt{\left(\frac{m}{n}\right)} + z_\epsilon \sqrt{\left(\frac{m+n}{n}\right)} \right\} \qquad (14)$$

B.3 Likelihood Approach: Prior Ignored Both in Predictions and in Analysis

Choi and Pepple (1989), Grieve (1991), Frei *et al.* (1987), and Hilsenbeck (1988) all discussed making predictions solely on the basis of the data so far. All effects of the prior can be removed by setting $n_0 = 0$ in either of expressions (13) or (11) to give

$$p_m^{\mathrm{ref}}\{P_{m+n}^{\mathrm{ref}}(\delta_1) < \epsilon\} = \Phi \left\{ z_m^{\mathrm{ref}}(\delta_1) \sqrt{\left(\frac{m+n}{n}\right)} + z_\epsilon \sqrt{\left(\frac{m}{n}\right)} \right\} \qquad (15)$$

which matches expression 4.6 of Jennison and Turnbull (1990). The interesting aspect of Eq. (15), noted in Spiegelhalter *et al.* (1993), is that it can be expressed solely in terms of the current standardized test statistic $z = z_m^{\mathrm{ref}}(\delta_1)$ and the fraction $f = m/(m + n)$ of the trial so far completed, to give the probability that the future tail area below δ_1 is less than ϵ:

$$p_m^{\mathrm{ref}}\{P_{m+n}^{\mathrm{ref}}(\delta_1) < \epsilon\} = \Phi \left\{ \frac{z + z_\epsilon \sqrt{f}}{\sqrt{(1 - f)}} \right\} \qquad (16)$$

This provides an exceptionally simple tool for those carrying out formal or informal interim analyses, who wish to make predictions based solely on the data so far. Often the consequences of such an analysis are depressing to investigators: if, say, they are halfway through a study ($f = 0.5$), and their sample mean is currently one standard error from the null hypothesis ($z = 1$), they may feel that they are well on the way to success, whereas Eq. (16) will tell them that there is only a $\Phi(\sqrt{2} - 1.96) = 29\%$ chance that they will eventually be able to report that the final tail area is less than 0.05. Graphical figures showing the predictive probability of the eventual tail area falling below 0.025 for a range of current z values and fractions of trial completed are given in Spiegelhalter *et al.* (1993).

Expression (16) can give rise to other useful predictive statements. Setting $\epsilon = \Phi^{-1}(-z)$ provides the predictive probability that the eventual tail area below δ_1 will be smaller than its current value $\Phi^{-1}(-z)$, whereas $\epsilon = 0.5$ gives the predictive probability that the sign of Z will stay the same. We see, for example, that for our investigator with $f = 0.5$ and $z = 1$ there is a 34% chance, based just on the data so far, that the tail area will be larger than its current value of 0.16, whereas there is an 8% chance that the final effect will point in the opposite direction. It is also straightforward to derive the full predictive distribution function for the future tail area by considering ϵ as a random variable in expression (15).

REFERENCES

AIMS Trial Study Group. (1988). Effect of intravenous APSAC on mortality after acute myocardial infarction: Preliminary report of a placebo-controlled clinical trial. *Lancet* 1: 545–549.

Armitage, P. A. (1985). The search for optimality in clinical trials. *Int. Statist. Rev.* 53: 15–24.

Armitage, P. A. (1989). Inference and decision in clinical trials. *J. Clin. Epidem.* 42: 293–299.

Becker, R. A., Chambers, J. M., and Wilks, A. R. (1990). *The New S Language*. Wadsworth and Brooks/Cole, Belmont, CA.

Berry, D. A. (1987). Interim analysis in clinical trials: the role of the likelihood principle. *Am. Statistn.* 41: 117–122.

BHAT Research Group. (1981). Beta-blocker heart attack trial design features. *Contr. Clin. Trials* 2: 275–285.

Box, G. E. P. (1980). Sampling and Bayes' inference in scientific modelling and robustness (with discussion). *J. R. Statist. Soc. A* 143: 383–430.

Brown, B. W., Herson, J., Atkinson, N., and Rozell, M. E. (1987). Projection from previous studies: A Bayesian and frequentist compromise. *Contr. Clin. Trials* 8: 29–44.

Byar, D. P., Simon, R. M., Friedewald, W. T., Schlesselman, J. J., DeMets, D. L., Ellenberg, J. H., Gail, M. H., and Ware, J. H. (1976). Randomized clinical trials: perspective on some recent ideas. *New Engl. J. Med.* 295: 74–80.

Byar, D. P. *et al.* (1990). Design considerations for AIDS trials. *New Engl. J. Med.* 323: 1343–1348.

Carlin, B. P., Chaloner, K., Church, T., Louis, T. A., and Matts, J. (1993). Bayesian approaches for monitoring clinical trials with an application to toxoplasmic encephalitis prophylaxis. *Statistician* 42, 355–368.

Carlin, J. B. (1992). Meta-analysis for 2 × 2 tables: A Bayesian approach. *Statist. Med.* 11: 141–158.

Chaloner, K., Church, T., Louis, T. A., and Matts, J. (1993). Graphical elicitation of a prior distribution for a clinical trial. *Statistician* 42, 341–354.

Choi, S. C. and Pepple, P. A. (1989). Monitoring clinical trials based on predictive probability of significance. *Biometrics* 45: 317–323.

Cornfield, J. (1966). Sequential trials, sequential analysis and the likelihood principle. *Am. Statistn.* 20: 18–23.

Crook, J. F. and Good, I. J. (1982). The powers and strengths of tests for multinomials and contingency tables. *J. Am. Statist. Ass.* 77: 793–802.

Dixon, D. O. and Simon, R. (1992). Bayesian subset analysis in a colorectal cancer clinical trial. *Statist. Med.* 11: 13–22.

Dunnett, C. W. and Gent, M. (1977). Significant testing to establish equivalence between treatments with special reference to data in the form of 2 × 2 tables. *Biometrics* 33: 593–602.

Easterbrook, P. J., Berlin, J. A., Gopalan, R., and Matthews, D. R. (1991). Publication bias in clinical research. *Lancet* 337: 867–872.

Errington, R. D., Ashby, D., Gore, S. M., Abrams, K. R., Myint, S., Bonnett, D. E., Blake, S. W., and Saxton, T. E. (1991). High energy neutron treatment for pelvic cancers: study stopped because of increased mortality. *Br. Med. J.* 302: 1045–1051.

Fleiss, J. L. (1981). *Statistical Methods for Rates and Proportions*, 2nd ed. Wiley, New York.

Fleming, T. R. and Watelet, L. F. (1989). Approaches to monitoring clinical trials. *J. Natn. Cancer Inst.* 81: 188–193.

Freedman, L. S. and Spiegelhalter, D. J. (1983). The assessment of subjective opinion and its use in relation to stopping rules for clinical trials. *Statistician* 32: 153–160.

Freedman, L. S. and Spiegelhalter, D. J. (1989). Comparison of Bayesian with group sequential methods for monitoring clinical trials. *Contr. Clin. Trials* 10: 357–367.

Freedman, L. S. and Spiegelhalter, D. J. (1992). Application of Bayesian statistics to decision-making during a clinical trial. *Statist. Med.* 11: 23–36.

Freedman, L. S., Lowe, D., and Macaskill, P. (1984). Stopping rules for clinical trials incorporating clinical opinion. *Biometrics* 40: 575–586.

Frei, A., Cottier, C., Wunderlich, P., and Ludin, E. (1987). Glycerol and dextran combined in the therapy of acute stroke. *Stroke* 18: 373–379.

Genest, C. and Zidek, J. V. (1986). Combining probability distributions: A critique and an annotated bibliography. *Statist. Sci.* 1: 114–148.

Greenhouse, J. B. (1992). On some applications of Bayesian methods in cancer clinical trials. *Statist. Med.* 11: 37–54.

Grieve, A. P. (1991). Predictive probability in clinical trials. *Biometrics* 47: 323–330.

Grossman, J., Parmar, M. K. B., Spiegelhalter, D. J., and Freedman, L. S. (1994) Unified hypothesis testing, point estimation and interval estimation for group sequential clinical trials. *Statist. Med.* 13: 1815–1826.

Halperin, M., Lan, K. K. G., Ware, J. H., Johnson, N. J., and DeMets, D. L. (1982). An aid to data monitoring in long-term clinical trials. *Contr. Clin. Trials* 3: 311–323.

Hilsenbeck, S. G. (1988). Early termination of a phase II clinical trial. *Contr. Clin. Trials* 9: 177–188.

Ingelfinger, J. A., Mosteller, F., Thibodeau, L. A., and Ware, J. H. (1987). *Biostatistics in Clinical Medicine*, 2nd ed. Macmillan, New York.

Jennison, C. and Turnbull, B. W. (1990). Statistical approaches to interim monitoring of medical trials: A review and commentary. *Statist. Sci.* 5: 299–317.

Kass, R. E. and Greenhouse, J. B. (1989). Comments on "Investigating therapies of potentially great benefit: ECMO" (by J. H. Ware). *Statist. Sci.* 4: 310–317.

Lachin, J. M. (1981). Introduction to sample size determination and power analysis for clinical trials. *Contr. Clin. Trials* 1: 13–28.

Laurie, J. A., Moertel, C. G., Fleming, T. R., *et al.* (1989). Surgical adjuvant therapy of large-bowel carcinoma: An evaluation of levamisole and the combination of levamisole and fluorouracil. *J. Clin. Onc.* 7: 1447–1456.

Lee, P. M. (1987). *Bayesian Statistics: An Introduction*. Arnold, London.

Lindley, D. V. and Singpurwalla, N. D. (1991). On the evidence needed to reach agreed action between adversaries, with application to acceptance sampling. *J. Am. Statist. Ass.* 86: 933–937.

Makuch, R. W. and Simon, R. (1982). Sample size requirements for comparing time-to-failure among k treatment groups. *J. Chron. Dis.* 35: 861–867.

McPherson, C. K. (1974). Statistics: The problem of examining accumulating data more than once. *New Engl. J. Med.* 290: 501–502.

Mehta, C. R. and Cain, K. C. (1984). Charts for the early stopping of pilot studies. *J. Clin. Onc.* 2: 676–682.

Meier, P. (1975). Statistics and medical experimentation. *Biometrics* 31: 511–529.

Moertel, C. G., Fleming, T. R., Macdonald, J. S., *et al.* (1990). Levamisole and fluorouracil for adjuvant therapy of resected colon carcinoma. *New Engl. J. Med.* 322: 352–358.

Moussa, M. A. A. (1989). Exact, conditional and predictive power in planning clinical trials. *Contr. Clin. Trials* 10: 378–385.

MRC Urological Working Party. (1985). Intravesical thiotepa for superficial bladder tumors: An MRC randomized study. *Br. J. Urol.* 57: 680–689.

MRC Working Party on Misonidazole in Cancer of the Cervix. (1983). The Medical Research Council Trial of misonidazole in carcinoma of the uterine cervix. *Br. J. Radiol.* 57: 491–499.

MRC Working Party on Misonidazole in Gliomas. (1982). A study of the effect of misonidazole in conjunction with radiotherapy for the treatment of grades 3 and 4 astrocytomas. *Br. J. Radiol.* 56: 673–682.

MRC Working Party on Misonidazole in Head and Neck Cancer. (1983). A study of the effect of misonidazole in conjunction with radiotherapy for the treatment of head and neck cancer. *Br. J. Radiol.* 57: 585–595.

O'Brien, P. C. and Fleming, T. R. (1979). A multiple testing procedure for clinical trials. *Biometrics* 35: 549–556.

Parmar, M. K. B., Spiegelhalter, D. J., and Freedman, L. S. (1994). The CHART trials: Bayesian design and monitoring in practice. *Statist. Med.*, 13: 1297–1312.

Peto, R., Pike, M. C., Armitage, P., Breslow, N. E., Cox, D. R., Howard, S. V., Mantel, N., McPherson, K., Peto, J., and Smith, P. G. (1976). Design and analysis of randomized clinical trials requiring prolonged observation of each patient: I. Introduction and design. *Br. J. Cancer* 34: 585–612.

Pocock, S. J. (1977). Group sequential methods in the design and analysis of clinical trials. *Biometrika* 64: 191–199.

Pocock (1983). *Clinical Trials: A Practical Approach.* Wiley, Chichester, UK.

Pocock, S. J. and Hughes, M. J. (1989). Practical problems in interim analyses, with particular regard to estimation. *Contr. Clin. Trials* 10: 209S–221S.

Pocock, S. and Spiegelhalter, D. J. (1992). Grampian region early anistreplase trial. *Br. Med. J.* 305: 1015.

Poon, M. A., O'Connell, M. J., Moertei, C. G., Wieand, H. S., Cullin, S. A., Everson, L. K., Krook, J. E., Mailliard, J. A., Laurie, J. A., Tschelter, L. K., and Wiesenfeld, M. (1989). Biochemical modulation of fluorouracil: Evidence of significant improvement of survival and quality of life in patients with advanced colorectal carcinoma. *J. Clin. Onc.* 7: 1407–1418.

Rosenbaum, P. R. and Rubin, D. (1984). Sensitivity of Bayes inference with data-dependent stopping rules. *Am. Statistn.* 38: 106–109.

Sasahara, A. A., Cole, T. M., Ederer, F., Murray, J. A., Wenger, N. K., Sherry, S. and Stengle, J. M. (1973). Urokinase pulmonary embolism trial: A national cooperative study. *Circulation* Suppl. 2: 1–108.

Schwartz, D., Flamant, R., and Lellouch, J. (1980). *Clinical Trials* (translated by M. J. R. Healy). Academic Press, London.

Spiegelhalter, D. J. and Freedman, L. S. (1986). A predictive approach to selecting the size of a clinical trial, based on subjective clinical opinion. *Statist. Med.* 5: 1–13.

Spiegelhalter, D. J. and Freedman, L. S. (1988). Bayesian approaches to clinical trials. In *Bayesian Statistics 3*, eds Bernardo, J. M., DeGroot, M. H., Lindley, D. V., and Smith A. F. M., pp. 453–477. Oxford University Press, Oxford, UK.

Spiegelhalter, D. J., Freedman, L. S., and Blackburn, P. R. (1986). Monitoring clinical trials: conditional or predictive power? *Contr. Clin. Trials* 7: 8–17.

Spiegelhalter, D. J., Freedman, L. S., and Parmar, M. K. B. (1993). Applying Bayesian thinking in drug development and clinical trials. *Statist. Med.* 12: 1501–1511.

Ten Centre Study Group. (1987). Ten centre study of artificial surfactant (artificial lung expanding compound) in very premature babies. *Br. Med. J.* 294: 991–996.

Tsiatis, A. A. (1981). The asymptotic joint distribution of the efficient scores test for the proportional hazards model calculated over time. *Biometrika* 68: 311–315.

United Kingdom Coordinating Committee on Cancer Research. (1989). Axis protocol. Technical Report, Medical Research Council Cancer Trials Office, Cambridge, UK.

Ware, J. H., Muller, J. E., and Braunwald, E. (1985). The futility index: An approach to the cost-effective termination of randomized clinical trials. *Am. J. Med.* 78: 635–643.

Whitehead, J. (1992). *The Design and Analysis of Sequential Clinical Trials.* Horwood, Chichester, UK.

Yusuf, S., Collins, R. and Peto, R. (1984). Why do we need some large and simple randomized trials? *Statist. Med.* 3: 409–420.

Yusuf, S., Peto, R., Lewis, J., Collins, R., and Sleight, P. (1985). Beta-blockade during and after myocardial infarction: An overview of the randomized trials. *Prog. Cardvasc. Dis.* 27: 335–371.

3
Bayesian Epidemiology

Deborah Ashby and Jane L. Hutton   University of Liverpool, Liverpool, England

ABSTRACT

Epidemiology progresses by an accumulation of evidence about the existence and the magnitude of effects of interest, so a Bayesian approach would appear to offer a natural framework for this process. This chapter outlines some of the main epidemiological study designs, specifically the analysis of routinely collected data, case-control studies, and cohort studies and gives examples of Bayesian work in these areas. Important issues in epidemiological work include measurement error, sample size determination, and combination of evidence. The potential contribution of a Bayesian approach to these issues is illustrated. Finally, some promising directions for future Bayesian work in epidemiology are outlined.

1 INTRODUCTION

Epidemiology can be defined as the numerical study of the distribution and determinants of disease in populations, both human and animal, with respect to time, place, and person. It provides information that can be used to guide laboratory-based research into disease mechanisms, on the one hand, and to inform preventative strategies and health service provision on the other.

In this chapter we consider human epidemiology. It is rarely possible to perform relevant experiments on humans, and the challenge is to piece together evidence from a wide range of observational data. These come from descriptive sources such as routinely collected data from birth and death notifications, which are typically used for hypothesis generation, from analytical studies, which are used to explore in detail the relationships between particular individual characteristics and risk of disease, and from intervention studies, which are experimental. Epidemiologists also refine methods of measurement of risk factors and consider modifications of study design and appropriate statistical models.

Traditionally, statistical analyses have been carried out at two levels. One level uses essentially "back-of-the-envelope" calculations of descriptive statistics such as measures of excess risk and accompanying tests of hypotheses of no excess risk. For fuller analysis a variety of methods are used, especially logistic and Poisson regression models. Typically, each study is analyzed separately. There are occasional reviews or meta-analyses which try to summarise estimates from several studies.

Berry and Stangl (Chapter 1) describe the essential components of Bayesian analysis. Briefly, existing evidence on the effects of interest may be available from other studies or from expert opinion. Assuming a suitable model, this evidence can be summarized quantitatively via a *prior distribution* for existing beliefs or evidence about the parameters of the model. The data from the study are used to form the *likelihood* function. The prior distribution and the likelihood function can then be formally combined by multiplying them to give a *posterior distribution* for the parameter of interest. This can be interpreted as the updated beliefs or evidence about the effects of interest. Epidemiology progresses by an accumulation of evidence about the existence and the magnitude of effects of interest, so a Bayesian approach would appear to offer a natural framework for this process.

Breslow (1990) described the impact that Bayesian methods are making in biostatistics generally, but in a review of statistical models used in epidemiology in the 1980s (Gail, 1991), only 5 of the 851 papers in the bibliography referred to Bayesian methods. Since then there has been considerable change, partly due to advances in computing. This chapter surveys current work in the area. Section 2 considers the analysis of routinely collected data and, in particular, spatial and temporal patterns. Section 3 looks at methods for case-control and cohort studies, and in Section 4 approaches to measurement error are considered. Section 5 outlines various Bayesian approaches to sample size determination. Section 6 looks at methods for combining evidence, and Section 7 points to likely directions for future developments in Bayesian epidemiology.

2 ANALYSIS OF SPATIAL AND TEMPORAL PATTERNS

Routinely collected data, such as birth and death notifications, are collected for administrative purposes but are an invaluable resource for epidemiologists. These data can be used either to get a broad picture of temporal and spatial patterns or to describe more local effects. Although the data are relatively simple, usually with few covariates, it is sensible to analyze routinely collected data before embarking on more specialized studies. Routine data sets are often large but can become sparse very quickly once the data are split by factors such as age groups, geographical region, calendar time, and birth cohorts. This means that the real effects in which we are interested can be masked by random noise, and, conversely, random noise can be mistaken for real effects. However, when analyzing such data, there is often a great deal of general knowledge about their structure which could be incorporated in the modeling. Example 1 makes use of expert opinion, and the other Bayesian approaches to routinely collected data outlined in this section all take account of known structure to produce smoothed estimates of the effects of interest.

2.1 Modeling Geographical Variation

The mapping of disease incidence, or its surrogate, disease mortality, is undertaken to gain clues to the etiology of disease or to test specific hypotheses about the effect of suspected risk factors, such as proximity to nuclear processing plants, climatic effects, or differences between urban and rural dwelling.

Example 1: Leukemia Near Nuclear Power Stations

An early Bayesian analysis of routine data was given by Clayton (1989), in his discussion of clusters of leukemia near a nuclear power station, Seascale. Four cases of childhood leukemia were observed, whereas only 0.25 would have been expected on the basis of national rates. Although a classical analysis gives a relative risk of 16, with a very extreme P-value, debate raged as to whether this was a real effect or not. There was criticism that the P-values should be adjusted upward because of post hoc hypothesis testing and the observed relative risk of 16 was thought by some too large to be plausible.

Clayton's analysis took a gamma prior distribution for θ, the relative risk, and a Poisson likelihood with mean θE, where E is the number of cases expected, for X, the number of cases observed. The prior distribution, gamma(ν, α), with shape parameter ν and scale parameter α is

$$f_{\Theta}(\theta) = \frac{\alpha^{\nu}\theta^{\nu-1}e^{-\alpha\theta}}{\Gamma(\nu)}$$

which has mean ν/α. The likelihood for X is $(\theta E)^{x}e^{-\theta E}/x!$, and, as this is a conjugate analysis, it is easily seen that this results in a gamma posterior distribution, gamma($\nu + x, \alpha + E$).

Clayton gave three analyses using improper prior distributions, which were uniform for log θ, θ and $\sqrt{\theta}$. Figure 1 shows the resulting posterior distributions. These yielded posterior means of 16.0, 20.0, and 18.0, respectively. The posterior probabilities of $\theta < 1$ are all less than .0001. Use of these improper priors gives results that are at best no more plausible than the classical estimate. Essentially, uniform distributions on θ and $\sqrt{\theta}$ give more weight to large values of θ than small values, so the posterior means are inflated. Clayton claims that "no epidemiologist would ever

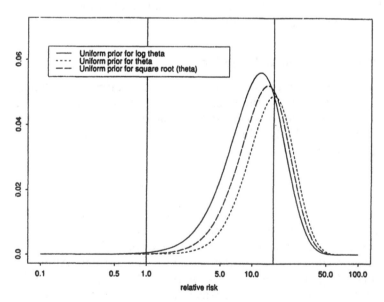

Figure 1   Clayton's analysis of the Seascale data: the resulting posterior distributions from using improper prior distributions.

have approached an ecological study with beliefs remotely like these improper priors.''

Analysis with three proper prior distributions, with mean 1 and 5% probability that the true relative risk exceeds 2, 1.57, or 1.35, were performed. Clearly, these prior distributions are very different from the uniform priors and express, with increasing conviction, a belief that large relative risks do not occur. Even the first of these prior distributions, gamma(3.5, 3.5), represents a person with considerable confidence in their own experience. A shape parameter of 3.5 represents having already observed 3.5 cases. In order to have a mean of 1, the shape and scale parameters are equated. A scale parameter of 3.5 implies, if one considers how the posterior is updated, a belief that the expected number of cases is 3.5, rather than the 0.25 based on national rates. The posterior distribution, gamma(7.5, 3.75), has mean 7.5/3.75 = 2.0, and the posterior probability of $\theta < 1$ is .06. Figure 2 illustrates this analysis. Although the posterior estimate of the relative risk is much reduced, there is clearly a strong possibility of an excess risk. The two other prior distributions, gamma(10, 10) and gamma(25, 25), overwhelm the data.

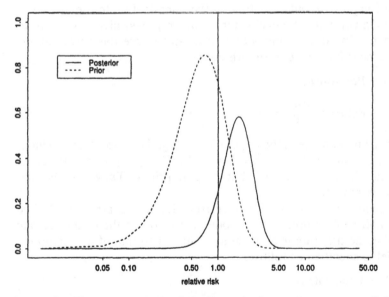

Figure 2 Clayton's analysis of the Seascale data, using a gamma (3.5, 3.5) as a prior distribution, and the resulting posterior distribution.

These analyses highlight the fact that when there is little information, one cannot easily attain a consensus in a scientific community, as personal beliefs will prevail.

Example 2: Mapping Lip Cancer in Scotland

Lip cancer rates in Scotland were first analyzed using an empirical Bayesian approach by Clayton and Kaldor (1987). Previously it had been noted that the spatial distribution of certain occupations bore a resemblance to the spatial distribution of lung cancer. They analyzed data from 56 regions and for each region they had the observed and expected number of cases, the percentage of the population engaged in agriculture, fishing, or forestry, and the position of each county expressed as a list of adjacent counties.

They produced estimates of the standardized mortality ratio (SMR) for each county, although they recognized that they did not fully account for the spatial autocorrelation. Breslow and Clayton (1993) reanalyzed these data using random-effects Poisson models, and Spiegelhalter *et al.* (in press) show how these analyses can be carried out using Bayesian graphical models. The use of such models is described in the clinical trials context in Chapter 15, but an outline of their application to the lip cancer data will be given here. Spiegelhalter *et al.* (in press) give two analyses of these data. The first assumes that, if O_i and E_i are the observed and expected cancer incidence in the ith county,

$$O_i \sim \text{Poisson}(\mu_i)$$

$$\log \mu_i = \log E_i + \frac{\alpha_1 x_i}{10} + b_i$$

where x_i is the percentage of the population engaged in agriculture, fishing, or forestry, $b_i \sim \text{normal}(\alpha_0, \tau)$, $\widehat{\text{SMR}}_i = 100\mu_i/E_i$, α_0, α_1, and τ, are given independent, "noninformative," but proper, priors. They call this the exchangeable model.

The second model allows for overdispersion and spatial correlation, using a conditional autoregressive model that smooths the estimated rate by assuming the true rate in any county is similar to that of its neighbors. This model is specified as follows:

$$O_i \sim \text{Poisson}(\mu_i)$$

$$\log \mu_i = \log E_i + \frac{\alpha_1 x_i}{10} + b_i$$

where $b_i \sim$ normal($\bar{b}_i$, τ_i), n_i = number of neighbors of i,

$$\bar{b}_i = \frac{1}{n_i} \sum_{j \in \text{neighbors}(i)} b_j$$

$\bar{\tau} = n_i \tau$, and $\widehat{\text{SMR}}_i = 100\mu_i/E_i$.

The latter model is shown graphically in Figure 3, where circles represent random quantities, squares observable data, and double squares contain known quantities. The authors use Gibbs sampling to estimate the parameters of the model, using their general-purpose software package BUGS (Spiegelhalter *et al.*, in press; 1994).

The parameter estimates for the two models are given in Table 1 and compare well with those obtained by Breslow and Clayton (1993), using penalized quasi-likelihood. The results for the two models are similar, each showing strong evidence for a covariate effect, although allowing for spatial homogeneity considerably reduces the estimated influence of the covariate. The authors give estimates of the SMRs of the different regions, and, as would be expected, the extreme values based on relatively few observations tend to be shrunk toward the overall mean. An advantage of the fully Bayesian approach over the penalized quasi-likelihood method is the ability to provide error estimates for the SMRs.

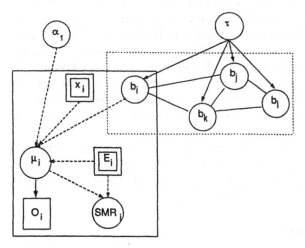

Figure 3 Graphical model for lip cancer example, assuming spatial smoothing of relative risks covered by a hyperparameter τ. (From Spiegelhalter *et al.*, 1994, with thanks.)

Table 1 Estimated Coefficients for Models of Maps of Lip Cancer in Scotland

Model	Constant (α_0)	$x/10(\alpha_1)$	σ
Exchangeable model	$-.46 \pm .16$	$.66 \pm .15$	$.61 \pm .08$
Spatial model	—	$.36 \pm .12$	$.76 \pm .13$

Several other authors have presented Bayesian approaches to mapping. Bernardinelli and Montomoli (1992) map geographical variation in breast cancer and Hodgkin's lymphoma in 22 districts in Sardinia, using both empirical Bayes and a fully Bayesian analysis. An empirical Bayes approach is often used to circumvent some of the computational problems of a Bayesian approach; it involves using an estimate of one or more parameters from the prior distribution, rather than fully accounting for uncertainty surrounding it. They follow Clayton and Kaldor's (1987) approach, modeling the area effect as the sum of a global mean and an area-specific effect, which they consider is a surrogate for unknown or unobserved variables that affect risk.

Much of statistical mapping activity, both traditional and Bayesian, has been from the field of cancer, but two applications from other areas are presented by Heisterkamp *et al.* (1993) and Lawson (1994). Heisterkamp *et al.* (1993) compare empirical Bayesian and fully Bayesian mapping approaches for several causes of death in the 64 health service districts in the Netherlands. Lawson (1994) models pneumonia mortality around a putative source of pollution in Scotland. He follows Clayton and Kaldor's approach, but whereas they simply specify that each area is "similar" to its immediate neighbors using adjacency matrices, Lawson specifies a covariance matrix that gives continuously varying weights to neighboring regions.

The big advantage of Bayesian approaches to mapping compared to the more conventional approaches is that the maps are smoother, which is more intuitively appealing, although there is always the danger of smoothing out a real but very localized effect.

2.2 Modeling Time Trends

Further understanding of the etiology of disease comes from the analysis of time trends, where the objective is often to separate out cohort effects from age and period (or secular) time effects. Cohort effects apply to a particular birth cohort or set of birth cohorts. A particular example is that infant immunization against measles is something that happened to cohorts born in the United Kingdom after about 1987 but not those born before. Insofar as the immunization or freedom from measles infection influences health, we may expect to see cohort effects in relevant health measures. Period, or secular, effects are those that apply to all those alive at a particular time period. The introduction of microwave ovens during the 1980s is an example of a secular trend, and any long-term influences on health would be referred to as secular effects. Clearly, the interaction between cohort and period effects may be complex. A further problem is that age, period, and cohort effects cannot be uniquely identified, so in practice good judgment needs to be exercised when fitting models to data.

Desouza (1991) uses empirical Bayes to look at cohort trends in malignant skin melanoma in Connecticut, Denmark, and Norway. She analyses incidence rates for the three countries stratified by period, cohort, and age groups. Her aim was to provide stable cohort effects even where numbers are low. She assumes a random-effects Poisson model, where the random effects are used to "borrow strength" across regions. She then uses empirical Bayes to estimate the geographic specific cohort effects, because of computing difficulties inherent in a fully Bayesian approach. The author found that she gained more stable estimates of the geographic specific cohort effect, as measured by the coefficient of variation.

Spiegelhalter *et al.* (in press) give an analysis of breast cancer rates in Iceland, using a Bayesian graphical modeling approach to fit a Poisson regression model, using an autoregressive structure for the cohort effects. Commenges and Etcheverry (1993) use an empirical Bayes approach to the problem of estimating the short-term incidence of acquired immunodeficiency syndrome (AIDS) in the United States, from back-calculation of the human immunodeficiency virus (HIV) curve. Back-calculation is based on the deconvolution of the AIDS incidence distribution into the HIV incident infection distribution and the incubation distribution. The main difficulty is in estimating the HIV incident infection distribution, and the authors tackle this using empirical Bayes with an informative prior distribution based on a simple epidemic model, which assumes that there is a high correlation between points close in time. Wild *et al.* (1993) provide a fully Bayesian analysis of these data.

3   ANALYSIS OF CASE-CONTROL AND COHORT STUDIES

In analytical epidemiology, two study designs, the case-control study and the cohort study, are very important. Both designs attempt to link potential risk factors to the onset of disease. In a cohort study, a group is enumerated at a particular time, possibly sampled with respect to the risk factor of interest, and is followed forward in time until the occurrence of the event of interest. In the case-control study, sampling is with respect to outcome. Cases of the disease of interest are taken, along with control subjects who do not have the disease. Information is then collected retrospectively on the potential risk factor. The design and analysis of these studies, together with variants such as the retrospective cohort and the nested case-control study, are described in classic monographs by Breslow and Day (1980, 1987).

The simplest form of data from either a case-control study or a cohort study is often presented as a two-by-two table. Data from an unmatched case-control study are given by classifying cases and controls by the presence or absence of a risk factor. The usual summary statistic of interest is the odds ratio for the presence of a factor in the cases relative to the presence of the factor in the controls. For a rare condition this gives a good approximation to the relative risk of developing the disease when the factor is present compared with when it is not. In a cohort study the table is similar in structure, but because of the method of sampling, one can work with the odds ratio or directly with the relative risk.

For a case-control study we estimate the probability of the risk factor given the disease classification, whereas for the cohort study we estimate the probability of disease given the exposure status. Nurminen and Mutanen (1987) give exact analyses for the odds ratio, or relative risk, and difference in risk of two proportions, and their analysis is discussed as Example 3. An alternative approach is to model the prior and posterior distributions of the relevant probabilities separately and then derive the distribution of the odds ratio, using a beta-binomial model. Another approach is to model directly the prior and posterior distributions for the odds ratio, using a normal-normal model. In the context of case-control studies these approaches are described by Marshall (1988) and Zelen and Parker (1986) and used by Ashby *et al.* (1993), and part of this application is described in Example 4.

3.1 Analyses of Case-Control Studies

Example 3: Occupational Exposure and Malignant Lymphoma

We concentrate here on an example, given in Nurminen and Mutanen (1987), which uses a uniform prior that summarizes vague beliefs. The reader who wishes to incorporate more specific prior knowledge is referred to the exact Bayesian analysis of the theoretical proportions or risks, R_i, of exposure to some factor in two groups, $i = 0, 1$, using a product of two beta distributions as the prior distribution for (R_0, R_1), in

Nurminen and Mutanen (1987). Briefly, as the data, the number of exposed individuals in the two groups, case and control, follow two independent binomial distributions, the posterior distribution for (R_0, R_1) is again a product of beta distributions. The distributions of interest are those of three summary statistics: the difference in risks, $R_1 - R_0$; the relative risk, R_1/R_0; and the odds ratio, $\{R_1(1 - R_0)\}/\{R_0(1 - R_1)\}$. The posterior distributions for these summary statistics involve convolutions of beta distributions and density functions, which are lengthy algebraically and tedious computationally.

Nurminen and Mutanen (1987) discuss data from a study of the association of occupational phenoxy acid and chlorophenol exposure with malignant lymphoma. The study subjects for this analysis had all had short-term employment in agriculture or forestry. There were 53 controls and 27 cases, and the results are summarized in Table 2.

Using a uniform prior and the posterior median as an estimate of the odds ratio, the authors obtain 2.41 for the odds ratio for the phenoxy acid–exposed group. This is the same as the classical estimate, using the conditional maximum likelihood estimate of the odds ratio, which allows matching of cases and controls. For this discussion, we quote the unconditional maximum likelihood estimate with a continuity correction,

$$\text{OR} = \frac{(a + 0.5)(d + 0.5)}{(b + 0.5)(c + 0.5)}$$

where a is the number of cases exposed, b is the number of cases not exposed, c is the number of controls exposed, and d is the number of controls not exposed, which is a typical back-of-the-envelope calculation.

For the phenoxy acid–exposed group, we have $(8.5 \times 44.5)/(9.5 \times 16.5) = 2.4$. The 95% posterior probability interval (or credible interval) of 0.82 to 7.15 differs slightly from the classical confidence interval (0.68, 8.50) but does not imply any substantive difference in the estimated size of the effect. For the chlorophenol group, the Bayesian analysis gives

Table 2 Case-Control Study of Malignant Lymphoma in Short-Term Workers in Forestry

	Not exposed	Exposed to phenoxy acids	Exposed to chlorophenols
Cases	16	3	8
Controls	44	3	0

an estimate of 14.4 with a 95% credibility interval of 1.64 to 117. The classical estimate of the odds ratio is quoted as infinity by the authors, because of the zero in one of the cells, and the 95% confidence interval is $(1.01, \infty)$. Use of the continuity correction yields OR = $(3.5 \times 44.5)/(0.5 \times 16.5)$ = 18.9, with a 95% confidence interval of (0.92, 385). The Bayesian analysis allows a calculation to be performed in a case in which the exact estimation in the classical framework is not feasible. However, the interpretation of the intervals is much the same: there is very little information about the magnitude of the association of chlorophenol and malignant lymphoma, but the evidence suggests there is an association. It is not obviously better to express uncertainty by an upper limit of 117 rather than infinity, especially as selection of cases and controls and decisions about whether or not they have been exposed might have introduced substantial biases. The possibility of such biases could be addressed in Bayesian inference by weakening the conclusions. As a vague prior distribution was used in these analysis, one could only relax the conclusions by somewhat arbitrarily widening the confidence intervals, tending towards the classical estimates.

Example 4: Risk of Leukemia Following Treatment for Hodgkin's Disease

In treating cancers, the primary objective is to maximize the survival of the patient, although considerations such as quality of life are also relevant. However, in cancers for which survival is good, there is a growing interest in the long-term effects of treatment. Recent research has focused on the occurrence of second primary cancers in these patients, with particular emphasis on how these are related to treatment. Hodgkin's disease, which is cancer of the lymph nodes, has a 5-year survival of about 80%, and survivors have excess risks of solid tumors, leukemias, and lymphomas (Swerdlow et al., 1992). Although these risks are very small compared with the benefits of treatment, there is interest in studying them, both to improve treatment and to shed light on causation of disease. Here we look specifically at two studies designed to estimate the risk of leukemia following Hodgkin's disease.

A cohort study (Kaldor et al., 1987) observed 62,654 people with Hodgkin's disease. In 137,401 person-years of follow-up, there were 106 cases of leukemia, compared with 10 cases expected. Although excess risk from treatment is clearly a possible explanation, no treatment data were available to explore this. The likelihood for the log of the relative risk may be approximated by a normal distribution, with mean $\log_e 10.6$ and variance $\frac{1}{106}$. This may be justified either by considering this a relative risk estimated from a two-by-two table with the total study population, the size of the control group, and the number of leukemias in the control group being so large that their contribution to the variance is negligible. Alternatively, this might be justified by considering that the number of leukemias follows a Poisson distribution with mean 106, and taking a lognormal approximation to the Poisson distribution.

Following this and other studies that confirmed the existence of an excess risk that seemed to be treatment related, an international case-control study (Kaldor et al., 1990) was set up to investigate further the effect of treatment. The results are shown in Table 3. From 149 cases, who had Hodgkin's disease followed by leukemia, and 411 matched controls, who had Hodgkin's disease but no subsequent leukemia, the authors demonstrated a ninefold risk for chemotherapy compared with radiotherapy and an eightfold risk for radiotherapy and chemotherapy combined compared with radiotherapy alone.

The odds ratios for chemotherapy and radiotherapy and chemotherapy versus radiotherapy only are very similar, so it seems reasonable to combine the groups into chemotherapy and no chemotherapy. As there are suggestions from elsewhere (Swerdlow et al., 1992) that radiotherapy carries little risk of leukemia for Hodgkin's disease patients, for the purpose of this analysis the odds ratio will be taken to refer to the risk of chemotherapy versus nothing, although clearly a fuller analysis would explore this in detail.

To illustrate a simple Bayesian approach, we give four analyses of the data from the case-control study, using noninformative priors, slightly informative prior information, and then using the cohort study as prior evidence. These illustrate the use of the normal-normal model. The use of the beta-binomial model on these and related data is also illustrated in Ashby et al. (1993). The results of these analyses are summarized in Table 4.

Table 3 Results from Case-Control Study of Leukemia Following Treatment for Hodgkin's Disease

Treatment	Cases	Controls	Odds ratio
No chemotherapy or radiotherapy	0	2	—
Radiotherapy only	11	158	1
Chemotherapy only	30	48	9.0
Radiotherapy and chemotherapy	108	203	7.6

Table 4 Bayesian Analyses of Case-Control and Cohort Studies of
Leukemia Following Treatment for Hodgkin's Disease

Prior evidence	Odds ratio	95% Credibility interval
Vague	8.0	4.2, 15.2
Slightly informative	7.5	4.1, 13.6
Cohort	10.4	8.6, 12.4
Downweighted cohort	8.5	4.9, 15.0

Case-Control Study with Vague Prior

If a vague prior distribution, in which all values of the parameters are
considered equally likely on the log-odds scale, is used, the estimated
odds ratio is 8.0. The posterior distribution is effectively a standardized
likelihood. The Bayesian 95% credibility interval is numerically the same
as the classical 95% confidence interval and goes from 4.2 to 15.2.

Case-Control Study with Slightly Informative Prior

As data are rarely analyzed from a position of complete ignorance, the
previous analysis may be criticised for failing to incorporate what is known
a priori. A belief that chemotherapy probably carries some excess risk,
which could be quite large, could be quantified as a "best guess" of 5,
with 95% certainty that the risk lies between 1 and 25. This is shown
graphically in Figure 4, along with the resulting posterior distribution from
Kaldor's case-control study. There is relatively little change from the reference analysis, which is appropriate with only slightly informative prior
beliefs.

Case-Control Study with Cohort Study as Prior

Large cohort studies often provide information about risk factors for cancers, but with little information on covariates. The information from
Kaldor's cohort study (Kaldor *et al.*, 1987) is used as an informative prior
distribution for the case-control study. Figure 5 shows that the posterior
distribution is strongly dominated by the prior distribution, as one would
expect given the relative sizes of the studies. The 95% credibility interval
of 8.6 to 12.4 does not include the "classical estimate" of the odds ratio
from the Kaldor study, which is 8.00.

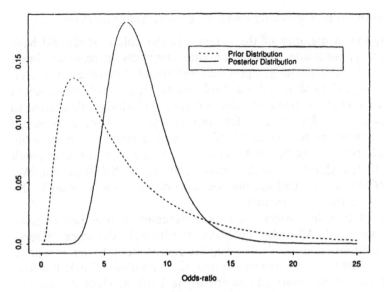

Figure 4 Analysis of Kaldor's case-control study using a slightly informative prior distribution and resulting posterior distribution for risk of leukemia following chemotherapy compared with radiotherapy for case-control study.

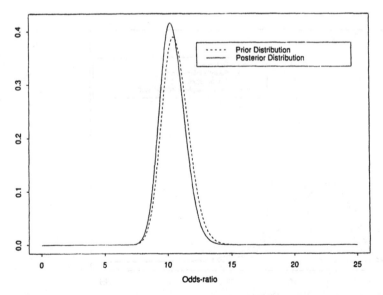

Figure 5 Analysis of Kaldor's case-control study using a prior distribution from Kaldor's cohort study and resulting posterior distribution for risk of leukemia following chemotherapy compared with radiotherapy for case-control study.

Case-Control Study with Downweighted Cohort Study as Prior

It is likely that a majority of the patients in the cohort study did have chemotherapy, but it may be putting undue weight on the results to interpret the relative risk as being a precisely estimated chemotherapy effect. More seriously, the risk is estimated relative to the general population in the cohort study and relative to other patients with Hodgkin's disease in the case-control study. We therefore downweight the prior distribution by assuming that the best estimate of 10.6 for the relative risk from the cohort study is the same but that it could be as little as a half or as much as double that, with a value in this range occurring with 95% probability. Because of the downweighting, the posterior distribution is closer to the likelihood from the study results.

Figure 6 shows the posterior distributions resulting from the four analyses just described. All of the analyses give broadly the same picture, with the exception of the analysis using the cohort study to derive a prior distribution, where the strong prior assumptions dominate the final analysis. The results are summarized numerically in Table 4. There is no doubt that there is a risk of leukemia associated with chemotherapy, of the order of 8, although it is imprecisely estimated and could be half or double this magnitude.

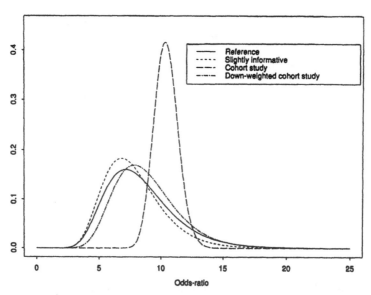

Figure 6   Analysis of Kaldor's case-control study using various prior distributions to obtain posterior distribution.

Further Issues in the Analysis of Case-Control Studies

For the analysis of a single study, there are several issues still to be addressed. The first is to note that the analysis given above can be cast as a logistic regression (Hughes, 1991). This gives the basis for extensions to fuller analyses. For a single continuous covariate O'Hagan (1990) shows how to carry out a logistic regression, and Bayesian methodology for generalized linear modeling is given by Dellaportas and Smith (1993).

The simple initial analysis of Example 4 has ignored matching. In fact, the matched and unmatched classical analyses differ only fractionally either in the estimate of the odds ratio or in its precision. Clearly, a fuller analysis should incorporate matching, and Skene and Wakefield (1990) mention that their hierarchical model is appropriate for a matched-pairs analysis.

A major problem with case-control studies is biases of design, due for example to the selection of controls or to differential recall between cases and controls. Because these biases have been studied, one might wish to try to model them.

The case-control analysis is conventionally modeled on the odds ratio, but in principle a Bayesian approach allows one to estimate the relative risk directly. This requires prior information on the prevalence of disease. If reasonable information is not available, it is sensible to work with the odds ratio.

The analyses given above each focus on one hypothesis, but one of the strengths of a case-control study is the ease with which data on several hypotheses can be collected simultaneously. This leads to multiple inference problems, and Greenland (1986) gives a semi-Bayes analysis of correlated multiple exposures in an occupational cancer mortality case-control study. Raftery and Richardson (Chapter 12) address the problem of model choice in a case-control study of nutritional risk factors for breast cancer, where they use Bayes factors to average across possible models in order to account fully for the uncertainties in inferences about the quantities of interest.

3.2 Analysis of a Cohort Study

In principle, the simplest analysis of a cohort study could be based on a two-by-two table and would proceed as outlined in Example 3 or 4. Schmid and Rosner (1993) investigate the effect of alcohol consumption for breast cancer. Gilks and Richardson (1992) consider the analysis of an occupational cohort study of bladder cancer. Both analyses use Bayesian logistic regression models, but as they also both consider the problem of measurement error, details are deferred until Section 4.

In practice, most cohort studies have a much more complex analysis, relying on either Poisson regression or survival models. At the moment, there are few actual examples of Bayesian analyses of cohort studies using these models. This situation should change, as current developments in Bayesian generalized linear modeling and survival analysis (Dellaportas and Smith, 1993) lay the groundwork for their analyses. Currently, many of the developments in Bayesian survival analysis are taking place in the clinical trials context. Carlin *et al.* (1993) and Abrams (1992), for example, describe models for survival data using Cox's proportional hazards model and a fully parametric approach, respectively, in the context of parallel group randomized trials. The example presented by Abrams *et al.* (Chapter 20) shows how these models can be extended, and work by Stangl (1995; Ch. 16) and Gray (1994) using hierarchical models to look at heterogeneity among centers in multicenter trials is clearly applicable in the epidemiological context. Work by Berzuini and Clayton (1994) on Bayesian analyses with multiple time scales is likely to prove particularly valuable in this context, as many cohort studies have to disentangle the effects of time since first exposure, total time exposed, and calendar time. Bayesian methods for the analysis of other kinds of longitudinal data have been developed by Gilks *et al.* (1993), and Nandram and Sedransk (1993).

4 MEASUREMENT ERROR

Because of its observational nature, epidemiology is bedeviled by measurement error problems. There is an explosion of literature on the subject, mainly from a classical perspective. However, Bayesian approaches are beginning to appear.

Example 5 Air Pollution and Respiratory Illness

Stephens and Dellaportas (1992) analyze some data from a study of exposure to nitrogen dioxide and risk of reported respiratory illness. The data from this study were given by Whittemore and Keller (1988), who analyzed them from a classical perspective, taking into account errors in the explanatory variable. The data are shown in Table 5. The model used for these data is a logistic regression with measurement error on the explanatory variable. The measurement error relationship is known from a calibration study to be

$$X_{aj} = \alpha + \beta X_{0j} + \epsilon_{xj}$$

where X_{aj} is the true value of the covariate and X_{0j} is the observed value,

Table 5 Respiratory Illness by Observed Nitrogen Dioxide Levels

Bedroom NO₂ level (ppb)	<20	20–39	40+	Total
Respiratory illness				
Yes ($Y = 1$)	21	20	15	56
No ($Y = 0$)	27	14	6	47
Total	48	34	21	103

with $X_0 = (X_{01}, X_{02}, X_{03}) = (10, 30, 50)$, $\alpha = 4.48$, $\beta = 0.76$, and ϵ_{xj} is a random element having zero mean and variance $\sigma_X^2 = 81.14$. This is a Berkson-type model, with the true value being modeled conditional on the observed value, as opposed to a measurement error model, where the observed value is modeled conditional on the true value.

Stephens and Dellaportas use the binomial generalized linear model (GLM) with logit link function and linear logistic predictor model proposed by Whittemore and Keller, which gives the likelihood for y, the presence of respiratory illness, as

$$l(y \mid \theta, X_a) \propto \exp\left\{ \sum_{j=1}^{3} [y_{1j}\eta_j - (y_{1j} + y_{2j}) \log(1 + \exp(\eta_j))] \right\}$$

where $\eta = \log(p_j(1 - p_j)) = \theta_1 + \theta_2 X_{aj}$ and p_j is the probability of the presence of respiratory illness.

Based on the calibration equation, they assume the prior density $p(\theta, X_a \mid X_0)$ is multivariate normal with expectation $\alpha + \beta X_0$ and covariance matrix $\sigma_X^2 I_3$. Assuming also that $p(\theta)$ is bivariate normal with mean μ and covariance matrix Σ, then

$$p(\theta, X_a \mid X_0)$$

$$\propto \exp\left\{ -\frac{1}{2}(\theta - \mu)^T \Sigma^{-1}(\theta - \mu) - \frac{1}{2\sigma_X^2}(X_a - \alpha - \beta X_0)^T(X_a - \alpha - \beta X_0) \right\}$$

Using Bayes' theorem, the joint posterior density for θ and X_a can then be evaluated.

Application of this measurement error model gives results with approximately the same mode for the parameter estimates of θ as the conventional model assuming no measurement error, but with larger variance and more skew. The posterior standard deviations are much larger than the standard errors given by the maximum-likelihood analysis of Whitte-more and Keller.

Stephens and Dellaportas show that the posterior distributions of interest may be obtained either by numerical integration (Naylor and Smith, 1982) or by using Gibbs sampling (Dellaportas and Smith, 1993). The former is efficient for low-dimensional problems such as this, whereas the latter should be able to cope with any number of parameters and, theoretically, any realistic posterior surface. This opens the door for Bayesian analyses accounting for measurement error in the kind of complex situations that typically occur in epidemiological studies. However, the practical challenge of quantifying the measurement error in a given situation remains.

Example 6:   Alcohol Consumption and Risk of Breast Cancer

Schmid and Rosner (1993) also consider measurement error in a logistic regression model in their study of alcohol consumption and breast cancer in a cohort of 89,538 female registered nurses in the United States. They incorporate measurement error in the alcohol consumption via a mixture distribution in their analysis and use Monte Carlo methods to estimate the parameter. Their analysis allows the measurement error distribution to change form with the observed exposure, and they also include a covariate, age. By adjusting for the underestimation of alcohol consumption, especially at higher levels, they show from their data an excess risk of breast cancer of 50% for a 25-g increase, compared with 34% from the uncorrected analysis.

The most promising general approach to measurement error problems is through Bayesian graphical models, which was outlined in Example 2 in the context of mapping. Gilks and Richardson (1992) use this approach in accounting for measurement error in the job exposure matrix in an occupational study of bladder cancer. Although their example uses simulated data, it is based on a real job exposure matrix and illustrates a potentially very important area of application for these models. Spiegelhalter et al. (in press) cite an example of a case-control study of herpes simplex virus (HSV) infection and the risk of cervical cancer, where the measurement error in the HSV is estimated on a subset of the population.

5   SAMPLE SIZE

In epidemiology, as in all research, good design is very important. Although it is unfortunate that how many subjects are needed is often the only statistical advice requested, the calculation of sample sizes is important, and a Bayesian philosophy allows several possible approaches to calculation.

Sample size calculations are well established within the classical framework. Desu and Raghavarao (1990) present the theory of sample size calculation for many study designs but make only a passing reference to Bayesian problems. Classical methods require the scientists to know or estimate the value and the variance of the parameters of interest and specify the type I error rate, in order to use a formula for the number of units necessary to obtain a given precision. The type II error rate must also be specified if a significance test comparing parameters is considered.

Instead of requiring a specified value and a variance, a Bayesian approach attempts to quantify the scientist's knowledge by defining a prior distribution for the parameter of interest. The sample size required to estimate a parameter can be determined by requiring the posterior distribution to have particular properties. This is a partial Bayesian approach—see Chapter 1. Comparison of different possible values of a parameter is dealt with by specifying relationships between the posterior distributions for the parameter, given different prior distributions. Using a fully Bayesian approach entails assessing utilities and viewing choice of sample size as a decision problem.

Although there is now considerable interest in using Bayesian methods in the analysis of health-related studies, there has been relatively little discussion of the use of Bayesian ideas in the design of clinical trials (Herson, 1979; McPherson, 1982; Freedman and Spiegelhalter, 1983; Spiegelhalter and Freedman, 1986) and even less in epidemiological work. Hutton and Owens (1993) focus on conflicting prior distributions and consider the number of subjects required for a prevalence survey or case-control study to achieve some consensus.

Example 7: Professionals' Beliefs about Child Sexual Abuse

Hutton *et al.* (1993) investigated professionals' beliefs about the prevalence of sexual abuse in children under 10 and about the prevalence of 11 signs of sexual abuse, including behavioral problems, urinary infections, sexually transmitted diseases, fear of toileting, and allegations by the child, in sexually abused and nonabused children. The study showed a wide range of beliefs. A further study of sexually abused and nonabused children is needed to establish which signs are useful, in the community and for use by referral agencies, in assessing whether a child has been sexually abused. To be certain that children in the group labeled "sexually abused" were so abused, one might choose children whose abusers have confessed. It is more difficult define a nonabused group with precision, and it might be appropriate to use a sample of all children instead. These substantive issues are not the focus of this example. We also do not discuss interactions between the signs. As both professionals and the general public often have strongly held beliefs about abuse that have important consequences, these data provide an appropriate illustration of the impact on the size of a study of allowing for prior beliefs.

The classical sample size technique is presented first, and we consider what revision of belief could be expected to result from a study of this size. We then consider three possible ways of defining a sample size within the Bayesian philosophy. The sample size needs to allow for the extremes of beliefs held by those who might reasonably change their practice as the result of the study. For example, a teacher who might have dismissed a child's allegation of abuse might need to change to regarding this as a reason for referring the child to a specialist agency. Although a layperson might expect all abused children to make an allegation, professionals are aware that such children are often threatened by the abuser with serious consequences if they disclose what has been done to them. In the study cited, half those questioned believed that no more than 15% of abused children would make an allegation of abuse.* One person believed that 30% of nonabused children would make allegations of abuse; such a person would be inclined to ignore an allegation, with the result that an abused child would be left at the mercy of their abuser. Of course, if many children do make false accusations, then many families would be subject to inappropriate investigation. Although one would aim to obtain sufficient information to improve the accuracy of professionals' beliefs, one must acknowledge that some people are unmoved by evidence.

As a first approach, the sample size can be determined by requiring the posterior credible interval to contain the true prevalence, θ_s, of allegations of sexual abuse among children who have been abused, assumed to have a known value for a given population. If there are limited data or a prior distribution on the prevalence, θ_s can be integrated out with respect to this distribution. Determining the sample size by taking the larger of the two numbers given by requiring the credible intervals for individuals with very different prior distributions to touch a fixed value for θ_s will obviously ensure that their posterior credible intervals overlap. An alternative approach is to calculate the sample size by the criterion that the posterior credible intervals of an individual whose prior estimate of prevalence is low should touch the lower limit of the individual whose prior

* One editor's query "Only 15%? Won't most children who have been abused say they've been, and most not say they've not?" illustrates the variety of opinion on this topic.

estimate is high. The formula are derived elsewhere (Hutton and Owens, 1993) and use beta prior distributions. Only illustrative results are given here.

As we do not have estimates of the precision or strength of the professionals' beliefs, an arbitrary number of cases observed by each professional of 100 is used in assigning parameters to the prior distributions. This is not an unreasonable number, as these were all people working in child sexual abuse.

For the classical calculation of the number of individuals, n, needed in a sample to give a $100(1 - \alpha)\%$ confidence interval of width δ for a proportion p, we use

$$n = \frac{z_\alpha^2 p(1 - p)}{\delta^2}$$

where z_α is the $\alpha/2$ percentage point of the standard normal distribution. We take the median estimated prevalences of allegations of sexual abuse among sexually abused and nonabused children, 15% and 1%, as the true proportions to be estimated. The classical calculation of the number of individuals, n, needed in a sample to give a 95% confidence interval of width 0.04 for proportions θ_s and θ_t, where θ_t is the prevalence of allegations of sexual abuse among children who have not been abused, gives sample sizes of 1225 and 95, respectively.

We wish to model the effect on a person's prior belief of observing a sign of sexual abuse by X_s out of N_s sexually abused children and X_t out of N_t nonabused children. Beta prior distributions, with parameters a and b,

$$\text{Beta}(a, b) = \frac{\theta_s^{a-1}(1 - \theta_s)^{b-1}}{B(a, b)}$$

where $B(a, b)$ is the standard beta function, are used to describe individual professional's beliefs, with a equal to the percentage given by the professional and $b = 100 - a$. The effect of observing allegations of sexual abuse by 184, 15%, of 1225 sexually abused children, and 1, 1%, of 95 nonabused children on the beliefs of two professionals with extreme beliefs is summarized using a beta($a + 184, b + 1241$) posterior distribution for the prevalence of allegations of sexual abuse among children who have been abused, and the corresponding distribution for nonabused children, to give the posterior means and 95% credible intervals.

One professional had the same prevalence of allegations, 0.5%, in both groups. The prior distributions of this individual for both groups are therefore beta(0.5, 99.5) and the posterior distributions are beta(184.5, 1140.5) for sexually abused children and beta(1.5, 193.5) for nonabused children. The posterior means and 95% credible intervals for

θ_s and θ_t are 14%, (12%, 16%) and 0.7%, (0.5%, 1.9%), respectively. The hypothesized "true" values lie within the credible intervals, so the individual's beliefs have been updated to conform to the hypothesized prevalence. In contrast, another person believed that no nonabused child would make an allegation of sexual abuse and that 50% of sexually abused children would do so. Using this prior information, the posterior means and 95% credible intervals for θ_s and θ_t are 19%, (17%, 21%) for sexually abused children and 0.5%, (0%, 1.5%) for nonabused children. The credible interval for θ_s lies above the true value, so this person's beliefs still do not conform with the hypothesized prevalence.

It is reasonable to plan for a sample of sufficient size to change the practice of moderately skeptical individuals. This suggests requiring the posterior credible intervals for such a skeptic to contain the true prevalence, θ_s. A first approach might assume a known value of θ_s for a given population. If we use a normal approximation to the posterior beta distribution, the upper limit of the $100(1 - \alpha)\%$ credible interval is $U = \mu + z_\alpha \times \sigma$, where $\mu = a + X_s/(a + b + N_s)$ and

$$\sigma^2 = \frac{(a + X_s)(b + N_s - X_s)}{(a + n + N_s)(a + b + N_s)}$$

Substituting the expected number showing the sign of children making allegations, $N_s\theta_s$, for X_s gives a quadratic equation for N_s.

In our example, with the two very different prior distributions for θ_s given above, that 0.5% and 50% of sexually abused children would make allegations, the sample sizes required for these individuals' posterior 95% credible intervals to include 15% are 413 and 1538, respectively. The upper limit of the first person's posterior interval and the lower limit of the second person's posterior interval would then be 15%. As the prior 95% credible intervals for the nonabused children already include the value 1%, by this criterion no further study is needed.

The individuals' point estimates for the prevalence of allegations of sexual abuse in sexually abused children span the range 0.5% to 90%. Sample sizes calculated by integrating over the observed distribution for θ_s reflect the lack of knowledge of the value of θ_s, and are 1459 and 1868, for the two professionals discussed above. For nonabused children, the point estimates range from 0% to 30%, and the sample sizes are 246 and 295 for the two people.

Allowing for the uncertainty in the parameters that we wish to estimate increases the required sample size.

There are several ways in which sample size calculations can be approached within a Bayesian framework. There is considerable scope in epidemiology for Bayesian reflections on sample sizes. In areas where there is marked disagreement, it is better to design a study to ensure that there will be agreement at the end. If the professionals who provide the prior distributions are in broad agreement, it is reasonable to work with a consensus derived by averaging their prior distributions and specify the properties of the posterior distribution to determine the sample size, as is done in the work in clinical trials (Spiegelhalter and Freedman, 1986). As there is increasing interest in the effective dissemination and implementation of research findings, the known conservativeness of individuals in updating their beliefs should be borne in mind (Slovic and Lichtenstein, 1968). This conservatism often results in people giving evidence less than full weight. If one expected many of the intended readership to down-weight evidence by half, one would need to quadruple the sample size in order to achieve a credible interval of specified width. Furthermore, we, in Examples 1 and 5, and other authors (Clayton, 1989) have suggested using prior distributions to reduce the strength of conclusions because of various sources of bias in epidemiological studies, which again suggests that sample sizes should be adjusted upward.

6 COMBINING EVIDENCE

Much of the work done in a classical context regarding aggregation of data from similar studies is relevant to combining evidence (Thompson and Pocock, 1991), and techniques for Bayesian meta-analysis for studies involving binary end points have been addressed by DuMouchel (1990) and Carlin (1992). Techniques for combining studies with survival end points have already been mentioned (Stangl, Chap. 16; Stangl, 1995; Gray, 1994). Smith and Spiegelhalter (1994) give an example of a meta-analysis where a proper prior distribution is used to encapsulate the knowledge that the true study effects are very unlikely to vary by more than two orders of magnitude. Nearly all of these examples involve meta-analysis of clinical trials, but the methodology is relevant to the epidemiological context. Rubin (1992) emphasizes the need for a meta-analysis to address the underlying scientific problem, rather than just to average effects from a collection of studies that happen to be available.

Quantification of expert opinion, perhaps based on physically motivated models, can also be addressed using a Bayesian approach. The strength of a Bayesian approach to epidemiological problems lies in the potential for summarizing evidence across a range of sources. If there are few data, then the posterior distribution will depend heavily on the quality of reasoning summarized in the prior distribution.

7 FUTURE PROSPECTS FOR BAYESIAN EPIDEMIOLOGY

There are some significant problems with drawing conclusions in epide-miological studies. These problems cannot be resolved by taking a particu-lar view of inference, although thinking through particular problems within a rigorous inferential framework may help clarify the issues. Bernardo and Smith (1994) give a development of Bayesian theory in a decision-making framework. Berry and Stangl (Chapter 1) emphasize the impor-tance of this approach in health care. Epidemiology is one of the key sciences on which public health decisions are based, so a Bayesian ap-proach to epidemiology leads naturally toward a more coherent approach to public health decision making. A particular area in which decisions often have to be made based on epidemiological evidence is in the field of pharmacoepidemiology, which looks at the safety of marketed medicines. Inevitably, decisions have to be made based on less evidence than one would ideally like, and formalizing this uncertainity may help to clarify the issues.

A Bayesian approach facilitates the downweighting of evidence, which is sometimes sensible because of the many sources of bias in epide-miological studies or because the evidence is only indirectly relevant. However, a fixed belief that nothing dramatic will ever be found could be damaging. Our uncertainty should include sufficient humility to enable serious events or mistakes to be detected.

Although Bayesian work in epidemiology is beginning to appear, it has made little impact on mainstream epidemiology. For that to change requires work at two levels. The first is for simple back-of-the-envelope Bayesian methods to be readily available. These are useful as an initial analysis of a study and also as a means of helping epidemiologists under-stand the Bayesian philosophy, so that they can appreciate the basis of more complex analysis. The second need is for accessible software for applied statisticians to carry out more complex modeling. Many of the illustrations in this chapter have been made possible by recent computa-tional developments, and indeed some of them have come from papers written primarily because of the statistical methodology.

There is currently considerable interest in Bayesian epidemiology, as the work surveyed in this chapter illustrates. As Bayesian methods for complex problems are developed further, this is likely to continue. A particularly fruitful area is likely to be the analysis of follow-up studies, whether looking at survival end points or other longitudinal data. The potential of Bayesian methods to deal flexibly with measurement error should yield a rich source of interesting problems, and the combination of evidence from different sources is another important area where Bayesian methods seem a natural resource.

Work still needs to be done on ways of expressing a Bayesian interpretation of studies to epidemiologists, doctors, and public health workers, so that the practical potential of Bayesian ideas in epidemiological research can be developed and assessed. The best work will come from the interaction between statisticians, clinicians, and other health workers who have thought deeply about the complex issues involved.

REFERENCES

Abrams, K. R. (1992). Bayesian Survival Models. Ph.D. thesis, University of Liverpool, Liverpool, UK.

Ashby, D., Hutton, J., and McGee, M. (1993). Simple Bayesian analyses for case-control studies in cancer epidemiology. *The Statistician* 42: 385–397.

Bernardinelli, L. and Montomoli, C. (1992). Empirical Bayes versus fully Bayesian analysis of geographical variation in disease risk. *Statistics in Medicine* 11: 983–1007.

Bernardo, J. and Smith, A. (1994). *Bayesian Theory*. Wiley, Chichester, UK.

Berzuini, C. and Clayton, D. (1994). Bayesian analysis of survival on multiple time scales. *Statistics in Medicine* 13: 823–828.

Breslow, N. (1990). Biostatistics and Bayes. *Statistical Science* 5: 269–298.

Breslow, N. and Clayton, D. (1993). Approximate inference in generalized linear mixed models. *Journal of the American Statistical Association* 88: 9–25.

Breslow, N. and Day, N. (1980). *Statistical Methods in Cancer Research I: The Analysis of Case-Control Studies*. Scientific Publication No. 32, International Agency for Research on Cancer, Lyon, France.

Breslow, N. and Day, N. (1987). *Statistical Methods in Cancer Research II: The Design and Analysis of Cohort Studies*. Scientific Publication No. 82, International Agency for Research on Cancer, Lyon, France.

Carlin, B., Chaloner, K., Church, T., Louis, T., and Matts, J. (1993). Bayesian approaches for monitoring clinical trials with an application to toxoplasmic encephalitis prophylaxis. *The Statistician* 42: 355–367.

Carlin, J. B. (1992). Meta-analysis for 2 × 2 tables: A Bayesian approach. *Statistics in Medicine* 11: 141–158.

Clayton, D. and Kaldor, J. (1987). Empirical Bayes estimates of age-standardized relative risks for use in disease mapping. *Biometrics* 43: 671–681.

Clayton, D. G. (1989). Discussion of 'Royal Statistical Society meeting on cancer near nuclear installations'. *Journal of the Royal Statistical Society Ser. A* 152: 365–384.

Commenges, D. and Etcheverry, B. (1993). An empirical Bayes approach to the estimation of the incidence curve of HIV infection. *Statistics in Medicine* 12: 1317–1324.

Dellaportas, P. and Smith, A. (1993). Bayesian inference for generalized linear and proportional hazards models via Gibbs sampling. *Applied Statistics* 42: 443–459.

Desouza, C. (1991). An empirical Bayes formulation of cohort models in cancer epidemiology. *Statistics in Medicine* 10: 1241–1256.

Desu, M. M. and Raghavarao, D. (1990). *Sample Size Methodology*. Academic Press, New York.

DuMouchel, W. (1990). Bayesian metaanalysis. In *Statistical Methodology in the Pharmaceutical Sciences*, ed. Berry, D., pp. 509–529. Marcel Dekker, New York.

Freedman, L. S. and Spiegelhalter, D. J. (1983). The assessment of subjective opinion and its use in relation to stopping rules for clinical trials. *The Statistician* 33: 153–160.

Gail, M. (1991). A bibliography and comments on the use of statistical models in epidemiology in the 1980's. *Statistics in Medicine* 10: 1819–1885.

Gilks, W. and Richardson, S. (1992). Ancillary risk factors with applications to job-exposure matrices. *Statistics in Medicine* 11: 1443–1463.

Gilks, W., Wang, C., Yvonnet, B., and Coursaget, P. (1993). Random-effects models for longitudinal data using Gibbs sampling. *Biometrics* 49: 441–453.

Gray, R. (1994). A Bayesian analysis of institutional effects in a multicenter cancer clinical trial. *Biometrics* 50: 244–253.

Greenland, S. (1986). A semi-Bayes approach to the analysis of correlated multiple associations, with an application to an occupational cancer-mortality study. *Statistics in Medicine* 11: 219–230.

Heisterkamp, S., Doorrnbos, G., and Gankema, M. (1993). Disease mapping using empirical Bayes and Bayes methods on mortality statistics in the Netherlands. *Statistics in Medicine* 12: 1895–1913.

Herson, J. (1979). Predictive probability early termination plans for phase II clinical trials. *Biometrics* 35: 775–783.

Hughes, M. D. (1991). Practical reporting of Bayesian analysis of clinical trials. *Drugs Information Journal* 25: 381–393.

Hutton, J. L. and Owens, R. G. (1993). Bayesian sample size calculations and prior beliefs about child sexual abuse. *The Statistician* 42: 399–404.

Hutton, J. L., Owens, R. G., Horne, L., and Glasgow, D. V. (1993). Professionals' beliefs about child sexual abuse'. *Health & Social Care in the Community* 1: 219–226.

Kaldor, J., Day, N., Band, P., Choi, N., Clarke, E., Coleman, M., Hakama, M., Koch, M., Langmark, F., Neal, F., Pettersson, F., Pompe-Kirn, V., Prior, P., and Storm, H. (1987). Second malignancies following testicular cancer, ovarian cancer and Hodgkin's disease: An international collaborative study among cancer registries. *International Journal of Cancer* 39: 571–585.

Kaldor, J., Day, N., Clarke, A., Van Leeuwen, F., Henry-Amar, M., Fiorentino, M., Bell, J., Pedersen, D., Band, P., Assouline, D., Koch, M., Choi, N., Prior, P., Blair, V., Langmark, F., Pompe-Kirn, V., Neal, F., Peters, D., Pfeiffer, R., Karjalainen, S., Cuzick, J., Sutcliffe, S., Somers, R., Pellae-Cosset, B., Pappagallo, G., Fraser, P., Storm, H., and Stovall, M. (1990). Leukaemia following Hodgkin's disease. *New England Journal of Medicine* 322: 7–13.

Lawson, A. (1994). Using spatial Gaussian priors to model heterogeneity in environmental epidemiology. *The Statistician* 43: 69–76.

Marshall, R. (1988). Bayesian analysis of case-control studies. *Statistics in Medicine* 7: 1223–1230.

McPherson, K. (1982). On choosing the number of interim analyses in clinical trials. *Statistics in Medicine* 1: 25–36.

Nandram, B. and Sedransk, J. (1993). Bayesian predictive inference for longitudinal sample surveys. *Biometrics* 49: 1045–1055.

Naylor, J. and Smith, A. (1982). Applications of a method for the efficient computation of posterior distributions. *Applied Statistics* 31(3): 214–225.

Nurminen, M. and Mutanen, P. (1987). Exact Bayesian analysis of two proportions. *Scandanavian Journal of Statistics* 14: 67–77.

O'Hagan, A. (1990). Practical Bayesian analysis of a simple logistic regression: Predicting corneal transplants. *Statistics in Medicine* 9: 1091–1101.

Rubin, D. (1992). Meta-analysis: Literature synthesis or effect-size surface estimation? *Journal of Educational Statistics* 17: 363–374.

Schmid, C. and rosner, B. (1993). A Bayesian approach to logistic regression models having measurement error following a mixture distribution. *Statistics in Medicine* 12: 1141–1153.

Skene, A. and Wakefield, J. (1990). Hierachical models for multicentre binary response models. *Statistics in Medicine* 9: 919–929.

Slovic, P. and Lichtenstein, S. (1968). Relative importance of probabilities and payoffs in risk taking. *Journal of Experimental Psychology* 78: 1–18.

Smith, T. and Spiegelhalter (1994). Proper priors for Bayesian meta-analysis. Report, Medical Research Council Biostatistics Unit, Cambridge, UK.

Spiegelhalter, D. J. and Freedman, L. S. (1986). A predictive approach to selecting the size of a clinical trial based on subjective clinical opinion. *Statistics in Medicine* 5: 1–13.

Spiegelhalter, D., Thomas, A., Best, N., and Gilks, W. (1994). BUGS examples, version 0.30 A. Technical report, Medical Research Council Biostatistics Unit, Cambridge, UK.

Spiegelhalter, D., Thomas, A., and Best, N. (in press). Computation on Bayesian graphical models. In *Bayesian Statistics 5*, eds. Berger, J., Bernardo, J., Dawid, A., and Smith, A. Oxford University Press, Oxford, UK.

Stangl, D. (1995). Modelling and decision making using Bayesian hierarchical models. *Statistics in Medicine* 14(20): 2173–2190.

Stephens, D. and Dellaportas, P. (1992). Bayesian analysis of generalised linear models with covariate measurement error. In *Bayesian Statistics 4*, eds. Bernardo, J., Berger, J., Dawid, A., and Smith, A., pp. 813–820. Oxford University Press, Oxford, UK.

Swerdlow, A., Douglas, A., Vaughan Hudson, G., Vaughan Hudson, B., Bennett, M., and MacLennan, K. (1992). Risk of second primary cancers after Hodgkin's disease by type of treatment: Analysis of 2846 patients in the British National Lymphoma Investigation. *British Medical Journal* 304: 1137–1143.

Thompson, S. G. and Pocock, S. J. (1991). Can meta-analyses be trusted? *Lancet* 338: 1127–30.

Whittemore, A. and Keller, J. (1988). Approximations for errors-in-variables regression. *Journal of the American Statistical Association* 83: 1957–1966.

Wild, P., Commenges, D., and Etcheverry, B. (1993). A hierarchical Bayesian approach to the back-calculation of numbers of HIV-infected subjects. *The Statistician* 42: 405–414.

Zelen, M. and Parker, R. (1986). Case-control studies and Bayesian inference. *Statistics in Medicine* 5: 261–269.

II
ASSESSING PROBABILITIES

4
Elicitation of Prior Distributions

Kathryn Chaloner   University of Minnesota, St. Paul, Minnesota

ABSTRACT

Research on methods for helping experts to specify subjective prior distributions is briefly reviewed and discussed. Specific methods for elicitation for clinical trials are also reviewed. Some suggestions are made and an example is given.

Keywords: probability assessment, subjective probability distribution

1 INTRODUCTION

There is a large volume of psychological literature on how people make judgments about uncertainty. This review is not comprehensive and only a few key references are given. Some general suggestions are provided in Section 5. These suggestions are based on my personal observational experiences working with physicians and researchers and are not based on scientific experiments or psychological theories. I would be delighted if better recommendations were developed. An example is given in the Appendix of eliciting a distribution for a human immunodeficiency virus/ acquired immunodeficiency syndrome (HIV/AIDS) clinical trial in progress.

2 OVERVIEW OF ELICITATION METHODS

Some of the initial work on elicitation can be found in Winkler (1967a,b, 1971) and Savage (1971). Hogarth (1975) gives an overview of early work incorporating much of the psychological and behavioral science literature into the discussion. Hogarth (1987) provides a more recent perspective, and Appendix B of his book is a useful tutorial on assessing probabilities. Other recent references on psychological and behavioral aspects of subjective probability assessment are found in Kahneman *et al.* (1982); this is a collection of papers by numerous authors about how people process uncertainty. Another important collection of papers is Kyberg and Smokler (1980). Wallsten and Budescu (1983) review the psychological aspects of subjective probability assessment and argue that measures of reliability and validity, as defined in measurement theory, should be applied to subjective probability assessments. They divide the literature into two parts: (1) studies of subjective probability assessment from nonexperts who have no expertise in either probability or the subject matter and (2) studies from experts, either subject matter experts or experts in probability or decision theory. They report that there is a lack of experiments investigating reliability and validity of experts. Von Winterfeldt and Edwards (1986, Chapter 4) also give a general discussion of elicitation and some of the basic issues.

Some probabilities are easier to assess than others. O'Hagan (1994, p. 107) gives the example that it is easy to assess the probability of a coin landing heads when tossed, but to assess the probability of 4 heads in 10 independent tosses is much harder. Ravinder *et al.* (1988) describe a technique they call *decomposition*. Rather than elicit a probability of an event A directly, they give circumstances in which it may be advantageous to elicit a series of marginal and conditional probabilities, $P(B)$ and $P(A \mid B_i)$ where the events B_i, $i = 1, \ldots, n$, form a partition of the sample space and

$$P(A) = \sum P(A \mid B_i)P(B_i)$$

Lindley (1985, pp. 39–41) calls this *extending the conversation* and also suggests its usefulness in elicitation.

Despite the large volume of psychological literature on probability assessment, few of the ideas, theories, and empirical results have been applied to develop operational methodology for eliciting prior distributions for specific statistical models and problems. The linear regression problem has received some attention: Kadane *et al.* (1980) suggested and implemented a method of elicitation based on specifying predictive distributions. They restricted beliefs to lie in the normal-gamma conjugate family and elicited quantiles of the predictive distributions at several values of the explanatory variables. They argue that predictive distributions on potentially observable quantities are easier to think about than distributions on unobservable parameters. Their method asks more than the minimum number of questions required and so any inconsistencies must be reconciled. An example of this method is given in Chapter 5, by Kadane and Wolfson. Garthwaite and Dickey extended the work of Kadane *et al.*; see Garthwaite and Dickey (1985, 1988, 1991) and Garthwaite (1992) for some successful elicitation methods for regression models. See also Dickey *et al.*, (1986). Other, related methods can be found in Laskey and Black (1989) and Black and Laskey (1989) for analysis of variance models, in Chapter 5, by Kadane and Wolfson, for exponential lifetime models, and in Chaloner and Duncan (1983, 1987) for binomial and multinomial problems. A novel way of evaluating elicitation procedures by adding random error to the values specified is given in Gavasakar (1988).

2.1   Problems

People are not typically very good probability assessors. They make mistakes and are inconsistent.

In elicitation methods such as that of Kadane *et al.* (1980) an assumption is made that the expert's beliefs follow a particular parametric family; the family is chosen for its convenience or conjugateness. Inconsistencies may arise because the subject's beliefs do not follow the chosen parametric family. For example, if an expert specifies a 5th, 50th, and 95th percentile of a distribution that is assumed to be normal, there may be no normal distribution with the specified percentiles. Or, an expert may specify a normal distribution by specifying a mean and a standard deviation and then when asked for the upper and lower quartiles give values that specify a different normal distribution.

Logical inconsistencies may also be present; for example, if a subject specifies a lower quartile of a distribution which is larger than an upper quartile specified, this is clearly an impossibility.

Kadane *et al.* (1980) suggest that probability distributions be elicited in several ways and the resulting inconsistencies reconciled. These methods are not particularly satisfactory and raise many difficult issues.

The elicitation technique known as the *device of imaginary results* has been advocated and used successfully (see Good, 1983). This technique requires subjects to give their beliefs after they are told about hypothetical data. From this, prior probabilities can be deduced, assuming that the subjects' beliefs obey Bayes' theorem. This method appears to have been successful, although Phillips and Edwards (1966) and others have demonstrated that people do not necessarily use Bayes' theorem to update their beliefs correctly in a situation in which Bayes' theorem should apply. Some experimental evidence indicates that people are conservative and do not adjust their beliefs as much as the rules of probability require. For a thoughtful discussion of the related experimental evidence and the corresponding psychological theories see von Winterfeld and Edwards (1986, Section 6.5 and 13.2).

3   ELICITATION FOR CLINICAL TRIALS

3.1   Why Is Elicitation Important?

Even without a Bayesian analysis it is important, before a trial begins, to document information and beliefs about the planned treatments. If the evidence is such that one treatment is firmly believed to be inferior, then the ethics of enrolling patients into the trial are questionable. As stated by, for example, Byar et al. (1990), "a trial should be open only to patients for whom the choice of recommended treatment remains substantially uncertain." Although Byar et al. do not quantify probabilistically what "substantially uncertain" means, it would clearly be helpful to document beliefs probabilistically in reviewing the ethical aspects of a trial.

Prior distributions can also used in the design of a clinical trial and, of course, prior distributions are also required for Bayesian monitoring and analysis of a clinical trial. Carlin et al. (1993, 1995) use the prior distributions elicited in Chaloner et al. (1993) to illustrate Bayesian monitoring for a toxoplasmosis prophylaxis trial.

Kadane (1986) suggests that subjective prior distributions be elicited from a number of experts, including a number of clinicians and patients, to form a "community" of prior distributions. Inferences can then be based on a consensus of posterior conclusions.

Spiegelhalter et al. (1994) recommend consulting a large number of experts and subsequently constructing a number of distributions, namely:

1.   A "clinical" prior distribution by averaging prior distributions elicited from a large number of experts
2.   A "vague" prior distribution leading to a posterior distribution proportional to the likelihood
3.   A "skeptical" prior distribution representing a clinician unenthusiastic about the new therapy centered at the new therapy having no effect

4.   An "enthusiastic" prior distribution centered at the new therapy
having a large effect

Greenhouse and Wasserman (1995) (following Huber, 1973, and
Berger, 1984) suggest that a single prior distribution π_0 be specified and
a class of prior distributions be considered that are close to π_0 in some
sense. A popular and tractable class is the "ϵ-contaminated class," which
are mixtures with probability $(1 - \epsilon)$ on π_0 and ϵ on some other distribution from a specified class. This approach cannot easily, however, reflect
a large amount of variability between opinions.

An excellent answer to the question "Why is elicitation important?"
was given by Garthwaite and Dickey (1991), who said that "expert personal opinion is of great potential value and can be used more efficiently,
communicated more accurately, and judged more critically if it is expressed as a probability distribution."

3.2   Elicitation for Clinical Trials

Freedman and Spiegelhalter (1983) describe a clinical trial of whether
using a particular drug immediately following surgery for superficial bladder cancer is an improvement over the standard treatment of not using
the drug. They elicit probability distributions from 18 clinicians on the
magnitude of the effect on probability of nonrecurrence for 2 years. They
ask clinicians about the difference in proportions in the control and treatment group. They describe an elicitation procedure in which clinicians
are asked for a mode and then upper and lower bounds that are thought
to be "very unlikely to be exceeded," and then the clinicians are asked
to assess the probability of the effect being larger than a series of intermediate points. Freedman and Spiegelhalter report a diversity of opinions
among the 18 clinicians. They also report results of asking the clinicians
for a "range of equivalence." These ranges are based on the belief that
before adopting a new treatment clinicians would demand not just that the
new treatment is efficacious but also that it is enough of an improvement to
counterbalance increased toxicity. The upper limit of the range is the
lowest treatment effect required for the clinician to adopt the drug as
standard care, and the lower limit of the range is the largest treatment
effect for which the clinician would not use the drug as standard care.
Within this range the treatments are, effectively, equivalent. The 18 clinicians reported a diversity of opinions. Freedman and Spiegelhalter suggest
that such a diversity may be a particular phenomenon of a multicenter
study.

Spiegelhalter and Freedman (1986, 1988) give an another example of a trial of chemotherapy in which seven consultant oncologists were interviewed and their prior distributions elicited using 20-minute structured interviews. Figure 2 of Spiegelhalter and Freedman (1988) gives the seven probability distributions, and there is a remarkable degree of consensus. The authors say "we note the consistency in judgement among "naive" subjects who had never undergone a similar exercise nor discussed it amongst themselves." As the opinions were so consistent and there were no clearly discordant opinions, taking a simple average seems reasonable. Spiegelhalter and Freedman (1988) also report the ranges of equivalence elicited for the treatments in this trial. The seven oncologists provide remarkably consistent beliefs with similar intervals.

Kadane (1994) describes a trial investigating two drugs designed to reduce hypertension after open heart surgery. As the response variable used was the deviation of arterial systolic blood pressure from a target of 75 mm Hg, regression models were used with several covariates. The elicitation method of Kadane *et al.* (1980) was used to elicit normal-gamma prior distributions from five experts. The experts had different beliefs: two preferred one drug for all types of patients considered and one expert always preferred the other.

Elicitation by interviewing experts individually is time consuming. Spiegelhalter *et al.* (1994) report an alternative approach using postal elicitation in several trials currently under way. The method is extremely simple and allows a large number of clinicians to report their beliefs. They ask clinicians for their probabilities that the effect falls in different intervals. Results from this method have yet to be reported.

Berry *et al.* (1992) take a different approach and describe subjectively assessing prior distributions for a large vaccine trial using primarily historical information. They use the collaborative expertise of the authors of the paper. Prior distributions are restricted to follow parametric distributions: gamma and F distributions. The authors mention that a computer program was written to represent graphically the parameters to be specified but provide little detail. They describe the prior distributions elicited as "reasonably open minded" and report the historical data and their experience upon which their prior distributions are based.

Chaloner *et al.* (1993) take yet another approach for a trial of a prophylactic treatment for toxoplasmosis in advanced HIV and AIDS patients. Like Freedman and Spiegelhalter (1983) they use a two-year probability: but this time it is the two-year probability of getting toxoplasmosis on the active treatment conditional on a placebo probability. They also make the assumption that the proportional hazards Cox regression model is appropriate. They describe eliciting beliefs from a group of five AIDS experts: three physicians, an epidemiologist, and a person working on AIDS research. The experts are asked to specify a guess for the 2-year probability on placebo, and then conditional on that probability a distribution is elicited on the probability for the active treatment arm. Initially a parametric distribution is specified, using upper and lower quartiles, and this distribution is shown to the expert on the screen of a workstation. The expert then adjusts the distribution by hand, using the mouse, to represent his or her beliefs. Like Freedman and Spiegelhalter (1983), Chaloner et al. (1993) report a wide diversity of opinions: Figure 1 of their paper is a plot of the five distributions. The distributions are not all unimodal and could not easily be represented by a simple parametric class. Chaloner et al. also report that although all five experts were enthusiastic about the treatment effect and had high probabilities on a large efficacious effect, the subsequent trial data proved them all wrong. The treatment had no effect in preventing toxoplasmosis and, in fact, those receiving the active drug had a higher death rate, possibly due to a harmful toxic effect of the drug. All the experts assigned this outcome little or no probability. Chaloner et al. (1993) describe a simple computer program, written in the xlispstat environment of Tierney (1990), using dynamic graphics and mouse input. They also describe repeating the elicitation for three of the five experts using 3-year rather than 2-year probabilities to give a check on the assumptions made. They do not suggest methods for reconciling inconsistencies.

By comparison with other problems, a variety of methods are available for elicitation for clinical trials and relatively wide experience. It would be valuable to compare the different methods and examine whether the more complicated method of Chaloner et al. is an improvement over the simple methods of Spiegelhalter et al.

An example is given in the Appendix of using the method of Chaloner et al. to elicit opinions from three experts about a trial of prophylaxis for Pneumocystis carinii pneumonia (PCP) in patients with HIV infection or AIDS.

4   OTHER BIOMEDICAL APPLICATIONS

4.1   Medical Diagnosis

In their review, Wallsten and Budescu (1983) include a section reviewing experiments on probabilistic assessments in medical diagnosis. See also Spiegelhalter (1987).

4.2 Design of Experiments

In designing experiments clinicians have often provided opinions as to the magnitude of the effects they expect. These point predictions have traditionally been used for sample size and power calculations. For example, it is often the practice that the sample size of a clinical trial be chosen to have 80% power to detect a specified effect: either a "smallest clinically meaningful effect," which might be the upper limit of Freedman and Spiegelhalter's range of equivalence, or an "expected" effect. As clinicians are familiar with providing guesses for the purpose of making sample size calculations, it is a natural step to provide probability distributions representing uncertainty so that uncertainty can be incorporated into the design process.

Methods for elicitation can also be used to elicit prior distributions to be used in design. Flournoy (1994) describes just such an example for the design of a phase I clinical trial. Her elicitation required a group of physicians to discuss and collectively provide a sketch, by hand, of an upper and lower response curve. She describes the historical information available to the physicians. She also describes in detail how the physicians "grappled together with their priors for this interval, sketching and re-sketching them" and describes how the final curves were agreed upon. She reports some inconsistencies such as an instance of an assessed 2.5th percentile above the 97.5th percentile. She also reports that although the physicians were asked for 50% probability intervals, it was clear from listening to them that they were describing something much closer to a 95% interval. Interestingly, the physicians also described it later as providing "maximum and minimum" values. Flournoy describes how a normal and a gamma prior distribution were specified using the resulting sketches.

This kind of detail is rare in the literature, and it is refreshing to see an elicitation processes described—and especially refreshing to see all the problems with the process laid out clearly and honestly. Only by sharing and describing these kinds of experiences will good methods for elicitation be developed.

5 GENERAL RECOMMENDATIONS

The following subjective recommendations are based on my experience, primarily as reported in the Chaloner et al. (1993) study.

1. *Interactive feedback*. Forming beliefs is an iterative process and subjects like to be able to change their minds and work toward

specifying a distribution. Presenting them with interactive feedback helps them formulate their ideas probabilistically. It could potentially enable the expert to reconcile inconsistencies subjectively, which seems preferable to having an automatic algorithm for doing so.

2. *Scripted interview*. A written script to structure the interview was found to be helpful in Chaloner *et al*. (1993). Freedman and Spiegelhalter (1983) also report using a "structured interview." In an interview experts will ask questions and it will be necessary to deviate from the script, but the script provides some uniformity across experts.

3. *Review*. The expert should be provided with the results of a systematic literature review.

4. *Percentiles*. Asking for quartiles is difficult. Like Flournoy (1994), Chaloner *et al*. (1993) report that even though experts were asked for 25th and 75th percentiles they interpreted these more like 2.5th and 97.5th percentiles to give intervals with probability content 0.95. A reasonable supposition is that experts are familiar with 95% confidence intervals, tend to interpret them as probability intervals, and tend to think about 95% probability intervals rather than upper and lower percentiles. It is probably better to ask for the 2.5th and 97.5th percentiles.

5. *Lots of experts*. Elicit opinions from as many experts as is practically feasible and from a variety of sources of expertise. Doctors, nurses, patients, scientists, and researchers all have opinions.

Other general comments are:

1. Postal elicitation (Spiegelhalter *et al*. 1994) seems promising but perhaps a little dangerous as so little is known about elicitation. There is no opportunity for discussion for clarification (such as when the expert does not understand what exactly a 75th percentile is) and no opportunity to deviate from the questions asked (such as when, as in Chaloner *et al*., 1993, an expert prefers to use a different end point definition than the one planned). It also eliminates the opportunity for interactive feedback. It does have the advantage of allowing a large number of clinicians to have their beliefs easily elicited.

2. Avoid restricting beliefs to parametric families. These families may be useful as a starting point or may even be necessary in high-dimensional problems but artificially constrain beliefs.

6 CONCLUSIONS

Elicitation not only enables the many potentially useful Bayesian methods to be used in practice but also aids discussion and provides valuable documentation of the expectations of a clinical experiment before it begins. There is an urgent need for operational methods for elicitation of subjective probability distributions.

ACKNOWLEDGMENT

I am grateful to George Duncan and Dalene Stangl for helpful comments and to Winston Cavert, Carlton Hogan, and Frank Rhame for their opinions. Support was provided in part by NIAID contract NO1-AI-05073.

APPENDIX

A description is now given of eliciting a prior distribution for a trial of PCP prophylactic therapy. The method of Chaloner *et al.* (1993) is used and the script was adapted from the script described in that paper. The method assumes that data from the trial will be analyzed using the Cox (1972) proportional hazards regression model and the prior distribution will be noninformative on all but the regression parameter corresponding to the treatment assignment. A probability distribution is elicited on the proportion of patients experiencing the end point on the treatment arm conditional on a guess for the proportion on the control arm. That is, the distribution elicited is a conditional subjective probability distribution.

The experts had available the protocol document of the trial in question: "A randomized, comparative, prospective study of daily trimethoprim/sulfamethoxazole (TMS) and thrice weekly TMS for prophylactic therapy against PCP in HIV-infected patients" (CPCRA 006). The protocol document contained a review of relevant information in the literature. The trial began enrollment in 1992 and at the time of elicitation in September 1994 results remain confidential.

In each of the three elicitations performed the experts asked many questions and entered into a lengthy discussion about what they thought the trial would show. Their distributions, together with a summary of the discussions, are given later. The script, which gave a structure to the process, is given below.

The Script

People with AIDS often develop *Pneumocystis carinii* pneumonia (PCP). Trimethoprim/sulfamethoxazole (TMS) has been shown to fight PCP but it is not without side effects. The PCP-TMS study is designed to show which dose of TMS is safest and which works best. Daily TMS has been shown to be effective in preventing a recurrence of PCP in people who have already had it. There is also evidence that taking TMS three times a week may be effective against PCP and may cause fewer side effects than a daily dose.

The purpose of this exercise is to quantify your beliefs about the efficacy of the two treatment arms in the PCP-TMS trial. One treatment is one double-strength (DS) tablet of TMS daily and the other treatment is one DS tablet three times a week (Monday, Wednesday, Friday). The protocol document describes the eligibility criteria for the trial as "designed to include all patients for whom a primary care physician would prescribe prophylaxis, while excluding some patients for safety reasons."

Do you need to know more about the trial? Or about the eligibility criteria? The protocol document provides a literature review of relevant studies on PCP prophylactic therapy and AIDS/HIV and is here if you want to study it or refer to it.

The end point of the trial is an episode of *Pneumocystis carinii* pneumonia (PCP). Deaths are not included in the end point. Some patients will reach an end point and some will not. Only patients experiencing PCP will reach an end point. Patients who die without experiencing PCP will be removed from the analysis.

Think about a large number of people enrolled in the trial for 2 years and think about the proportion of people who will reach the PCP end point.

What is your best guess of the percentage of people assigned to the daily group who will experience PCP 2 years after enrollment? Please give your best guess—you will have uncertainty about this guess—but if you had to make a guess what would your guess be? Choose the value you think most likely. Your guess should relate to people similar to those expected to be enrolled in the trial—some of whom may supplement their study drug or reduce their dosage, some may fail to comply with the treatment, and some may develop intolerance to TMS—but think of the entire group of people assigned to an arm.

Now suppose your guess of this proportion turns out to be correct—the percentage of people on placebo experiencing PCP is exactly your best guess. Think about the people on the thrice weekly arm and think about an interval estimate for what you would expect for the percentage of people on the thrice weekly TMS arm who will experience PCP in two years *given* that the proportion experiencing PCP on the daily TMS arm is what you guessed.

Please specify an interval, by an upper number and a lower number, within which you think that the percentage of people experiencing PCP on the three times a week arm will lie. The interval should be such that you have probability 0.95 that the proportion will lie in the interval: probability 0.025 that the percentage will be higher than the upper limit and probability 0.025 that the percentage will be lower than the lower limit. In other words, the upper and lower limits are the 2.5th and 97.5th percentiles of your probability distribution on the percentage.

You will now be shown a plot of a probability distribution with the properties you have given. It is a smooth estimate of a distribution with the percentiles you specified. You can adjust the curve smoothly using the slider-dialogs to something that you think is reasonable. The curve is constrained to be of a particular shape.

You can now adjust the distribution to more accurately reflect your beliefs using the hand adjustment by changing the mouse mode to hand drawing. (If at any point you want to start again from the beginning then we can do so.)

The Results

Beliefs were elicited from three experts: two infectious disease MDs treating AIDS patients (experts A and C) and a person involved in the running and design of AIDS clinical trials (expert B). The three values for the best guess of the proportion experiencing PCP on the daily dose arms were 0.07, 0.15, 0.10. The probability distribution, conditional on this percentage, of the corresponding proportion of patients experiencing PCP on the three times a week arm is plotted on the top line of Figure 1. On each of the three plots the expert's best guess of the proportion for the daily arm is indicated by a cross ($\times$). The corresponding distribution on the regression coefficient in a proportional hazards regression model was calculated for each expert and is plotted on the second row of Figure 1.

Expert A thought that either dose would be completely effective if the patient complied with the dose. That is, the only patients experiencing PCP would be those who either did not comply with treatment or became intolerant to TMS and could not comply. This expert thought that if a patient on the three times a week arm forgot to take a dose, the patient might become susceptible to PCP. On the daily dose, however, forgetting to take one dose occasionally would probably not increase susceptibility. Because of this, expert A typically prescribes TMS to be taken daily,

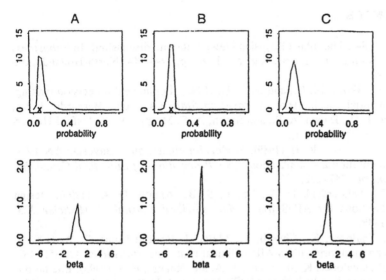

Figure 1 Prior distributions on the probability (top row) and on the regression coefficient (bottom row).

although expert A believes more people will develop intolerance on the daily dose. Expert A's guess for the proportion of end points in 2 years on the daily arm was 7% with a corresponding initial conditional interval for the three times a week arm of 3% to 20%. Expert A adjusted the initial plot to give the plot shown in Figure 1.

Expert B had the opinion that the two arms were equivalent. This expert guessed that the proportion of patients experiencing PCP on the daily arm would be about 15% and had a 95% interval of 11.25 to 18.75 (which is 15% ± 25% of 15%) for the percentage on the thrice weekly arm conditional on 15% on the daily arm. Although this expert claimed he had great uncertainty, his distribution was, in fact, the least variable of the three.

Like expert B, expert C also believed the two arms to be equivalent and indeed, in clinical practice, typically gives patients a choice of either daily TMS or TMS three times a week. Unlike expert A, expert C believes that developing intolerance to TMS will occur equally often on the two doses. Expert C's guess for the daily arm was 10% with an initial conditional interval of 5% to 20% for the thrice weekly arm.

All three experts had uncertainty about which of the two treatments would be best. This should be reassuring to the designers of the trial, as clinical trials should be designed to answer questions about which there is uncertainty.

154

Chaloner

REFERENCES

Berger, J. (1984). The robust Bayesian viewpoint (with discussion). In *Robustness in Bayesian Statistics*, ed. Kadane, J. B., pp. 64–144. North-Holland, Amsterdam.

Berry, D. A., Wolff, M. C., and Sack, D. (1992). Public health decision making: A sequential vaccine trial. In *Bayesian Statistics 4*, eds. Bernardo, J. M., Berger, J. O., Dawid, A. P., and Smith, A. F. M., pp. 79–96. Oxford University Press, Oxford, UK.

Black, P. and Laskey, K. B. (1989). Models for elicitation in Bayesian ANOVA: Implementation and application. *ASA Proceedings of Statistical Computing Section*, pp. 247–252.

Byar, D. P., Schoenfeld, D. A., Green, S. B., Amato, D. A. (1990). Design considerations for AIDS trials. *New England Journal of Medicine* 323: 1343–1348.

Carlin, B., Chaloner, K., Church, T., Matts, J. P., and Louis, T. A. (1993a). Bayesian monitoring of an AIDS clinical trail. *The Statistician* 42: 355–367.

Carlin, B. P., Chaloner, K. M., Louis, T. A. and Rhame, F. S. (1995). Elicitation, monitoring and analysis of an AIDS clinical trial. In *Case Studies of Bayesian Statistics in Science and Industry, 2*, eds. Gatsonis, C., Hodges, J., Kass, R., and Singpurwalla, N., pp. 48–89, Springer-Verlag, New York.

Chaloner, K. M. and Duncan, G. T. (1983). Assessment of a beta prior distribution: PM elicitation. *Statistician* 32: 174–180.

Chaloner, K. M. and Duncan, G. T. (1987). Some properties of the Dirichlet-multinomial distribution and its use in prior elicitation. *Communications in Statistics A* 16: 511–523.

Chaloner, K., Church, T., Matts, J. P., and Louis, T. A. (1993). Graphical elicitation of a prior distribution for an AIDS clinical trial. *The Statistician* 42: 341–353.

Cox, D. R. (1972). Regression models and life-tables (with discussion). *J. Royal Statistical Soc. Ser. B* 34: 187–220.

Dickey, J. M., Dawid, A. P. and Kadane, J. B. (1986). Subjective-probability assessment methods for multivariate-t and matrix-t models. In *Bayesian Inference and Decision Techniques: Essays in Honor of Bruno de Finetti*, eds. Goel, P. K., and Zellner, A. pp. 177–195. North-Holland, Amsterdam.

Flournoy, N. (1994). A clinical experiment in bone marrow transplantation: Estimating a percentage point of a quantal response curve. In *Case Studies in Bayesian Statistics*, eds. Gatsonis C., pp. 324–336, Springer-Verlag, New York.

Freedman, L. S. and Spiegelhalter, D. J. (1983). The assessment of subjective opinion and its use in relation to stopping rules for clinical trials. *Statistician* 32: 153–160.

Freedman, L. S. and Spiegelhalter, D. J. (1992). Application of Bayesian statistics to decision making during a clinical trial. *Statistics in Medicine* 11: 23–35.

Garthwaite, P. H. (1992). Preposterior expected loss as a scoring rule for prior distributions. *Communications in Statistics—Theory Meth.* 21(12): 3601–3619.

Garthwaite, P. H. and Dickey, J. M. (1985). Double- and single-bisection methods for subjective probability assessment in a location-scale family. *Journal of Econometrics* 29: 149–163.

Garthwaite, P. H. and Dickey, J. M. (1988). Quantifying expert opinion in linear regression models. *Journal of the Royal Statistical Society Ser. B* 50: 462–474.

Garthwaite, P. H. and Dickey, J. M. (1991). An elicitation method for multiple linear regression models. *Journal of Behavioral Decision Making* 4: 17–31.

Gavasakar, U. (1988). A comparison of two elicitation methods for a prior distribution for a binomial parameter. *Management Science* 34: 784–790.

Good, I. J. (1983). *Good Thinking.* University of Minnesota Press, Minneapolis.

Greenhouse, J. B. and Wasserman, L. (1995). Robust Bayesian Methods for Monitoring Clinical Trials. *Statistics in Medicine* 14: 1379–1391.

Hogarth, R. M. (1975). Cognitive processes and the assessment of subjective probability distributions. *Journal of the American Statistical Association* 70: 271–294.

Hogarth, R. M. (1987). *Judgement and Choice*, 2nd ed. Wiley, New York.

Huber, P. (1973). The use of Choquet capacities in statistics. *Bull. Internat. Statist. Inst.* 45: 181–191.

Kadane, J. B. (1986). Progress toward a more ethical method for clinical trials. *The Journal of Medicine and Philosophy* 11: 385–404.

Kadane, J. B. (1994). An application of robust Bayesian analysis to a medical experiment (with discussion). *Journal of Statistical Planning and Inference* 40: 221–232.

Kadane, J. B., Dickey, J. M., Winkler, R. L., Smith, W. S. and Peters, S. C. (1980). Interactive elicitation of opinion for a normal linear model. *Journal of the American Statistical Association* 75: 845–854.

Kahneman, D., Slovic, P., and Tverskey, A., eds. (1982). *Judgement under Uncertainty: Heuristics and Biases.* Cambridge University Press, Cambridge, UK.

Keeney, R. and Raiffa, H. (1976). *Decisions with Multiple Objectives: Preferences and Value Tradeoffs.* Wiley, New York.

Kyberg, H. E. Jr. and Smokler, H. E., eds. (1980). *Studies in Subjective Probability.* Krieger, New York.

Laskey, K. B. and Black, P. (1989). Models for elicitation in Bayesian analysis of variance. *Proceedings of Computer Science and Statistics: 21st Annual Symposium on the Interface*, Florida pp. 242–247.

Lindley, D. V. (1985). *Making Decisions*, 2nd ed. Wiley, New York.

O'Hagan, A. (1994). *Kendall's Advanced Theory of Statistics*, Vol. 2B, *Bayesian Inference.* Arnold, London.

Phillips, L. D. and Edwards, W. (1966). Conservatism in simple probability inference tasks. *Journal of Experimental Psychology* 72: 346–357.

Ravinder, H. V., Kleinmuntz, D., and Dyer, J. S. (1988). The reliability of subjective probabilities obtained through decomposition. *Management Science* 34: 186–199.

Savage, L. J. (1971). Elicitation of personal probabilities and expectations. *Journal of the American Statistical Association* 66: 783–801.

Spiegelhalter, D. J. (1987). Probability expert systems in medicine: practical issues in handling uncertainty (with discussion). *Statistical Science* 2: 25–34.

Spiegelhalter, D. J. and Freedman, L. S. (1986). A predictive approach to selecting the size of a clinical trial, based on subjective clinical opinion. *Statistics in Medicine* 5: 1–13.

Spiegelhalter, D. J. and Freedman, L. S. (1988). Bayesian approaches to clinical trials. In *Bayesian Statistics 3*, eds. Bernardo, J. M., DeGroot, M. H., Lindley, D. V., and Smith, A. F. M., pp. 453–477. Oxford University Press, Oxford, UK.

Spiegelhalter, D. J., Freedman, L. S., and Parmar, M. K. (1993). Applying Bayesian ideas in drug development and clinical trials. *Statistics in Medicine* 12: 1501–1511.

Spiegelhalter, D. J., Freedman, L. S., and Parmar, M. K. (1994). Bayesian approaches to randomized trials. *J. Royal Statistical Soc. Ser. A*, 157: 357–416.

Tierney, L. (1990). *LISP-STAT, an Object-Oriented Environment for Statistical Computing and Dynamic Graphics*. Wiley, New York.

von Winterfeldt, D. and Edwards, W. (1986). *Decision analysis and behavioral research*. Cambridge University Press, Cambridge, UK.

Wallsten, T. S. and Budescu, D. V. (1983). Encoding subjective probabilities: A psychological and psychometric review. *Management Science* 34: 186–199.

Winkler, R. L. (1967a). The assessment of prior distributions in Bayesian analysis. *Journal of the American Statistical Association* 62: 776–800.

Winkler, R. L. (1971). Probabilistic prediction: Some experimental results. *Journal of the American Statistical Association* 66: 675–685.

Winkler, R. L. (1967b). The quantification of judgment: Some methodological suggestions. *Journal of the American Statistical Association* 62: 1105–1120.

5

Priors for the Design and Analysis of Clinical Trials

Joseph B. Kadane   Carnegie Mellon University, Pittsburgh, Pennsylvania

Lara J. Wolfson   University of Waterloo, Waterloo, Ontario, Canada

ABSTRACT

This chapter discusses methods available for eliciting opinion in the form of prior distributions. A program is available in Statlib for elicitation in the context of a normal linear model with a conjugate prior. An example is presented using this method. The discussion of elicitation methods also focuses on survival models, as these two classes represent the models most commonly used in the design and analysis of clinical trials.

1 ATTITUDES TOWARD PRIOR DISTRIBUTIONS

Prior distributions mean different things to different statisticians. They are necessary in the Bayesian context to compute posterior distributions. To many Bayesians, they are also necessary as a way of expressing important beliefs about the parameter and/or summarizing what was learned in previous studies.

In the most minimal sense of priors, there are Bayesians who try to find "reference" or "noninformative" prior distributions. The intention is to find a prior distribution that incorporates little knowledge and is minimally controversial (for a review see Kass and Wasserman, 1993). Each attempt along these lines has its troubles. For example, priors that are noninformative in one parameterization can be highly informative in

another (Box and Tiao, 1973, Chapter 1). The presence of nuisance parameters can change the form of a noninformative prior. The use of the term
"reference" is intended to obviate some of these concerns by renaming,
but the questions remain about what distinguishes such a prior and why
a person should use it.

Specification of informative priors may depend on the availability of
relevant data. Sometimes a prior can be estimated using data from another
source judged to be identical in some respects (i.e., Wolfson *et al.*, 1994).
Sometimes informative priors can be obtained using maximum-entropy
methods, which maximize a measure of lack of information subject to
constraints derived from the data. So-called empirical Bayes methods
amount to using classical estimation on the highest level in a hierarchical
model (see Deely and Lindley, 1981).

The approach used in this chapter regards prior distributions as an
important modeling component. Priors allow available information to be
stated in a form that others can examine and compare with their own
beliefs. Some object to this use of prior distributions because not everyone
need agree about them; that is, they are personal or subjective. Although
this is true, the same may be said of the likelihood function and the loss
structure. Agreement cannot be coerced; in such an environment, the best
that analysis can do is to make explicit the areas of relative agreement
and disagreement. In this spirit, the use of prior distributions, admitting
their subjectivity, is merely a step toward honesty and the open disclosure
of assumptions.

2 USES OF PRIOR DISTRIBUTIONS

The standard use of a prior distribution is to combine it with a likelihood
function and data to compute the resultant posterior distribution. A good
review of such Bayesian work in the context of clinical trials is given by
Spiegelhalter *et al.* (Chapter 2). Although common, this is not the only
use of priors.

A second common use is in the design of experiments. In the simplest
case, a Bayesian chooses a design, obtains data, and makes a decision.
The same prior is used both to analyze the data after they are collected
and to determine the expected utility of possible designs before data collection (for a review see Lindley, 1971).

This conception of experimental design seems much too constrained
to be an accurate representation of how designs are, or ought to be,
thought of. They are a good model for what Kadane and Seidenfeld (KS)
(1990) have called *experiments to learn*, i.e., experiments that involve

only one person, who wants to learn as much as possible from the experiment, but it omits the audience, others who may also want to learn. Typically, pilot experiments are experiments to learn.

In contrast to experiments to learn, KS propose *experiments to prove*, which involve people and views other than those of the experimenter. In designing an experiment to prove something to someone else, that "someone else" becomes an actor in the design, with a prior conceptually separate and distinct from that of the designer of the experiment. Thus there would be a designer and an estimator, each of whom would have a subjective prior. Here the designer would find the resulting expected loss from the decision the estimator would make after seeing each possible data set. The expected utility of the design would depend on the use that would be made of it by the estimator and the resulting expected utility of that use to the designer. Etzioni and Kadane (1993) use just such a framework for simple normal linear model design problems. Additional work along these lines is done in Lodh (1993). Berry *et al.* (1992) provide an interesting example of an experiment conducted to prove the effectiveness of a vaccine to the Food and Drug Administration (FDA).

A third use of elicited prior distributions is to employ the opinions to protect patients against the least advantageous treatments to which randomization might expose them. This is the spirit of Kadane and Sedransk (1980) and Kadane (1986, 1995).

3 ELICITATION METHODS IN GENERAL

In beginning any elicitation process, there are three phases of probability encoding (Spetzler and Staël von Holstein, 1975). These include the deterministic phase, in which the form of the problem is specified; the probabilistic phase, in which an expert assessor answers questions; and the informational phase, in which the answers obtained in the probabilistic phase are verified. The deterministic phase is perhaps the most difficult, because it involves choosing functional forms for any model, and although we consider this to be an important topic for future research, currently no formal methods are available for extracting this type of information from experts. Instead, the statistician substitutes his or her "expert" judgment, choosing a conjugate pair for mathematical convenience in the case of the two methods we discuss here.

In the probabilistic phase, there are various ways to elicit information from the expert. Some of the most common types of questions asked by psychologists are "indirect response with external (or internal) reference" (Wilson, 1994). An expert is asked to choose which of two alternatives is more likely, either indirectly by comparing the quantity of interest with a reference quantity, as in the question "Which do you consider more likely, that it will rain tomorrow, or that student X will score less than 50 on the test?", or directly by comparing two possible outcomes for the quantity of interest, as in the question "Is it more likely that student X will score less than 50 or greater than 70 on the test?" We dislike the first type of question for two reasons: (1) we consider our experts to be experts in a particular subfield, not in any other field, and (2) this is not very effective in taking advantage of covariate information. The second type of question gives only a probability inequality and hence is a weak form of information gathering. Rather, asking the expert assessor to specify the quantiles, where he or she is basically placing a bet, is more in line with the type of information we believe our experts are able to communicate.

So the type of question that we believe optimal in the sense of extracting the most information from an expert assessor is to ask for quantiles, from which the center and spread of a distribution are estimated. For example, a typical question might be to ask an assessor for the 75th percentile. In the case of a survival model, the question could be postulated as "What is the time at which (for a given level of the covariates) you expect that 25% of the patients will still be alive?" These can be phrased as questions comparing the probability of a certain medical outcome with the probability of drawing a ball of a certain color from an urn of specified makeup.

Given that quantiles are the target, it remains to justify the choice of posterior predictive quantiles. This choice basically stems from the fact that even statisticians have trouble thinking about the distribution of parameters. Both experts and statisticians can (we hope!) think about the predictive distribution of a single observation. So it makes sense to ask such questions. Returning to the previous example of a survival model, take the model to have a specific parametric form of an exponential likelihood and a prior that follows a gamma distribution with hyperparameters α and β. If one attempts to elicit the prior distribution directly (and ask about quantiles of that distribution), then there is no direct extrapolation to asking about survival times. Instead, the question would be "What is the 75th percentile of the prior distribution?" If, however, one asks about quantiles of the posterior predictive distribution (in this case, a shifted Pareto distribution), then a question is posed that has direct meaning to the expert, as stated in the preceding paragraph.

In the informational encoding phase the responses are verified by informing the expert of any assessments that are "outliers," or to which the fit is extremely sensitive, giving the expert a chance to reconcile any discrepancies, should he or she so choose.

Probability assessors often have difficulty assessing the tails of the distribution (Hora *et al.*, 1992; Alpert and Raiffa, 1982; Chaloner, Chapter 4). Wallsten and Budescu (1983) note that experts are fairly coherent in the center of the distribution. In cases in which the posterior distribution may be fairly sensitive to the tails of the prior distribution, it is important to account for the uncertainty the expert may have in assessing that part of the distribution.

Seidenfeld (1985) reviews some of the literature on "calibration." There are two ways in which an expert might be "calibrated." Both methods assume that the expert should have perfect predictive ability. In the first method, the opinion of the expert is elicited for the problem of interest, data are collected, and then the expert's opinion is adjusted so that the prior that is used has perfect predictive performance. The second method requires the elicitation of the expert's opinion about another problem, assessing the expert's predictive performance for that problem, and then using that information to adjust the elicited opinion for the problem of interest. Both methods violate the spirit of expert elicitation, since no expert is infallible, and adjusting the elicited responses to yield perfect predictive performance results in the prior distribution being something other than a subjective, personal probability (Savage, 1954). The second method has an additional flaw, that an expert on medical treatments is not a weather forecaster and the abilities to make judgments about treatments and about the weather are different.

4 METHODS AVAILABLE FOR ELICITATION IN CLINICAL SETTINGS

Having explained the background for the methods we advocate using in this chapter, we review the programs that can be used for practical elicitation. From the standpoint of obtaining expert opinion for the design and analysis of clinical trials, there are two types of models for which elicitation methods are currently available: the normal linear regression model (Kadane *et al.*, 1980; Garthwaite and Dickey, 1988, 1992); a Cox proportional hazards model (Chaloner *et al.*, 1993; Chaloner, Chapter 4); and simple exponential, lognormal, and gamma survival models (Wolfson, 1995). In addition, some case-specific examples of elicitation are given in DuMouchel (1988) for multiple comparisons of parameters measured on the same scale in the context of a normal linear model; in Kadane and Schum (1992), for eliciting prior probabilities with regard to legal evidence; in Berry *et al.* (1992) for use in the design of a sequential vaccine trial; and in Kadane and Hastorf (1988) for paleoethnobotany. Chaloner and

Duncan (1983) and Gavasakar (1988) propose methods for the elicitation of the prior distribution in the beta-binomial model.

The normal linear model, often known as a regression model, is frequently used in statistics when it is believed that there is a linear relationship between some dependent variable (usually denoted Y) and some independent variables that may be used to predict the value of the dependent variable. These independent variables are usually referred to as covariates, and in the regression model, for each value of the dependent variable Y, there is a row in the design matrix X with the corresponding values of the independent variables.

A program for the elicitation of the normal linear regression model is available in Statlib. The program is called *elicit-normlin.f* and runs in FORTRAN, using various IMSL subroutines. Later in this section, we introduce an example of how the program can be used to elicit a prior distribution. The results of the example are documented in Appendix A, and we relate the steps of the method to the example in the appendix. The methods in the program are based on the paper by Kadane *et al.* (1980), and they allow an expert to formulate his or her opinion based on up to four covariates. The basic idea of this method is to elicit quantiles of the posterior predictive distribution, in the belief that experts can more easily express opinions about the predictive quantity, which they experience, than about the parameters, which they do not experience. Thus the question of interest is: given the predictive distribution and the normal linear likelihood, how can (conjugate) distributions be estimated for the prior on the parameters?

The formulation of the model, with the usual conjugate prior, is that the likelihood of the dependent variable follows a normal distribution, with mean given by $X^T\beta$, and variance $\sigma^2 I$. Conditional on σ^2, the vector β also follows a normal distribution, with prior mean b and prior variance-covariance matrix $\sigma^2 R^{-1}$. The precision ($1/\sigma^2$) follows a scaled gamma distribution, with parameters as given below:

$$Y \mid X, \beta, \sigma^2 \sim N(X^T\beta, \sigma^2 I)$$

$$\beta \mid \sigma^2 \sim N(b, \sigma^2 R^{-1})$$

$$\frac{1}{\sigma^2} \sim \frac{1}{w\delta} \Gamma\left(\frac{\delta}{2}, 2\right)$$

So the elicitation program must elicit four hyperparameters: b, R, w, and δ. Kadane *et al.* (1980), first ask the expert to specify the range of the independent variables (step A in the Appendix). The expert is then offered a series of design points (specific values of the independent variables) at which to assess 50th, 75th, and 90th percentiles (step B in the Appendix). The design points are the median ±30% of the range for each variable. The expert may reject any design point and must specify percentiles for a minimum number of design points to allow estimation of the hyper-parameters. The assessor is allowed to assess design points beyond this minimum, and it is recommended that if the elicited prior is to be used in a problem in which few or no data are to be collected, the expert assesses as many points as possible. The expert is allowed to reconcile incoherences or change answers after seeing the estimated values for the hyper-parameters b (the prior mean for β) and δ, the degrees of freedom parameter (step C in the Appendix). The next stage of the program elicits conditional 50th and 75th percentiles from the expert, where the "conditional" part comes from the fact that the expert is shown hypothetical data points at particular design points (calculated from the hyperparameters b and δ) (step D in the Appendix). At no point is the expert ever asked to "forget" hypothetical data points, so the hypothetical data set increases as more design points are added. After being given the opportunity to reconcile incoherences, this information is used to assess the hyper-parameters w and R, by constructing a matrix of assessments of the variances and covariances of the conditional assessments, from which estimates of w and R can be derived. In practice, if the expert fails to truly "condition" on the hypothetical data and the conditional percentiles are too close to their unconditional percentiles, then a reasonable estimate of w is hard to find. The program provides one, but with a warning message, encouraging the expert to reassess the conditional percentiles. Finally, the program displays the hyperparameters and other information such as highest posterior density regions (step E in the Appendix).

Garthwaite and Dickey (1988, 1992) extend these methods to deal with the problem of variable selection. This is applicable to clinical trials particularly, because it may not be known a priori which variables have an effect on treatment efficacy.

In Chaloner et al. (1993) a graphical elicitation method for the Cox proportional hazards model is given in the context of a clinical trial for treatment to prevent toxoplasmosis in patients with AIDS. There are three treatments (including a placebo) in this trial. The expert is asked to assess a "best guess" for the probability of patients experiencing toxoplasmosis. This can also be thought of as the proportion of the AIDS group of patients who will be afflicted with toxoplasmosis. Conditional on this probability, the bivariate distribution for the probability of disease based on the other two treatments can be elicited.

The expert is asked to specify (marginal) upper and lower quartiles for the other two probabilities, and this is used to provide initial estimates for generating plots of the marginal distributions, assuming independence of the two treatments. In addition, a contour plot of the joint distribution is shown, along with sliders that allow the assessor to adjust both the marginal distributions (by changing the means and variances for P_A and P_B) and the correlation. To begin changing the correlation, the expert specifies a probability that both P_A and P_B are larger than their marginal respective medians. The correlation is constrained to lie between 0.25 and 0.5.

This method has the advantage of providing immediate feedback for any changes that occur. However, it has the disadvantage of asking experts to deal with actual probability distributions directly, and the jury is still out on whether nonstatistical experts can do that effectively. Chaloner's experts were able to do the elicitations effectively, which is supportive. As the dimensionality of the parameters space increases, however, the tasks may become very difficult for them.

Wolfson (1995) presents a method for eliciting the conjugate prior for the exponential survival model. The method entails expert assessments of quantiles of the posterior predictive distribution. These assessments are used to provide initial estimates of the hyperparameters α and β of the gamma distribution, which is the prior. In addition, an estimate of the uncertainty in the assessment, σ, is calculated. The expert is then shown quantiles of the posterior predictive distribution but with a twist: it is assumed that there is some uncertainty in the specification of the prior distribution. The expert assesses conditional probabilities associated with the hypothetical quantiles, and the assessments of the hyperparameters are updated based on this information. The example shown in Wolfson (1995), eliciting an engineer about the post dialing delay experienced by users of long-distance services on particular telephone networks, illustrates the additional information that explicit modeling of uncertainty provides. This method is easily generalizable to other univariate survival models, such as the gamma, Weibull, and lognormal models.

5 AN EXAMPLE OF ELICITATION

To illustrate the use of elicitation, a statistician working with data in the health care field was elicited for her prior for a normal linear model, using the methods of Kadane *et al.* (1980). On January 1, 1994, 10 of 41 area programs in North Carolina began a waiver program [under the authority of section 1915 (b) of the Social Security Act] for children who are Medicaid eligible and in need of mental health services. The waiver program, called Carolina Alternatives, will test the impact of capitation funding within a public-sector managed care system. A grant from the Kate B. Reynolds Health Care Trust was used to fund the development of the program.

During phase I of the waiver program, local mental health centers (called area programs), which are in charge of managing care, will be given a capitated (lump sum) amount for inpatient services for all children falling within their catchment area. The area programs will be trying to channel children into outpatient rather than inpatient services via community-based diversion and step-down services. Diversion services will prevent children from entering hospitals, and step-down services work to speed up the discharge process for children who do end up in the hospital. The incentive for the area programs to restrict the use of inpatient services is that they will be allowed to keep savings from the capitation allotment and use these savings to develop community-based alternatives.

One aspect of the evaluation of this program will be comparing length of hospital stay prior to the implementation of the program to length of stay during the program. So the response variable is CHANGE in length of hospital stay, and the independent variables are CA, an indicator variable for whether the child was in an area that was participating in the waiver program, PCCA, a variable ranging from 0 to 1 that measures community-based alternative capacity within the child's area program, and an interaction between the two. PCCA is important because if the community services are not available, the program will not have much impact on length of hospitalization. The interaction between PCCA and CA is included because it is thought that PCCA would have a stronger effect on children from participating area programs.

A condensed version of the elicitation script for this example is in Appendix A. For explicatory purposes, we have identified five "stages" of the process: step A, where the model is specified; step B, where initial assessments are made; step C, where estimates of b and δ are calculated; step D, where conditional assessments are made; and step E, where the hyperparameters w and R^{-1} are calculated and some summary statistics are shown.

The estimates obtained for the hyperparameters of the prior distribution are:

$$b = \begin{bmatrix} -2.0 \\ 4.0 \\ 6.0 \end{bmatrix}, \quad R^{-1} = \begin{bmatrix} 17.1 & -20.0 & -16.0 \\ -20.0 & 120.7 & 67.0 \\ -16.0 & 67.0 & 157.6 \end{bmatrix}$$

$w = 1.11$

$\delta = 1.2$

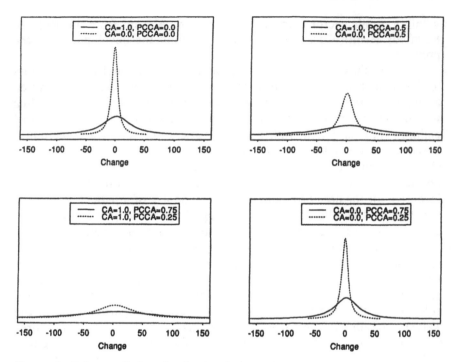

Figure 1   Prior predictive distribution for change.

Figure 1 shows the elicited predictive distribution for various scenarios, with children being either in or out of participating area programs and at varying levels of the community-based alternatives within that program. This predictive distribution is a t-distribution, with degrees of freedom given by δ, center given by b, and spread given by

$$w\left(\frac{1}{3 + \delta}\, xR^{-1}x^{T} + I\right)$$

where x represents the values taken by the independent variables.

Figure 1 shows that the predictive distribution is always sharper when the child is not in a participating area program and when the level of the community-based alternatives is low. This is consistent with what is expected; because the expert has experience with what happens when there is no waiver program, she is more certain of that part of the distribution.

6 DISCUSSION

This chapter reviews the uses of elicitation in the design and analysis of clinical trials and the methods available to accomplish it. Elicitation is still in an early stage of development, and more needs to be done. As progress continues to be made on Bayesian computation, other problems, such as what to use as inputs to the computation and using posterior distributions in decision making, will become more acute. Further experience with the use of elicitation in clinical settings would be helpful in guiding methodological advances. A community of applied and methodological "knowledge engineers" would help elicitation become more practically available and useful.

APPENDIX

Introduction

> Specification of a prior distribution for the parameters for the standard normal linear regression model

This program will assist you in assessing the parameters of your prior distribution for the normal linear model. The prior distribution has three components (known as hyperparameters):

1. A vector indicating your "prior mean" of the regression coefficients.
2. The degrees of freedom and the scale factor of your distribution of the variance of the error: jointly, these two hyperparameters specify the center and spread of the distribution.
3. A matrix measuring the dispersion of your distribution of the regression coefficients.

To determine your prior distribution for these parameters the procedure will begin by asking you to specify your 50th, 75th, an 90th predictive percentile values of the dependent variable conditional on specific values for the independent variables.

A Specifying the Model

> Type the number of the independent variables (max = 4), not including the intercept.   2
> Enter the name of the response variable:
> max length of each is 8 characters.
>   CHANGE

Enter names for each of the independent (predictor) variables. The maximum length of each is 9 characters.

The name of variable 1 is   CA

The name of variable 2 is   PCCA

Now specify the smallest and the largest values of the independent variables for which you can comfortably make predictions about CHANGE

Enter the smallest value of CA   0

Enter the largest value of CA   1

Enter the smallest value of PCCA   0

Enter the largest value of PCCA   1

B Initial Assessments

We will now ask you to enter 50th, 75th, and 90th predictive percentile of CHANGE for various joint values of CA and PCCA.

Think of the 50th percentile of CHANGE as the value for which you would accept an even bet that CHANGE would be less than or greater than that value.

Think of the 75th percentile of CHANGE as the value for which you would accept a 3-to-1 bet that CHANGE would be less than that value or a 1-to-3 bet that CHANGE would be greater than that value.

Think of the 90th percentile of CHANGE as the value for which you would accept a 9-to-1 bet that CHANGE would be less than that value or a 1-to-9 bet that CHANGE would be greater than that value.

It is now necessary to consider for what values of the independent variables you will be asked to provide predictive assessments. The program will initially generate a total of 25.00 points by taking the maximum and minimum values for each independent variable and divide the distance between these end points into four equal intervals. The program will offer each point sequentially. At each point you can

1. Accept and assess the predictive percentiles for this point.
2. Reject this point (skip) and continue to the subsequent point.
3. Review or change any previous assessed percentiles.

After you have provided sufficient information to fit a distribution for the parameters, you may stop the assessment. However the more points you provide the more thoroughly we can check the coherence of your beliefs and therefore guide you to a more accurate assessment.

The program will now offer you points at which you will be asked to assess percentiles.

If you wish to specify your own points, please enter '1'. Otherwise, enter '0'.   1

You will now be asked to enter values of the independent variables:
The value of the intercept is always 1.

    Please enter a value for CA   :   0
    Please enter a value for PCCA   :   0
    Consider the distribution of CHANGE for the specified values of the independent variables:
    Variable Name   Value
    CA   .000
    PCCA   .000
    Type '1' if you want to assess percentiles for CHANGE
    Type '2' to enter a new design point.
    Type '3' to see the previous assessed percentiles
                                                          1
    Enter 50th percentile of CHANGE   0
    Enter 75th percentile of CHANGE   2.1
    Enter 90th percentile of CHANGE   6
    You will now be asked to enter values of
    The independent variables:
    The value of the intercept is always 1.
    Please enter a value for CA   :1
    Please enter a value for PCCA   :0

The way in which further points are elicited is similar: we will just show the results:

Here are the predictive percentiles you have specified for CHANGE at this time:

	Independent variables		Predictive percentiles		
	CA	PCCA	50th	75th	90th
1	.0	.0	.0	2.1	6.0
2	1.0	.0	.0	3.5	10.0
3	.0	.5	1.0	3.1	7.0
4	1.0	.5	5.0	8.5	15.0
5	.0	1.0	2.0	4.1	8.0

At this time, you can change the value of any percentiles you have previously assessed. To change any entry, type the number of the entry that you wish to change.

    If you do not wish to change any entries,
    Type '0'   0

You have now entered 5 points. This is sufficient to fit a distribution. If you intend to do a posterior analysis based on at least 5 real data points, there is no need to enter more points.

However, if you now wish to address a problem with less than 5 real data points, or with no data, it would be useful to provide at least 5 additional points. For sensitive problems, a total of 10 points would be desirable.

To enter additional points,   type '1'
To stop collecting points,   type '2'

1

You will now be asked to enter values of the independent variables: the value of the intercept is always 1.

Please enter a value for CA   : 1
Please enter a value for PCCA   : 1

Consider the distribution of CHANGE for the specified values of the independent variables:

Variable Name	Value
CA	1.000
PCCA	1.000

Type '1' if you want to assess percentile for CHANGE.
Type '2' to enter a new design point.
Type '3' to see the previous assessed percentiles.

1

Enter 50th percentile of CHANGE   10
Enter 75th percentile of CHANGE   13.5
Enter 90th percentile of CHANGE   20
To enter additional points,   type '1'
To stop collecting points,   type '2'   2

Here are the predictive percentiles you have specified for CHANGE at this time:

Independent variables		Predictive percentiles		
CA	PCCA	50th	75th	90th
1	.0	.0	.0	2.1	6.0
2	1.0	.0	.0	3.5	10.0
3	.0	.5	1.0	3.1	7.0
4	1.0	.5	5.0	8.5	15.0
5	.0	1.0	2.0	4.1	8.0
6	1.0	1.0	10.0	13.5	20.0

At this time, you can change the value of any percentile you have previously assessed. To change any entry, type the number of the entry that you wish to change.

If you do not with to change any entries,
Type '0'   0

Thank you for assessing the percentiles.

C Estimates of *b* and *δ*

The hyperparameters for the mean of the regression coefficients and the degrees of freedom will now be assessed. If there are any percentiles you entered that appear to be incoherent, you will be informed and given the opportunity to reassess them, should you so choose.

Variable	Fitted value
Intercept	−2.000
CA	4.000
PCCA	6.000

Please type the number of the option you want.

1. Accept the fitted values
2. Change the specified percentiles
3. Display the residuals from the assessed median   3

Here are the residuals from the least squares fit. Those residuals that are large in absolute value are marked with a '1' in the flag column, because they may be out of line. Please consider if you want to change any of the assessed medians in light of these residuals.

	CA	PCCA	Residual	Flag
1	.0	.0	2.0	1.0
2	1.0	.0	−2.0	1.0
3	.0	.5	.0	.0
4	1.0	.5	.0	.0
5	.0	1.0	−2.0	1.0
6	1.0	1.0	2.0	1.0

Please type the number of the option you want.

1. Accept the fitted values
2. Change the specified percentiles
3. Display the residuals from the assessed median   1

The estimated degrees of freedom for the predictive T distribution for CHANGE given CA PCCA is 1.200. This value was obtained by averaging the tail ratios for the 6 sets of assessments you provided.

These tail ratios are obtained as the ratio of the differences between the 90th and 50th percentiles and the 75th and 50th percentiles of CHANGE.

To continue the assessment,   type '1'
To see the tail ratios,   type '2'   2

If flag = 1, then the ratio for the tail distribution is large. You may want to consider reassessing that point.

Independent Variables

	CA	PCCA	Ratio	Flag
1	.00	.00	2.86	.00
2	1.00	.00	2.86	.00
3	.00	.50	2.86	.00
4	1.00	.50	2.86	.00
5	.00	1.00	2.86	.00
6	1.00	1.00	2.86	.00

If you would like to change any of your assessments,   type '1'
To continue the elicitation,   type '2'   2

D Conditional Assessments

You have now given sufficient information to estimate a vector indicating the central tendency of your distribution of the regression coefficients and degrees of freedom. Your answers to the next set of questions will allow the estimation of the remaining parameters, a scale factor of your distribution of the variance of the error, and a matrix measuring the dispersion of your distribution for the regression coefficients.

We now wish to determine how observed data would affect your conditional judgments about CHANGE. After being given some hypothetical data, you will be asked to provide new 50th and 75th percentiles of CHANGE given the dependent variables.

You are asked to provide the number of points used in your previous assessment for which you will specify new 50th and 75th percentiles of CHANGE. This must be a number not less than 4 and no more than 6. A larger number of points provides more thorough coherence testing.

If you are asked about

4 points, you must answer	14 questions.
5 points, you must answer	20 questions.
6 points, you must answer	27 questions.

Enter the number of points you will use   4

Suppose CHANGE has the following values at the indicated points.
Point Number 1

CA   .00
PCCA   .00

Assessed		Fitted	New
50th	75th	50th	Obs.
.00	2.10	−2.00	−.88

Conditional on the new observation(s) enter your new 50th and 75th percentiles for CHANGE at the indicated values of the independent variables
Point Number 2

CA   1.00
PCCA   .00

Assessed		Fitted	New	
50th	75th	50th	50th &	75th
.00	3.50	2.00	−.5	3

For some further points now enter your new 50th percentile of CHANGE
at the indicated values of the independent variables
Point Number 1

CA   .00
PCCA   .00

Assessed		Fitted	New
50th	75th	50th	50th
.00	2.10	−2.00	−.75

Point Number 3

CA   .00
PCCA   .50

Assessed		Fitted	New
50th	75th	50th	50th
1.00	3.10	1.00	.25

Point Number 4

CA   1.00
PCCA   .50

Assessed		Fitted	New
50th	75th	50th	50th
5.00	8.50	5.00	4.5

Here are the conditional percentiles that you specified.
Conditional on
Point Number 1. CHANGE = −.88

Assessed		Fitted	Conditional	
50th	75th	50th	50th	75th

Point Number 1

| CA | .00 |
| PCCA | .00 |

.00	2.10	−2.00	−.75

Point Number 2

| CA | 1.00 |
| PCCA | .00 |

.00	3.50	2.00	−.50	3.00

Point Number 3

| CA | .00 |
| PCCA | .50 |

1.00	3.10	1.00	.25

Point Number 4

| CA | 1.00 |
| PCCA | .50 |

5.00	8.50	5.00	4.50

Do you wish to change any of conditional percentiles that you have specified given the hypothetical data?
If so, you may change the percentiles just given based on the hypothetical value given for Point Number 1   CHANGE = −.88
Enter '0' if you do not wish to change any points, or the number of the point you wish to change otherwise   0

Suppose CHANGE has the following values at the indicated points.
Point Number 1

CA   .00
PCCA   .00

	Assessed		Fitted	New
	50th	75th	50th	Obs.
	.00	2.10	− 2.00	− .88

Point Number 2

CA   1.00
PCCA   .00

	Assessed		Fitted	New
	50th	75th	50th	Obs.
	.00	3.50	2.00	1.37

Conditional on the new observation(s) enter your new 50th and 75th percentiles for CHANGE at the indicated values of the independent variables.
Point Number 3

CA   .00
PCCA   .50

	Assessed		Fitted	New	
	50th	75th	50th	50th & 75th	
	1.00	3.10	1.00	1.5	3.6

For some further points now enter your new 50th percentile of CHANGE at the indicated values of the independent variables
Point Number 2

CA   1.00
PCCA   .00

Assessed		Fitted	New
50th	75th	50th	50th
.00	3.50	2.00	2.5

Point Number 4

CA	1.00
PCCA	.50

Assessed		Fitted	New
50th	75th	50th	50th
5.00	8.50	5.00	6

All of the conditionally assessed percentiles were shown and the assessor given the opportunity to change their assessments.
Suppose CHANGE has the following values at the indicated points.
Point Number 1

CA	.00
PCCA	.00

Assessed		Fitted	New
50th	75th	50th	Obs.
.00	2.10	−2.00	−.88

Point Number 2

CA	1.00
PCCA	.00

Assessed		Fitted	New
50th	75th	50th	Obs.
.00	3.50	2.00	1.37

Point Number 3

CA	.00
PCCA	.50

Assessed		Fitted	New
50th	75th	50th	Obs.
1.00	3.10	1.00	2.62

Conditional on the new observation(s) enter your new 50th and 75th percentiles for CHANGE at the indicated values of the independent variables.
Point Number 4

CA	1.00
PCCA	.50

Assessed		Fitted	New	
50th	75th	50th	50th &	75th
5.00	8.50	5.00	5.5	9

For some further points now enter your new 50th percentile of CHANGE at the indicated values of the independent variables.
Point Number 3

CA	.00
PCCA	.50

Assessed		Fitted	New
50th	75th	50th	50th
1.00	3.10	1.00	2.5

. . . assessor is shown assessed points . . .

Suppose CHANGE has the following values at the indicated points.
Point Number 1

CA	.00
PCCA	.00

Assessed		Fitted	New
50th	75th	50th	Obs.
.00	2.10	−2.00	−.88

Point Number 2

CA   1.00
PCCA   .00

Assessed		Fitted	New
50th	75th	50th	Obs.
.00	3.50	2.00	1.37

Point Number 3

CA   .00
PCCA   .50

Assessed		Fitted	New
50th	75th	50th	Obs.
1.00	3.10	1.00	2.62

Point Number 4

CA   1.00
PCCA   .50

Assessed		Fitted	New
50th	75th	50th	Obs.
5.00	8.50	5.00	7.37

Conditional on the new observation(s) enter your new 50th and 75th percentiles for CHANGE at the indicated values of the independent variables.
Point Number 5

CA   .00
PCCA   1.00

Assessed		Fitted	New
50th	75th	50th	50th
2.00	4.10	4.00	2.5

For some further points now enter your new 50th percentile of CHANGE at the indicated values of the independent variables.
Point Number 4

CA   1.00
PCCA   .50

Assessed		Fitted	New
50th	75th	50th	50th
5.00	8.50	5.00	6.5

E Estimation of w and R^{-1}

The estimate for the hyperparameter w is $w = 1.11$.
    The estimated variance-covariance matrix (the hyperparameter R) is given by:

	R		
	1	2	3
1	.07358	.01052	.00299
2	.01052	.01235	−.00418
3	.00299	−.00418	.00843

The distribution of the variance of the error is an inverse chi-square variable on 1.20 degrees of freedom with the scale factor 1.11 and

25th Percentile = .5064
50th Percentile = 1.0344
75th Percentile = 2.5750
50 Percent HDR   .0000 to   .8423
75 Percent HDR   .0000 to   2.1709
95 Percent HDR   .0000 to   5.7604

The distribution of the regression coefficients (the parameter vector beta) is a 3-variate T variable with 1.2000 degrees of freedom and the following parameters:

        Center

intercept = −2.0000
CA   = 4.0000
PCCA   = 6.0000

	Spread Matrix		
	1	2	3
1	.07358	.01052	.00299
2	.01052	.01235	−.00418
3	.00299	−.00418	.00843

If you would like the explanations of what the center and spread of the t-distribution are, type '1', otherwise, type '0'.
1

The standard multivariate t variable with g degrees of freedom is distributed as the product of a standard multivariate normal variable and the square root of g divided by an independent chi-square variable with g degrees of freedom. Call this random variable Y.

Then the general multivariate t variable with g degrees of freedom has the generic vector

$$Z = A + B \cdot Y$$

where A is a constant vector and B a constant matrix. We say that:

$C(Z) = A$ is the center of Z
$S(Z) = B \cdot B'$ is the spread of Z

When the degrees of freedom g is greater than 1, so that the mean exists, the center equals the mean. When g is greater than 2, so that the variance exists, the spread times the constant $g/(g - 2)$ is the variance.

Here is a summary of the parameters of your distribution.

1. A vector for the central tendency

intercept	−2.000
CA	4.0000
PCCA	6.0000

2. The degrees of freedom and the scale factor.

D.F. =	1.2000
Scale factor =	1.1117

3. A matrix for the dispersion of the regression coefficients.

	Dispersion Matrix		
	1	2	3
1	.07358	.01052	.00299
2	.01052	.01235	−.00418
3	.00299	−.00418	.00843

These results can be saved in this order for your use in a posterior analysis.
Please enter the name of the file you wish to use
results.1
Please enter a 2nd file name for saving your assessments into.
results.2
Goodbye!

ACKNOWLEDGMENT

The authors were supported in part by the National Science Foundation under NSF grant DMS-9303557. They thank Dalene Stangl for doing the sample elicitations reported in Section 5 and Juana Sanchez for helpful comments.

REFERENCES

Alpert, M. and Raiffa, H. (1982). A progress report on the training of probability assessors. In *Judgment Under Uncertainty: Heuristics and Biases*, eds. Kahneman, D., Slovic, P., and Tversky, A., pp. 294–305. Cambridge University Press, Cambridge, UK.

Berry, D. A., Wolff, M. C., and Sack, D., (1992). Public health decision making: A sequential vaccine trial. In *Bayesian Statistics 4*, eds. Bernardo, J. M., Berger, J. O., Dawid, A. P., and Smith, A. F. M., pp. 79–96. Oxford University Press, Oxford, UK.

Box, G. E. P. and Tiao, G. C. (1973). *Bayesian Inference in Statistical Analysis*. Wiley, New York.

Chaloner, K. M. (1996). The elicitation of prior distributions. In *Bayesian Biostatistics*, eds. Berry, D. A. and Stangl, D. K., Marcel Dekker, New York.

Chaloner, K. M., Church, T., Louis, T. A., and Matts, J. P. (1993). Graphical elicitation of a prior distribution for a clinical trial. *The Statistician* 42: 341–353.

Chaloner, K. M. and Duncan, G. T. (1983). Assessment of a beta prior distribution: PM elicitation. *The Statistician* 32: 174–180.

Deely, J. J. and Lindley, D. V. (1981). Bayes empirical Bayes. *Journal of the American Statistical Association* 76: 833–841.

DuMouchel, W. (1988). A Bayesian model and a graphical elicitation procedure for multiple comparison. In *Bayesian Statistics 3*, eds. Bernardo, J. M., De-Groot, M. H., Lindley, D. V., and Smith, A. F. M., pp. 127–145. Oxford University Press, Oxford, UK.

Etzioni, R. and Kadane, J. (1993). Optimal experimental design for another's analysis. *Journal of the American Statistical Association* 88: 1404–1411.

Garthwaite, P. H. and Dickey, J. M. (1988). Quantifying expert opinion in linear regression problems. *Journal of the Royal Statistical Society Ser. B* 50: 462–474.

Garthwaite, P. H. and Dickey, J. M. (1992). Elicitation of prior distributions for variable-selection problems in regression. *The Annals of Statistics* 20: 1697–1719.

Gavasakar, U. (1988). A comparison of two elicitation methods for a prior distribution for a binomial parameter. *Management Science* 34: 784–790.

Hora, S. C., Hora, J. A., and Dodd, N. G. (1992). Assessment of probability distributions for continuous random variables: A comparison of the bisection and fixed-value methods. *Organizational Behavior and Human Decision Processes* 51: 133–155.

Kadane, J. (1986). Progress toward a more ethical method for clinical trials. *Journal of Medicine and Philosophy* 11: 385–404.

Kadane, J. (1995). *Bayesian Methods and Ethics in a Clinical Trial Design*, Wiley, New York, in press.

Kadane, J. B. and Hastorf, C. A. (1988). Bayesian paleoethnobotany. In *Bayesian Statistics 3*, eds. Bernardo, J. M., DeGroot, M. H., Lindley, D. V., and Smith, A. F. M., pp. 243–259. Oxford University Press, Oxford, UK.

Kadane, J. B. and Schum, D. A. (1992). Opinions in dispute: The Sacco-Vanzetti case. In *Bayesian Statistics 4*, eds. Bernardo, J. M., Berger, J. O., Dawid, A. P., and Smith, A. F. M., pp. 79–96. Oxford University Press, Oxford, UK.

Kadane, J. and Sedransk, N. (1980). Toward a more ethical clinical trial. In *Bayesian Statistics*, eds. Bernardo, J. M., DeGroot, M. H., Lindley, D. V., and Smith, A. F. M., pp. 329–338. University Press, Valencia.

Kadane, J. and Seidenfeld, T. (1990). Randomization in a Bayesian perspective. *Journal of Statistical Planning and Inference 25*: 329–345.

Kadane, J. and Seidenfeld, T. (1995). Statistical Issues in the Analysis of Data Gathered in the New Designs. In *Bayesian Methods and Ethics in a Clinical Trial Design*, ed. Kadane, J. Wiley, New York, in press.

Kadane, J., Dickey, J., Winkler, R., Smith, W., and Peters, S. (1980). Interactive elicitation of opinion for a normal linear model. *Journal of the American Statistical Association 75*: 845–854.

Kass, R. E. and Wasserman, L. A. (1993). "Formal Rules for Selecting Prior Distributions: A Review and Annotated Bibliography." Technical Report 583, Department of Statistics, Carnegie-Mellon University, Pittsburgh, PA.

Lindley, D. V. (1971). *Bayesian Statistics, a Review*. SIAM, Philadelphia.

Lodh, M. (1993). Experimental Studies by Distinct Designer and Estimators. Ph.D. dissertation, Department of Statistics, Carnegie-Mellon University, Pittsburgh, PA.

Savage, L. J. (1954). *Foundations of Statistics*, Wiley, New York.

Seidenfeld, T. (1985). Calibration, coherence, and scoring rules. *Philosophy of Science 52*: 274–294.

Spetzler, C. S. and Staël von Holstein, C. S. (1975). Probability encoding in decision analysis. *Management Science 22*: 340–358.

Wallsten, T. S. and Budescu, D. V. (1983). Encoding subjective probabilities: A psychological and psychometric review. *Management Science 29*: 151–173.

Wilson, A. G. (1994). Cognitive factors affecting subjective probability assessment." Discussion paper 94-02, Institute of Statistics and Decision Sciences, Duke University, Durham, NC.

Wolfson, L. J. (1995). Elicitation of Priors and Utilities for Bayesian Analysis. Ph.D. thesis, Department of Statistics, Carnegie-Mellon University, Pittsburgh, PA.

Wolfson, L. J., Kadane, J. B., and Small, M. J. (1994). Statistical Decision Theory for Environmental Remediation. Technical Report 594, Department of Statistics, Carnegie-Mellon University, Pittsburgh, PA.

III
DECISION PROBLEMS

6

A Weibull Model for Survival Data: Using Prediction to Decide When to Stop a Clinical Trial

Jiang Qian Abbott Laboratories, Abbott Park, Illinois
Dalene K. Stangl and Stephen George Duke University, Durham, North Carolina

ABSTRACT

In 1984, the Cancer and Leukemia Group B (CALGB) opened a phase III clinical trial for patients with stage III non–small cell lung cancer (NSCLC). The original fixed sample size design called for randomly assigning 240 patients to either radiotherapy alone or chemotherapy followed by radiotherapy. Analysis of results was to occur after 190 deaths. Instead, early in the trial it was decided to apply group sequential concepts using a truncated O'Brien-Fleming stopping rule, implemented via a Lan-DeMets α-spending function. A study monitoring committee stopped the trial at the fifth interim analysis after only 155 eligible patients had been entered. This chapter presents a Bayesian alternative to the standard frequentist approach of assessing stopping times in clinical trials. It demonstrates how the use of posterior and predictive distributions provides a natural way to address many of the issues raised in monitoring clinical trials.

1 INTRODUCTION

Carcinoma of the lung is the leading cause of cancer deaths in the United States (Ginsberg *et al.*, 1993). For treatment purposes, lung cancer is generally divided into two categories, based on the cell type presenting at diagnosis. The first type, small cell (or oat cell) anaplastic carcinoma, is strongly associated with smoking and accounts for roughly 20% of all lung cancers (Ihde *et al.*, 1993). The second type, non–small cell lung cancer (NSCLC), which includes adenocarcinoma, squamous cell carcinoma, and large cell anaplastic carcinoma, is also strongly associated with smoking and accounts for the remaining 80% of lung cancer cases. Cancers are usually classified further according to the extent or stage of disease so that therapies may be tailored to the particular disease stage. Patients with stage III non–small cell lung cancer are those who have no demonstrable distant metastases but do have locally extensive or invasive disease or involvement of mediastinal lymph nodes. Stage III disease is rarely cured by surgery alone because of the association of these findings with micrometastatic disease.

For many years, radiation therapy (RT) has been the treatment of choice for such patients, although the practice has been questioned. Thoracic radiation to the primary tumor and the hilar and mediastinal lymph nodes can usually control local symptoms. However, the overall impact of radiation alone on survival of stage III NSCLC patients in minimal at best, producing median survivals of approximately 9 months, a 2-year survival of approximately 15% and a 3-year survival of approximately 5%.

In attempts to improve survival in these patients, clinical researchers in the early 1980s considered the possibility that RT alone might not be sufficient to eradicate micrometastatic disease. At that time there was also some evidence that platinum-based chemotherapy (CT) increased survival in patients with more advanced disease. Therefore, various systematic approaches to treatment of stage III patients, which included a combination of CT and standard RT, were proposed. The study addressed in this chapter was designed to compare the standard treatment (RT only), consisting of RT delivered over 6 weeks to the original tumor volume and involved regional lymph nodes, to an experimental treatment (CT + RT), which employed 5 weeks of cisplatin plus vinblastine prior to the RT. The RT on the experimental arm was identical in dose, schedule, and volume to that for the standard treatment.

1.1 Design of the Study

We consider a randomized phase III clinical trial for patients with stage III NSCLC. The trial was opened in 1984 by the Cancer and Leukemia

Group B (CALGB). It was designed to compare survival rates on standard treatment (RT alone) to survival rates on an experimental treatment regimen of two courses of combination chemotherapy given prior to the standard radiotherapy (CT + RT). The original fixed sample size for this trial was 240 patients (120 patients per treatment arm). This sample size was calculated to provide 80% power to detect a hazard ratio of 1.5:1 using a Cox (1972) proportional hazards model and assuming that the log-rank test would be used at a two-sided significance level of $\alpha = 0.05$. This hazard ratio represents a 50% difference in median survival between the two treatment groups.

Shortly after the trial began, it was decided to apply group sequential concepts by using a truncated O'Brien-Fleming (1979) stopping rule, implemented via a Lan-Demets (1983) α-spending function. A data monitoring committee (DMC) was established to review the analyses as they were produced. The population of patients for this study was limited to patients with documented regional stage III NSCLC. The patients eligibility criteria included no prior CT, RT, or total resection (tumor removal), performance status (PS) of 0 or 1 (patient either fully active or ambulatory and capable of light work), and weight loss of less than 5% in the 3-month interval prior to study entry. Standard CALGB eligibility criteria regarding laboratory values, other diseases, and so on were also used. All eligible patients who began treatment and for whom follow-up data were attained were included in the analysis. Eight percent of the patients in this study failed to meet these criteria and were excluded from the analysis.

Recruitment began in May 1984. Accrual proceeded at about one-quarter the anticipated rate. Hypothesized reasons for the slow recruitment included lack of investigator enthusiasm for the design of the study and the restrictive selection criteria. The study was stopped at the fifth interim analysis in May 1987 after 155 eligible patients had been entered, and follow-up data were available for only 105 patients.

1.2 Preliminary Analysis of the CALGB Trial

From May 1984 to May 1987, five interim analyses were performed and presented to the DMC. Table 1 gives specific information about the number of eligible patients accrued and numbers of deaths on each arm at each interim analysis. Here, n_i is the total number of patients registered with follow-up information on treatment i, and d_i is the number of deaths observed among these patients ($i = 1, 2$). Treatment 1 refers to RT, and treatment 2 refers to CT + RT. It also gives observed p-values, boundary significance levels, and the decision of the DMC about whether to close the trial or to keep it open. Initial accrual was slow, and the first interim analysis was performed in September 1985. The sample size at this time

Table 1 Summary of Statistics and Boundary Significance Levels Used by the DMC

Analysis	d_1	n_1	$\sum_1^{n_1} t_{1j}$	d_2	n_2	$\sum_1^{n_2} t_{2j}$	Logrank p-value	Truncated O'Brien-Fleming boundary significance level	Pocock boundary significance level	Decision
Sept. 1985	7	25	135.9	3	25	174.7	—[a]	.0013	.0041	Keep open
Mar. 1986	12	41	279.5	4	38	340.6	0.0210	.0013	.0034	Keep open
Aug. 1986	20	41	279.5	14	47	407.3	0.0071	.0013	.0078	Keep open
Oct. 1986	24	46	356.0	18	49	431.9	0.0015	.0013	.0061	Keep open
Mar. 1987	32	51	406.1	24	54	517.4	0.0015	.0013	.0081	Close

[a] This value is missing because the database cannot be reconstructed to the state of September 1985, and this value was never recorded.

was too small to allow any meaningful comparisons of survival by treatment. The O'Brien-Fleming boundary (O'Brien and Fleming, 1979) truncated at 3.0 standard deviations (boundary significance level of 0.0013) was selected as the boundary to be used for all monitoring of the study. We have included the Pocock boundary value (Pocock, 1982) to illustrate how the decision process would have differed if another boundary had been used.

By the October 1986 analysis, the observed significance for the comparison of survival by treatment had decreased to $p = .0015$, which almost reached the truncated O'Brien-Fleming boundary significance level of .0013. Despite the close proximity to the boundary, the DMC again decided that there was insufficient evidence to recommend closure of the study and that the next analysis would be performed as scheduled in March 1987. The March 1987 interim analysis led to the decision to close the trial to further accrual. At this time, 180 of the projected 240 patients had been accrued and 155 of the 180 patients were eligible. Follow-up data were available for 105 patients, 56 of whom had died. The observed p-value was .0015 with a total of 56 failures. A comprehensive analysis including use of the Cox (1972) proportional hazards model was undertaken. The prognostic factors examined were comparable in the two treatment arms. The results of the Cox analysis, which controlled for various prognostic factors, gave $p = .0008$ for the treatment comparison. This analysis reaffirmed that the observed difference represented a real treatment effect. This adjusted p-value crossed the truncated O'Brien-Fleming boundary, and the DMC unanimously voted to close the study. In Figure 1, Kaplan-Meier plots are displayed for each interim analysis and as of

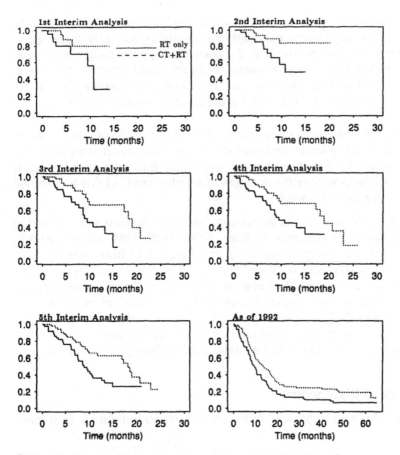

Figure 1   Kaplan-Meier plots at each interim analysis and as of 1992.

June 1992. In the early interim analysis, large differences between treatments occurred. As time went on, the difference between survival curves narrowed, although patients receiving CT + RT still had longer survival.

2   A BAYESIAN WEIBULL SURVIVAL MODEL

In this section we describe a Bayesian approach to analyzing data from the CALGB trial. We assume that the Bayesian analyses were done at the same time points as described earlier for the actual analyses. The Bayesian approach has several advantages in clinical trials:

Inferences are flexible. In a Bayesian analysis there is no penalty for analyzing interim data with the possibility for early stopping.

Focus is on the probability of a particular hypothesis given the data rather than the probability of results more extreme than those we observed given the null hypothesis. The probability of the hypothesis that CT + RT prolongs life over RT alone is the focus of this analysis.

A Bayesian analysis can provide probabilities for future observations and incorporate those probabilities into a decision analysis framework. In this chapter predictive probabilities are calculated that will help answer the question of whether the CALGB trial should have been stopped.

For further advantages and discussion, interested readers are referred to Spiegelhalter *et al.* (Chapter 2), Berry (1985, 1987, 1988, 1989a, 1991a,b, 1993), Berry and Stangl (Chapter 1), and Berger and Berry (1988).

Parallel to the work of Stangl (1991, p. 429), an exponential survival model was used by Li (1994) and George *et al.* (1994) for analyzing the CALGB trial. In this chapter we use a Weibull survival model as was used in Qian (1994). Because the Weibull includes the exponential as a special case, it is more general and yet allows us to compare our results with those from the exponential model used in George *et al.* (1994).

2.1 Weibull Distribution

A lifetime T is said to have a Weibull(θ, α) distribution if its hazard function is

$$h(t; \theta, \alpha) = \theta \alpha t^{\alpha - 1} \qquad (1)$$

The Weibull density function is

$$f(t; \theta, \alpha) = \theta \alpha t^{\alpha - 1} \exp(-\theta t^\alpha) \qquad (2)$$

and its survival function is

$$S(t; \theta, \alpha) = \exp(-\theta t^\alpha) \qquad (3)$$

where θ is a scale parameter, α is a shape parameter, and $\theta > 0$, $\alpha > 0$. When $\alpha < 1$, the hazard rate decreases with time, and when $\alpha > 1$, the hazard rate increases with time. When $\alpha = 1$, the hazard rate is constant and the Weibull corresponds to an exponential distribution with parameter θ.

2.2   The Model

Let subscript 1 denote standard treatment (RT only) and subscript 2 denote experimental treatment CT + RT in the CALGB trial; t_{ij} is the survival time of the jth patient in the ith treatment, where $i = 1, 2, j = 1, 2, \ldots, n_i$. Here n_i is the total number of patients receiving treatment i. Let d_i denote the number of deaths in treatment i at any particular time (for example, at an interim analysis time point). Furthermore let $j = 1, 2, \ldots, d_i$ and $j = d_i + 1, \ldots, n_i$ denote the patients who died and survived, respectively, on treatment i. This notation requires reordering patient indices at each interim analysis.

Now we are ready to use the Weibull distribution to specify a Bayesian hierarchical model. At the first stage of the hierarchy, assume observations t_{ij}, conditional on (θ_i, α_i), are independent and identically distributed with Weibull (θ_i, α_i). This stage provides a parametric model for lifetime t_{ij}. At the second stage, prior distributions are given for unknown parameters (θ_i, α_i) $i = 1, 2$. Following the parameterization of Stangl (p. 429) and Li (1994), we let $\eta = \log(\theta_2/\theta_1)$. Solving for θ_2 we have

$$\theta_2 = \theta_1 e^{\eta} \tag{4}$$

This parameterization is convenient because when $\alpha_1 = \alpha_2 = 1$, the survival distributions for both treatments are exponential, with θ_1 the hazard rate for patients on treatment 1 and η the log hazard ratio of treatment 2 to treatment 1.

We have chosen to place a gamma prior on the α_i and θ_1 and a normal prior distribution on η. Our model assumptions give the following hierarchical model (*iid* means independently identically distributed):

Stage I:

$$t_{ij} \mid (\theta_i, \alpha_i) \stackrel{iid}{\sim} \text{Weibull}(\theta_i, \alpha_i), \qquad i = 1, 2; j = 1, 2, \cdots, n_i \tag{5}$$

Stage II:

$$\begin{aligned} \alpha_i \mid (u_\alpha, v_\alpha) &\stackrel{iid}{\sim} \text{gamma}(u_\alpha, v_\alpha) \\ \theta_1 \mid (u_\theta, v_\theta) &\sim \text{gamma}(u_\theta, v_\theta) \\ \eta \mid (\mu_\eta, \sigma_\eta^2) &\sim \text{normal}(\mu_\eta, \sigma_\eta^2) \end{aligned} \tag{6}$$

Using this parameterization, we have the following likelihood function:

$$L(\theta_1, \eta, \alpha_1, \alpha_2) = \prod_{i=1}^{2} \left\{ \prod_{j=1}^{d_i} f(t_{ij}; \theta_i, \alpha_i) \prod_{j=d_i+1}^{n_i} S(t_{ij}; \theta_i, \alpha_i) \right\}$$

$$= \prod_{i=1}^{2} \left\{ (\theta_i\alpha_i)^{d_i} \left(\prod_{j=1}^{d_i} t_{ij} \right)^{\alpha_i - 1} \exp\left(-\theta_i \sum_{j=1}^{n_i} t_{ij}^{\alpha_i} \right) \right\} \qquad (7)$$

$$= (\theta_1\alpha_1)^{d_1} \left(\prod_{j=1}^{d_1} t_{1j} \right)^{\alpha_1 - 1} \exp\left(-\theta_1 \sum_{j=1}^{n_1} t_{1j}^{\alpha_1} \right)$$

$$(\theta_1 e^\eta \alpha_1)^{d_2} \left(\prod_{j=1}^{d_2} t_{2j} \right)^{\alpha_2 - 1} \exp\left(-\theta_1 e^\eta \sum_{j=1}^{n_2} t_{2j}^{\alpha_2} \right)$$

2.3 Prior for (θ_1, η, α_1, α_2)

We chose gamma priors for both α_1 and α_2 with a mode equal to 1. We varied the variance of the gamma prior to assess the sensitivity of results to this distribution. Review of the literature on NSCLC before 1984 showed the median survival time for NSCLC patients to be approximately 9 months, with a 2-year survival of approximately 15% and 3-year survival of approximately 5%. A gamma prior for θ_1, gamma(2, 20), approximates this information, with a median survival of 8.3 months. Under an exponential model η represents the log-hazard ratio of treatment 2 to 1. A normal(0, 1) prior was chosen to represent our prior belief about η. This prior reflects the belief that each treatment was equally likely to be more effective.

2.4 The Posterior

Given the prior $\pi(\theta_1, \eta, \alpha_1, \alpha_2)$ and the likelihood $L(\theta_1, \eta, \alpha_1, \alpha_2)$, the posterior $p(\theta_1, \eta, \alpha_1, \alpha_2)$ is

$$p(\theta_1, \eta, \alpha_1, \alpha_2) \propto L(\theta_1, \eta, \alpha_1, \alpha_2) \times \pi(\theta_1, \eta, \alpha_1, \alpha_2)$$

$$\propto L(\theta_1, \eta, \alpha_1, \alpha_2) \times \theta_1^{u_\theta - 1} e^{-v_\theta \theta_1} \times e^{-\eta^2/2}$$

$$\times \alpha_1^{u_\alpha - 1} e^{-v_\alpha \alpha_1} \times \alpha_2^{u_\alpha - 1} \times e^{-v_\alpha \alpha_2}$$

$$\propto (\theta_1\alpha_1)^{d_1} \left(\prod_{j=1}^{d_1} t_{1j} \right)^{\alpha_1 - 1} \exp\left(-\theta_1 \sum_{j=1}^{n_1} t_{1j}^{\alpha_1} \right) \qquad (8)$$

$$(\theta_1 e^\eta \alpha_2)^{d_2} \left(\prod_{j=1}^{d_2} t_{2j} \right)^{\alpha_2 - 1} \exp\left(-\theta_1 e^\eta \sum_{j=1}^{n_2} t_{2j}^{\alpha_2} \right)$$

$$\times \theta_1^{u_\theta - 1} e^{-v_\theta \theta_1} \times e^{-\eta^2/2}$$

$$\times \alpha_1^{u_\alpha - 1} e^{-v_\alpha \alpha_1} \times \alpha_2^{u_\alpha - 1} e^{-v_\alpha \alpha_2}$$

We used Gibbs sampling to calculate the posterior distributions of parameters at each interim analysis. Because many other chapters (Albert and Chib, Ch. 22; Clyde *et al.*, Ch. 11; Dellaportas *et al.*, Ch. 23; Joseph *et al.*, Ch. 24, Racine-Poon *et al.*, Ch. 13; and Smith *et al.*, Ch. 15) give examples and explanations of Gibbs sampling, we will not present details here.

2.5 Treatment Comparison

When survival times under both treatments are exponential, assessing the posterior distribution of η is a convenient way to summarize treatment effects; however, when the models deviate from exponential, the posterior distributions of model parameters are less intuitive summary measures of treatment differences. There are many alternative ways to describe the treatment effects using functions of the parameters. The two used throughout the rest of this chapter are described here.

1. The first measure used in this chapter is the predictive probability that the survival probability of treatment 1 is smaller than that on treatment 2 at time t:

$$D(t) = \Pr(S_1(t) < S_2(t) \mid \text{Data}) \qquad (9)$$

Here "Data" means the data available at the time of the analysis. $S_1(t)$ and $S_2(t)$ are the survival functions under treatments 1 and 2, respectively. If α_1 and α_2 were both equal to 1, then $D(t)$ would be constant in t.

2. The second measure used in this chapter is the predictive mean difference between the survival probabilities on treatments 1 and 2 at time t:

$$d(t) = E(S_1(t) - S_2(t) \mid \text{Data}) \qquad (10)$$

3 APPLICATION OF THE WEIBULL MODEL TO THE CALGB TRIAL

In this section we apply the Weibull model proposed to the CALGB trial. Results are presented and contrasted for two sets of priors, the first of which places high prior probability on models close to exponential survival models and the second of which does not. Results are also presented to assess the sensitivity of results to selection of the variance parameter in the prior for η.

3.1 Results: Interim, Final, and as of Current (1992)

The priors chosen for θ_1 and η were independent gamma(2, 20) and normal(0, 1) distributions, respectively. The first prior distributions chosen for the shape parameters α_1 and α_2 were independent gamma distributions with the same shape and scale parameters for each, gamma(10001, 10000). This prior distribution has small variance: $10001/10000^2 \approx 10^{-4}$, and hence the Weibull model under this prior distribution corresponds closely to the simple exponential case considered in George *et al.* (1994).

Figure 2 shows the prior and posterior distributions of parameters and predictive characteristics $D(t)$ and $d(t)$. The first plot shows the density for θ_1. The smallest mode of approximately .04 occurs at the second interim analysis, the largest mode of approximately .075 at the fifth interim analysis, and the mode is approximately .06 as of 1992. These modes represent survival times of 16 to 25 months. The second plot shows the density of η. Here posterior modes range from -1.2 to $-.6$. Exponentiating, these numbers correspond to hazard rate ratios of .3 and .55, meaning longer survival times under the combination treatment. Zero is clearly in the upper tail (95th or greater percentile) of each interim density. The posterior densities for the shape parameters are presented in the second row. There is little difference between posterior and prior distributions for α_1 and α_2, because our prior is so peaked (almost a point mass on 1) and the sample size is moderate. The plot of $D(t)$, the probability of greater survival on the combination treatment over radiotherapy alone, also shows little deviation from an exponential model, because it is nearly constant over time. It also shows probabilities greater than .98 for all but the first interim analysis. The last plot displays the predictive plot of $d(t)$, the expected difference in survival rates between the treatment groups across time. As of 1992, the expected difference at 20 months is about .18. The expected difference at 20 months was the lowest, approximately $-.35$, at the second interim analysis.

Comparing these distributions with the posterior distribution of θ_1 and η in George *et al.* (1994), there are only slight differences. This is expected because the posteriors of α_1 and α_2 are peaked about 1, implying an exponential distribution. Comparing the posterior modes as of 1992 with the maximum-likelihood estimates, similar results are also seen. The maximum-likelihood estimates as of 1992 are

$\hat{\theta}_1 = 0.076$

$\hat{\eta} = -0.500$

$\hat{\alpha}_1 = 0.936$

$\hat{\alpha}_2 = 0.939$

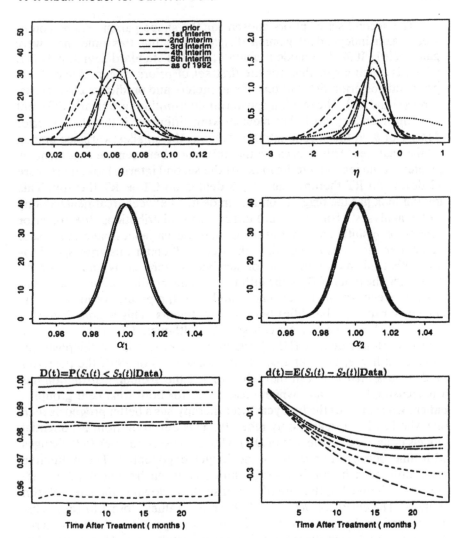

Figure 2 Posterior distributions of $(\theta_1, \eta, \alpha_1, \alpha_2)$ and predictive characteristics when the prior is $\alpha_i \sim \text{gamma}(10001, 10000)$, $\theta_1 \sim \text{gamma}(2, 20)$, $\eta \sim \text{normal}(0, 1)$.

The second set of priors chosen for the shape parameters α_1 and α_2 were again independent gamma distributions, but this time they were gamma(101, 100). The mode is again 1, but the standard deviation is 0.1, which is much larger than for the first set of priors. Figure 3 shows the prior and posterior distributions of parameters and predictive characteristics $D(t)$ and $d(t)$. The resulting posterior distributions are clearly different from those under prior 1. The most striking difference is that the posterior density of θ_1 at the second interim look has the smallest variance among all the analyses. This is because the difference between the treatments is greatest at the second interim look. At the second interim look, there were 12 deaths on RT therapy and only 3 deaths on CT + RT therapy. This interim look has the biggest contrast in deaths between treatments among all the analyses. Using the more diffuse prior distributions, the posterior density of η shifts slightly toward 0, and the variances have increased. Much more area lies in the tail to the right of 0 under this prior specification. What this means in terms of survival is more easily understood by examining the plots of $D(t)$, the difference in the predictive survival probability between treatments at time t, and $d(t)$, the predictive mean difference in survival probabilities between treatments. This is done below.

For both α_1 and α_2, we can see that although 1 is within the 95% highest posterior density (HPD) region of the α_i's, it is not the posterior mode of either α_1 or α_2 at 1992. In both cases, as of 1992, the posterior mode is less than 1. Recall that when $\alpha_i < 1$ the Weibull distribution has a decreasing failure rate, which means that for both treatment groups a patient who survives the first year after therapy has a better prognosis than one who has been treated only recently. This deviation from exponential models is also seen in the plot of $D(t)$. Under the second prior $D(t)$ changes with time because our model is no longer exponential. The predictive mean difference in survival probabilities between treatments, $d(t)$, decreases with time but the rate of decrease slows down in the later interim analysis. The curve for $d(t)$ as of 1992 shows values closest to zero. This is expected as the posterior for η for this time period is closest to zero. At 5 months after treatment $d(t)$ is approximately $-.1$. This means that the difference in expected survival probability between the two treatments is 10% at 5 months after treatment. By 20 months the difference is about 15%.

3.2 Further Sensitivity Analysis

To see further how different priors on η would affect our conclusion about this trial, 10 more values were chosen for σ, the standard deviation of η, the log hazard ratio. The values chosen were 0.01, 0.1, 0.2, 0.3, 0.4, 0.5,

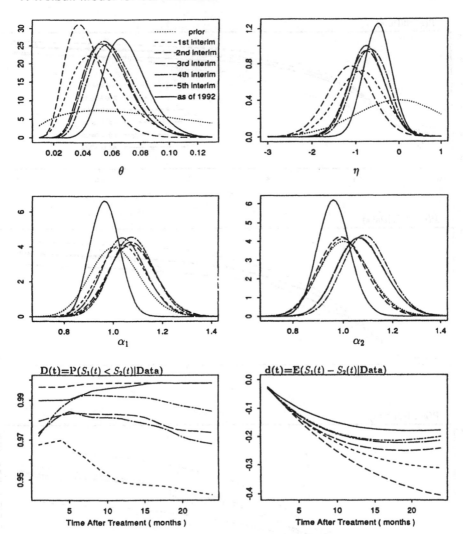

Figure 3 Posterior distributions of $(\theta_1, \eta, \alpha_1, \alpha_2)$ and predictive characteristics when the prior is $\alpha_i \sim$ gamma(101, 100), $\theta_1 \sim$ gamma(2, 20), $\eta \sim$ normal(0, 1).

0.6, 0.7, 0.8, and 0.9. Figure 4 shows the effect of changing σ on posterior probabilities of η. As the information is accumulating (along the interim analyses), the probabilities of $\eta < 0$, $\eta < -0.25$, $\eta < -0.5$ change less for values of $\sigma > 0.5$. Except for the third interim look and very small σ, the $P(\eta < 0)$ is always greater than .5. For $\sigma < 0.3$ the $P(\eta < -0.25)$ is less than .5 for all interim looks, and for $\sigma > 0.3$ the $P(\eta < -0.25)$ is

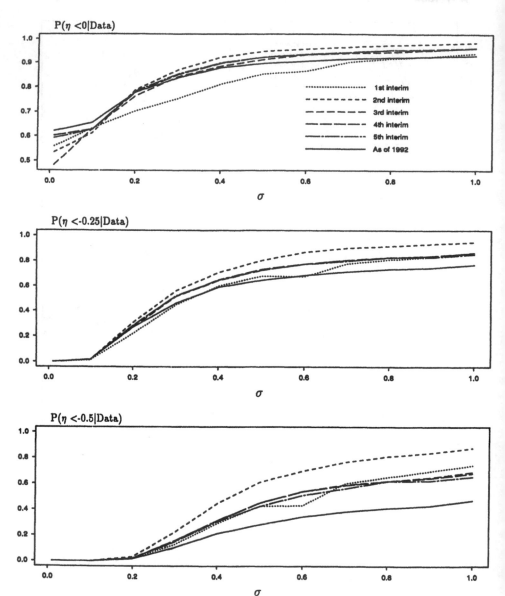

Figure 4   Posterior probabilities of η as a function of σ.

greater than .5 for all interim looks. Thus, how compelling an advantage CT + RT provides over RT depends on the prior specification of this parameter. It must be remembered that this prior was centered around 0, so larger σ allows a greater probability of a treatment effect in either direction.

4 USING PREDICTION TO EVALUATE STOPPING TIMES IN THE CALGB TRIAL

Early stopping of the CALGB trial was controversial. The original plan was to accrue a total of 240 patients, with 120 per treatment. The trial was stopped in April 1987 prior to achieving the accrual goal because the DMC viewed the data as providing strong support of the superiority of CT + RT. The total accrual in April 1987 was 180 patients, of whom only 155 were eligible: 77 in the RT treatment and 78 in the CT + RT treatment. Thus, another 85 patients would have been required to fulfill the original plan. We can now look back and ask what would have happened had accrual to the trial not stopped in 1987. To what degree would these extra 85 patients have influenced the study's conclusion and impact? To examine this question, a simulation study was carried out. First, we determined the actual accrual rate from the first interim look time to the end of the trial. The accrual rate was about 6 patients per month. Next we simulated accrual of patients after April 1987 at the same rate. Lifetimes were generated based on all the information available as of April 1987. Observations were simulated according to Weibull distributions. The parameters of the Weibull were sampled from the posterior distribution of the parameters given all the data as of April 1987.

$$t_{ij} \sim \text{Weibull}(\theta_i, \alpha_i)$$

$$i = 1: j = 78, \ldots, 120 \tag{11}$$

$$i = 2: j = 79, \ldots, 120$$

$$(\theta_1, \eta, \alpha_1, \alpha_2) \sim p(\theta_1, \eta, \alpha_1, \alpha_2 \mid \text{Data as of 04/87})$$

Repeating this simulation procedure 500 times makes it clear that accruing these extra 85 patients would not have changed the 1987 conclusion, at least not very much. This can be seen from Figure 5. Figure 5 shows the histogram of $d(t)$, the predictive mean difference in survival probabilities between treatments at times $t = 6, 12, 18,$ and 24 months for these 500 simulations. The small triangle at the base of each plot denotes the estimated $d(t)$ value as of April 1987. For each of these plots the mean of the histogram is approximately the same as the April 1987 estimated value.

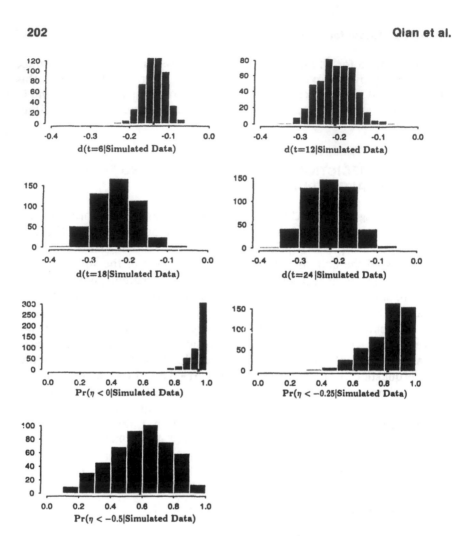

Figure 5   Histograms of $d(t)$, $P(\eta < 0)$, $P(\eta < -0.25)$, $P(\eta < -0.5)$ based on 500 simulations if the trial had not been stopped at fifth interim look.

Our uncertainty about the location of the mean will diminish with an increased sample, but the question of whether the certainty gained justifies the further randomization must be answered.

Figure 5 also shows the histograms for $P(\eta < 0)$, $P(\eta < -0.25)$, $P(\eta < -0.5)$ for the 500 simulations. The black triangles on the plots denote the estimated values as of April 1987. Each of these plots also shows that conclusions about the direction or size of the treatment effect would not have changed by further patient accrual. Here the means of the histograms are to the right of the April 1987 estimates. This shows that our uncertainty about the treatment effect would have diminished with further accrual.

5 CONCLUSIONS AND DISCUSSION

This chapter presents a Bayesian method for analyzing clinical trials and assessing stopping times in clinical trials. Based on data accrued up to the time of a particular analysis, posterior distributions of parameters or functions of parameters can be calculated. No penalty is imposed for frequent interim analysis. These posterior distributions can be used to simulate future observations and to calculate predictive distributions of quantities of interest. Posterior distributions, based on the data observed, can be compared to predictive distributions based on the data observed and further simulated observations to determine what is to be gained from further patient accrual. In addition, these distributions can be used in a decision theoretic framework to do a cost-benefit analysis of the consequences. The impact of alternative decisions on individuals' survival and quality of life as well as on health policy such as clinical guidelines must be weighed. In the CALGB clinical trial, despite clear evidence of a treatment effect (lifetimes were increased by about 11 months), the study was discounted by many practitioners and hence has had little impact on clinical practice. Most patients are still treated with RT alone. While the discounting of the current study may be partially due to the early stopping and partially due to the evidence provided by other studies that showed less positive results for chemotherapy, deciding whether a trial should be stopped must include judgments about whether the results will be convincing to the medical community and hence influence practice. Had practitioners been exposed to the predictive distributions for the full CALGB sample, perhaps they would not have discounted the study.

In the analysis presented here, the clinical trial had already been stopped, and the question of whether the trial should have been stopped was addressed. The same methods would be useful in making the decision whether to stop. Data monitoring committees could look at posterior distributions and assess how predictive distributions would compare based on the accrual of more patients. Also, these committees must address the decision problem by incorporating the impact of the study on clinical practice into their utility functions.

In our example, the number of future patients was chosen to match the initial design of the trial; however, this need not be the case. Predictive distributions for alternative numbers of future accruals could be carried out. If a sequential design is planned from the outset, this method could be used at various time points to determine the appropriateness of accruing further patients.

ACKNOWLEDGMENT

The authors would like to thank the Cancer and Leukemia Group B (CALGB) for providing the data for this analysis and Mark Green for providing useful comments on the manuscript. This research was supported in part by grant CA 33601 from the National Cancer Institute.

REFERENCES

Berger, J. O., and Berry, D. A. (1988). The relevance of stopping rules in statistical inference (with discussion). *Statistical Decision Theory and Related Topics IV 1*, ed. Berger, J. O. and Gupta, S., pp. 29–72. Springer-Verlag, New York.

Berry, D. A. (1985). Interim analysis in clinical trials: Classical vs. Bayesian approaches. *Statistics in Medicine* 4: 521–526.

Berry, D. A. (1987). Interim analysis in clinical trials: The role of the likelihood principle. *American Statistician* 41: 117–122.

Berry, D. A. (1988). Interim analysis in clinical research. *Cancer Investigation 5*: 469–477.

Berry, D. A. (1989a). Monitoring accumulating data in a clinical trial. *Biometrics* 45: 1197–1211.

Berry, D. A. (1991a). Bayesian methodology in Phase III trials. *Drug Information Association Journal* 25:345–368.

Berry, D. A. (1991b). Experimental design for drug development: Bayesian approach. *Journal of Biopharmaceutical Statistics* 1:18–101.

Berry, D. A. (1993) A case for Bayesianism in clinical trials (with discussion). *Statistics in Medicine* 12:1377–1393.

Cox, D. R. (1972). Regression models and life-tables (with discussion). *Journal of the Royal Statistical Society Series B* 34; 187–220.

George, S., Li, C., Berry, D. A., and Green, M. (1994). Stopping a clinical trial early: Frequentist and Bayesian approaches applied to a CALGB trial in non-small-cell lung cancer. *Statistics in Medicine* 13: 1313–1327.

Ginsberg, R. J., Kris, M. G., and Armstrong, J. G. (1993). Cancer of the lung. In *Cancer: Principles and Practice of Oncology*, eds. DeVita, V. T., Jr., Hellman, S., and Rosenberg, S. A., p. 673. J. B. Lippincott, Co., Philadelphia.

Ihde, D. C., Pass, H. I., and Glatstein, E. J. (1993). Cancer of the lung. In *Cancer: Principles and Practice of Oncology*, eds. DeVita, V. T., Jr., Hellman, S., and Rosenberg, S. A., p. 723. J. B. Lippincott, Philadelphia.

Kaplan, E. L. and Meier, P. (1958). Nonparametric estimation from incomplete observations. *Journals of American Statistical Association* 53: 457–481.

Lan, K. K. G. and Demets, D. L. (1983). Discrete sequential boundaries for clinical trials. *Biometrika* 70: 659–663.

Li, C. (1994). Comparing Survival Data for Two Therapies: Nonhierarchical and Hierarchical Bayesian Approaches. Ph.D. dissertation, ISDS, Duke University, Durham, NC.

O'Brien, P. C. and Fleming, T. R. (1979). A multiple testing procedure for clinical trials. *Biometrics* 35: 549–556.

Perez, C. A., Pajak, T. F., Rubin, P., *et al.* (1987). Long-term observations of the patterns of failure in patients with unresectable non–oat cell carcinoma of the lung treated with definitive radiotherapy. Report by the Radiation Therapy Oncology Group. *Cancer* 59: 1874–1881.

Pocock, S. J. (1982). Interim analyses for randomized clinical trials: The group sequential approach. *Biometrics* 38:153–162.

Propert, K. J. and Kim, K. (1992). Group sequential methods in multi-institutional cancer clinical trials. In *Biopharmaceutical Sequential Statistical Applications*, ed. Peace, K. E., pp. 133–153. Marcel Dekker, New York.

Qian, J. (1994). A Bayesian Weibull Survival Model. Ph.D. dissertation, ISDS, Duke University, Durham, NC

Stangl, D. K. (1991). Modeling Heterogeneity in Multi-center Clinical Trials Using Bayesian Hierarchical Survival Models. Ph.D. thesis, Carnegie-Mellon University, Pittsburgh, PA.

7

Decision Models in Clinical Recommendations Development: The Stroke Prevention Policy Model

Giovanni Parmigiani, Marek Ancukiewicz,
and David Matchar   Duke University, Durham, North Carolina

ABSTRACT

The goal of this chapter is to give a general overview and a practical illustration of the role of Bayesian methods in large interdisciplinary studies for clinical recommendation development. These studies have a strongly decision-oriented nature, they draw from several sources of information, and they require a probabilistic assessment of the uncertainty of the answers. These are among the features that make them an ideal terrain for the application of Bayesian methods.

The chapter consists of two parts. The first is a broad overview of issues, with pointers to some of the relevant literature. The second is an illustration in the context of the Stroke Prevention Policy Model, being developed by the Stroke Prevention Patient Outcome Research Team. We discuss in detail selected aspects of the study. We focus on general goals and structure of the decision model, computing framework for predictive distributions of the outcomes of interest, and combining information by simulation and resampling methods.

1 INTRODUCTION

Guidelines and recommendations for clinical practice can be developed by systematically evaluating and comparing alternative intervention

strategies based on their predicted health outcomes and costs. Matchar
(1994) discusses general decision analysis tools useful to this purpose.
From a statistical point of view, the general framework for the comparison
of alternative interventions is, to varying degrees of formalization, that
of Bayesian decision theory: evaluation proceeds by deriving a joint predictive distribution of the outcomes under each strategy and by using it
to determine expectations of patient utilities, costs, and so forth. This
approach accommodates the expected utility paradigm for decision making (see Bernardo and Smith, 1994).

Determining the appropriate predictive distributions, utilities, and
costs requires complex modeling and estimation, revolving around what
is usually called a decision model. The decision model serves as a central
organizing focus, helping to choose data sets and determine which projects
and analyses are of primary interest. It is also a policy tool designed to
assess the relative merits of various strategies in the sense described
above. Bayesian methods can provide a general unifying framework and
effective implementation techniques for building the decision model and
making it operational.

As a case study, we consider the Stroke Prevention Policy Model
(SPPM) developed by the Stroke Prevention Patient Outcome Research
Team (PORT). The goal of the SPPM is to identify the most appropriate
and cost-effective clinical strategies for secondary and tertiary prevention
of stroke. We focus on selected aspects of the SPPM: the general structure
of the model, the specification of the components of the utility function,
the computation of predictive distributions for the outcomes of interest,
and the probabilistic sensitivity analysis. For each of the components of
the decision model, we briefly review the role of Bayesian methods. The
study is still in progress as this chapter is being written, and final results
are not yet available. However, the current state of the project provides
an adequate illustration of some important points and in many ways reflects the state of the art in clinical policy modeling.

2 DECISION MODEL

In this section we give a general overview of the building blocks of the
type of decision model used in guideline development. We then review
very briefly the role of Bayesian methods in implementing the decision
model.

The goals of a decision model are to integrate the evidence regarding
alternative interventions and to provide predictions of expected clinical
outcomes for different patient/intervention scenarios. Once the model is

operational, the inputs are a patient profile and a specific intervention strategy, and the output is a joint predictive distribution for various outcomes of interest, such as patient's utilities, costs, and survival times. From the predictive distribution one can then derive the expectations that are typically used in practice for comparing interventions. For ease of exposition, we distinguish the following elements of a decision model:

Health states
Transitions
Duration distributions
Intervention effects
Health outcomes and costs

We comment briefly on each of these elements.

2.1 Health States

The core of the model is a representation of the progress of the disease, in the absence of clinical intervention. This begins with a list of relevant health states: an exhaustive and mutually exclusive classification of a patient's health at a given time, or age. Figure 1 illustrates a hypothetical elementary example: health states are represented by the ovals. Specifying the possible health states is a crucial element of modeling. Initially, it is based on informal incorporation of published results and expert opinion. Subsequently, the list and configuration of states may change as the result of estimation of other parts of the models. In particular, distinguishing between clinically different health states is important only if there is also a difference in terms of patient's prognosis, utilities, or costs.

2.2 Transitions

The health states provide a static representation of an individual's health. A model for the progress of the disease over time requires transitions between states. In Figure 1, these are represented by the arrows and are mimicking a typical pattern for a chronic disease. The resulting model is a stochastic compartment model. Manton and Stallard (1988) give an indepth discussion. Wakefield *et al.* (1994) illustrate Bayesian estimation for a related class of models.

2.3 Duration Distributions

The time at which transitions occur is usually modeled in terms of the patient's age rather than chronological time. This in turn may be discrete or continuous. A convenient way of specifying the probabilistic structure

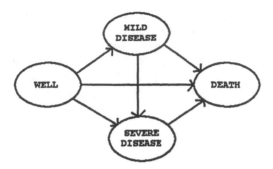

Figure 1   Disease model for a hypothetical disease.

of the model is to assign a duration distribution to each of the states. If the duration distributions are memoryless, the model is a Markov decision model. See Puterman *et al.* (1994) and Van der Duyn Schouten (1983) for a general discussion of Markov decision theory.

Estimation of the duration distribution is typically based on registries of major epidemiologic studies. From a statistical viewpoint, there are several challenges. First, this is the stage in which it is necessary to model patient- and provider-specific features. This is typically implemented by generalized linear models, or by proportional hazard models, with numerous covariates. Second, due to the high number of duration distributions and covariates, estimation must often resort to information derived from different databases or studies. In some cases, little is lost by focusing on one good study or data set properly addressing each of the questions of interest. Often, limitations of individual data sets make it necessary to combine formally information from different sources.

Bayesian methods are well suited to address both problems within the framework of hierarchical models, which we briefly outline at the end of the section. The general theory of hierarchical models is discussed in Bernardo and Smith (1994). Thomas *et al.* (1992) and Spiegelhalter *et al.* (1993) give further examples and provide software. Berry (1990), Zeger and Karim (1991), Morris and Normand (1992), and Gatsonis *et al.* (1994b) consider methodological issues in the health sciences. Gatsonis *et al.* (1993, 1994a) address geographical variation in the specific context of guideline development. Also, Bayesian methods can be employed for handling missing covariates across different studies. Methods based on data augmentation are reviewed by Tanner (1991). In this book hierarchical models are used in DuMouchel *et al.* (Chapter 17), Racine-Poon and Wakefield (Chapter 13), Smith *et al.* (Chapter 15), Albert and Chib (Chapter 22), Stangl (Chapter 16), Simon *et al.* (Chapter 21), Christiansen and Morris (Chapter 18).

2.4 Intervention Effects

The model components described so far can be used to evaluate the implications of no intervention. To determine the implications of other available clinical strategies, an effective option is to modify the incidence rates in the natural history model, using intervention-specific risk ratios. These are obtained by reviewing and analytically synthesizing published research and other information on clinical strategies. Ideally, evaluation of risk ratios should be based on randomized clinical trials.

In this phase there is clear opportunity to apply Bayesian meta-analysis. When meta-analysis is based on hierarchical modeling it can be cast in a unified framework with the estimation of the natural history model. An important advantage of Bayesian meta-analysis in this context is that it provides a formal assessment of the uncertainty in the evaluation of the risk ratio. Sources of uncertainty that can be incorporated include sampling variation within each study, between-study variability, and selection bias. References to Bayesian methods in meta-analysis of clinical trials include Berry (1990), DuMouchel (1990), Eddy (1992), Eddy *et al.* (1992), and Morris and Normand (1992).

2.5 Health Outcomes and Costs

Consequences of clinical interventions are usually evaluated based on a number of different outcomes. Formally the problem is probably best captured by a multiway utility function, whose dimensions may include survival times, patient's individual utilities, financial costs, and so on. This can in turn be used to perform various kinds of comparisons, such as cost-utility analysis (Kamlet, 1991), cost-effectiveness analysis, or one-dimensional utility maximizations by linear combination of components. The term utility is used here to indicate anything that is an objective of the decision-making process. This is standard in statistical decision theory but not so in the medical decision-making literature, where utility typically designates patients' individual preferences for health states.

In the context of a stochastic compartment model, costs and individual utilities can be incorporated by attaching a value to each health state and transition. The cost value represents the cost of remaining in the given health state for one unit of time, or the cost of making a certain transition. Similarly, individuals are assigned a utility value depending on their health state at each point in time. A utility value of 1 is assigned to "full health" and a utility value of 0 to "death." Intermediate values are interpreted accordingly. The quality-adjusted life year (QALY) measure for a health intervention is then determined by integrating this utility weight over an individual's life subsequent to the health intervention. Weinstein *et al.* (1980) discuss the formal conditions under which such an outcome measure is consistent with subjective expected utility theory.

Cost estimation is based on administrative data sets and is discussed in Lave *et al.* (1994). Practical evaluation of individual utility is based on direct elicitation by interview of patients, physicians, or the general public. Kamlet (1991) and Sackett and Torrance (1978) give further details.

In summary, in the context of decision modeling, the Bayesian approach provides:

Theoretical foundations for individual and policy decision making
Effective computing strategies for prediction and uncertainty assessment
Methods for combining information from various data sources

Before providing more detail on the latter two points we will introduce some basic definitions and notation.

2.6 Notation

The clinical interventions available are denoted by $i = 1, \ldots, I$; x is the vector of covariates defining patient profiles; y are the data, including the databases used for estimating duration distribution, costs, and risk ratios; (λ, θ) is the vector of unknown parameters of the natural history model, where λ is the subvector of the rates associated with the duration distributions, and θ includes all the other parameters; κ_i is the vector of risk ratios for intervention i; finally, ω is the vector of outcome measures. The parameter vectors will typically be structured hierarchically, but we do not need to account for this explicitly in the notation of this section.

2.7 Outcome Prediction

The natural history model can be identified with the distribution:

$$p(\omega \,|\, x, \lambda, \theta) \tag{1}$$

defined implicitly by the assumptions on duration distributions, costs, and utilities. Also, Bayes' rule leads to a posterior distribution, denoted by

$$p(\lambda, \theta \,|\, x, y) \tag{2}$$

Outcome predictions for each patient profile can then be obtained by the equation

$$p(\omega \mid x, y) = \int p(\omega \mid x, \lambda, \theta)p(\lambda, \theta \mid x, y) \, d\lambda \, d\theta \tag{3}$$

Similarly, intervention-specific predictions are derived by

$$p_i(\omega \mid x, y) = \int p(\omega \mid x, \lambda\kappa_i, \theta)p(\lambda, \kappa_i, \theta, \mid x, y) \, d\lambda \, d\theta \, d\kappa_i,$$
$$i = 1, \ldots, I \tag{4}$$

where $\lambda\kappa_i$ denotes pointwise multiplication of the elements in the vectors.

The Bayesian approach to prediction, based on averaging over unknown values of the parameters in Eq. (3) and Eq. (4), can be contrasted to the ad hoc approach of estimating the model by replacing point estimates for the parameters in Eq. (1). The latter is easier computationally, but it leads to predictive distributions that are underdispersed, as no accounting is taken of uncertainty about the parameter estimates. Analysis based on replacing point estimates in Eq. (1) are often accompanied by sensitivity analysis. The Bayesian prediction paradigm produces automatically a probabilistic sensitivity analysis in which the parameter distributions are the posterior distributions given the data.

A further important feature of Eq. (3) is that the joint distribution of the outcomes can be used to assess the strength of the dependence between outcomes, as a function of clinical interventions.

3 COMBINING INFORMATION

In this section we briefly outline the general strategy for combining information. For ease of exposition we divide the discussion into two parts: combining information for different subpopulations using hierarchical models and incorporating expert opinion via the prior distributions.

3.1 Hierarchical Models

In combining information from different studies, it is helpful to model both between-study variation and within-study variation. This leads naturally to the idea that each study pertains to a specific subpopulation (say the patient population of a certain clinic, study, cohort, geographic area, etc.), randomly chosen from a more general population (say the patient population of interest for the SPPM). Combining information amounts to learning about the general population, and it is done by deriving posterior distributions for population parameters, conditional on the subpopulation data.

The basic framework is as follows. We concentrate on estimating the duration distribution for a specific health state, based on several studies. For expository purposes we assume that there are S studies with the same type of survival and covariate information. For each study, we may assume a proportional hazard model with study-specific parameters. Using the notation of the previous section:

$$\lambda_s = \lambda_0 \exp\{\beta_s x\}, \qquad s = 1, \ldots, S$$

This set of models constitutes the first stage of the hierarchy. The second stage consists of assuming that the first-stage parameters β_s are draws from a wider population, in our case a population of studies. In particular, it is assumed that

$$\beta_s \sim \pi(\cdot \mid \theta)$$

where π is a parametric probability distribution and θ is a vector of unknown parameters. This can be extended to the baseline hazards as well. Inference on θ produces transition probabilities that incorporate information from all the studies. Such inference can also provide uncertainty assessment to be used in sensitivity analysis. Prevailing classical estimation methods are based on the so-called empirical Bayes technique (Morris, 1983). Computational developments have made available fully Bayesian techniques as well. In the specific context of proportional hazard models, recent work includes Zeger and Karim (1991) Dellaportas and Smith (1993), Stangl (1995), and Stangl (Chapter 16).

Extensions of this framework can account for missing covariates in some of the data sets, for study-specific baseline rates, and so forth. A special treatment is necessary if estimates rather than raw data are available for each of the sites. One approach consists of interpreting estimates as sufficient statistics for some probability model. Another approach based on bootstrap resampling of the study-specific parameter estimates is discussed later in this chapter. A further feature that can be incorporated in hierarchical models is discounting of older or less reliable studies whose evidence is not considered of the same strength as that of current studies. Likelihood adjustments for such cases are discussed by Berry and Stangl (Chapter 1) and Wolpert and Mengersen (1995).

Expert Judgment

A complex decision model will inevitably require specification of parameters that are not adequately estimated based on the available data sets alone. In such cases it will be necessary to complement data-based information with clinicians' input. This is best done by modeling data-based information via the likelihood function and clinicians' input via the a priori distribution. These can then be combined using Bayes' rule.

A crucial step in this endeavor is the elicitation of the a priori distribution, that is, the translation of a physician's knowledge into the formal statement necessary to apply Bayes' rule. A thorough discussion of several simple and efficient strategies is provided by Berger (1985), Chaloner (Chapter 4), and Kadane and Wolfson (Chapter 5). Some of the alternatives are similar to the specification of bounds for sensitivity analysis, to which many clinicians are already accustomed.

Even though the model may contain a large number of parameters, the process of incorporating expert information can be done selectively, by using experts' opinion only when data-based information is inadequate. An effective strategy is to proceed sequentially: begin by estimating parameters based on data alone. Identify components of the models that are difficult to estimate, for example, by monitoring the variability of the parameters. Such sections will typically correspond to combinations of covariates for which the number of patients is small. At that point one can elicit experts' opinion, reestimate, and so on.

3.2 Some Difficulties with Combining Data

We now discuss in more detail two of the implementation steps that proved especially challenging in the case study of Section 5: handling different sets of covariates in different studies and eliciting expert opinion.

As we indicated, (for the synthesis of information) the first stage of the hierarchical model is typically either a proportional hazard model or a generalized linear model. The coefficients of this model are random and modeled at the second stage of the hierarchy. Such a modeling approach works best when a similar type of survival and covariate information is available for the different data sources. Unfortunately, studies synthesizing useful medical knowledge often use data sources registering different covariates. Then, covariate coefficients for different sources may have different interpretations. Carrying out properly the full hierarchical estimation requires either the unrealistic assumptions of orthogonality of the design matrices, or the assumption that the omitted variables have little or no effect, or a reformulation of the problem to treat missing covariates as missing data. In principle, augmentation algorithms can be used to this end. In practice, however, this cannot be done reliably in studies with a high number of covariates and sparse observation in most of the cells, even for the studies that do report most of the covariates of interest.

Regarding elicitation of expert opinion, it can be difficult to elicit opinions on model parameters for very complex and high-dimensional parameterizations in which parameters do not always have an immediately intuitive interpretation. Experts can be more comfortable, and therefore more reliable, when thinking in terms of model outcomes such as survival rates or costs.

Strategies to alleviate both of these difficulties can be developed by approaching the problem from a predictive viewpoint. First, different data sources may use different covariates but are likely to provide their outcome predictions on comparable scales. For example, data for the natural history model are derived from epidemiologic studies that are all designed to provide projections of survival. Likewise, experts will often be able to predict outcomes of the model based on their experience and data available in the literature. Elicitation of prior hyperparameters via the specification of predictive distributions is commonplace in Bayesian statistics (see Berger, 1985). Implementation of complex models may be difficult, and approximate solutions may be necessary. We will touch on these again in the next section.

4 COMPUTING FRAMEWORK

Practical implementation of decision models on a scale such as that of the SPPM requires substantial computing effort. In this section we describe a general simulation-based framework for the computations. This approach addresses jointly several steps that are often thought of as separate. These are (1) estimating model parameters, perhaps by combining multiple sources of information; (2) predicting outcomes; and (3) performing a sensitivity analysis.

4.1 Monte Carlo Simulations

In realistic decision models, the distributions in Eqs. (1), (2), and (3) have very high dimensionality or are intractable analytically and must be determined numerically. Currently, the most efficient and flexible numerical methods are based on simulation.

Estimation of the model distribution in Eq. (1) usually requires Monte Carlo methods. The prevalent strategy for computing Eq. (2) in hierarchical models involves simulation methods based on Markov chain Monte Carlo [see Spiegelhalter *et al.* (1993) for a tutorial, software and further references]. Briefly, one generates a sample of parameter values from Eq. (2) by iteratively updating a subgroup of parameters each time. Each iteration provides a value of $(\lambda, \theta, \kappa_1, \ldots, \kappa_I)$. The collection of iterations gives a sample from the desired posterior distribution.

In decision modeling, the most important distribution is the predictive distribution, Eq. (3). We can regard the model equation (1) and the posterior distribution of Eq. (2) as intermediate steps, and $(\lambda, \theta, \kappa_1, \ldots, \kappa_I)$ as nuisance parameters. The most efficient strategy for generating simulated samples of outcomes measures, and incorporating the uncertainty about the model parameters, is to set up a single simulation scheme to perform estimation, combination of information, and prediction. The basic idea is to sample from the joint distribution of ω and $(\lambda, \theta, \kappa_1, \ldots, \kappa_I)$, given x and y, and then use only the simulated values for the outcome, which constitute a draw from Eq. (3). In practice this can be implemented by nesting a prediction step within each Markov chain Monte Carlo iteration. Each iteration generates $(\lambda, \theta, \kappa_1, \ldots, \kappa_I)$; given these one can use a Monte Carlo step to simulate one draw from Eq. (1).

Using the resulting sample of outcomes for decision making is then a simple matter of summarization, which can be done using standard graphical tools such as boxplots. Important summaries such as cost-effectiveness ratios can be evaluated by Monte Carlo averages. Predictions incorporate variability of model parameters $(\lambda, \theta, \kappa_1, \ldots, \kappa_I)$. Probabilistic sensitivity analysis with respect to other quantities relevant to the decision model can be performed by simulating a value from the postulated distribution of such quantities at each step of the simulation scheme.

4.2 Resampling

The full implementation of the Markov chain Monte Carlo approach requires direct access to full data sets. In practice, in complex studies that use many data sources, such as the SPPM, it may prove impossible or impractical to amass all useful data in one central computer and raw data may have to be replaced by partial summaries. For example, some participating collaborators may provide a series of parameter estimates rather than a copy of their databases, perhaps due to the proprietary nature of some of the data sources. In the SPPM, we handled this problem using a bootstrap-based resampling technique, which makes it easy to incorporate partial site data into the general computing framework. We first discuss how to incorporate bootstrap estimates into a Bayesian framework and then move to combining bootstrap samples from different studies.

We focus on a specific study having parameter vector ψ. In the SPPM this includes coefficients of proportional hazard models as well as baseline survival functions and is of very large dimension in most studies. Assume that a bootstrap sample of parameter estimates $\hat{\psi}_1, \ldots, \hat{\psi}_B$ is available.

The principle of bootstrap—an idea introduced in a classical setting by Efron (1979)—is to approximate the data distribution $F(\cdot)$ by the empirical distribution $F_n(\cdot)$ and then to derive an estimated distribution of statistics by sampling from this distribution. Efron's original bootstrap uses B bootstrap repetitions, obtained through resampling with replacement of the original data. The statistic $\hat{\psi}$ is computed for each bootstrap repetition. The resulting estimates $\hat{\psi}_1, \ldots, \hat{\psi}_B$ approximate the distribution of $\hat{\psi}$.

Applications of similar ideas in Bayesian inference began with Rubin (1981), who proposed an algorithm in which each bootstrap replication generates a posterior probability for each data point (points not observed are given zero posterior probability). Press (1989) describes the following version of the bootstrap in a Bayesian context: (1) obtain B bootstrap repetitions, each time drawing b points; (2) for each repetition, calculate a statistic $\hat{\psi}$, estimating a parameter ψ; (3) compute an empirical probability density $\hat{p}(\hat{\psi} \mid \psi)$, and (4) use the following estimator for the posterior probability:

$$\hat{p}(\psi \mid y) = \frac{\hat{p}(\hat{\psi} \mid \psi)\pi(\psi)}{\int \hat{p}(\hat{\psi} \mid \psi)\pi(\psi)\, d\psi} \tag{5}$$

This relies crucially on the choice of the empirical density $\hat{p}(\hat{\psi} \mid \psi)$; standard assumptions include location and location-scale families to relate $\hat{\psi}$ and ψ. Convergence of Eq. (5) to the posterior density $p(\psi \mid y)$ is discussed, for example, by Boos and Monahan (1986). One advantage of this approach over the full hierarchical model is ease of implementation of nonparametric or semiparametric assumptions (see also Muliere and Secchi, 1994).

4.3 Combining Bootstrap Predictions

Assume now that there are S studies available, each providing a bootstrap sample of the study-specific parameters. Using $\hat{p}(\hat{\psi}_s \mid \psi_s)$ as the study-specific likelihood function and assuming, as in the previous section, that $\psi_s \sim \pi(\cdot \mid \theta, \lambda)$, it is possible to set up a hierarchical model for combining the S studies. As we indicated, handling missing covariates would be difficult but not impossible. For example, a missing covariate in study s can be handled either as a missing value (if the covariate was not recorded at all) or as a zero value (if the covariate was recorded but not included in the study's model) of the corresponding coordinate of ψ_s in the bootstrap sample.

Using the assumption that the studies are conditionally independent given θ, we can write

$$p(\theta, \lambda \mid y) \propto p(\theta, \lambda) \prod_{s=1}^{S} \int \hat{p}(\hat{\psi}_s \mid \psi_s) \pi(\psi_s \mid \theta, \lambda) \, d\psi_s \tag{6}$$

where both $\hat{\psi}_s$ and ψ_s are vectors of dimensionality k_s. Based on this, we can generate predictions from Eq. (3).

In the SPPM, the ψ_s's are of very large dimensionality (up to one thousand in the larger studies), because bootstrap replications include baseline hazard functions. This makes the choice of $\hat{p}$ and π very difficult. Also, evaluating Eq. (6) becomes a challenging and time-consuming computing problem.

The current implementation of the SPPM uses an ad hoc, but effective, strategy for combining studies, based on operating directly on the distributions of the predicted outcomes. The technique consists of two steps: (1) using bootstrap samples of site-specific parameters to generate outcomes from approximate site-specific predictive distributions; and (2) combining the resulting simulated outcome using a weighting scheme approximating the random effects model of Eq. (6).

Assume that the bootstrap sample size B is common to all studies. For a fixed covariate profile x, we simulate I predicted outcomes ω_{sbi} from $p_s(\omega \mid x, \psi_{bs})$, for each bootstrap replication $b = 1, \ldots, B$, and each study $s = 1, \ldots, S$. If we assume that the bootstrap sample is approximately a sample from the study-specific posterior distribution $p(\psi_s \mid x, y)$, we can consider $\omega_{sbi}, b = 1, \ldots, B, i = 1, \ldots, I$ as an approximate sample from Eq. (3) for study s. The bootstrap estimates are replacing the Markov chain Monte Carlo step.

We can now generate combined predictions according to the linear pooling rule:

$$p(\omega \mid x, y) = \sum_{s=1}^{S} w_s p_s(\omega \mid x, y)$$

where w_s are weights. We determine the values of w_s by mimicking the weights in the estimation of a common mean from several samples, under a random effects model (see also Box and Tiao, 1973). More precisely, think of the following model for the simulated outcome values:

$$\omega_{sb\cdot} = \theta + \alpha_s + \beta_{sb} \tag{7}$$

where $\omega_{sb\cdot} = (1/I) \sum_{i=1}^{I} \omega_{sb}$. Moreover, let $E(\alpha_s) = E(\beta_{sb}) = 0$ and $\mathrm{VAR}(\alpha_s) = \tau^2$ and $\mathrm{VAR}(\beta_{sb}) = \tau_s^2$. To determine heuristically the weights w_s we think of this as estimating the common mean θ based on S normal samples each with variance $\tau_s^2 + \tau^2$. This leads to

$$w_s \propto \left[\frac{1}{S-1} \sum_{s=1}^{S} (\omega_{s..} - \omega_{...})^2 + \frac{1}{B-1} \sum_{b=1}^{B} (\omega_{sb.} - \omega_{s..})^2 \right]^{-1}$$

Variants of this rule can be constructed by assuming that some prior information on the variance components is available. In particular, it may be helpful to incorporate prior knowledge on τ^2.

In summary, combining bootstrap prediction presents several attractive features in the context of the SPPM:

1. The dimensionality of the problem is substantially reduced. There are typically around 10 outcome variables as opposed to hundreds of dimensions in the parameter space. This is perhaps the most important advantage of combining predictions.

2. By combining predictions we need much weaker modeling assumptions on the likelihood function for the parameters in each of the studies. The bootstrap sample of parameter values provides this directly.

3. The form of the predictive distribution is not restricted by any distributional assumption. It is only the weight given to each of the studies that is based on normality. The study-specific distribution remains general.

4. Combining different studies accounts in an intuitively appealing way for the different sources of variability. The weights depend on study-to-study variability and on bootstrap variability but, for sufficiently large I, are independent of the variability in the predictive distribution.

5   THE STROKE PREVENTION POLICY MODEL

In this final section we illustrate some of the issues discussed so far using the Stroke Prevention Policy Model. The SPPM is being developed by the Stroke Patient Outcome Research Team with the goal of comparing the efficacy of various stroke prevention interventions. For a general discussion see Matchar *et al.* (1993). The model incorporates four components:

A stochastic model of natural history of cerebrovascular disease
A model for prevention strategies, or intervention model
A model for assessment of patient preferences
A model for assessment of costs

The general strategy reflects the discussion of Section 2. To compare interventions, we first developed a natural history model, using data from major epidemiologic studies: the Framingham study, the Cardiovascular Health Study, and the Olmsted County Registry. The natural history model represents the implications of the "no intervention" strategy. To determine the implications of other stroke prevention strategies, we formed intervention models by modifying incidence rates in the natural history model, using risk ratios obtained from the PORT's literature review and meta-analysis.

5.1 Outcomes

Model outcomes are captured as a vector ω whose components include:

Expected survival time (life years, LY)
Expected quality-adjusted life years (QALY)
Expected lifetime accumulated direct medical costs (MC)

These are the end results used in the evidence tables summarizing the results of the study for clinicians. Additional outcomes, such as the expected lifetime stroke rate or myocardial infarction rate, can be determined easily. In general, call $H(t)$ and $C(t)$ the processes recording, respectively, health states and rate of accrual of cumulative direct medical costs at time t. Then any (possibly multivariate) stochastic functional defined on the patient's clinical history $H(\cdot)$ and the direct health care costs $C(\cdot)$ can be regarded as a model outcome.

A joint distribution of the above outcomes can be used for decision making. Alternatively, marginal cost-effectiveness ratios and quality-adjusted marginal cost-effectiveness ratios can be estimated to compare interventions (Kamlet, 1991).

5.2 Events and Health States

The model takes into account clinical events whose relative importance and cost were deemed to influence the choice of stroke prevention policy. In general, these include stroke by severity and type, myocardial infarction, other complications of treatment (primarily gastrointestinal bleeding), and death.

We define health state at time t by the latest prior clinical event. Health states of our model are listed in Table 1. The distinction between minor and major stroke states is based on the Rankin score, a disability measure, assessed 30 days after the event. Strokes with scores 1–3 are assumed to be minor and strokes fatal or with scores 4 and 5 are assumed to be major.

Table 1   Health States in the SPPM

No.	Symbol	Description
1	AS	No history of cardiovascular or cerebrovascular events
2	TIA	Transient ischemic attack
3	μIS	Minor ischemic stroke
4	IS	Major ischemic stroke
5	MI	Myocardial infarction
6	DT	Death (all causes)
7	μHS	Minor hemorrhagic stroke
8	HS	Major hemorrhagic stroke
9	OC	Nonstroke, non-MI complications of treatment

5.3 Natural History Model

The natural history of disease can be described as a process $H(t)$, recording health state as a function of time. At any time, the probability of transition to another state is assumed to depend on:

1. The present state
2. The occurrence of certain important events (strokes, TIAs, MIs) in the past
3. The state to be visited next
4. The time spent in the present state (in some cases this needs to be replaced by the transition time from another previous disease state)
5. The patient-specific covariates

The predictions of the model may depend on some *covariates*, that is, patient-specific or setting-specific features, like patient age, gender, geographic location. Table 2 lists covariates used in the SPPM and shows their purpose.

    The distribution of the time T spent in the state s_i before moving to s_j can be described by the transition-specific hazard function $\lambda_{ij}(t)$

$$\lambda_{ij}(t) = \lim_{\Delta t \to 0} \frac{\Pr(t \le T < t + \Delta t, j \mid T \ge t, i)}{\Delta t}$$

assuming that the limit exists. Specifically, if patient covariates are x and the current state s_i is a disease state, then the hazard $\lambda_{ij}(t \mid x)$ depends solely on

Table 2 Covariates Used by the SPPM

Covariate	Used for modeling of			
	Hazards	Costs	Utilities	Interventions
Age (years)	•	•	•	•
Gender	•	•	•	•
Race (white/other)	•	•		
Smoking status (never/ever)	•			
Systolic blood pressure (mm Hg)	•			
Hypertensive medications	•			
Cervical bruit	•			
Evidence of carotid artery stenosis	•			•
Diabetes mellitus	•			
Coronary artery disease	•			
Atrial fibrillation	•			
Congestive heart failure	•			
Peripheral vascular disease	•			
Valvular heart disease	•			
History of past TIA or IS	•	•	•	•
History of past MI	•	•		
History of major IS		•	•	
Geographic location (state)		•		
Type of hospital (teaching/other)		•		

Next state s_j
Sojourn time
Occurrence of disease states in the past

We assume that occurrence of HS or OC affects exclusively the risk of death; that is, all other transition-specific hazard functions are determined by the other previous state. We could ignore HS and OC, accounting for a somewhat higher hazard of death when the patient is in one of remaining states (similarly, accounting for somewhat higher costs and lower preferences). We decided to keep the present structure of the model for reasons that are indicated later in this section.

An additional modification of the model could be a redefinition of health states such that transition-specific hazard functions would depend only on the present state, the state to be visited next, and the sojourn time. This would be easily accomplished by defining the states as vectors $(h_{TIA}, h_{IS}, h_{MI}, s)$. Here h_{TIA}, h_{IS}, and h_{MI} are indicator variables recording the occurrence of, respectively, TIA, IS, and MI; s denotes the recent clinical event from the list in Table 1. The stochastic process denoting a redefined health state of a patient indexed by time is a semi-Markov process (Pyke, 1961).

Therefore, the stochastic process $H(t)$, which records the health state as a function of time, is equivalent to a semi-Markov process. For simplicity, we have chosen a non-Markov representation, with HS and OC as separate states and with transition-specific hazards dependent upon past states. The main advantage of such a representation is the low number of states and, consequently, easier implementation and interpretation of the model. Also, the delineation of separate complication states (HS and OC) simplifies modeling of mortality, cost, and disutility of treatment complications. An obvious disadvantage of not using a semi-Markov representation is the loss of a well-developed analytical framework developed for semi-Markovian processes. However, this has little importance in the SPPM, where the estimation is based on simulations rather than analytic expressions.

5.4 Estimation

The transition-specific hazard functions were estimated under assumptions of

Proportional hazards
Multiplicative covariate effects
Conditional independence of transitions to different states given a set of covariates

using the Cox proportional hazard (PH) model:

$$\lambda_{ij}(t \mid x) = \lambda_{0,ij}(t) \exp \beta_{ij} x$$

Here $\lambda_{0,ij}(t)$ is the baseline hazard at time t, for the transition between states s_i and s_j; β_{ij} is the vector of regression coefficients specific to the transition from i to j. No prior information on the coefficients is incorporated. Estimation is based on the partial likelihood (Cox, 1972). The baseline hazard function $\lambda_{0,ij}(t)$ was estimated according to method of Kalbfleisch and Prentice (1980, p. 85). Noninformative censoring is assumed throughout the estimation process. This means that if we are estimating

$\lambda_{ij}(t \mid x))$, only the transitions to state s_j are informative. Other transitions, which censor observations for the purpose of estimating the transition-specific hazard function $\lambda_{ij}(t \mid x)$, are noninformative. This assumption is restrictive but not completely unreasonable, as we are conditioning on a number of covariates, including past history.

Simulation

The outcomes are estimated from transition-specific hazard functions by Monte Carlo simulations. In each realization, the simulation generates times of clinical events, starting from time zero, until patients' death or until a fixed maximum age.

To simulate the time-to-next-transition and the next state, we used the inversion algorithm from Devroye (1985) for generating a nonhomogeneous Poisson variate. This is then used to generate the transition times for the various transition-specific hazard functions (which depend on patient characteristics, time, and past events). Finally, the transition with the shortest time is selected, with ties broken at random.

A specially designed algorithm counts the amount of time spent in each state, costs, and utilities (discounted, if needed) relevant for model outcomes. The expectation is approximated by the average from all realizations.

5.5 Intervention Models

We assumed that both costs and hazards of the natural history model are modified in the presence of various interventions. Intervention models quantify the influence of modeled prevention strategies on hazards of transitions and on medical costs.

SPPM allows for an unlimited number of prevention strategies, which, however, consist of only three basic components:

Diagnostic test. This may be either an invasive or a noninvasive test. Each test is characterized further by its cost, sensitivity, specificity, and by a vector of probabilities of the following complication: ischemic stroke, hemorrhagic stroke, myocardial infarction, and death.

Carotid endarterectomy. This is a surgical procedure. It is characterized by its cost, complication probabilities, effectiveness, and effectiveness duration. Complication probabilities are described by a vector, the components of which denote the probability of transition to ischemic stroke, hemorrhagic stroke, myocardial infarction, and death. Similarly, effectiveness is characterized by a matrix of relative risks. These are multipliers that are applied to the hazard functions, as in Eq. (4).

Medical therapy. The medical therapy is characterized by its duration, monthly average cost, effectiveness (again a matrix of multipliers for the hazard functions), and complications (a matrix of hazard functions). Effectiveness and complications may depend on time. Duration may be fixed in advance or the treatment may last until a particular clinical event (for example, until the first TIA). A special case of medical therapy, with relative risks of one and with zero costs, corresponds to a "no intervention" strategy.

Figure 2 shows the admissible sequences of the basic treatment components. Intervention strategies consist of a sequence of these components, with possibly different characteristics. It is assumed that complications of tests and surgery occur instantaneously.

The data used for intervention models were obtained from hospital records (McCrory, 1994), from meta-analysis of data on major carotid endarterectomy trials, from meta-analysis of published results (Matchar *et al.*, 1994; Goldstein and Matchar, 1994) and from Medicare fee schedules.

5.6 Utilities

We assume that patient preferences about health states depend on the current health state, on demographic covariates (i.e., age, sex, race), and on the history of MI and strokes. Evidence about patient's utilities is provided by a patient survey conducted especially for this purpose by the PORT. Individual preferences are combined by simply averaging utilities for each combination of covariates.

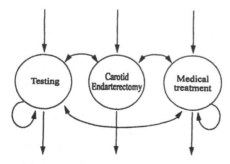

Figure 2   Elements of the intervention strategies and their relationship.

5.7 Cost Models

In the SPPM we found it convenient to estimate the cost model separately from the natural history model and to combine the two models at the simulation stage, when generating outcome predictions.

To make this more precise we need to introduce some further notation. We represent the complete clinical history, including events and time-to-events, by a vector h of finite dimension. Each component of this vector may correspond, for example, to the state occupied in each day or month. Similarly, we represent the rate of accrual of cumulative medical costs by a finite-dimensional vector c. Again, components correspond to costs per day or month.

Model outcomes can be computed based on the joint distribution of the cost stream c and of clinical history h, given the covariates x. This joint probability can be written as

$$\Pr(c, h \mid x) = \Pr(c \mid h, x)\Pr(h \mid x)$$

Our approach is to estimate the two factors on the right side of this equation separately. The first factor, which is the conditional distribution of the cost stream c given the patient covariates x and the complete clinical history h, will be termed the cost model. In the SPPM the cost model is estimated from administrative records, mostly consisting of insurance claim files. The second factor is the natural history model, possibly modified as the result of an intervention. We assume here that the clinical history h given patient characteristics does not depend on cost c, so that the natural history model can be adequately estimated from epidemiologic data.

In practice, our cost model included only a limited number of patient covariates and a limited part of the clinical history. In particular, the cost stream in the present state was modeled based on the two last states and the next state. This was motivated by model simplicity and by the absence of quality data linking clinical states and cost.

We estimated the distribution of the hospitalization costs associated with each clinical event by the corresponding empirical distribution, as the administrative data on this are extensive. We assumed that such hospitalization costs (termed acute costs) occurred instantaneously.

Chronic costs, that is, costs incurred after discharge from the hospital following a clinical event, require more complex modeling. We began by dividing the time horizon into months. The rate of accrual of chronic costs in the jth month since entry into state i, called C_{ij}, was assumed to be constant within each month. The cost stream model can then be written as

$$\Pr(C_{i1}, \ldots, C_{iJ} \mid h, x) = \prod_{j=1}^{J} \Pr(C_{ij} \mid C_{i1}, \ldots, C_{ij-1}, h, x)$$

The distribution of C_{ij} was modeled using a stratified proportional hazard model (Kalbfleisch and Prentice, 1980):

$$\Pr(C_{ij} > c \mid z) = C_{0ij}(c)^{\exp(\beta_i z)}$$

Here z denotes a vector of covariates comprising patient features x, data from clinical history h, and data about costs in previous months. The baseline cost function $C_{0ij}(c)$ describes the distribution of costs for a patient in state s_i, in the jth month since the last clinical event, given zero covariates. Observations are censored if the patient experienced a transition to a new state before the jth month was completed. Because a patient typically stays in the same state for several months and thus contributes many uncensored observations for each censored one, the assumption of noninformative censoring seems to be a good approximation in this case. Both acute and chronic costs for each health and complication state were discounted and integrated over time and health states. This integration was done using Monte Carlo methods.

5.8 Example

To conclude, we illustrate using the decision model to compare the two specific prevention strategies defined below.

> *Screening.* Patients are given a noninvasive test, known as the duplex Doppler ultrasound test, to detect severe carotid artery stenosis. If the test is positive, it is followed by carotid angiography—a more accurate, although more expensive and risky, diagnostic test. If the invasive test is also positive, then carotid endarterectomy is performed, followed by administration of aspirin for the lifetime. If any of the tests is negative, patients are not treated at all.

> *Watchful waiting.* Patients are followed until a first TIA or a minor ischemic stroke. Then they are tested for the presence of stenosis of at least 70%. If the test is positive, carotid endarterectomy is performed, followed by aspirin therapy. If the test is negative, patients are given aspirin for the lifetime.

The flowchart in Figure 3 outlines the logic used for simulating model outcomes for the two strategies.

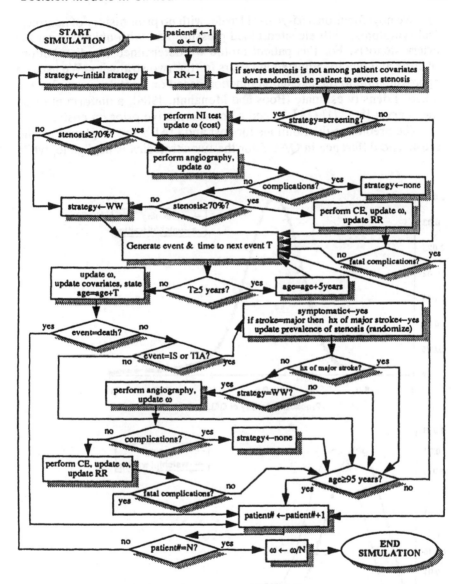

Figure 3 Flowchart for simulation of the "screening" and "watchful waiting" stroke prevention strategies.

We now focus on a 65-year-old male, with no prior history of neurological symptoms, with elevated blood pressure and with unknown carotid artery stenosis. For this patient profile we determined the distributions of costs and quality-adjusted life years for the two strategies. We used an approximation to the likelihood function based on bootstrap samples and a kernel density estimate (Boos and Monahan, 1986), a uniform prior on the interval [8.5, 11] QALY, and a noninformative prior on costs.

The predictive distributions for cost and QALY are displayed in Figure 4. The difference in QALY for the two strategies is negligible, while

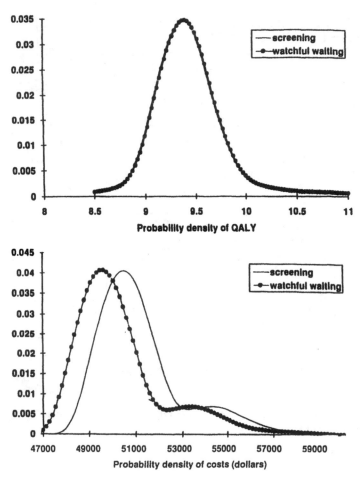

Figure 4   Simulated implications of "screening" and "watchful waiting": marginal densities of quality-adjusted life years and costs.

the distributions of costs are definitely different. The bimodal form of the distribution of costs reflects the fact that multiple sources of information are contributing to the prediction.

At this stage in the project, patient utilities are not yet available. In lieu of empirical evidence, we used utility estimates based on clinicians' input, assuming a utility of 0.4 for stroke and 0.7 for myocardial infarction.

The example illustrates several of the points highlighted in this chapter. First, the model permits a relatively detailed clinical description of a complex health problem. Second, the interventions can be modeled in a flexible way, with future decisions dependent on the actual unfolding events. Third, the model permits outputs that reflect the uncertainty in the inputs.

In conclusion, the SPPM offers both fertile terrain and important challenges for the application of Bayesian methodology. Although full implementation of the Bayesian paradigm in every aspect of the project has not been possible, the general outlook has provided a useful framework for assembling diverse submodels and information sources into one cohesive decision-making tool.

REFERENCES

Berger, J. O. (1985). *Statistical Decision Theory and Bayesian Analysis*. Springer, New York.

Bernardo, J. M. and Smith, A. F. M. (1994). *Bayesian Theory*. Wiley, New York.

Berry, D. A. (1990). A Bayesian approach to multicenter trials and meta-analysis. *ASA Proceedings of Biopharmaceutical Section*, American Statistical Association, Alexandria, VA. pp. 1–10.

Boos, D. D. and Monahan, J. F. (1986). Bootstrap methods using prior information. *Biometrika* 73: 77–83.

Box, G. E. P. and Tiao, G. C. (1973). *Bayesian Inference in Statistical Analysis*. Addison-Wesley, Reading, MA.

Cox, D. R. (1972). Regression models and life tables (with discussion). *Journal of the Royal Statistical Society*, Series B. 34: 187–220.

Dellaportas, P. and Smith, A. F. M. (1993). Bayesian inference for generalized linear and proportional hazard models via Gibbs sampling. *Applied Statistics* 27: 443–460.

Devroye, L. (1986). Non-uniform random variate generation, Springer-Verlag, New York.

DuMouchel, W. (1990). Bayesian meta-analysis. In *Statistical Methodology in the Pharmaceutical Sciences*, ed. Berry, D. A., pp. 509–529. Marcel Dekker, New York.

Eddy, D. M. (1992). *A Manual for Assessing Health Practices and Designing Practice Policies*. American College of Physicians, Philadelphia.

Eddy, D. M., Hasselblad, V., and Shachter, R. (1992). *Meta-Analysis by the Confidence Profile Method*. Academic Press, New York.

Efron, B. (1979). Bootstrap methods: Another look at the jackknife. *Annals of Statistics* 7: 1–26.

Gatsonis, C., Normand, S. L., Liu, C., and Morris, C. (1993). Geographic variation of procedure utilization: A hierarchical model approach. *Medical Care* 31(5 Suppl): YS54–59.

Gatsonis, C., Epstein, A. M., Newhouse, J. P., Normand, S. L., and McNeil, B. J. (1994a). Variations in the Utilization of Coronary Angiography for Elderly Patients with an Acute Myocardial Infarction. An Analysis Using Hierarchical Logistic Regression. Technical Report, Harvard Medical School, Boston.

Gatsonis, C., Liu, C., and Normand, S. L. (1994b) Hierarchical Regression Models for Health Care Utilization Data: Bayesian Methods for Hierarchical Logistic and Normal Models. Technical Report, Harvard Medical School.

Goldstein, L. B. and Matchar, D. B. (1994). Clinical assessment in the evaluation of stroke. *Journal of the American Medical Association* 271: 1114–1120.

Goldstein, L. B., Matchar, D. B., Stroke PORT Investigators, and The Stroke PORT. (1994). Secondary and Tertiary Prevention of Ischemic Stroke *Health Reports* 6(1): 154–159.

Kalbfleisch, J. D. and Prentice, R. L. (1980). *The Statistical Analysis of Failure Time Data*. Wiley, New York.

Lave, J. R., Anderson, G., Brailer, D., Bubolz, T., Conrad, D., Freund, D., Keeler, E., Lipscomb, J., Luft, H., Pashos, C., Provenzano, G., and McCrory, D. (1994). Impact of risk factors on stroke rate: A metaanalysis. *Medical Care* 32(7 Suppl): J577–J589.

Kamlet, M. S. (1991). *A Framework for Cost-Utility Analysis of Government Health Care Programs*. U.S. Department of Health and Human Services, Washington, DC.

Manton, K. G. and Stallard, E. (1988). *Chronic Disease Modelling*, Oxford, New York.

Matchar, D. B. (1994). Application of Decision Analysis to Guideline Development. In *Methodology Manual for Guideline Development*. AHCPR, Washington, DC.

Matchar, D. B., Duncan, P., Samsa, G., Whisnant, J., DeFriese, G., Ballard, D., Paul, J., Witter, D., and Mitchell J. (1993). The stroke prevention Patient Outcomes Research Team: Goals and methods. *Stroke* 24(12): 2135–2142.

McCrory, D. C. (1994). Prediction of Complications of Carotid Endarterectomy. Master's thesis, Biometry Training Program, Duke University Medical Center, Durham, NC.

Morris, C. N. (1983). Parametric empirical Bayes inference: Theory and applications (with discussion). *Journal of the American Statistical Association* 78: 47–65.

Morris, C. and Normand, S. L. (1992). Hierarchical models for combining information and for meta-analyses. In *Bayesian Statistics* ed. J. M. Bernardo, J. O. Berger, A. P. Dawid and Smith, A. F. M., pp. 321–335. Oxford University Press, New York.

Muliere, P. and Secchi, P. (1994). Exchangeability, Predictive Sufficiency and Bayesian Bootstrap. TR-579, Department of Statistics, University of Minnesota, Minneapolis, MN.

Press, S. J. (1989). Bayesian Statistics: Principles, models and applications, Wiley, N.Y.

Pyke, R. (1961). Markov renewal process: Definitions and preliminary properties. *Annals of Mathematical Statistics* 32: 1231–1242.

Sackett, D. L. and Torrance, G. W. (1978). The utility of different health states as perceived by the general public. *Journal of Chronic Disease* 31: 697–704.

Spiegelhalter, D., Thomas, A., and Gilks, W. (1993). BUGS, Bayesian Inference using Gibbs Sampling. Technical Report, MRC Biostatistics Unit, Cambridge, UK.

Stangl, D. (1995). Modeling and decision making using Bayesian hierarchical models. *Statistics in Medicine* 14(20): 2173–2190.

Tanner, M. A. (1991). *Tools for Statistical Inference–Observed Data and Data Augmentation Methods*, Lecture Notes in Statistics, 67. Springer-Verlag, New York.

Thomas, A., Spiegelhalter, D., and Gilks, W. (1992). BUGS, a program to perform Bayesian inference using Gibbs sampling. In *Bayesian Statistics 4*, ed. J. M. Bernardo, J. O. Berger, A. P. Dawid and Smith, A. F. M. Oxford University Press, New York.

van der Duyn Schouten, F. A. (1983). *Markov Decision Processes with Continuous Time Parameter*. Mathematisch Centrum, Amsterdam.

Wakefield, J. C., Smith, A. F. M., Racine-Poon, A., and Gelfand, A. (1994). Bayesian analysis of linear and non-linear population models using the Gibbs sampler. *Applied Statistics* 43: 201–221.

Weinstein, M. C., Fineberg, H. V., Elstein, A. S., Frazier H. S., Neuhauser, D., Neutra, R. R., and McNeil B. J. (1980). *Clinical Decision Making*. Saunders, Philadelphia.

Wolpert, R. L. and Mengersen, K. L. (1995). Adjusted and synthetic likelihood for combining empirical evidence. Discussion paper, Institute of Statistics and Decision Sciences. Duke University, Durham, NC.

Wolpert, R. L. and Warren-Hicks, R. (1992). Bayesian hierarchical logistic models for combining field and laboratory survival data. In *Bayesian Statistics 4*, ed. J. M. Bernardo, J. O. Berger, A. P. Dawid and Smith, A. F. M. Oxford University Press, New York.

Zeger, S. L. and Karim, M. R. (1991). Generalized linear models with random effects: A Gibbs sampling approach. *Journal of the American Statistical Association* 86: 79–86.

8

Dose-Response Analysis of Toxic Chemicals

Vic Hasselblad   Duke University, Durham, North Carolina
Annie M. Jarabek   U.S. Environmental Protection Agency,
Research Triangle Park, North Carolina

1 INTRODUCTION

In 1983, the National Academy of Sciences (NAS) published a report entitled *Risk Assessment in the Federal Government: Managing the Process* (National Research Council, 1983). The NAS had been charged with evaluating the process of risk assessment as performed at the federal level in order to determine the "mechanisms to ensure that government regulation rests on the best available scientific knowledge and to preserve the integrity of scientific data and judgments" so that controversial decisions regulating chronic hazards could be avoided. The NAS recommended a framework that separated the scientific aspects of risk assessment from the policy aspects of risk management.

Dose-response assessment, the estimation of the relation between the magnitude of exposure and the occurrence of the adverse health effects or toxicity end points in question, is a key component of risk assessment. The term "dose" in this context refers to an applied external concentration and not an internal measure of exposure. Thus, inhalation exposure concentration and dose are used interchangeably herein. Usually the dose-response relationship is determined from data from experiments in which the exposures were controlled. Once the dose-response estimate is calculated, the estimate is compared to the exposures measured or estimated for the assessment scenario of interest, e.g., environmental levels at a hazardous waste site. This comparison between the estimated exposure judged to be associated with a minimum of adverse risk (dose-response estimate) and an exposure measure or estimate is called risk characterization. The risk characterization is then used as a basis for risk management decisions. For example, if the environmental exposure is well above the risk estimate, control technology may be requested to mitigate the exposure.

Current approaches to dose-response assessment of both noncancer and cancer toxicity are briefly reviewed here. Case examples of typical toxicity data are then analyzed using the various approaches and their results are contrasted with those of a proposed Bayesian procedure.

2 CURRENT DOSE-RESPONSE ANALYSIS APPROACHES

At this time, different approaches are used for noncancer versus cancer data analysis based on assumptions regarding thresholds for toxicity. As the mechanistic determinants underlying the pathogenesis of different disease end points become understood, this artificial distinction between the approaches may dissolve. The current distinction between approaches is really based on the type of data obtained and the level of detail in reporting of standardized testing protocols.

2.1 Noncancer Toxicity

Noncancer toxicity refers to adverse health effects or toxic end points, other than cancer and gene mutations, that are due to the effects of environmental agents on the structure or function of various organ systems. Most chemicals that produce noncancer toxicity do not cause a similar degree of toxicity in all organs but usually demonstrate major toxicity to one or two organs. These are referred to as the target organs of toxicity for that chemical (Klaassen, 1986). Generally, based on understanding of homeostatic and adaptive mechanisms, most dose-response assessment procedures operationally approach noncancer health effects as though there is an identifiable threshold (both for the individual and for the population) below which effects are not observable. This concept assumes that a range of exposures from zero to some finite value can be tolerated by the organism without adverse effects. For example, there could be a large number of cells that perform the same or similar function whose population must be significantly depleted before an adverse effect is seen.

This threshold will vary from one individual to another, so that there will be a distribution of thresholds in the population. As a result it is difficult to model this threshold, and nonthreshold models are often used to fit noncancer toxicity data. Also, many of the end points used for analysis have an approximately linear relationship with dose. The threshold determination must be based on a toxicologist's scientific judgment as to what constitutes an adverse effect or one with minimal or acceptable severity. Because sensitive subpopulations (i.e., individuals with low thresholds) are frequently of concern in setting exposure standards, dose-response analysis procedures are aimed at estimating levels at which these sensitive individuals would not be expected to respond.

2.2 Noael/Loael Approach

The most common approach to derivation of dose-response estimates for noncancer toxicity is to determine the NOAEL (no-observed-adverse-effect level) or LOAEL (lowest-observed-adverse-effect level). These are determined for the specified adverse effect from the exposure levels of a given individual study. An NOAEL is defined as the highest exposure level tested at which neither a biologically nor a statistically significant increase in the frequency or severity of the specified adverse effect is produced and is therefore, by definition, a subthreshold level (Klaassen, 1986). An LOAEL is defined as the lowest exposure level that produces a biologically or statistically significant increase in the frequency or severity of the specified adverse effect and is by definition, therefore, a level above threshold. The entire array of NOAELs and LOAELs available for a chemical is then analyzed and an NOAEL or LOAEL that is most representative of the threshold for critical adverse effects is chosen for the basis of the derivation of a risk estimate. Details of how uncertainty factors for extrapolation of the chosen NOAEL or LOAEL are then used to derive human dose-response estimates such as oral reference doses (RfDs) or inhalation reference concentrations (RfCs) are provided elsewhere (Jarabek et al., 1990).

The NOAEL/LOAEL approach has several problems. Because the designation of the effect levels of individual studies is restricted to the exposure levels tested, the dose spacing used in the experiments can influence the resultant NOAEL and LOAEL values dramatically. The designation of an effect or not at a given level does not readily account for the number of animals or subjects tested. Although the slope of the dose-response curve is an important consideration in many applications, it is generally ignored and not computed statistically. Finally, the estimate is presented as a single number that does not adequately summarize the information in a particular study. Calculation of a confidence interval is precluded and therefore there is no inherent reflection of the variability of the data underlying the resultant estimate.

2.3 Benchmark Dose Approach

Crump (1984) and others recognized the problems of the NOAEL/LOAEL approach and proposed using a benchmark dose (BMD) approach. A benchmark dose is calculated from a prespecified (adverse) health effect. For example, a 10% increase in the incidence of a specific developmental abnormality (e.g., missing limb) in exposed rats versus control rats is commonly judged to be abnormal and thus specified as an adverse health effect. An estimate of the dose that produces this specified effect is calculated (usually using maximum-likelihood estimation) and commonly called the BMD_{10}. The associated lower confidence limit is also calculated and is called the $BMDL_{10}$. This lower confidence limit is the benchmark dose and could be used in a manner analogous to the NOAEL described above. The issue of whether a $BMDL_{10}$ should serve as an NOAEL or LOAEL for the derivation of dose-response estimates remains unresolved (Barnes *et al.*, 1994). Resolution awaits systematic study of the approach to different end points (e.g., respiratory tract, liver, and kidney toxicities), different outcome measures (e.g., quantal or continuous measures), and peculiarities of different types of data sets (e.g., thresholds or nonzero backgrounds). Nevertheless, the approach is a promising improvement on the NOAEL/LOAEL approach and addresses many of the limitations described above. Because the use of the model allows interpolation between or below the experimental levels to estimate an adverse effect level, additional testing to determine adversity levels may be obviated. This is a particular advantage in that fewer animals may need to be sacrificed and saved monies may be devoted to competing priorities.

The concept of a specified health effect is not new and is related to the concept of "relative potency." Finney (1962) defines relative potency in his description of a direct assay. A direct assay is one in which "that doses of the standard and test preparations sufficient to produce a specified response are directly measured. The ratio between these doses estimates the potency of the test preparation relative to the standard." The choice of a "specified response" is key to the definition. More specific terminology has been devised for quantal assays. The terms median effective concentration (EC_{50}) and median lethal concentration (LC_{50}) refer to concentrations that will produce an effect or death, respectively, in 50% of the subjects. Several different designs are available for estimating an EC_{50} or an LC_{50} (Kalish, 1990). As discussed by Finney, these effects need not be at the median—for example, the 10th percentile may be of interest and this would be referred to as the EC_{10} or LC_{10}. Again, these are based on predetermining a specified response of interest.

Calculating an EC_{10} or an EC_{50} requires choosing a model. For dichotomous outcome measures, Crump (1984) uses a quantal polynomial regression model (also known as a multistage model):

$$P(x) = \gamma + (1 - \gamma) \left(1 - \exp\left(- \sum_{j=1}^{k} \beta_j(x - \alpha)^j \right) \right) \tag{1}$$

where $\alpha > 0$ and $\beta_j > 0, j = 1, \ldots, k$. A second model used by Crump (1984) is the quantal Weibull model:

$$P(x) = \gamma + (1 - \gamma)(1 - \exp(-\alpha x^\beta)) \tag{2}$$

where $\alpha > 0$ and $\beta > 0$. The third model described by Crump (1984) is the log-probit model or the log-logistic model:

$$P(x) = \gamma + (1 - \gamma) \Phi(\alpha + \beta \log(x)) \tag{3}$$

where $\Phi(x)$ is the cumulative normal distribution or the cumulative logistic distribution. Once the specified response and the appropriate model have been selected, the corresponding concentration can be calculated using maximum-likelihood methods. Finally, the lower 95% confidence limit (BMDL) can be calculated based on likelihood ratios, and it is this value that is usually reported as the benchmark dose (see Crump and Howe, 1985). Studies with smaller sample sizes tend to have lower BMDL values because of the uncertainty in the estimate.

Although the BMD approach eliminates many of the problems associated with the NOAEL/LOAEL approach, it has some problems of its own. The benchmark dose is a single number and there is no obvious method for comparing or combining benchmark doses calculated from similar studies using the same specified health effect. Sometimes there are difficulties in the estimation of the BMDL resulting from using the asymptotic normality of maximum-likelihood estimates, and these difficulties were recognized by Crump and Howe (1985). Confidence intervals for the lower limits can be incorrect by orders of magnitudes. As an alternative, Crump and Howe (1985) suggest constructing the confidence limits using the asymptotic distribution of the likelihood ratio. This is accomplished by reparameterizing the likelihood function using the BMD as a variable. The likelihood function is then calculated for different values of the BMD while maximizing the other (nuisance) parameters. The lowest value of the BMD producing a likelihood ratio test chi-squared value that is just significant at the specified level (e.g., 0.05) is the BMDL.

3 CANCER TOXICITY

The identification of a threshold distinguishes approaches for noncancer toxicity from those for cancer. Cancer is typically addressed as a nonthreshold process. Guidelines on how to perform a cancer dose-response assessment, such as information on how to select the appropriate data for analysis based on considerations of data quality, relevance to human exposure, biological relevance, and exposure route, are provided elsewhere (U.S. Environment Protection Agency, 1987a; Jarabek and Farland, 1990).

The U.S. Environmental Protection Agency (EPA) has developed a system for stratifying the weight of the evidence that a given chemical is carcinogenic based on the available human and animal data. Agents that are classified in group A (carcinogenic to humans) or B (probably carcinogenic to humans; B1 indicates that limited human data are available to warrant this classification and B2 indicates sufficient evidence in laboratory animals with inadequate data in humans) in the weight-of-evidence stratification are regarded as suitable for quantitative risk assessment.

The quantitative problem is to discern the real dose-response curve in the dose (exposure) range of inference (Jarabek and Farland, 1990). Different mathematical models used for extrapolation may fit the observed data and yet lead to large differences in the predicted risk at low doses. The choice of the low-dose extrapolation model is governed by consistency with current understanding of the mechanism of carcinogenesis and is not solely based on goodness of fit to the observed tumor data. When data are limited, and when uncertainty exists regarding the mechanism of carcinogenic action, the linearized multistage model is commonly used by the U.S. Environmental Protection Agency (1988).

The linearized multistage model is an exponential model approaching 100% risk at high doses with a shape at low doses described by a polynomial function. When the polynomial is of first degree, the model is equivalent to a one-hit or linear model, so called because at low doses it produces an approximately linear relationship between dose and cancer risk. Other models that could be used include the Weibull, probit, logit, and gamma multihit models. Except for the one-hit model, these models all tend to give the characteristic S shapes of many biological experiments, with varying curvature and tail length. The lower bounds tend to parallel the curvature of the models themselves unless a procedure has been devised (such as the linearized multistage) to provide otherwise.

The slope factor is the slope of the straight line from the lower bound at zero dose to the dose producing a lower bound at 1% (per mg/kg/d). In order to estimate the concentration of air or water associated with certain designated levels of lifetime risk, a quantity known as the unit risk is calculated. This is the risk associated with one unit of drinking water (μg/L) or one unit of air breathed (μg/m^3). The unit risk is proportional to the inverse of the slope factor. The standardized duration of exposure is understood to be continuous lifetime exposure. Default assumptions for ventilation rates and water consumption are used (U.S. EPA, 1988). In current practice, calculation of the geometric mean of the unit risks for a number of studies is sometimes performed to combine data. However, neither slope factors nor unit risk estimates are parameter estimates in the classical sense; rather they are lower confidence limits of parameter estimates. As such, they have no statistical properties that would allow their combination. Thus, this calculation of a geometric mean represents an ad hoc attempt at data combination.

3.1 Bayesian Approach

Given the difficulties of using classical estimation methods to describe lower bounds for concentrations producing specified health effects, it is not surprising that Bayesian methods have been proposed by Hoadley (1970) and Buonaccorsi and Gatsonis (1988). Their approach is to place a prior distribution on the concentration producing a specified effect so that the posterior of the dose is integrable. Although this method solves the problems of negative slope estimates, it does not appear to be the natural way to define the prior distribution.

As an alternative to the above method, a Bayesian approach using the confidence profile method as described by Eddy *et al.* (1992) is proposed. The confidence profile method (CPM) is a very general method for combining virtually any kind of evidence about various parameters, as long as those parameters can be described in some model. Models consist of three elements: (1) basic parameters, (2) functional parameters, and (3) likelihood functions relating evidence to basic or functional parameters. As applied to the problem of estimating the concentration producing a specified health effect, the parameters of the dose-response models of Crump (1984) are the basic parameters. These basic parameters (basic because they are not functions of other parameters in the model) must have priors defined, and we usually choose these to be standard noninformative priors. For example, the log-logistic model [Eq. (3)] without a nonzero background would use a two-dimensional uniform prior for α (the intercept) and β (the slope). Because most risk assessments are made assuming no positive benefit from a potentially toxic substance, the prior for beta is restricted to be positive. Nonzero background rates (γ) can have a Jeffries (Berger, 1985) or uniform prior. The sample size for the zero dose is usually quite large and the choice of a prior on the background rate has a minimal effect. The parameter of interest, the concentration producing a specified health effect, is a functional parameter and is determined from the priors and likelihood function. For a discussion of likelihood functions see Berger and Wolpert (1984). The marginal posterior distribution for this parameter can be calculated, and this distribution represents our belief about the concentration that could cause this specified health effect. Percentiles of the distribution can be compared with those calculated from other specified health effects. Multiple studies of the same health effect can be combined assuming conditional independence in order to compute a posterior distribution based on all studies.

The models used by the BMD approach and the Bayesian approach are the same, and so there are similarities between the two approaches. However, the BMD approach results in a single number that summarizes an experiment, whereas the Bayesian approach results in a posterior distribution describing each experiment. Visual display of the distribution provides much information about the usefulness of the data for quantitative dose-response estimation. Skewness and nonnormality of the posterior distribution are readily observed, whereas these properties are not observed when using tabular or single summary statistics. Finally, only the Bayesian approach uses formal statistical methodology to combine information.

4 CHEMICAL CASE STUDIES

The advantages that the Bayesian approach offers to dose-response analysis of noncancer toxicity and cancer data will be illustrated using case examples of different chemicals. The examples were chosen because the data illustrate cases in which non-Bayesian approaches for analysis are inadequate. The chemicals chosen have either undergone quantitative dose-response assessment performed by the U.S. EPA or have data typical of the type routinely available for such analyses.

4.1 Noncancer Dose-Response Assessment: Ingestion of Acrylamide

All acrylamide in the environment is anthropogenic. It is released primarily into the environment in wastewater during its production and during its use in the synthesis of dyes or in the manufacture of polymers, adhesives, paper, paperboard, and textile additives. Acrylamide may also be released into water from liquid-solid separation for processing minerals in mining or the treatment of wastewater with polyacrylamide as a flocculating agent. Other releases may result from the disposal of the solid on land or its leaching as a residue from polyacrylamides. Human exposure is primarily occupational via dermal contact and inhalation, although oral exposure of the general public has resulted from water contaminated as described.

Acrylamide is toxic to the peripheral nervous system and chronic exposure can result in polyneuritis with sensory changes in the limbs, weakness, and ataxia. Typical data on peripheral neurotoxicity of acrylamide that might be used for dose-response analysis are those of Johnson *et al.* (1986). Male and female Fischer 344 rats were maintained with drinking water containing acrylamide at dosages of 0.01, 0.1, 0.5, or 2.0 mg/kg body weight (BW) per day for 13 weeks. These levels were chosen because another study had shown clinical effects at 20 mg/kg BW/d and pathologic alterations in nerves visible on electron microscopy at 1.0 mg/kg BW/d. Degeneration of the tibial nerve was observed on histopathology. The administered dose levels and respective incidences of toxic response are shown in Table 1.

For these data, designation of the LOAEL and NOAEL would be difficult. No statistical significance was achieved at any dose level. Based on the severity (e.g., "moderate" versus "slight" tibial nerve degeneration), the LOAEL would likely be assigned to the 2.0 mg/kg/d level and the NOAEL at 0.5 mg/kg/d.

These data do show, however, a marginally significant trend increase in the number of cases of tibial nerve degeneration as a function of the dose. For example, the likelihood ratio test for an effect assuming a log-logistic model with a nonzero background gives a chi-squared value of 5.084 for 2 degrees of freedom ($p = .0787$). Using Fisher's exact test, comparing the control group with the highest exposure group gives a two-sided p-value of .18 and a one-sided p-value of .09. Since there is a nonzero background rate, a log-logistic function with a nonzero background [Eq.

Table 1 Effect of Acrylamide on Tibial Nerve Degeneration in Male Rats

Exposure (mg/kg BW/d)	Incidence of changes	Severity	
		Moderate	Severe
0.00	9/60	8	1
0.01	6/60	5	1
0.10	12/60	12	0
0.50	13/60	13	0
2.00	16/60	12	4

Source: Data of Johnson *et al.* (1986).

(3)] is a logical choice for a model to use the BMD approach. The concentration to produce a 10% increase in rats with nerve damage, X, is given by

$$\log(X) = \frac{-\log(.1/.9) - \alpha}{\beta} \tag{4}$$

A 10% increase in response is chosen for illustration as a typical candidate response level for the BMD approach to quantal data (Barnes *et al.*, 1995).

If background, γ, is fixed at its maximum-likelihood estimate (0.131), then the joint likelihood for α and β is as shown in Figure 1. A significant fraction of the likelihood for β extends below zero. It is a relatively straightforward process to obtain the maximum-likelihood estimates using a modified Gauss-Newton method as described by Berndt *et al.* (1974). The estimated benchmark dose is 0.584 (mg/kg/d) and the lower 95% confidence limit for X using the methods of Crump and Howe (1985) is 0.0308 (mg/kg/d).

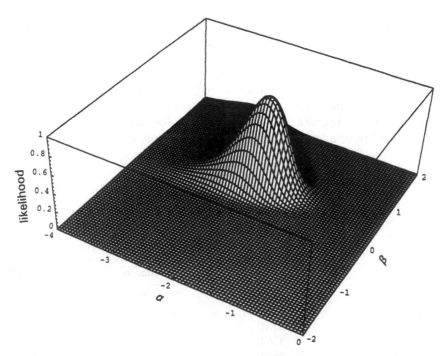

Figure 1   Likelihood for data of Johnson *et al.* (1986) as a function of α and β with γ fixed at 0.131.

Using the Bayesian approach with the same model, the three natural parameters are α, β, and γ. These are the basic parameters of the problem and the concentration (producing a 10% increase in tibial nerve damage), X, is a functional parameter. A uniform prior over the interval (0, 1) was used for γ. A natural prior distribution for (α, β) is the uniform prior over the half-plane where β is positive. The posterior distribution for X was calculated by reparameterizing the likelihood function and integrating out the remaining nuisance parameter. This posterior distribution is shown in Figure 2. The distribution illustrates the great uncertainty in using these data to establish the specified health effect and its lower limit. The distribution has a mode at zero and 95% credible set limits of 0.000012 to 27.6 (mg/kg/d). The distribution does not have a finite mean or variance. Thus, the Bayesian approach clearly demonstrates the uncertainty that may not be apparent when the raw data are used to designate a NOAEL or when the BMD approach is used to calculate a surrogate. Additional experiments with larger sample sizes could be expected to increase the lower limit.

4.2 Noncancer Dose-Response Assessment: Inhalation of *n*-Hexane

n-Hexane is produced during the cracking and fractional distillation of crude oil and is used in such applications as printing of laminated products;

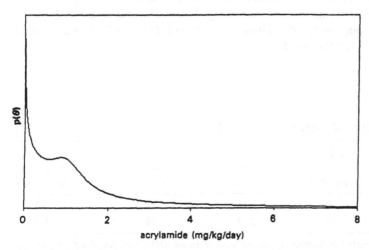

Figure 2 Posterior density for concentration of acrylamide producing a 10-percent increase in tibial nerve damage for data of Johnson *et al.* (1986).

vegetable oil extraction; as a solvent in glues, paints, varnishes, and inks; as a diluent in the production of plastics and rubber; and as a minor component of gasoline. An estimated 2.5 million workers have been exposed occupationally. n-Hexane is known to cause peripheral neuropathy. The initial symptoms are symmetrical sensory numbness and parasthesias of distal portions of the extremities. Recovery usually occurs after removal from exposure but severe cases may retain some neurological deficits. Outbreaks of this type of neurotoxic syndrome have been reported in the laminating, pharmaceutical, shoe making, furniture production, and adhesive bandaging manufacture industries.

An inhalation reference concentration (RfC) was derived for n-hexane based on adverse health effects observed in two different studies (U.S. EPA, 1990). The first study was a human occupational study of Sanagi *et al.* (1980) and the second was an inhalation study in mice as reported by Dunnick *et al.* (1989). The data for these studies were subjectively combined as the basis for the RfC derivation.

Sanagi *et al.* (1980) conducted a cross-sectional occupational investigation of neurotoxicity in two age-matched cohorts. Both groups underwent clinical neurological examinations, and nerve stimulation studies were performed with a surface electrode and included a number of parameters. Recordings were made in a temperature-controlled room and subjects were acclimated 30 minutes prior to examinations. A decrement in mean motor nerve conduction velocity (MNCV) is indicative of neurotoxicity and considered to be an adverse effect. The measured mean MNCV values (m/sec) for the two cohorts are shown in Table 2. The mean MNCV for the exposed group was significantly decreased compared to that for the control group ($p < .05$). Thus, the value of 73 mg/m^3 was designated as an LOAEL for this study.

Table 2 Effect of n-Hexane on MNCV

Cohort	HEC exposure[a] (mg/m^3)	Sample size	Mean MNCV (m/sec)	Standard deviation
Controls	0	14	48.3	2.1
Exposed	73	14	46.6	2.3

[a] Exposure converted to human equivalent concentration (HEC) as described by Jarabek and Hasselblad (1992).
Source: Data of Sanagi *et al.* (1980).

The observed decrease in mean MNCV (1.7 m/sec) between control and exposed cohorts in the Sanagi *et al.* (1980) study was a change of 4% and, as mentioned above, was statistically significant. Thus, this change can be specified as the adverse health effect because decrements in MNCV are considered adverse and this was a statistically significant change. A linear model is reasonable to assume, as dose-response curves are usually linear in the low dose range. The slope of the line is then estimated by the difference in mean MNCV values divided by the difference in exposure (73 − 0 = 73 mg/m³). Because the effect of *n*-hexane was assumed not to be beneficial, a flat prior over the negative half-line was used, resulting in the posterior shown in Figure 3. Finally, the posterior for that concentration producing a 4% decrease in MNCV is shown in Figure 4. The mode is 54 mg/m³ and the 2.5 and 97.5 percentiles are 36 and 442 mg/m³, respectively.

The second study used for dose-response analysis of hexane was an inhalation study in B6C3F1 mice (10/sex/concentration) exposed to 0, 500, 1000, 4000, and 10,000 ppm *n*-hexane 6 hours/day, 5 days/week for 13 weeks. Histopathologic changes included mild inflammatory, erosive, and regenerative lesions of the olfactory epithelium of the nasal cavity in a concentration-related manner in female mice. The summary of these results in female mice is shown in Table 3.

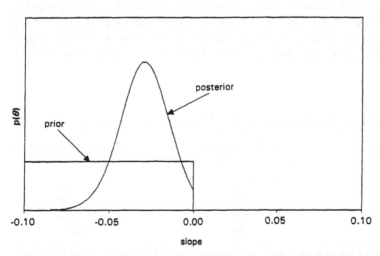

Figure 3 Prior and posterior densities for the slope of MNCV per-unit change in *n*-hexane based on Sanagi *et al.* (1980).

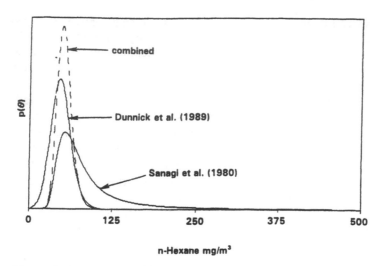

Figure 4 Posterior distributions for concentration of *n*-hexane producing a specified health effect.

The specified health effect for the Dunnick *et al.* (1989) study is designated as a 10% increase in the rate of nasal lesions. The 10% increase in
response was chosen for illustration as a typical candidate level for the
BMD approach to quantal data (Barnes *et al.*, 1994). The model was the
log-logistic model used earlier [Eq. (3)]. The resulting posterior for the
concentration of *n*-hexane producing the specified health effect is also
shown in Figure 4. The resulting distribution has a mode of 40 mg/m^3,

Table 3   Effect of *n*-Hexane on Nasal Lesions in Female Mice

Exposure (ppm)	HEC^a (mg/m^3)	Incidence of changes
0	0	0/10
500	38	2/10
1,000	77	1/10
4,000	307	9/10
10,000	768	10/10

^{a} Exposure converted to human equivalent concentration (HEC) as described by Jarabek
and Hasselblad (1992).
Source: Data of Dunnick *et al.* (1989).

and the 2.5 and 97.5 percentiles are 24 and 78 mg/m^3, respectively. The shape of this posterior distribution, in contrast to that of the Sanagi *et al.* (1980) study, clearly illustrates that these data generated from an investigation with an adequate number and test concentrations produce a much narrower distribution with reduced variance.

Although the two end points are different, the concentrations producing a specified health effect can be combined. In this case, the two different end points (neurotoxicity versus respiratory toxicity in humans versus laboratory animals) gave overlapping posterior distributions. The marginal posterior distribution for the concentration producing the specified health effect was calculated by numerical integration for each experiment. The combined posterior distribution was calculated numerically as the product of the posterior distributions assuming independence. The calculated posterior distribution has a mode of 47 mg/m^3, and the 2.5 and 97.5 percentiles are 33 and 76 mg/m^3, respectively. The resultant posterior distribution for the combined evidence is not drastically different from the individual distributions from which it was derived (see Figure 4). This may be due to the fact that both investigations of *n*-hexane measured very sensitive end points that are probably sentinel for more serious effects that might occur at higher concentrations. Target organs for noncancer toxicity are often not concordant across species, whereas the severity (e.g., magnitude of dysfunction) often is. Heywood (1981, 1983) showed poor correlation of target organ toxicity between rodent and nonrodent species for a number of compounds. The differences may be due to differences in types of assays performed across the species or to species differences in pharmacokinetics and/or pharmacodynamics. When data are not comparable with respect to assayed end points, but individual studies investigate very sensitive end points (i.e., near the threshold or subthreshold region), the combination of these data may provide a more likely estimate of the region of concern. The EPA used the combination of the same two studies to derive the inhalation RfC for *n*-hexane. The quantitative estimate was based on the LOAEL of the Sanagi *et al.* (1980) study. The results comparing the Bayesian approaches with the other approaches are shown in Table 4.

4.3 Cancer Dose-Response Assessment: Ingestion of Aldrin

Aldrin (1,2,3,4,10,10-hexachloro-1,4,4α,5,8,8α-hexahydro-1,4-*endo,exo*-5,8-dimethanonaphthalene) and dieldrin (1,2,3,4,10,10-hexachloro-6,7-epoxy-1,4,4α,5,6,7,8,8α-octahydro-1,4-*endo,exo*-5,8-dimethanonaphthalene) are the common names of two structurally similar insecticides. Aldrin was formerly used against termites and soil-dwelling pests such as

Table 4 HEC^a (mg/m^3) of n-Hexane Producing Specified
Health Effects

			Benchmark		Bayesian	
Study	NOAEL	LOAEL	BMDL	BMD	5%ile	Mode
Sanagi et al. (1980)	—	73	36	66	40	54
Dunnick et al. (1989)	38	77	20	43	24	46
Combined	—	75	29	54	37	51

^{a} Exposure converted to human equivalent concentration (HEC) as described by Jarabek
and Hasselblad (1992).

ants and grubs. Its presence in the environment has resulted from these
uses as an insecticide or from leaking storage containers at waste sites.
Although moderately persistent, aldrin converts into dieldrin, so many of
these past releases may now be present as dieldrin. For most people,
exposure to aldrin or dieldrin occurs when they eat contaminated foods.
Contaminated foods might include fish or shellfish from contaminated
waters, root crops, dairy products, or meats from animals that grazed on
contaminated sites. Exposure might also occur via breathing or touching
contaminated soil at contaminated sites.

Aldrin and dieldrin cause similar adverse health effects. These effects
include nervous system disorders such as convulsions. Laboratory animal
data show adverse changes in the kidney, liver, and male reproductive
systems. Some evidence suggests that it causes developmental effects.
Because aldrin has also been shown to cause liver tumors in both female
and male mice, it has been classified as B2—a probable human carcinogen.
A quantitative assessment of the cancer risk of aldrin was based on two
studies of liver tumors in mice (U.S. EPA, 1987b). The study of Davis
(1965) gave results for both males and females, and the National Cancer
Institute (1978) study was restricted to males. The results for the two
studies are shown in Table 5.

These studies used so few dose levels that models like the Weibull
and log-logistic could not be fitted to the data. Instead, a quantal linear
model [Eq. (1) with $k = 1$] was fitted to the data. The estimated posterior
distribution for the concentration of aldrin producing a 10% increase in
liver tumors is shown in Figure 5. The results of analyzing the three studies
using various approaches are shown in Table 6. Note the similarity between the BMD and Bayesian results.

Table 5 Studies of the Effect of Aldrin on Liver Tumors in Mice

Study	Administered dose (ppm)	Human equivalent dose[a] (mg/kg/d)	Tumor incidence
Davis (1965), females	0	0.000	2/53
	10	0.104	72/85
Davis (1965), males	0	0.000	22/73
	10	0.104	75/91
NCI (1978), males	0	0.00	3/20
	4	0.04	16/49
	8	0.08	25/45

[a] Human equivalent dose calculated as described in U.S. EPA (1987b).

The liver tumors were all similar in these three studies. In addition, five compounds structurally similar to aldrin (dieldrin, chlordan, heptachlor, heptachlor epoxide, and chlorendic acid) have induced malignant liver tumors in mice. Chlorendic acid has also induced liver tumors in rats. For these reasons, and because the slope factors for the studies were

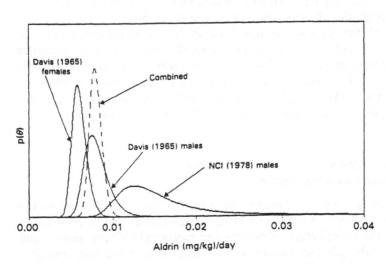

Figure 5 Posterior densities for concentration of aldrin producing a specified health effect.

Table 6 Exposure to Aldrin Producing Specified Health Effect

Study	Slope factor[a]	Benchmark[b]		Bayesian[b]	
		BMDL	BMD	5%ile	Mode
Davis (1965), females	23	.0048	.0060	.0047	.0058
Davis (1965), males	18	.0061	.0079	.0060	.0075
NCI (1978), males	12	.0096	.0139	.0103	.0132
Combined	17	.0066	.0087	.0068	.0078

[a] Slope factor in units per mg/kg/d.
[b] Concentrations in mg/kg/d.

similar, EPA combined the data by taking a geometric mean to calculate the quantitative slope factor estimate at 17 per mg/kg/d. The corresponding unit risk value is 0.00049 mg/kg/d. The value producing a 10% risk (0.0049 mg/kg/d) can then be calculated and compared with the benchmark and Bayesian values.

The resulting posterior for combining the data using the Bayesian approach is shown as a dashed line in Figure 5. The mode of the combined posterior distribution is .0078 and the lower 5th percentile of the combined posterior is .0068 mg/kg/d. This value is somewhat larger than that calculated by combining the data using a geometric mean.

Although the results from the three studies were reasonably similar in comparison with results for other chemicals, there was some evidence of a lack of homogeneity (see Figure 5). A natural extension of the models presented in this chapter would be the use of hierarchical models. In particular, the reader should refer to the chapters by Clyde *et al.* (Chapter 11), DuMouchel, Waternaux, and Kinney (Chapter 17), and Stangl (Chapter 16).

4.4 Cancer Dose-Response Assessment: Ingestion of Acrylonitrile

Acrylonitrile is a high-production chemical that may be released to the environment as fugitive emissions or in wastewater during its production and use in the manufacture of acrylic and modacrylic fibers, resins, and other chemicals. It is also found in automotive exhaust. Residual amounts may be released from clothing, furniture, etc. made with polyacrylic fiber or leached from polyacrylonitrile containers. Acrylonitrile is also released during the burning of plastic containers and is found in cigarette smoke. Humans are exposed primarily in the workplace via inhalation. The general public may be exposed via the air due to dispersal of acrylonitrile from industrial sources, outgassing from acrylic fibers, auto exhaust, and cigarette smoke and via food in contact with plastic containers from which the chemical has leached. Acrylonitrile has been found in water and soil of some hazardous waste sites, so there is also a potential for exposure by breathing contaminated air or water at such sites.

Significant increases in the incidence of lung cancer in exposed workers and the observation of tumors in laboratory rats exposed by various routes (drinking water, gavage, and inhalation) have resulted in the classification of acrylonitrile as a B1—a probable human carcinogen. A quantitative cancer assessment was based on three studies of various kinds of tumors (brain and spinal cord astrocytomas, zymbal gland carcinomas, and stomach papillomas/carcinomas) in rats dosed orally with acrylonitrile in drinking water (U.S. EPA, 1987c). Two of the studies were done by Biodynamics (1980a,b) on two different strains of male rats. A third study, by Quast *et al.* (1980), was also done on male rats. The results are shown in Table 7.

Table 7 Effect of Acrylonitrile in Drinking Water on Various Tumors in Male Rats

Study	Administered dose (ppm)	Human equivalent dose[a] (mg/kg/d)	Tumor incidence
Biodynamics (1980a),	0	0.00	6/100
Sprague-Dawley rats	1	0.02	6/98
	100	1.36	36/98
Biodynamics (1980b),	0	0.00	5/200
Fischer 344 rats	1	0.01	4/100
	3	0.04	5/100
	10	0.14	7/100
	30	0.43	20/100
	100	1.39	34/100
Quast *et al.* (1980),	0	0.00	4/80
Sprague-Dawley rats	35	0.58	18/47
	100	1.46	36/48
	300	3.62	45/48

[a] Human equivalent dose calculated as described in U.S. EPA (1987c).

Table 8 Exposure to Acrylonitrile Producing Specified Health Effect

Study	Slope factor[a]	Benchmark[b]		Bayesian[b]	
		BMDL	BMD	5% tile	Mode
Biodynamics (1980a)	0.40	0.26	0.34	0.29	0.38
Biodynamics (1980b)	0.40	0.27	0.36	0.27	0.33
Quast et al. (1980)	0.99	0.11	0.13	0.11	0.13

[a] Slope factor in units per mg/kg/d.
[b] Concentrations in mg/kg/d.

Because the slope factors for these studies differed by only approximately twofold, the EPA chose to combine these data using a geometric mean in order to derive a quantitative estimate. The combined slope factor is 0.54 per mg/kg/d and the corresponding unit risk value is 1.85 per mg/kg/d. The estimate of the dose to produce a 10% risk is 0.19 mg/kg/d.

Bayesian estimates of the concentration corresponding to a specified health effect are shown in Table 8, and a graph of the posterior distributions is shown in Figure 6. The results of Quast et al. (1980) are clearly quite different from those of the two Biodynamics (1980a,b) studies. The

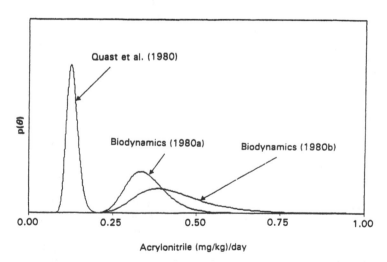

Figure 6 Posterior densities for concentration of acrylonitrile producing a specified health effect.

Quast *et al.* (1980) data have a narrower confidence interval, probably because of the wider range of administered dose levels. The differences between the studies may represent interlaboratory variability. The same strain of rats was used in the Biodynamics (1980a) and Quast *et al.* (1980) studies, and thus strain differences are unlikely to be the reason for the observed variability. Quast *et al.* (1980) identified a larger range of tumors, possibly indicating a more thorough histopathologic examination. The range of tumor types observed in all three studies, however, cannot be linked to similar mechanisms of action. Without this biologic plausibility, and given the difference in qualitative characteristics of the studies, it makes little sense to combine the results. Using the Quast *et al.* (1980) data alone results in an estimate for the dose producing a 10% risk of 0.11 mg/kg/d.

5 DISCUSSION

These results demonstrate the ability of the Bayesian approach to address the problems of the determination of dose-response estimates for toxic substances. The Bayesian approach produces posterior distributions that can be compared readily, both visually and statistically. Risk assessors routinely subjectively synthesize data when attempting to evaluate the overall data array or toxicity profile of a given toxic agent. In situations in which the biologic motivation to combine data exists, the Bayesian approach is a straightforward method for combining the data. Combining results gives narrower limits on an estimate, thereby increasing accuracy.

Case examples have illustrated specific attributes of the Bayesian approach. In the case of acrylamide, the Bayesian approach graphically shows the uncertainty with some data sets that is not apparent when using the other two approaches. The case of *n*-hexane illustrated the efficacy of the approach for comparing and combining data. The data represented both continuous and quantal outcome measures. The Bayesian method has the advantage that it describes the results in probability distributions that can be combined formally. The NOAEL/LOAEL and BMD approaches do not allow formal combination. The two case examples assessing cancer data, for aldrin and acrylonitrile, further illustrate issues involved in data combination that become apparent only using the Bayesian approach. In the case of aldrin, when data can be combined with some biologic motivation to do so, the Bayesian approach reflects the greater confidence in the accuracy of the estimate with a narrower confidence range on the combined data. This increase in accuracy of the estimate is not achieved with the current EPA approach, which combines data on the basis of the geometric mean. The case example of acrylonitrile illustrates the advantage of the graphical display of the data provided by the Bayesian approach. In this case, the display clearly indicates that it is inappropriate to combine studies, a fact that is not apparent when comparing slope factors alone.

Other methods have been suggested for combining studies of toxic effects of a compound. Some risk assessments (U.S. EPA, 1987b) have combined the slope factors by calculating a geometric mean. Because the slope factors are inversely proportional to the BMDLs, this method is equivalent to combining BMDLs using geometric means. Although this can be calculated easily, it is not a logical computation to make. Consider a hypothetical situation in which three studies were of the same size and by chance gave the same results for both the BMDs and BMDLs. The combined value would be the same as the individual values, and yet logic would suggest that the spread between the BMD and its lower limit should shrink as more information becomes available. This method also does not assign larger weights to studies with larger sample sizes. Neither of these problems exists when using Bayesian techniques.

Two other statistical approaches were given by DuMouchel and Harris (1983) and Vater *et al.* (1993). The method of DuMouchel and Harris (1983) uses a hierarchical Bayesian model, but it assumes that all studies can be reduced to a slope estimate and that the distribution of these slopes is normal. For many examples presented earlier this is clearly an unreasonable assumption. Vater *et al.* (1993) assume that studies are similar enough so that they can be described by similar models. When this assumption is met, the models of Vater *et al.* (1993) can be used with the Bayesian method described in this chapter.

In conclusion, the Bayesian approach offers significant advantages over current approaches used for dose-response assessment. Specific advantages include the graphical display of variability in the concentrations producing specified health effects and the ability to combine such data formally, regardless of type of outcome measure, when determined appropriate. Additional research to develop guidance or criteria for establishing biological motivation for data combination (e.g., a tiered approach to frame mechanistic data in order to gain insight into the toxic action of a chemical) is warranted. Research on statistical criteria for data combination is also required, as standard statistical tests of homogeneity may be insufficient.

REFERENCES

Barnes, D. G., Daston, G. P., Evans, J. S., Jarabek, A. M., Kavlock, R. J., Kimmel, C. A., Park, C., and Spitzer, H. L. (1995). Benchmark Dose Workshop: Criteria for use of a benchmark dose to estimate a reference dose. *Regul. Toxicol. Pharmacol.* 21: 296–306.

Berger, J. O. (1985). *Statistical Decision Theory and Bayesian Analysis.* 2nd ed. Springer-Verlag, New York.

Berger, J. O. and Wolpert, R. L. (1984). The likelihood principle. In Lecture Notes—Monograph Series Volume 6, ed. S. S. Gupta. Institute of Mathematical Statistics, Hayward, CA.

Berndt, E., Hall, B., Hall, R., and Hausman, J. (1974). Estimation and inference in nonlinear structural models. *Ann. Econ. Soc. Measure.* 3: 655–665.

Biodynamics, Inc. (1980a). A Twenty-four Month Oral Toxicity/Carcinogenicity Study of Acrylonitrile Administered to Spartan Rats in the Drinking Water, Vols. 1 and 2. Prepared by Biodynamics Inc., Division of Biology and Safety Evaluation, East Millstone, NJ, for Monsanto Company, St. Louis, MO.

Biodynamics, Inc. (1980b). A Twenty-Four Month Oral Toxicity/Carcinogenicity Study of Acrylonitrile Administered to Fischer 344 Rats in the Drinking Water, Vols. 1 and 2. Prepared by Biodynamics Inc., Division of Biology and Safety Evaluation, East Millstone, NJ, for Monsanto Company, St. Louis, MO.

Box, G. E. P. and Tiao, G. C. (1973). *Bayesian Inference in Statistical Analysis.* Addison-Wesley, Reading MA.

Buonaccorsi, J. P. and Gatsonis, C. A. (1988). Bayesian inference for ratios of coefficients in a linear model. *Biometrics* 44: 87–101.

Crump, K. S. (1984). A new method for determining allowable daily intakes. *Fundamental and Applied Toxicology* 4: 854–871.

Crump, K. and Howe, R. (1985). A review of methods for calculating confidence limits in low dose extrapolation. In *Toxicological Risk Assessment*, ed. Krewski D. CRC Press, Boca Raton, FL.

Davis, K. J. (1965). Pathology Report on Mice Fed Dieldrin, Aldrin, Heptachlor, or Heptachlor Epoxide for Two Years. Internal FDA Memorandum to Dr. A. J. Lehrman, July 19, 1965.

DuMouchel, W. H. and Harris, J. E. (1983). Bayes method for combining the results of cancer studies in humans and other species. *J. Am. Stat. Assoc.* 78: 293–315.

Dunnick, J. K., Graham, D. G., Yang, R. S., Haber, S. B., and Brown, H. R. (1989). Thirteen-week toxicity study of *n*-hexane in B6C3F1 mice after inhalation exposure. *Toxicology* 57: 163–172.

Eddy, D. M., Hasselblad, V., and Shachter, R. D. (1992). *The Statistical Synthesis of Evidence: Meta-Analysis by the Confidence Profile Method.* Academic Press, San Diego.

Fieller, E. C. (1940). *Journal of the Royal Statistical Society Supplement* 7: 1–64.

Finney, D. J. (1948). *Probit Analysis.* The University Press, Cambridge, UK.

Finney, D. J. (1962). *Statistical Methods in Biological Assay.* Hafner, New York.

Heywood, R. (1981). Target organ toxicity. *Toxicology Letters* 8: 349–358.

Heywood, R. (1983). Target organ toxicity II. *Toxicology Letters* 18: 83–88.

Hoadley, B. (1970). A Bayesian look at inverse linear regression. *Journal of the American Statistical Association* 65: 356–369.

Jarabek, A. M. and Farland, W. H. (1990). The U.S. Environmental Protection Agency's risk assessment guidelines. *Toxicology and Industrial Health* 6: 199–216.

Jarabek, A. M. and Hasselblad, V. (1991). Inhalation reference concentration methodology: Impact of dosimetric adjustments and future directions using the confidence profile method, *Proceedings of the 84th Annual Meeting of the Air & Waste Management Association*, Vancouver, Canada.

Jarabek, A. M., Menache, M. G., Overton, J. H., Jr., Dourson, M. L., and Miller, F. J. (1990). The U.S. Environmental Protection Agency's inhalation RfD methodology: Risk assessment for air toxics. *Toxicol. Ind. Health* 6(5): 279–301.

Johnson, K., S. Gorzinski, K. Bodner, *et al.* (1986). Chronic toxicity and oncogenicity study on acrylamide incorporated in the drinking water of Fischer 344 rats. *Toxicol. Appl. Pharmacol.* 85: 154–168.

Kalish, L. A. (1990). Efficient design for estimation of median lethal dose and quantal dose-response curves. *Biometrics* 46: 737–748.

Klaassen, C. D. (1986). Principles of Toxicology, In Klaasen, C. D., Amdur, M. O., and Doull, J., eds., *Casarett and Doull's Toxicology: The Basic Science of Poisons*, 3rd ed., Macmillan, New York.

National Cancer Institute. (1978). Bioassays of Aldrin and Dieldrin for Possible Carcinogenicity. DHEW Publication No. (NIH) 78-821, NCI Carcinogenesis Terchnnical Report Series No. 21, NCI-C6-TR-21.

National Research Council. (1983). *Risk Assessment in the Federal Government: Managing the Process*. National Academy Press, Washington, DC.

Quast, J. F., Wade, C. E., Humiston, C. G., Carreon, R. M., Hermann, E. A., Park, C. N., and Schwetz, B. A. (1980). A two-year toxicity and oncogenicity study with acrylonitrile incorporated in the drinking water of rats. Prepared by the Toxicology Research Laboratory, Health and Environmental Sciences, Dow Chemical USA, Midland, MI, for the Chemical Manufacturers Association, Washington, DC.

Sanagi, S., Seki, Y., Sugimoto K., and Hirata, M. (1980). Peripheral nervous system functions of workers exposed to *n*-hexane at a low level. *Int. Arch. Occup. Health* 47: 69–70.

Tsutakawa, R. K. (1982). Statistical methods in bioassay. In *Encyclopedia of Statistical Sciences*, Vol. 1, ed. Kotz, S. and Johnson, N. L. pp. 236–243. Wiley, New York.

U.S. Environmental Protection Agency. 1987a. The Risk Assessment Guidelines of 1986. Office of Research and Development, Office of Health and Environmental Assessments, EPA Report No. EPA/600/8-87/045, Washington, DC.

U.S. Environmental Protection Agency. 1987b. Integrated Risk Information System (IRIS). Carcinogenicity Assessment for Lifetime Exposure to Aldrin. Online. (Verification date 03/22/87). Office of Health and Environmental Assessment, Environmental Criteria and Assessment Office, Cincinnati, OH.

U.S. Environmental Protection Agency. 1987c. Integrated Risk Information System (IRIS). Carcinogenicity Assessment for Lifetime Exposure to Acrylonitrile. Online. (Verification date 02/11/87). Office of Health and Environmental Assessment, Environmental Criteria and Assessment Office, Cincinnati, OH.

U.S. Environmental Protection Agency. 1988. EPA Approach for Assessing the Risks Associated with Chronic Exposures to Carcinogens. Integrated Risk Information System (IRIS). Online. Intra-Agency Carcinogen Risk Assessment Verification Endeavor (CRAVE) Work Group, Office of Health and Environmental Assessment, Environmental Criteria and Assessment Office, Cincinnati, OH.

U.S. Environmental Protection Agency. 1990. Integrated Risk Information System (IRIS). Reference Concentration for Inhalation Exposure for *n*-Hexane. Online. (Verification date 04/19/90). Office of Health and Environmental Assessment, Environmental Criteria and Assessment Office, Cincinnati, OH.

Vater, S. T., McGinnis, P. M., Schoeny, R. S., and Velazquez, S. F. (1993). Biological considerations for combining carcinogenicity data for quantitative risk assessment. *Reg. Tox. Pharm.* 18: 403–418.

9
Expected Utility as a Policy-Making Tool: An Environmental Health Example

Lara J. Wolfson   University of Waterloo, Waterloo, Ontario, Canada
Joseph B. Kadane and Mitchell J. Small   Carnegie Mellon University, Pittsburgh, Pennsylvania

ABSTRACT

Statistical decision theory can be a valuable tool when considering the results of a statistical analysis for health policy decisions. In this chapter we review briefly the history of decision theory in statistics and introduce methods for eliciting loss functions to be used when two parties with possibly conflicting interests need to reach a decision. We illustrate the methods with examples concerning public exposure to potential health risks from environmental contamination.

1 INTRODUCTION

Decision making under uncertainty has long been an aim of statistical inference. Much work has been done on quantifying uncertainty for the purpose of constructing statistical models of various phenomena. However, when it comes to translating those statistical models into the framework of making a decision that is optimal in some sense, the problem of accounting for the evaluation of intangibles, such as goodwill or a willingness to compromise, is very difficult. In this chapter we discuss how

Bayesian models can be used, along with elicited utilities, to provide a framework for two parties (under certain conditions) to negotiate an optimal compromise in the context of maximizing expected utility, without requiring explicit evaluation of intangibles. This method thus allows for consideration of some of the more qualitative aspects of decision making.

The outline of the Chapter is as follows. Section 2 outlines a historical perspective of decision theory. In Section 3, we discuss some of the mechanics of eliciting loss functions. Section 4 presents two examples of optimal decision making in the context of environmental remediation. The first example illustrates a situation in which a prior opinion derived from empirical information is used in a Bayesian model, but the two parties who need to reach a decision have different utilities. In the second example, empirical prior information is combined with the subjective opinions of each party, resulting in each party having not only different utilities, but different posterior distributions. In Section 5, we discuss the implications of compromise, as well as discussing how an elementary loss function can serve as a point of departure for a more complex decision-making process.

2   A BRIEF HISTORY OF OPTIMAL DECISION MAKING

The mathematical theory of probability arose as much from the analysis of gambling as from any other source. Generally the expectation of a random quantity of money was used as an indication of its worth. One important early contributor was Pascal (1623–1662), who combined mathematical interests with a deep theological attachment. In the latter connection, he wrote concerning whether to believe in God, as follows:

> Let us weigh the gain and loss in wagering that God is. Let us estimate these two chances. If you gain, you gain all; if you lose, you lose nothing. Wager, then, without hesitation that He is. (Pascal, Pensée 233, 1958, p. 67)

Thus Pascal's argument is that you have a two-decision problem (to believe or not to believe) and two states of the world (God exists or does not). Since you have infinite utility if God exists and you believe and finite utility otherwise, Pascal recommends belief.

A second major step was taken by D. Bernoulli (1700–1782) in response to the St. Petersburg paradox. (Jorland, 1987). The problem is that a fair coin is flipped; if the first head occurs on the nth trial, A pays B 2^{n-1} coins. How much should B pay for this? The expected number of coins A will pay B is

$$\sum_{n=1}^{\infty} \left(\frac{1}{2^n} \right) (2^{n-1}) = \infty$$

which is more than any reasonable person would pay. Bernoulli developed the idea of moral expectation or utility, so that the game is worth

$$\sum_{n=1}^{\infty} \left(\frac{1}{2^n}\right) U\left(2^{n-1}\right)$$

where Bernoulli took U to be logarithmic, leading to a finite sum. In later developments, U was recognized to be quite flexible, but probability was defined only in terms of equally likely events.

The first "modern" essay in this area is by Frank Ramsey (1931), although his ideas about expected utility are presented without emphasis and seem to have attracted little notice at the time. The route that attracted statisticians to this theory largely started with Fisher's ideas about tests of significance, with emphasis on only the null hypothesis. Neyman and Pearson proposed tests of hypotheses in which the alternative and the probability of rejection under the alternative also matter. Then came Wald's (1950) ideas about statistical decision functions, in which the expected losses of different statistical procedures were compared. When a second procedure's expected losses were everywhere (in the parameter space) at least as high, and sometimes higher than a first procedure's, the first procedure is inadmissible. Because only procedures that are Bayes (based on a prior distribution) and certain limits of them are admissible, this led to interest in Bayesian ideas. Furthermore, Wald's decision theory did not lend itself to choice among admissible procedures, whereas the Bayesian approach does. This development finally led to Savage's (1954) unification of decision theory with Bayesian ideas.

Several points should be made concerning modern Bayesian decision theory. First, it is normative, in that it prescribes how a person should make decisions under uncertainty (and be rational), not how people actually do make decisions. Second, it is subjective, or as Savage would have put it, personal. The utilities and probabilities used represent the decision maker's values and beliefs, respectively, and not necessarily anyone else's. Third, the theory applies only to a single individual. Attempts to make theories of rational decision making for groups have so far not been very successful. The interested reader should refer to Raiffa (1968), Berger (1985), DeGroot (1970), and Smith (1988) for statistical treatments of modern decision theory and to von Winterfeldt and Edwards (1986) and Edwards (1992) for discussion of the more qualitative issues.

3 ELICITING LOSS FUNCTIONS

Making decisions based on expected utility is synonymous with making decisions that minimize expected loss. Determining expected loss means constructing loss functions that consist of outcomes based on a quantity of interest, Q, and the loss associated with each outcome. The loss can be fixed or random. In general, there is a set A of possible actions that can be taken, and for each element of A, there are losses which may or may not depend on Q.

In this chapter we consider that there are two parties involved in the decision-making process. Both parties should be involved in the elicitation of the loss functions, since the ideal solution would be that the loss functions of the two parties coincide. In this section, we describe a procedure for eliciting the utilities; in the next section, we illustrate the procedure through two examples.

The first stage of the elicitation process should be to choose the action set A and the quantity of interest Q. The set of actions should be well defined and must contain at least two elements. These actions are usually dependent on some quantity of interest. For example, if the action set consists of assigning patients to two different treatments in a clinical trial, then the quantity of interest might be some quantification of the state of the patient's health. In choosing the quantity of interest, it is important to make sure that Q is a quantity that incorporates the uncertainty inherent in the problem being addressed. Generally, Q will be a random quantity defined by a statistical model, and as such, it can have several dimensions. In this chapter, however, we address primarily the case in which Q is one-dimensional.

Once the action set A and the quantity of interest Q have been determined, the second stage of the elicitation is to construct loss functions based on them. This is the most difficult stage of the elicitation process. Each party will construct these separately and must give careful thought to the consequences inherent for each action in A for any value of Q.

In general, there will be losses associated with any action. These losses can be monetary costs, or they can represent the relative benefits at a particular value of Q of one course of action versus another. Specifying the losses can be tricky; it is often advisable to construct several loss functions for each party. Each loss function represents the values being placed by an individual (or group) on various costs and benefits.

Some examples of the elements of a loss function, particularly examples related to health and environmental issues, are monetary costs associated with actions, loss or gain of goodwill, increase in life expectancy, medical problems associated with an action, quality of life, and potential threat of litigation. This is not an exhaustive list but merely an indication of the type of elements to consider.

The loss function can be specified parametrically, rather than assigning specific values at the outset to the elements of the loss functions. It is often the case, as is demonstrated in this chapter, that the optimal decision will depend on the relationship between elements of the loss function rather than the absolute value of the elements. A key point of this chapter is that loss functions specified in this manner allow the investigation of values of the parameters of the loss function that will give rise to different decisions.

4 QUESTIONS OF ENVIRONMENTAL REMEDIATION

In this section, we present two examples of how decision theory can be used in the decision-making process. Both examples deal with issues of environmental policy. Both address the perspective of a regulatory agency charged with protection of the public interests in human health and the environment (e.g., the U.S. Environmental Protection Agency, or EPA) and the regulated community (e.g., industry).

In the first example, the decision is to determine when and where the current owners of a former battery recycling plant are responsible for remediation to reduce soil lead concentrations in the surrounding residential area. Loss functions are illustrated that, under the model, allow each party to understand the implications of particular choices of decision rules for the interests of the other party. The second example concerns whether or not radioactive waste has contaminated a residential water supply and whether further monitoring and remediation are therefore needed. In the second instance, because the two parties have different prior beliefs, as well as different utilities, compromise between the two may be difficult.

The management of hazardous waste sites in the United States falls under legislation known as the Comprehensive Environmental Response, Compensation, and Liability Act (CERCLA) of 1980 (amended in 1986). Sites designated to be "Superfund" sites are required to have a phased program of investigation, design, and remedial action. A high priority is placed on remediation, which typically involves cleanup and restoration of impacted areas, when the surrounding community is at risk from the hazardous waste. In the example dealing with lead contamination in soil, the risk is in ingesting lead particles, so remediation entails replacing all the carpets in a home, thoroughly cleaning the home, and replacing the topsoil and relandscaping the yard. For the radioactive groundwater contamination, it entails providing the community with an alternative water supply.

Environmental remediation is a particularly appropriate source of examples of the use of expected utility theory to make decisions. Most problems in this area have a great degree of uncertainty inherent in the modeling process and have many intangible quantities involved in the utilities.

4.1 Example 1: Lead Contamination in Soil

When toxic contamination occurs in the United States, remediation is usually mandated by the U.S. EPA or by state environmental agencies to whom such authority has been delegated. The determination of who bears the costs is generally the product of negotiations between the parties responsible for the contamination and the EPA. When the contaminant is one that could have come from several sources, the negotiations become much more complicated.

In this section, we describe a potential solution to one such environmental remediation problem. Small *et al.* (1995) and Wolfson *et al.* (1994) consider the case of a former battery recycling plant that had caused toxic lead contamination in the surrounding community. Remediation near the site is mandated wherever the lead concentration in soil exceeds 500 ppm, but the problem is complicated by the fact that there are other "background" sources of lead contamination, such as lead-based paint used on homes and vehicular emissions from past use of leaded gasoline. The question is, how far from the plant should the current owners of the plant be held responsible for the costs of remediation?

In modeling the lead concentration in soil, the total observed soil lead concentration, L, was assumed in Wolfson *et al.* (1994) to be equal to $B + P$, where B is a random variable representing the soil lead contributed from background sources and P is a random variable representing soil lead contributed from the former battery recycling plant. In this case, there are two actions to be considered, remediation and no remediation, for homes whose lead concentrations exceed the 500-ppm limit, as set by the regulations at the time the study was done. The quantity of interest is

$$R = \frac{B}{B + P}$$

where R represents, in some sense, the proportion for which the plant is not responsible for exceeding the level at which remediation is mandated.

Empirical information in the form of lead concentrations from a nearby city located upwind of the plant is available for use in constructing a prior distribution for background lead concentrations. The background lead was modeled as following a gamma distribution with mean parameter μ_b and shape parameter β. The posterior mean and variance of these parameters estimated by using a reference prior (Berger, 1985; Box and

Tiao, 1983) are used as estimates of the hyperparameters for the prior distribution on B:

$$\mu_b \sim N(268, 22^2) \tag{1}$$

$$\beta \sim N(294, 35^2) \tag{2}$$

with μ_b and β assumed to be independent. Lead from the plant, P, is modeled as following a gamma distribution with mean μ_p and shape parameter β. The spatial trend of lead concentrations from the plant (P) decreasing with distance from the plant is included by setting $\mu_p = \exp(\theta_1 + \theta_2 XY + \theta_3 X^2 + \theta_4 Y^2)$, where X and Y described the north-south and east-west (respectively) directions of a particular sample, relative to the plant. Wolfson *et al.* (1994) showed that for this model, the posterior distribution of R is well approximated by a beta distribution, with parameters α_b, α_p given by

$$\alpha_b = \frac{\mu_b}{\beta} = \frac{274}{305}$$

$$\alpha_p = \frac{\mu_p}{\beta} = \frac{\exp(7.42 - 300XY - 191X^2 - 253Y^2)}{305}$$

Thus, the plant contribution to soil lead concentrations is a function of distance from the plant, with the mean value ranging from 1200 ppm near the plant to 5 ppm for homes a great distance from the plant.

The two parties involved in the negotiation are the EPA and the "potentially responsible parties" or PRPs (in this case, the current owners of the former battery recycling plant). To illustrate the methods described in the previous section, we construct two loss functions, one from the perspective of each party. Then we discuss how they can be compared to help formulate policy (in this case, which houses to remediate).

From the perspective of the PRP, there are three elements in the loss function; dollars, goodwill, and lawsuit risk. If the PRP remediates there is a fixed monetary cost; call it m. If it does not remediate when responsible, it loses goodwill in the community and opens itself to potential lawsuits. Call this loss g. If it does not remediate when it is not responsible, it loses nothing. Defining q as a threshold value for R that indicates whether the plant is responsible, we get the corresponding loss function L_{PRP}

$$L_{PRP}(R, \text{no remediation}) = \begin{cases} 0 & \text{if } R \geq q \\ g & \text{if } R < q \end{cases} \tag{3}$$

$$L_{PRP}(R, \text{remediation}) = m$$

R is different for each home, but m, g, and q are set to be the same for all homes. The action that minimizes expected loss is to remediate when

$$\frac{m}{g} < P(R \leq q)$$

In this loss function, it was not necessary to specify m and g directly, since it turns out that it is the *ratio* of the two, and not their absolute value, that is important for decision making. This is rather convenient, as it is often difficult to assess the dollar value associated with an intangible such as g. However, choosing a ratio is effectively choosing a value for g, given m.

A loss function from the perspective of the EPA looks somewhat different. Suppose that the EPA's stance is that there is no loss to the public if remediation does not occur when the proportion of background lead is high, but there is a loss when remediation does not occur and the proportion of background lead is low. This loss, h, can be viewed as the costs associated with underremediation (i.e., when homes are not remediated, but should be), such as increased health risks in the surrounding community, as the EPA is charged with protecting the public interests. When the action chosen is remediation, then the EPA experiences no losses if the PRP remediates and the proportion of background lead to total lead is below the cutoff point q, because the PRP is then remediating when it is indeed responsible; however, if the PRP is remediating when the proportion of background lead to total lead is high, and thus above the cutoff point q, then the EPA's relationship with the PRP is damaged, and the cost associated with forcing the PRP to overremediate is k. The EPA is not going to bear a direct financial burden, so a plausible loss function might be

$$L_{\text{EPA}}(R, \text{not remediating}) = \begin{cases} 0 & \text{if } R \geq q \\ h & \text{otherwise} \end{cases}$$

$$\tag{4}$$

$$L_{\text{EPA}}(R, \text{remediating}) = \begin{cases} 0 & \text{if } R < q \\ k & \text{otherwise} \end{cases}$$

which implies that the action that minimizes expected loss is to remediate when

$$\frac{k}{h} < \frac{P(R \leq q)}{1 - P(R \leq q)}$$

Once again, the values for k and h need not be assessed directly; it is their relative values that are of import. The loss functions L_{EPA} and L_{PRP}

provide simple, yet plausible loss representations for both the tangible and intangible policy factors of the decision-making process under uncertainty.

Now that a loss function has been constructed for each party, it is possible to compare the two, to facilitate the decision-making process. For any value of q, $L_{PRP} = L_{EPA}$ when

$$\frac{m}{g} = \frac{k}{k+h} \tag{5}$$

In practical terms, this translates to the following. For any choice of cutoff point q, suppose that the costs the EPA associates with underremediation are valued at twice the costs the EPA associates with overremediation, corresponding to a value of $k/h = 0.5$. Then under the PRP's loss function, this corresponds to $m/g = 0.33$, so that for the PRP to agree to this particular decision rule, it is valuing their financial costs at one-third the value it places on the risks associated with underremediation.

Figure 1 shows several remediation contour lines and the associated values of L_{EPA} and L_{PRP} when $q = 0.67$. The homes shown on the graph that fall within a given contour line would be eligible for remediation by the PRP under the associated decision rule. The shapes of the contour lines are determined by several factors: prevailing wind patterns, spatial trend of lead depositions in soil, and the fact that in the lower left-hand corner of the graph, there was a cemetery (which was not considered for

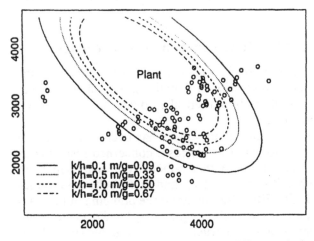

Figure 1   Remediation contour lines for different choices of k/h and m/g.

remediation) and in the upper right-hand corner of the graph there is a housing development that was established after the battery recycling operation ceased.

When Eq. (5) is not satisfied, disagreements arise on the appropriate extent of remediation. Typically, m/g exceeds $k/(k + h)$, with the result that the EPA wishes to extend the area of remediation farther from the site than is desired by the PRP. Both loss functions presented here are somewhat simplistic; a loss function where losses are proportional to the amount of contamination contributed by the plant, and likely more representative of the actual utilities of the parties involved in the soil lead contamination example, is

$$L^*_{PRP}(R, \text{not remediating}) = g(1 - R)$$
$$L^*_{PRP}(R, \text{remediating}) = m \tag{6}$$

$$L^*_{EPA}(R, \text{not remediating}) = h(1 - R)$$
$$L^*_{EPA}(R, \text{remediating}) = kR \tag{7}$$

Under these loss functions, the PRPs would remediate whenever $1 - E[R] \leq m/g$, and the EPA's optimal decision would be to have remediation occur whenever $1 - E[R] \leq k/(k + h)$. This is the same as the equivalence relationship shown previously, which means that values of m, g, k, and h satisfying Eq. (5) would lead to an agreement between the PRP and the EPA. In particular, this implies that the decisions made under the loss functions specified in Eqs. (3) and (4) will be identical to the decisions made under the loss functions specified in Eqs. (6) and (7), illustrating that the relationship between m, g, k, and h is very important in the decision-making process.

4.2   Example 2: Radioactive Contamination of Groundwater

This example illustrates a situation in which two parties involved in the negotiation process have different prior probabilities, as well as different utility functions. The example is hypothetical in detail but is based on experience at Superfund sites.

In the area surrounding a landfill site, there has been some concern on the part of various stakeholders (in this case, concerned residents as well as some public interest groups) that radioactive waste had been dumped in the landfill when it was operative, and as a result, the groundwater was contaminated. The EPA did not authorize dumping of radioactive waste in the landfill, but some illicit dumping may have occurred, and residents of the area claim to have seen trucks with radioactive symbols on them dumping waste during the night on several occasions.

To determine if there has indeed been radioactive contamination of the groundwater supply, the EPA authorized testing that reports whether or not a sample is "above" or "below" threshold but not the amount above or below. If there had been a moderate level of contamination, one would expect, on the average, about 40% of the samples taken from the water supply to have radioactive contamination levels above the threshold level for detection, but if there is no contamination one would expect approximately 10% of samples to be above threshold, due to some background contamination, with some variability in each case. Let $\mathscr{C}$ be the set of possible states of nature, such that $\mathscr{C} \in \{C, NC\}$, where C indicates there is some contamination, defined as the proportion of samples above threshold being 20% or greater, and NC is no contamination, defined as the proportion of samples being below 20%.

Let E represent the evidence collected. In this case, E is the number of samples whose radioactive level exceeded a set threshold value (this tends to vary depending on current legislation and understanding of the health risks involved) and E follows a binomial distribution with parameters n (number of samples collected) and p (probability of a particular sample being above threshold). Based on prior sample sizes of 4 and the historical expected proportion of samples above threshold given the presence or absence of contamination, the conjugate prior distributions under the two possible states of nature are both truncated Beta(1, 3) distributions, shown in Table 1.

Because $\pi(p \mid C)$ and $\pi(p \mid NC)$ differ only by a constant, they will combine to form a single Beta(1,3) distribution if and only if the weights on the two pieces are

$$P(NC) = P(p^* \leq 0.2 \mid p^* \sim \text{Beta}(1, 3)) \quad \text{and}$$

$$P(C) = P(0.2 \leq p^* \leq 1 \mid p^* \sim \text{Beta}(1, 3))$$

The stakeholders have a high prior probability that contamination has occurred and a low probability on no contamination, whereas the converse is true for the EPA, with the results shown in Table 2.

Table 1 Empirical Prior Distributions for p, Conditional on Contamination Levels

Contamination level ($\mathscr{C}$)	$\pi(p \mid \mathscr{C})$	$E(p \mid \mathscr{C})$
No contamination (NC)	$\propto (1 - p)^2 I(0 \leq p < .2)$	.09
Contamination (C)	$\propto (1 - p)^2 I(.2 \leq p \leq 1)$	.4

Table 2 Prior Probabilities on Contamination Levels

Contamination level ($\mathscr{C}$)	Range	Stakeholders	EPA
No contamination (NC)	$0 \le p < .2$	$P(NC) = .05$	$P(NC) = .95$
Contamination (C)	$.2 \le p \le 1$	$P(C) = .95$	$P(C) = .05$

Given a binomial sampling model and the prior distributions on contamination of both the stakeholders and the regulatory agency, one can use Bayes' formula to compute the posterior probability of contamination:

$$P(C \mid E) = P(0.2 \le p \le 1 \mid E) = \frac{P(E \mid C)P(C)}{P(E)}$$

where, for $\mathscr{C} \in \{C, NC\}$, $\alpha = 1$, $\beta = 3$,

$$P(E \mid p, \mathscr{C}) \sim \text{Binomial}(n, p)$$

$$P(p \mid \mathscr{C}) = \frac{p^{\alpha-1}(1-p)^{\beta-1}}{\int_{\mathscr{C}} p^{\alpha-1}(1-p)^{\beta-1} \, dp}$$

$$P(E \mid \mathscr{C}) = \int_{\mathscr{C}} P(E \mid p, \mathscr{C})P(p \mid \mathscr{C}) \, dp$$

$$\propto \int_{\mathscr{C}} \frac{p^{\alpha+x-1}(1-p)^{\beta+n-x-1}}{\int_{\mathscr{C}} p^{\alpha-1}(1-p)^{\beta-1} \, dp} \, dp$$

$P(\mathscr{C})$: From Table 2

$$P(E) = \sum_{\mathscr{C} \in \{C, NC\}} P(E \mid \mathscr{C})P(\mathscr{C})$$

and x is the number of the n samples that are above threshold. Figure 2 shows that the prior predictive probability distribution of x/n will be concentrated in the region $x/n < 0.2$ for the EPA, and for the stakeholders it will be concentrated in the region $0.2 \le x/n$.

Suppose that the utilities of the stakeholders are expressed by the following loss function:

$$L_S(p, \text{not remediating}) = v_1 p$$
$$L_S(p, \text{remediating}) = v_2(1 - p) \qquad (8)$$

Thus v_1 is the value placed by the stakeholders on the effect of contamination, and v_2 is the value they place on public spending by the EPA when it may not be necessary. Thus the stakeholders would choose to remediate when $v_2/(v_1 + v_2) < E_S[p \mid E]$.

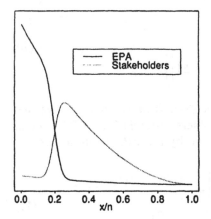

Figure 2   Prior predictive distribution of x/n.

The EPA's utility can be expressed by a loss function that says there is a fixed loss, v_3, when remediation occurs (the cost of the remediation), and there will be a loss proportional to p when remediation does not occur. The value v_4 represents the value placed on underremediation by the EPA:

$$L_{\text{EPA}}(p, \text{not remediating}) = v_4 p$$
$$L_{\text{EPA}}(p, \text{remediating}) = v_3 \tag{9}$$

The EPA would choose to remediate when $v_3/v_4 < E_{\text{EPA}}[p \mid E]$. Unlike the previous example, because of the different prior opinions, it is not the case that we can express a relationship between v_1, v_2, v_3, and v_4 that will result in the decision made by both parties being the same.

Lindley and Singpurwalla (1991), Etzioni and Kadane (1993), and Lodh (1993) consider the problem of selecting the optimal sample size in this type of situation, in which the two parties have both differing prior opinions and different utility functions. Lindley and Singpurwalla (1991) refer to this as an "adversarial" relationship. In the problem just described, the EPA will choose the sample size. It can be shown that the sample size the EPA would choose using L_{EPA} would be substantially less than the sample size they would choose using L_S. To illustrate, if the EPA were the only party involved in making the decision, and the cost of sampling is v_5 per unit, then the sample size n would be chosen to minimize the EPA's expected loss:

$$v_5 n + v_3 P_{\text{EPA}} \left[\frac{v_3}{v_4} < E_{\text{EPA}}[p \mid E] \right]$$

$$+ v_4 E_{\text{EPA}} \left\{ p * P_{\text{EPA}} \left[\frac{v_3}{v_4} \geq E_{\text{EPA}}[p \mid E] \right] \right\} \tag{10}$$

This "preposterior" analysis can also be done from the perspective that the decision will be made by the stakeholders instead of the EPA. In that situation, the sample size would be chosen by the EPA to minimize

$$v_5 n + v_3 P_{\text{EPA}} \left[\frac{v_2}{v_1 + v_2} < E_{\text{S}}[p \mid E] \right]$$

$$+ v_4 E_{\text{EPA}} \left\{ p * P_{\text{EPA}} \left[\frac{v_2}{v_1 + v_2} \geq E_{\text{S}}[p \mid E] \right] \right\} \tag{11}$$

The sample sizes in Eqs. (10) and (11) can be determined numerically, for particular values of $v_1, \ldots v_5$, and the software for doing this in S-Plus is available on StatLib. As an example, suppose that the stakeholders chose $v_2/(v_1 + v_2) = 0.1$, implying that the loss for underremediation is nine times that of overremediation. Further, suppose that the EPA valued the penalty for underremediation at four times the cost of remediation, so that $v_3/v_4 = 0.25$, and the cost of a single example is $v_5 = 1/4000$. Then the sample size chosen by the EPA under Eq. (10) would be 30, and the sample size chosen by the EPA under Eq. (11) would be 96. If $v_3/v_4 = 0.33$, and v_1, v_2, and v_5 remain the same, then under Eq. (10) only one sample would be taken, because as Figure 2 shows, the EPA's prior predictive probability that $x/n <= 0.33$ is very small. Under Eq. (11), the optimal sample size would be 45. If v_3/v_4 were decreased to 0.2, keeping all other values the same, then the EPA would choose a sample size of 40 under L_{EPA} and a sample size of 116 under L_{S}.

The diagonal curves in Figure 3 show the posterior expected value of p for the stakeholders and the EPA for samples of size 30 and 96, over the range of possible values that could be obtained for x. For the stakeholders, this "curve" is actually a straight line, whereas it is a curve for the EPA. The reason is that given the stakeholders' prior opinion, their posterior opinion about p is that p increases linearly as a function of the number of samples observed above threshold. For the EPA, the number of samples above threshold does not have as large an influence on their posterior opinion when the data are somewhat uninformative. In the plots, four regions are defined. In region A, under L_{EPA}, the stakeholders would conclude that remediation is mandated, but the EPA would not, and in

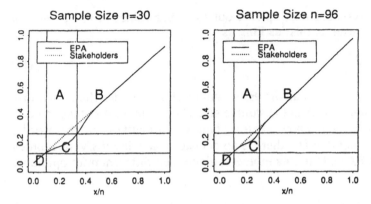

Figure 3 Posterior expected value of *p* as a function of possible outcomes.

region C, under L_S, both parties would agree that remediation is mandated. In regions B and D, both parties would reach the same decision regardless of whose utility is being used. As the sample size increases, the *proportion* of samples above threshold that would lead to disagreement by putting a decision maker in the region A or C is declining. Table 3 gives exact numbers, which summarized imply that with a sample of size 30, 27% of the possible outcomes will put the decision maker in the indeterminate region, and with a sample of size 96, only 19% of the possible outcomes will do this. As the sample size gets larger, the area of regions A and C will decrease.

Table 3 Remediation Rules[a]

	Prior		
Loss	EPA	Stakeholders	
L_{EPA}	11	8	$n = 30$
	28	24	$n = 96$
L_S	3	3	$n = 30$
	10	10	$n = 96$

[a] Each cell represents the minimum number of samples that would have to be above threshold for the decision to be made to remediate under given Loss function and Prior.

This analysis could be carried out for other values of $v_1 \ldots v_5$, but the essential result is this: as long as

$$\frac{v_3}{v_4} \geq \frac{v_2}{v_1 + v_2} \tag{12}$$

then the sample size that the EPA would choose under their own loss functions will be less than the sample size chosen under the stakeholders' utilities. Unless the samples taken are highly informative—in other words, they fall in region B or D—there is no decision that will satisfy the utilities of both the EPA and the stakeholders, given their different prior opinions.

5 DISCUSSION

The examples in this chapter illustrate the implications of utility functions combined with posterior distributions for practical decision making. In the soil lead contamination example, the PRP and the EPA agreed to a distribution based on empirical data, and thus despite their having different utilities, it is possible to see room for compromises.

It is useful that the loss functions for the two competing parties be functionally comparable in the initial formulation. There are two parts to the decision-making process as presented in this chapter. The first part is to settle on relative values of some intangible quantities, such as perceived risk, goodwill, etc. The second part of the decision-making process is to use those relative values to make a decision. For the first part, it is convenient for the loss functions of the two parties to be of a form that allows one to express a relationship between the utility quantities of each party in the negotiation process, such as in Eq. (5). The decision-making process can be an iterative process. Once a relationship between the intangible quantities has been developed, and relative values agreed upon, a more representative loss function can be employed.

The second example illustrates that when the two parties involved in the negotiation process have both differing prior opinions and differing utilities, unless the data are highly informative there is no agreement that maximizes the expected utility of both.

The approach we have taken to analyzing this type of environmental health problem is suitable when the contaminants are known to have harmful effects, and it is known at approximately what level they are harmful. When the question at hand is not the extent of the environmental contamination, but rather the impact of different levels of contamination on environmental health, analysis of the type presented in Hasselblad and Jarabek (Chapter 8) may be more appropriate.

The methods described in this chapter are applicable in any kind of policy-oriented decision-making process, where the quantities that affect decisions can be stated in a reasonable fashion. This type of quantitative analysis can also account for the additional uncertainty that is present in decision making when intangible quantities must be incorporated.

REFERENCES

Berger, J. O. (1985). *Statistical Decision Theory and Bayesian Analysis*. Springer-Verlag, New York.

DeGroot, M. H. (1970). *Optimal Statistical Decision*. McGraw-Hill, New York.

Edwards, W., ed. (1992). *Utility Theories: Measurements and Applications*. Kluwer Academic Publishers, Boston.

Etzioni, R., and Kadane, J. B. (1993). Optimal experimental design for another's analysis. *Journal of the American Statistical Association* 88: 1404–1411.

Jorland, G. (1987). The St. Petersburg paradox, 1713–1937. In *The Probabilistic Revolution*, Vol. 1: *Ideas in History*, ed. Kruger, L., Daston, L., and Heidelberger, M., pp. 157–190. MIT Press, Cambridge, MA.

Lindley, D. V., and Singpurwalla, N. D. (1991). On the evidence needed to reach agreed action between adversaries, with application to acceptance sampling. *Journal of the American Statistical Association* 86: 933–937.

Lodh, M. (1993). Experimental Studies by Distinct Designer and Estimators. Ph.D dissertation, Department of Statistics, Carnegie-Mellon University, Pittsburgh.

Pascal, B. (1958). *Pensées, with an Introduction by T. S. Eliot*. Dutton, New York.

Raiffa, H. (1968). *Decision Analysis: Introductory Lectures on Choices under Uncertainty*. Addison-Wesley, Reading, MA.

Ramsey, F. P. (1931). Truth and probability. In *The Foundations of Mathematics and other Logical Essays by Frank Plumpton Ramsey*, ed. Braithwaite, R. B., pp. 158–198. Routledge and Kegan Paul, London.

Savage, L. J. (1954). *Foundations of Statistics*. Wiley, New York.

Seidenfeld, T., Kadane, J. B., and Schervish, M. (1989). On the shared preferences of two Bayesian decision makers. *Journal of Philosophy* 76(5): 225–244.

Small, M. J., Nunn, A. B., Forslund, B. L., and Daily, D. A. (1995). Source attribution of elevated residential soil lead near a battery recycling site. *Environmental Science and Technology* 29, 883–895.

Smith, J. Q. (1988). *Decision Analysis: A Bayesian Approach*. Chapman & Hall, London.

von Winterfeldt, D., and Edwards, W. (1986). *Decision Analysis and Behavioral Research*. Cambridge University Press, Cambridge.

Wald, A. (1950). *Statistical Decision Functions*. Wiley, New York.

Wolfson, L. J., Kadane, J. B., and Small, M. J. (1994). Statistical Decision Theory for Environmental Remediation. Technical Report 594, Department of Statistics, Carnegie-Mellon University, submitted to *Technometrics*.

IV
DESIGN

10
Bayesian Hypothesis Testing: Interim Analysis of a Clinical Trial Evaluating Phenytoin for the Prophylaxis of Early Post-Traumatic Seizures in Children

Roger J. Lewis UCLA School of Medicine, Los Angeles, and Harbor-UCLA Medical Center, Torrance, California

1 CLINICAL BACKGROUND

Head injury, both accidental and inflicted, results in approximately 12,000 physician visits, 200 hospitalizations, and 10 deaths per 100,000 children in the United States each year (Bruce, 1990; Kraus *et al.*, 1986; Annegers, 1983). Neurologic damage due to head injury occurs both at the time of the initial injury (primary injury) and during the period of time that follows (secondary injury) (Bruce, 1990). Reduced blood flow to the brain, due to increased intracranial pressure, is responsible for most secondary injury. The primary goal of the initial care of children with significant head trauma is the prevention of secondary injury (Bruce, 1990; James, 1986).

Post-traumatic seizures occur in up to 15% of all children with head trauma and up to 40% of children with severe head trauma (Hahn *et al.*, 1988a, b; Hendrick and Harris, 1968; Jennett, 1973; Jennett, 1969; Lewis *et al.*, 1993). Seizures that occur within 1 week of the initial injury are defined as early post-traumatic seizures (Hendrick and Harris, 1968; Jennett, 1969). The majority of early post-traumatic seizures occur within 24 to 48 hours of the initial injury (Hahn *et al.*, 1988b; Hendrick and Harris, 1968; Lewis *et al.*, 1993). Early post-traumatic seizures are fundamentally different from the long-term seizure disorders that may follow significant head injury.

In a retrospective study of 937 children with head trauma, Hahn *et al.* (1988b) observed an incidence of post-traumatic seizures of 9.8% in all children and an incidence of 35% in patients with severe head injury, as defined by a Glasgow Coma Scale (GCS) value less than 9 (Teasdale and Jennett, 1974). Similarly, a retrospective study of children seen at our institution showed that those with a depressed GCS had a markedly increased risk of post-traumatic seizures. Of the patients with GCS scores of 3 to 8, 38.7% (12/31) suffered post-traumatic seizures, whereas only 3.8% (6/159) of patients with GCS scores of 9 to 15 suffered post-traumatic seizures (Lewis *et al.*, 1993).

Early post-traumatic seizures cause a rapid increase in the cerebral metabolic rate. Post-traumatic seizures may also increase the intracranial pressure directly, thus interfering with cerebral blood flow. Because secondary injury is most likely when increased cerebral metabolism occurs in the presence of increased intracranial pressure and decreased blood flow to the brain, prevention of post-traumatic seizures may reduce secondary neurologic injury and improve neurologic outcome. The anticonvulsant medication phenytoin is often administered in an attempt to prevent post-traumatic seizures in children with moderate to severe closed head injury (Lewis *et al.*, 1993; Young *et al.*, 1983; Temkin *et al.*, 1990).

Several prospective studies have addressed the question of whether or not prophylactic phenytoin is effective in preventing early post-traumatic seizures, with conflicting results. In one randomized, double-blind, placebo-controlled trial involving 244 patients, 5 of 136 (3.7%) patients treated with phenytoin and 4 of 108 (3.7%) patients receiving placebo suffered early post-traumatic seizures (Young *et al.*, 1983). This negative study, because of the low rate of early post-traumatic seizures observed, the small number of children (only 34 of 244 patients), and the fact that the results were not separated by age group, gives little information on the efficacy of phenytoin in children.

Temkin *et al.* (1990) published a prospective, randomized, placebo-controlled trial of phenytoin for the prevention of both early and late post-traumatic seizures in adults with serious head trauma. Four hundred and four patients were randomized, with 208 receiving phenytoin and 196 receiving placebo. The cumulative risk of a post-traumatic seizure occurring within the first week was determined by Kaplan-Meier analysis (Kaplan and Meier, 1958) and was found to be 14.2% in the placebo-treated group and 3.6% in the phenytoin-treated group (classical *P* value less than 0.001).

Although Temkin *et al.* (1990) provide evidence that phenytoin is probably effective for the prophylaxis of early post-traumatic seizures in adults, this conclusion cannot be reliably extended to children. The child's brain reacts differently from the adult's to acute trauma, and the pathogenesis of post-traumatic seizures in children and adults may be very different.

2 THE CLINICAL TRIAL

The clinical trial considered here is a prospective, double-blind, randomized placebo-controlled trial of intravenous phenytoin for the prophylaxis of early post-traumatic seizures in children who have suffered moderate to severe blunt head injury. The primary end point is the occurrence of a clinically apparent or electroencephalogram (EEG)-documented seizure within 48 hours of the child's arrival in the emergency department.

Pediatric patients who have suffered blunt head injury are eligible if their age is less than 16 years, they have a heart rate greater than 60 beats per minute, and they have a markedly depressed coma scale. Children are enrolled in six stratification groups with three age levels and two coma scale score levels. Within each stratification group, children are allocated in randomized permuted blocks to receive phenytoin or placebo in a 1:1 ratio. An initial dose of the study medication (phenytoin or placebo) is given within 1 hour of the patient's arrival in the emergency department, and maintenance doses of the study drug are given for the next 48 hours.

Upon the child's arrival in the emergency department, a "blinded" 48-hour observation period begins, during which the child is observed for seizure activity. This observation period is terminated prior to 48 hours if the child experiences a clinically apparent or EEG-documented seizure (the primary end point), experiences an adverse reaction that is attributable to the study drug, or dies. If the child receives placebo and suffers a seizure, the clinicians have the option of administering phenytoin in an attempt to prevent further seizures.

3 FREQUENTIST INTERIM ANALYSIS

The design of most clinical trials includes planned interim analyses of the accumulating data, to allow termination of the trial should the accumulating data demonstrate that a clear difference exists between the treatments. If a clear difference exists, continuing the trial would unnecessarily expose some patients to the less efficacious therapy, whichever one that turns out to be, and delay applying the better treatment to patients not enrolled in the study.

Many frequentist procedures for interim analysis have been suggested (Pocock, 1977; O'Brien and Fleming, 1979; Lan and DeMets, 1983; Kim and DeMets, 1987). These procedures have the characteristic that they control the overall type I error rate, usually keeping it at 0.05. The possibility of stopping the clinical trial at each interim analysis "spends" a portion of the type I error risk, and the different classical designs are different plans for spending the type I error risk.

The Pocock design (Pocock, 1977; Lewis, 1993; Geller and Pocock, 1987) uses the same significance level at each interim analysis. For example, a three-look Pocock design (two interim analyses and one final analysis) with an overall α of 0.05 uses a nominal significance level of 0.0221 for each of the three analyses. The O'Brien-Fleming design spends less of the error risk and is more conservative at the early analyses, compared with the Pocock design (O'Brien and Fleming, 1979; Geller and Pocock, 1987; Lewis, 1993). The O'Brien-Fleming design is less conservative, however, in the later and final analyses, should the trial not be stopped earlier. For a three-look O'Brien-Fleming design and an overall $\alpha = 0.05$, the three significance levels are 0.0005, 0.0141, and 0.0451.

A larger number of analyses gives more opportunities for early stopping and decreases the mean sample size should the treatment effect be large. On the other hand, increasing the number of analyses can actually increase the expected number of patients required for the trial under the null hypothesis, because the significance levels must be adjusted downward to maintain the overall type I error rate. This "penalty" for additional interim analyses decreases the probability of stopping at any particular early analysis. In addition, when a large number of interim analyses are planned, the maximum sample size must be adjusted upward to obtain a given power because the terminal significance level is decreased. The optimum number of interim analyses depends on the uncertainty in the size of the treatment effect (McPherson, 1982). Between 2 and 10 interim analyses usually lead to a minimum mean sample size, with a larger number of analyses being better when there is great uncertainty in the magnitude of the treatment effect. In situations analogous to most clinical trials, two to five interim analyses are usually optimal.

Based on previous data (Lewis et al., 1993), we assumed the placebo group would experience a 25% to 40% incidence of post-traumatic seizures. The trial was designed to have a classical power of 95% to detect a treatment efficacy of 50%, that is, a decrease in the observed incidence of seizures from 25% to 12.5% with phenytoin therapy. The high power was chosen because of the uncertainty in the true seizure rate in the placebo group and the desire that the trial should have adequate power should the observed seizure rate in the placebo group be lower than expected.

During the initial planning of the trial, a classical group-sequential design was considered. A classical four-look Pocock design, assuming a one-tailed α of 0.05 and a power of 95%, requires a sample size of 124 patients per block to achieve the required power, for a maximum sample size of 496 patients. The classical group-sequential design of Pocock was selected over that of O'Brien and Fleming because, under the assumption that phenytoin is efficacious, the Pocock design has a lower expected sample size than the O'Brien-Fleming design.

There are two conceptual difficulties with the decision to terminate a clinical trial based on a classical interim analysis. First of all, the decision is based on a P value, which is the probability of obtaining the observed pattern in the data, or one more inconsistent with the null hypothesis, assuming the null hypothesis is true. This probability is not directly related to the probability of interest, the probability that the null hypothesis is in fact false (Diamond and Forrester, 1983; Lewis and Wears, 1993). Thus the decision to terminate the trial is based not on a calculation of the probability that one therapy is superior to the other but instead on a loosely related probability. In contrast, a Bayesian analysis allows the direct calculation of the probability that one therapy is superior to another (Berry, 1985).

The second conceptual difficulty arises because the classical decision to terminate a clinical trial does not take into account the cost and potential benefit of continued patient enrollment. The decision to continue a clinical trial should be based on the knowledge that the likely benefit, in terms of a reduced risk of an erroneous final conclusion, is greater than the expected cost of enrolling the additional patients. Conversely, the decision to terminate a clinical trial at an interim analysis should be based on the knowledge that continued patient enrollment is unlikely to reduce a final error risk to a degree that warrants the additional cost. No such comparison of expected benefit and expected cost is included in a classical interim analysis. In a Bayesian decision theoretic approach to interim analysis, however, the decision to continue or terminate a clinical trial after an interim analysis is based on an explicit comparison of the expected cost and benefit of continued patient enrollment.

4 BAYESIAN INTERIM ANALYSIS

When a Bayesian approach is used for interim data analysis, the probability density function (pdf) for the treatment effect is continually updated using Bayes' theorem and, at any time, an interim analysis may be performed by using the current pdf (Lewis and Wears, 1993; Freedman and Spiegelhalter, 1989; Berry, 1985). The likelihood principle implies that the interpretation of the data does not depend on the number of inspections of the pdf that have been performed or the stopping rule for the trial (Berry, 1985, 1987). Thus no penalty is paid for frequent interim analyses. This advantage of the Bayesian approach, i.e., no penalty for frequent interim analyses, is realized whether a simple pdf-based stopping rule is used (Freedman and Spiegelhalter, 1989; Berry, 1989; Lewis and Wears, 1993), or a more complex decision-theoretic approach is taken (Berry and Ho, 1988; Berry et al., 1992; Lewis and Berry, 1992, 1994).

4.1 pdf-Based Bayesian Interim Analysis

The first approach to the Bayesian interim analysis of this clinical trial that we consider is an approach which relies on the subjective interpretation of the pdf for the difference in efficacy of the two treatments (Lewis and Wears, 1993). The pdf for the difference in efficacy is used to determine the probability that phenytoin is better than placebo, based on the available data. This probability can then be used by the medical personnel involved in the trial to determine whether or not the trial should be terminated.

The advantage of this Bayesian approach over the classical approach is that physicians are likely to interpret the Bayesian probability correctly, as a true predictive probability, whereas they usually *misinterpret* the classical P value as a predictive probability, which it is not. Thus, while the actual decision-making process used by the clinicians is subjective and ill-defined in this approach to Bayesian interim analysis, at least it is based on an accurate interpretation of the results of the statistical analysis.

This type of simple pdf-based interim analysis can be used for relatively informal "safety analyses" of clinical trial data. Such analyses may be motivated by unexpected trends in the data or by new medical information that suggests the original criteria for stopping the trial may need to be changed. For the clinical trial considered here, this type of informal interim analysis was motivated by a trend toward a higher rate of seizures in the phenytoin-treated group, a completely unanticipated possibility.

At the time the informal Bayesian safety analysis was conducted, 14 patients had been enrolled in the trial. None of the seven patients (0%) given the placebo had suffered a seizure, but two of the seven (29%) given phenytoin had seizures. The goal of the analysis was to determine (1) the probability that phenytoin was more efficacious than placebo for preventing post-traumatic seizures and (2) the sensitivity of this probability estimate to changes in the assumed prior information.

We now will set up the Bayesian framework for the analyses that follow. We consider a clinical trial comparing two treatments for a disease with a binary outcome, loosely termed success and failure. The disease is closed head injury and the treatment failure is the occurrence of a seizure. The true rates of success are denoted p_i, $i = 1, 2$, and the difference in efficacy is $\delta = p_2 - p_1$. The index $i = 2$ corresponds to the phenytoin treatment and $i = 1$ corresponds to the placebo treatment. The null hypothesis is $\delta < \delta_0$ where δ_0 is the minimum (positive) difference in efficacy sought by the trial. The alternative hypothesis is $\delta > 0$. These are overlapping hypotheses.

Patients are enrolled into the trial in randomized permuted blocks of size $2N$, with N patients in each block receiving each therapy. After the jth block of patients has been studied, jN patients have been allocated to the ith treatment, with s_i successes and $f_i = jN - s_i$ failures. The trial may be terminated after any complete block and the null hypothesis accepted (A) or rejected (R) in favor of the alternative hypothesis.

The prior pdfs for p_i are beta(a_i, b_i), $i = 1, 2$, where the beta pdf is defined as usual,

$$\text{beta}(p \mid a, b) = \frac{\Gamma(a + b)}{\Gamma(a)\Gamma(b)} p^{a-1}(1 - p)^{b-1}$$

Because of the conjugate nature of beta distributions, the p_i continue to have beta distributions and are independent throughout the trial. The state or pattern of information is given by the number of blocks of patients treated, j, and the parameters of these beta distributions (x_i, y_i). If the state of information is (j, x_1, y_1, x_2, y_2) then the state after the next block is $(j + 1, x_1 + \Delta s_1, y_1 + \Delta f_1, x_2 + \Delta s_2, y_2 + \Delta f_2)$, where Δs_i and Δf_i are the numbers of successes and failures observed with the ith treatment in that block.

Based on previous studies, the incidence of post-traumatic seizures in the placebo group was considered most likely to be between 25% and 40% (a success rate of 60% to 75%). The large uncertainty in the expected incidence of post-traumatic seizures in the placebo group, despite the substantial number of studies published on post-traumatic seizures in children, occurs because there are important but poorly characterized differences between the different pediatric head-trauma populations that have been studied. This is manifested by the large variance in reported rates of post-traumatic seizures in different studies.

We consider three sets of prior pdfs, which we call wide, pessimistic, and optimistic. These prior pdfs are shown in Figure 1. The wide priors are $p_i = \text{beta}(2, 2)$ $i = 1, 2$. The pessimistic priors are $p_i = \text{beta}(7, 3)$, $i = 1, 2$. This set of prior pdfs is termed pessimistic because it implies that phenytoin is not more effective than placebo and the prior pdfs have

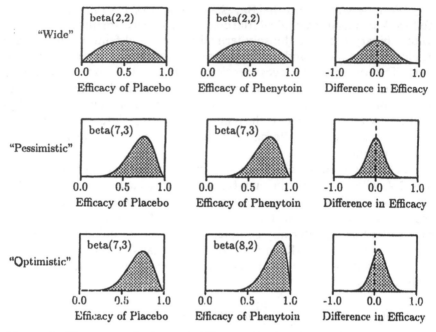

Figure 1 Three sets of prior probability density functions (pdfs) used in a pdf-based interim analysis of the phenytoin prophylaxis study. The first row shows a relatively wide set of priors for the efficacy of the placebo treatment, the efficacy of phenytoin, and the difference in efficacy, $\delta = p_2 - p_1$. The next row shows a "pessimistic" set of priors, in which the true efficacy of the placebo is better defined, and the prior for phenytoin is the same as the prior for the placebo. The last row shows an "optimistic" set of priors, in which the true efficacies of both the placebo and phenytoin are better defined than in the case of the "wide" priors, and there is some belief that phenytoin is more effective than placebo.

smaller variance than the wide priors. The optimistic priors are $p_1 =$ beta(7, 3) and $p_2 =$ beta(8, 2), suggesting a mild tendancy toward believing that phenytoin is more effective than placebo. Each pair of prior pdfs for the p_i imply a prior pdf for the difference in efficacy, δ. These prior pdfs for δ are shown in the right-hand column of Figure 1. Compared to "informative" priors used in other applications, all of the priors used in the present work are relatively "noninformative."

Each of the prior pdfs for δ, shown in the right-hand column of Figure 1 and the left-hand column of Figure 2, can be updated using the data for

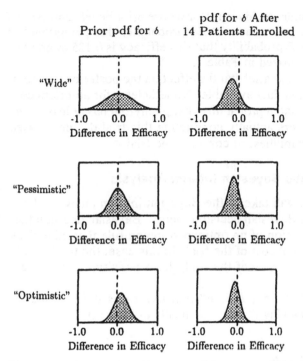

Figure 2 Three prior pdfs for the difference in efficacy between phenytoin and placebo, each paired with the resulting updated pdf using the data available after 14 patients were enrolled in the study.

the 14 patients enrolled in the study at the time of the informal interim analysis. The resulting pdfs for δ are shown in the right-hand column of Figure 2. These updated pdfs yield the quantitative predictive probabilities that are used to determine whether the clinical trial should be continued.

Consider the pdf for δ that results when the wide priors are updated. This posterior pdf, shown in the upper right-hand panel of Figure 2, implies that the probability that $\delta > 0.125$ is 0.044 and the probability that $\delta > 0$ is 0.152. In other words, an investigator whose prior beliefs correspond to the wide priors has, based on the new data, a 15.2% probability that phenytoin has some efficacy but only a 4.4% probability that this efficacy is as big as 0.125 (a reduction in the incidence of post-traumatic seizures from 25% to 12.5%, for example). Similarly, an investigator whose prior beliefs correspond to the pessimistic set of priors has a 19.7% probability that phenytoin has some efficacy but only a 4.2% probability that this efficacy is 0.125 or greater. An investigator whose prior beliefs correspond to the optimistic set of priors has a 32.7% probability that phenytoin has some efficacy and a 8.4% probability that this efficacy is 0.125 or greater. These results are summarized in Table 1.

The choice of priors has a substantial effect on the posterior probabilities, reflecting the limited data available. Nonetheless, for all priors considered, there is a substantial probability that phenytoin has at least some efficacy. The investigators actually conducting this trial decided, based on these posterior probabilities, to continue the trial.

4.2 Decision Theoretic Bayesian Interim Analysis

The other approach we will take to the Bayesian interim analysis of the clinical trial is based on Bayesian decision theory. Bayesian decision theory allows the determination of an optimal stopping rule that minimizes the expectation of the total cost of the trial. In this case, the total cost is defined as the final sample size of the trial plus a penalty, L, if a type I or type II error occurs.

The terminal loss function, $L(\delta, \text{action})$, expresses the hypothesis-testing focus of the trial and is shown graphically in Figure 3. It is defined by

$$L(\delta, A) = \begin{cases} 0 & \text{if } \delta \leq \delta_0 \\ K & \text{if } \delta > \delta_0 \end{cases}$$
$$L(\delta, R) = \begin{cases} K & \text{if } \delta < 0 \\ 0 & \text{if } \delta \geq 0 \end{cases} \tag{1}$$

This loss function implies a "zone of indifference" that extends from 0 to δ_0. If the true difference in efficacy lies in this region, no penalty is associated with accepting or rejecting the null hypothesis. This is equiva-

Table 1 pdf-Based Bayesian Interim Analysis
(After 14 Patients Enrolled)

Prior type	Priors π_1	Priors π_2	Prob $\delta > 0.0$	Prob $\delta > 0.125$
Wide	beta(2, 2)	beta(2, 2)	0.152	0.044
Pessimistic	beta(7, 3)	beta(7, 3)	0.197	0.042
Optimistic	beta(7, 3)	beta(8, 2)	0.327	0.084

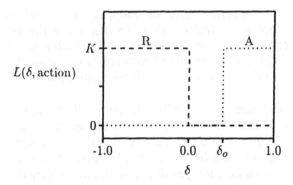

Figure 3 The terminal decision loss function for the action to accept (A) or reject (R) the null hypothesis $\delta < \delta_0$, as a function of the true value of δ.

lent to the "range of equivalence" used by Freedman and Spiegelhalter in their pdf-based approach to the development of Bayesian stopping rules (Freedman and Spiegelhalter, 1983, 1989; Freedman *et al.*, 1984).

Although Bayesian decision theoretic designs for clinical trials have been suggested for a wide variety of loss functions, we approached the design of this clinical trial from an unusual Bayesian perspective. Specifically, we use a loss function that explicitly addresses the classical problem of hypothesis testing. In this manner, we will obtain a class of Bayesian designs that can be compared directly with classical group-sequential trial designs on frequentist grounds.

The unit of cost is the cost of enrolling a patient into the study, which is assumed constant. The total cost for a trial that is terminated after the jth block of patients is the sample size plus the terminal loss, or $2jN + L(\delta, \text{action})$. The constant K in the terminal loss function [Eq. (1)] is the cost of committing a classical error, either type I or type II. A large value of K implies that the accuracy of the final conclusion is important relative to the cost of patient enrollment. Thus clinical trial designs developed with larger values of K will have larger average sample sizes and lower error rates.

In most decision analysis problems, the loss function is chosen to represent as closely as possible the true costs incurred for various actions and true states of nature. Here, because we wish to create a Bayesian decision theoretic trial design that is directly comparable to the classical group-sequential design, a value of K will be used which is found to yield type I and type II error rates near 0.05.

After any block j, the overall cost of stopping the trial is $W_{stop}(j, x_1, y_1, x_2, y_2) = 2jN + \min[EL(\delta, A), EL(\delta, R)]$ where the expectation E is with respect to the current distribution of δ, which is given by $\pi_\delta = \int \pi_2(p_1 + \delta)\pi_1(p_1) \, dp_1$. Thus, during a "forward" calculation the value of $W_{stop}(j, x_1, y_1, x_2, y_2)$ may be directly calculated for all possible clinical trial results.

We will use a truncated design, meaning that the trial will be stopped after at most M blocks of patients. The value of M is chosen to yield a maximum sample size similar to that used by the classical Pocock design.

To initiate the backward induction (Berger, 1985; DeGroot, 1970), we determine the costs associated with each possible trial outcome after the Mth block of patients. A terminal action, either to accept or reject the null hypothesis, will be necessary should sampling continue through block M, so the cost at that block is $W_{stop}(M, x_1, y_1, x_2, y_2)$. The possible values of x_i, y_i are restricted by $(x_1, y_1, x_2, y_2) = (a_1 + s_1, b_1 + f_1, a_2 + s_2, b_2 + f_2)$ where $s_1 + f_1 = s_2 + f_2 = MN$.

The remainder of the function W is defined and calculated recursively, starting at $j = M$ and proceeding backward to $j = 0$. The fundamental backward induction equation is

$$W(j, x_1, y_1, x_2, y_2)$$
$$= \min[W_{stop}(j, x_1, y_1, x_2, y_2), W_{cont}(j, x_1, y_1, x_2, y_2)]$$

for $j = M - 1, M - 2, \ldots, 0$, where W_{stop} is defined above and

$$W_{cont}(j, x_1, y_1, x_2, y_2)$$
$$= EW(j + 1, x_1 + \Delta s_1, y_1 + \Delta f_1, x_2 + \Delta s_2, y_2 + \Delta f_2)$$

Here, the expectation E refers to the predictive distribution of the numbers of successes, Δs_i, and failures, $\Delta f_i = N - \Delta s_i$, in block $j + 1$. This predictive distribution is given by the product of two beta-binomial distributions (Ferguson, 1967),

$$\prod_{i=1}^{2} \binom{N}{\Delta s_i} \left[\frac{\Gamma(x_i + y_i)\Gamma(x_i + \Delta s_i)\Gamma(y_i + \Delta f_i)}{\Gamma(x_i)\Gamma(y_i)\Gamma(x_i + y_i + N)} \right]$$

As the function W is calculated, the action that results in the minimum W for each possible interim and final trial result is recorded. This pattern of actions is the stopping and decision rule for the trial.

The final calculation in this backward induction gives $W(0, a_1, b_1, a_2, b_2)$. In performing the actual calculations, the contributions of the terminal loss function and the number of patients enrolled to the function $W(0, a_1, b_1, a_2, b_2)$ are kept separate. This allows the determination of the expected sample size for the trial design and, when the terminal loss contribution is divided by K, the expected overall error rate.

5 NUMERICAL SIMULATIONS

Characteristics of the Bayesian clinical trial design can be determined by Monte Carlo simulation. Two types of simulations were performed, differing in the distributions of the true values of the p_i. In Bayesian simulations, the true values of the p_i used for each simulated trial were sampled from beta(a_i, b_i). Thus there was exact agreement between the prior pdf assumed and the true distribution of the p_i. The mean sample size under these conditions is denoted $\bar{n}$ and the rate of type I and type II errors is termed the Bayesian error rate.

In the second set of simulations, the p_i were given fixed values, allowing the determination of classical error rates. Classical type I error rates were determined with $p_1 = p_2 = 0.750$ and are denoted α. The mean sample size under these conditions is $\bar{n}_\alpha$. Classical type II error rates were determined with $p_1 = 0.750$ and $p_2 = 0.875$ and are denoted β. The corresponding mean sample size is $\bar{n}_\beta$.

The mean cost of the Bayesian trial $\bar{c}$ is $\bar{n}$ plus ($K \times$ Bayesian error rate) and, for comparison, the same definition was used for the mean cost of a classical group-sequential design, despite the fact that the constant K has no relevance to the frequentist. The value of K used to determine the mean cost of the classical design was that used to create the Bayesian design.

One-tailed classical group-sequential clinical trials of Pocock (1977) and O'Brien and Fleming (1979) were simulated as well. Three interim analyses and one final analysis were used with the overall $\alpha = 0.05$. Group sizes ($2N$) were chosen to yield classical type II error rates close to 0.05.

Table 2 shows the Bayesian and classical error rates and mean sample sizes of the Bayesian trial design and the classical designs. The Bayesian design uses up to 32 blocks of 16 patients each. The prior pdfs are beta(2, 2), the "wide" priors used earlier. The value of K is 20,000. As expected, the Bayesian design has a substantially lower mean cost $\bar{c}$ than the classical designs. It is more surprising to compare the mean sample sizes for the classical simulations, in which the values of the p_i were fixed. The Bayesian design has a lower mean sample size than the classical designs, despite having similar classical error rates.

The sample size difference shown in Table 2 can be investigated by examining the dependence of the mean sample size on the p_i for the differ-

Table 2 Comparison of Bayesian and Classical Group Sequential Designs

Design type	2N	Max n	Bayesian Error rate	$\bar{n}$	$\bar{c}$	α	$\bar{n}_\alpha$	β	$\bar{n}_\beta$
Pocock	124	496	0.00509	364.1	465.9	0.0482	484.7	0.0496	253.6
OBF[a]	106	424	0.00481	335.6	431.7	0.0489	421.3	0.0488	298.7
Bayesian	16	512	0.00347	131.6	201.0	0.0499	257.5	0.0482	242.9

[a] OBF = O'Brien-Fleming.

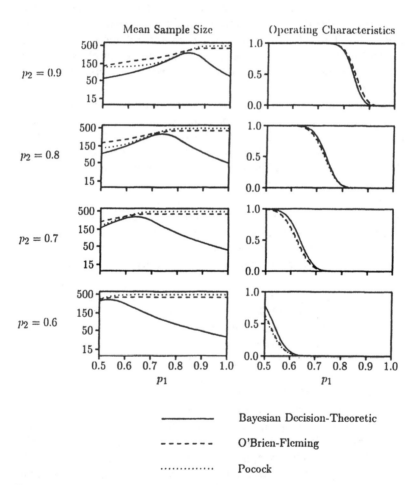

Figure 4 Mean sample sizes (left-hand panels) and operating characteristics (right-hand panels) of the Bayesian clinical trial design and two classical clinical trial designs.

ent trial designs. The left-hand panels of Figure 4 show the mean sample size of each clinical trial design, as a function of the p_i. For all values of the p_i studied, the Bayesian design had a lower mean sample size than the classical designs. The right-hand panels of Figure 4 show the operating characteristics of the same trial designs, also as a function of the p_i. In general, the operating characteristics are very similar, which is not surprising given that the values of K and the sample sizes of the classical designs were chosen to give similar classical type I and type II error rates.

6 APPLICATION TO THE PHENYTOIN PROPHYLAXIS TRIAL

As can be seen from Table 2, a Bayesian decision theoretic design using $K = 20,000$ and beta(2, 2) priors has appropriate characteristics for use with the phenytoin prophylaxis trial. The first interim analysis with this Bayesian design occurred after 16 patients had been enrolled. At that point there were seven of eight treatment successes in the placebo group and six of eight treatment successes in the phenytoin group.

Given these interim data, the action that minimizes the expected cost of the trial is to continue patient enrollment, and the probability that phenytoin has at least some efficacy is 0.318. The probability that the efficacy is at least 0.125 is 0.121. The expected number of additional patients to be enrolled, if the trial is conducted according to the optimal decision rule, is 139.9 patients and the probability that a type I or type II error will ultimately be made is 0.004. It is the decrease in expected error rate, made possible through the enrollment of additional patients, that justifies this additional enrollment. If one were forced to stop the trial and draw a conclusion from the limited data, the best conclusion would be that phenytoin is not efficacious and the probability of a type II error would be 0.121. Note that $K \times (0.121 - 0.004) \geqslant 139.9$. In other words, the expected decrease in cost due to the decreased chance of error is larger in magnitude than the expected cost of patient enrollment. Thus the optimal decision is to continue the trial.

The Bayesian decision theoretic approach to the design of a group-sequential clinical trial is potentially superior to the classical approach for several reasons. First, the interim analyses are performed using predictive probabilities instead of P values. Second, there is no penalty in error risk or sample size for frequent interim analyses. The effect of incorporating many interim analyses into a Bayesian design is generally a reduction in the expected sample size (Lewis and Berry, 1994). The sample-size savings of the Bayesian design over the classical designs is explained in part by the additional interim analyses allowed with the Bayesian approach.

Third, even when a "one-tailed" hypothesis-testing approach is taken, a Bayesian trial may be terminated early when the test treatment appears ineffective, whereas in a classical one-tailed group sequential trial, sampling must continue through the final block before the null hypothesis can be accepted (Lewis and Berry, 1994).

The Bayesian design presented here allows interpretation of the final results along either Bayesian or frequentist lines. To the Bayesian, the design has the advantages of minimizing the expectation of the total cost and allowing the direct calculation of the pdf for the difference in efficacy, δ. To the frequentist, the design has well-characterized classical type I and type II error rates and generally leads to a reduction in the mean sample size relative to commonly used classical group-sequential designs (Lewis and Berry, 1994).

REFERENCES

Annegers, F. (1983). The epidemiology of head trauma in children. In *Pediatric Head Trauma*, ed. Shapiro, I., pp. 1–10. Futura, Mount Kisco, NY.

Berger, J. O. (1985). *Statistical Decision Theory and Bayesian Analysis*. Springer-Verlag, New York.

Berry, D. A. (1985). Interim analysis in clinical trials: Classical vs. Bayesian approaches. *Statistics in Medicine* 4: 521–526.

Berry, D. A. (1987). Interim analysis in clinical trials: The role of the likelihood principle. *American Statistician* 41: 117–122.

Berry, D. A. (1989). Monitoring accumulating data in a clinical trial. *Biometrics* 45: 1197–1211.

Berry, D. A., and Ho, C. H. (1988). One-sided sequential stopping boundaries for clinical trials: A decision-theoretic approach. *Biometrics* 44: 219–227.

Berry, D. A., Wolff, M. C., and Sack, D. (1992). Public health decision making: A sequential vaccine trial. In *Bayesian Statistics 4*, ed. Bernardo, J. M., Berger, J. O., Dawid, A. P., and Smith, A. F. M., pp. 79–92. Oxford University Press, Oxford.

Bruce, D. A. (1990). Head injuries in the pediatric population. *Curr. Probl. Pediatr.* 20: 61–107.

DeGroot, M. H. (1970). *Optimal Statistical Decisions*. McGraw-Hill, New York.

Diamond, G. A. and Forrester, J. S. (1983). Clinical trials and statistical verdicts: Probable grounds for appeal. *Ann. Intern. Med.* 98: 385–394.

Ferguson, T. S. (1967). *Mathematical Statistics: A Decision Theoretic Approach*, pp. 100 and 104. Academic Press, New York.

Freedman, L. S., and Spiegelhalter, D. J. (1983). The assessment of subjective opinion and its use in relation to stopping rules for clinical trials. *Statistician* 32: 153–160.

Freedman, L. S. and Spiegelhalter, D. J. (1989). Comparison of Bayesian with group sequential methods for monitoring clinical trials. *Controlled Clinical Trials* 10: 357–367.

Freedman, L. S., Lowe, D., and Macaskill, P. (1984). Stopping rules for clinical trials incorporating clinical opinion. *Biometrics* 40: 575–586.

Geller, N. L. and Pocock, S. J. (1987). Interim analyses in randomized clinical trials: Ramifications and guidelines for practitioners. *Biometrics* 43: 213–223.

Hahn, Y. S., Chyung, C., Barthel, M. J., Bailes, J., Flannery, A. M., and McLone, D. G. (1988a). Head injuries in children under 36 months of age: Demography and outcome. *Child's Nerv. Syst.* 4: 34–40.

Hahn, Y. S., Fuchs, S., Flannery, A. M., Barthel, M. J., and McLone, D. G. (1988b). Factors influencing post-traumatic seizures in children. *Neurosurgery* 22: 864–867.

Hendrick, E. B. and Harris, L. (1968). Post-traumatic epilepsy in children. *J. Trauma* 8: 547–556.

James, H. E. (1986). Neurologic evaluation and support in the child with an acute brain insult. *Pediatric Annals* 15: 16–22.

Jennett, B. (1973). Trauma as a cause of epilepsy in childhood. *Develop. Med. Child Neurol.* 15: 56–62.

Jennett, W. B. (1969). Early traumatic epilepsy. *Lancet* 1: 1023–1025.

Kaplan, E. L. and Meier, P. (1958). Nonparametric estimation from incomplete observations. *J. Am. Stat. Assoc.* 53: 457–481.

Kim, K. and DeMets, D. L. (1987). Design and analysis of group sequential tests based on the type I error spending rate function. *Biometrika* 74: 149–154.

Kraus, J. F., Fife, D., Cox, P., Ramstein, K., and Conroy, C. (1986). Incidence, severity, and external causes of pediatric brain injury. *AJDC* 140: 687–693.

Lan, K. K. G. and DeMets, D. L. (1983). Discrete sequential boundaries for clinical trials. *Biometrika* 70: 659–663.

Lewis, R. J. (1993). An introduction to the use of interim analyses in clinical trials. *Annals of Emergency Medicine* 22: 1463–1469.

Lewis, R. J. and Berry, D. A. (1992). A comparison of Bayesian and classical group-sequential clinical trial designs. *Annals of Emergency Medicine* 21: 641.

Lewis, R. J. and Berry, D. A. (1994). Group-sequential clinical trials: A classical evaluation of Bayesian decision theoretic designs. *J. Am. Stat. Assoc.* 89: 1528–1534.

Lewis, R. J. and Wears, R. L. (1993). An introduction to the Bayesian analysis of clinical trials. *Annals of Emergency Medicine* 22: 1328–1336.

Lewis, R. J., Yee, L., Inkelis, S. H., and Gilmore, D. (1993). Clinical predictors of post-traumatic seizures in children with head trauma. *Annals of Emergency Medicine* 22: 1114–1118.

McPherson, K. (1982). On choosing the number of interim analyses in clinical trials. *Statistics in Medicine* 1: 25–36.

O'Brien, P. C. and Fleming, T. R. (1979). A multiple testing procedure for clinical trials. *Biometrics* 35: 549–556.

Pocock, S. J. (1977). Group sequential methods in the design and analysis of clinical trials. *Biometrika* 64: 191–199.

Teasdale, G. and Jennett, B. (1974). Assessment of coma and impaired consciousness: A practical scale. *Lancet* 2: 81–84.

Temkin, N. R., Dikmen, S. S., Wilensky, A. J., Keihm, J., Chabal, S., and Winn, H. R. (1990). A randomized, double-blind study of phenytoin for the prevention of post-traumatic seizures. *N. Engl. J. Med.* 323: 497–502.

Young, B., Rapp, R. P., Norton, J. A., Haack, D., Tibbs, P. A., and Bean, J. R. (1983). Failure of prophylactically administered phenytoin to prevent early post-traumatic seizures. *J. Neurosurg.* 58: 231–235.

11

Inference and Design Strategies for a Hierarchical Logistic Regression Model

Merlise Clyde, Peter Müller, and Giovanni Parmigiani   Duke University, Durham, North Carolina

ABSTRACT

This chapter focuses on Bayesian inference and design in binary regression experiments. As a case study, we consider heart defibrillator experiments, in which the number of observations that can be taken is limited and it is important to incorporate all available prior information. In particular, by modeling the individual-to-individual variation in the appropriate defibrillation setting, we can use information on past patients in formulating a sensible prior distribution for designing experiments for current patients. The first part illustrates the use of hierarchical models to obtain such prior distributions. The second part of the chapter considers design strategies. An important advantage of a Bayesian technique is that it is conceptually easy to adapt to information that accrues sequentially. This is particularly desirable when early stopping of the experimentation is of interest. In general, analytic expressions for optimal sequential solutions are not available, and a combination of approximation techniques and numerical computation must be used. Here we focus on finding optima within restricted sets of strategies. We compare an adaptive strategy based on fixed percentage changes in the energy levels and variable sample size with a strategy in which all levels are chosen optimally but the sample size is fixed.

1   INTRODUCTION

The goal of this chapter is to give a practical illustration of Bayesian inference and design in binary regression experiments. In general, formalization of prior information and explicit consideration of objectives in a decision theoretic framework are important aspects of most Bayesian design problems. In addition, two important features often arise in applications:

> The experiment that is being designed will be performed on a certain unit, subject, batch etc., and information is available in the form of similar experiments on exchangeable units, subjects, or batches.
> Analytic expressions for the optimal solutions are not available, and a combination of approximation techniques and numerical computation techniques must be used to determine the optimal design.

These can be addressed successfully within the Bayesian approach. This chapter is a case study in which both issues appear, and, accordingly, it has two parts. The first part illustrates the use of hierarchical models to obtain the prior distribution for the next unit. The prior distribution is derived as the distribution of the next draw of unit-specific parameters, conditional on the units already observed. The second part considers design strategies. We illustrate two numerical procedures that have been developed (Chaloner and Larntz, 1989; Müller and Parmigiani, 1995).

We consider designs for heart defibrillator experiments. An automatic implantable cardioverter defibrillator is a device that detects fibrillation in a patient's heart and responds by discharging a specified pulse of energy to the heart to restore the normal rhythm. The probability of successful defibrillation depends on the energy level and can be modeled as in dose-response using, for example, logistic regression. The required energy level, however, is patient specific. Using the highest possible setting may lead to 100% successful defibrillation, but it is not acceptable because of the risk of myocardial damage and premature battery failure in the defibrillator. Therefore, while implanting a defibrillator, the surgeon fibrillates the patient's heart several times at different test strengths to estimate the strength that will be necessary for successful defibrillation of this patient's heart. Typically, tests are carried out at successively lower strengths until defibrillation does not occur, with the penultimate test strength (or some function of it) used as the setting for the device. Bourland et al. (1978) and Manolis et al. (1989) discuss general issues related to implantable defibrillators.

The number of observations that can be taken on an individual patient's heart is limited. Therefore, the choice of test strengths is extremely important in efficiently learning about the appropriate setting for the patient's defibrillator. Incorporating all available prior information is essential. In particular, modeling the patient-to-patient variation in the appropriate defibrillation setting allows for using information on past patients in formulating a sensible prior distribution. While tests are carried out sequentially on a patient's heart, the problem of determining the optimal sequence of test strengths is computationally very demanding. It is therefore of interest to find fast and yet efficient ways of choosing the test strengths and the setting. In this chapter we discuss a Bayesian decision theoretic approach that stresses these two aspects of the problem.

The material in the remainder of this chapter is organized as follows. Section 2 describes the hierarchical logistic regression model and the Markov chain Monte Carlo methods used for inference. Bayesian analysis of hierarchical logistic regression requires special care, as not all the so-called full conditional distributions are available in closed form. We describe an efficient simulation strategy. Section 3 considers optimal design problems and optimization strategies. We address both fixed and adaptive designs. In particular, we consider the approximately optimal fixed design of Clyde et al. (1995) and compare it to a computationally simpler alternative based on equal spacing of test strengths on the log scale. We also illustrate two different computational tools: for the Clyde et al. (1994) design we use an analytic approximation. For the log-uniform design we use a Monte Carlo–based scheme, which can be efficiently integrated with the methods of Section 2. Finally, we briefly review some further results of Clyde et al. on the efficiency of certain adaptive designs.

Applications of Bayesian techniques in defibrillator design problems are described by Malkin et al. (1993) and Clyde et al. (1995). Further references on design for logistic regression include Tsutakawa (1972, 1980), Chaloner and Larntz (1989), Chaloner (1993), and Flournoy (1993). The defibrillation experiments and models have many aspects in common with problems in phase I clinical trials and dose-response experiments for bioassay; see, for example, Zacks (1977), Freeman (1983), Storer (1989), O'Quigley et al. (1990), Gatsonis and Greenhouse (1992), Durham and Flournoy (1993) and Wakefield (1994). The book by Govindarajulu (1988) discusses up-down designs and other sequential approaches in bioassay problems. Chaloner and Verdinelli (1994) give an excellent review of recent research in Bayesian experimental design.

2   HIERARCHICAL MODEL

2.1   Model and Prior Distributions

To allow for patient-specific settings, we fit the following hierarchical model for the current data on J patients. Patient j is tested at energy levels

$x_{1j}, \ldots, x_{I_j j}$ and at each design point x_{ij}, the experimenter observes a Bernoulli response variable y_{ij}, where $y_{ij} = 1$ indicates that defibrillation was successful at test level x_{ij} and $y_{ij} = 0$ indicates that defibrillation was unsuccessful. The response variable y is assumed to be related to the energy level x and unknown parameters θ by the probability $P(y = 1 \mid x, \theta) = p(\theta, x)$.

Many choices for $p(\theta, x)$ are discussed and compared in the defibrillation literature (Gliner *et al.*, 1990; McDaniel and Schuder, 1987; Davy *et al.*, 1987; Davy, 1984). We use the logistic regression model

$$p(y = 1 \mid \theta, x) = \frac{1}{1 + \exp\{-\beta(x - \lambda) - \log(.95/.05)\}} = p(\theta, x) \qquad (1)$$

where $\theta = (\beta, \lambda)$. All of the methods presented here can be adapted to other probability models. Conditional on the parameter vector θ and on the energy level x, successive responses are assumed to be independent. In general, there is enough time between successive tests so that any carryover effect is assumed to be negligible.

The goal is to estimate the implantation strength, which we take to be the effective strength that defibrillates 95% of the time (the ED_{95}). In this parameterization, the ED_{95} corresponds to λ and we use λ and ED_{95} interchangeably in the remainder of the chapter. By replacing $\log(.95/.05)$ with $\log(q/(1 - q))$, we can easily consider design and analysis for other quantiles of interest.

For the hierarchical model, we assume that each patient has a separate logistic regression, with parameters (β_j, λ_j), for $j = 1, \ldots, J$. We find it useful to reparameterize (1) using $\phi = (\log(\beta), \log(\lambda))$. The ϕ parameterization is more convenient, because in the decision problem we will be primarily interested in the marginal posterior distribution of $\log(\lambda)$. Also, the log posterior in this parameterization is better approximated by a quadratic function. This improves the accuracy of the normal approximations used in finding the optimal nonsequential designs and can improve convergence rates of the Markov chain methods used in obtaining the posterior distributions.

To complete the hierarchical model setup, the ϕ_j's are assumed to come from a normal distribution with mean μ and covariance matrix V, $N(\mu, V)$, representing the distribution of the population of patients. We also used conjugate hyperpriors for μ and V; that is, μ is normally distributed with mean m and covariance matrix B and V^{-1} has a Wishart distribution with scalar parameter q and matrix parameter $(qQ)^{-1}$, denoted by $W(q, (qQ)^{-1})$. In this representation, $E(V^{-1}) = Q^{-1}$.

In summary, if $i = 1, \ldots, I_j$ indexes the test strengths and $j = 1, \ldots, J$ indexes the patients, we have the following distributions for the hierarchical model:

$$y_{ij} \mid \theta_j \sim \text{Bernoulli}(p(\theta_j, x_{ij})), \qquad \theta_j = (\beta_j, \lambda_j)$$

$$\phi_j = (\log \beta_j, \log \lambda_j) \mid \mu, V \sim N(\mu, V)$$

$$\mu \sim N(m, B)$$

$$V^{-1} \sim W(q, (qQ)^{-1})$$

2.2 A Markov Chain Monte Carlo Scheme

The posterior distribution cannot be obtained analytically. However, a simulated sample from the posterior distribution suffices for making the required inferences. Such a sample can be generated by Markov chain Monte Carlo (MCMC) methods. For an overview of MCMC in a variety of models see, for example, Smith and Roberts (1993), Gelfand and Smith (1990), and Gilks *et al.* (1993). Albert and Chib (Chapter 22) also consider MCMC methods for a related model based on a probit regression model. We now describe our particular MCMC implementation used in the logistic regression framework.

We generate a posterior Monte Carlo sample by simulating a Markov chain described as follows. Assume that $(\phi_1, \ldots, \phi_J, \mu, V)$ is the current state of the simulated Markov chain. To move to the next state we go through the following three steps.

1. We generate a new value for μ from the full conditional posterior distribution,

$$\mu \mid \phi_1, \ldots, \phi_J, V, y \sim N(m^*, B^*)$$

where

$$\bar{\phi} = \sum_{j=1}^{J} \frac{\phi_j}{J}$$

$$B^* = (JV^{-1} + B^{-1})^{-1}$$

$$m^* = B^*(JV^{-1}\bar{\phi} + B^{-1}m)$$

2. To obtain a new value for V we draw from the conditional posterior distribution,

$$V^{-1} \mid \phi_1, \ldots, \phi_J, \mu, y \sim W\left(q + J, \left(qQ + \sum_{j=1}^{J} (\phi_j - \mu)(\phi_j - \mu)^T\right)^{-1}\right)$$

3. The ϕ's are updated using a Metropolis step (Smith and Roberts,

1993) as follows. Candidate values ϕ_j^* are generated by

$$\phi_j^* \sim N(\phi_j, D), \qquad j = 1, \ldots, J$$

where D is a two-by-two diagonal matrix with $D_{ii} = V_{ii}$. Let $p_{\phi_j}(\phi \mid \mu, V)$ denote the normal prior density evaluated at ϕ and use $p(\exp(\phi), x)$ to denote the Bernoulli probability defined by Eq. (1). Each of the candidates ϕ_j^* is accepted with probability $\min[1, \rho_j]$, where

$$\rho_j = \frac{p_{\phi_j}(\phi_j^* \mid \mu, V) \prod_{i=1}^{l_j} p(\exp(\phi_j^*), x_{ij})^{y_{ij}}(1 - p(\exp(\phi_j^*), x_{ij}))^{1-y_{ij}}}{p_{\phi_j}(\phi_j \mid \mu, V) \prod_{i=1}^{l_j} p(\exp(\phi_j), x_{ij})^{y_{ij}}(1 - p(\exp(\phi_j), x_{ij}))^{1-y_{ij}}}$$

If the candidate value ϕ^* is accepted, then the current value of ϕ_j is replaced by ϕ_j^* and otherwise it remains unchanged. Alternative to the Metropolis step described here, an independence chain step, based on a normal approximation for the complete conditional $p(\phi_j \mid \mu, V, y)$, could be implemented. Details are described, for example, in Müller and Rosner (1994) or Gamerman (1994).

By iterating steps 1–3, we generate an approximate sample from the posterior distribution, $p(\phi_1, \ldots, \phi_J, \mu, V \mid y)$.

To illustrate the methods, we have used data from a defibrillator experiment on 10 dogs with 304 tests provided by Mark Kroll, AngeMed, Inc. The hyperparameters are chosen to obtain relatively disperse distributions for μ and V^{-1}, but the choice is such that virtually all of the mass of the predictive distribution for a new patient is within known physical limits. In particular, we used $q = 10$, $m = (0, 2)$ and

$$Q = \begin{bmatrix} 0.44 & -0.12 \\ -0.12 & 0.14 \end{bmatrix}, \qquad B^{-1} = \begin{bmatrix} 25 & 5 \\ 5 & 25 \end{bmatrix}$$

The Markov chain was run for 10,000 iterations. Excluding the first 1000 iterations as a burn-in period, we kept every 10th sample for 900 simulated values to estimate the posterior quantities of interest. Figure 1 shows the estimated posterior densities for μ_2 and $\lambda_1, \ldots, \lambda_{10}$ and the dog-to-dog variation.

2.3 Utility for Estimation

Throughout our discussion we will assume that the goal of the analysis is estimation of the ED_{95}. In implantation problems, setting the patient-specific level at the ED_{95} can be thought of as a crude resolution of the trade-

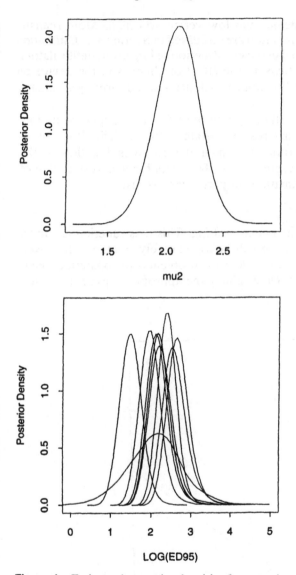

Figure 1 Estimated posterior densities for μ_2, prior mean of the λ_i's (log(ED$_{95}$)), and for $\lambda_1, \ldots, \lambda_{10}$.

off between setting the device too low versus too high. More realistic formulations of the utility function are discussed in Section 3.2. Estimation of the ED_{95} is a realistic goal in the construction of improved defibrillators, where it is of interest to measure the effect of changes in the device on quantiles of the response function, which are related directly to energy consumption.

Typically, there is uncertainty about the ED_{95} after experimentation. A convenient and somewhat realistic strategy for modeling the consequences of estimation errors is a squared error loss function in the $\log(ED_{95})$ scale. This corresponds to higher costs for underestimation of the ED_{95} than for overestimation. In particular, we use

$$l(\phi, d) = \{\log(\lambda) - d\}^2$$

where d (for decision) is the chosen setting. This loss function is shown in Figure 2. It is realistic while still being relatively tractable. It is well known that the Bayes rule for this decision problem is the posterior mean, $E(\log(\lambda) \mid y)$ (Berger, 1985). Determining the appropriate patient-specific

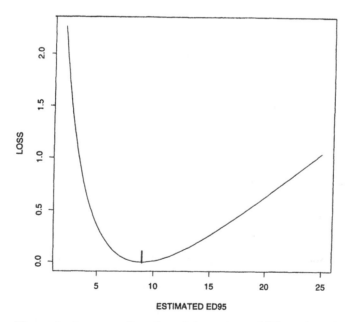

Figure 2   Loss as a function of the estimated ED_{95} when the true value is at λ = 9.0 (marked by the vertical bar). The steep penalty for underestimation reflects the asymmetry of the decision problem.

Table 1 Posterior Mean of the $\log(ED_{95})$
and Expected Losses for the 10 Dogs[a]

Setting	$E[\log(ED_{95}) \mid Y]$	$E(\text{loss} \mid Y)$
4.48	1.50	0.0386
7.28	1.98	0.0448
8.22	2.10	0.4364
8.68	2.16	0.0516
9.10	2.20	0.0456
9.11	2.21	0.0470
9.61	2.26	0.0637
11.11	2.40	0.0280
13.42	2.59	0.0650
14.72	2.68	0.0532

[a] The final optimal setting using this loss function is exp($E[\log(ED_{95}) \mid Y]$).

defibrillation shock strength reduces to calculating the posterior mean of the $\log(ED_{95})$ given the responses from the test data, and then transforming to the original scale. The posterior means for $\log(\lambda_j)$ are estimated by taking the average from the simulated posterior sample. The results for the 10 dogs are given in Table 1.

In the next section, we consider how to design a defibrillator experiment for a new patient. This design takes into account the information on the current sample of patients.

3 EXPERIMENTAL DESIGNS FOR NEW PATIENTS

3.1 Criteria

To determine the optimal choice of design for a new patient, in Clyde *et al.* (1995), we used a utility function combining the cost of observation with the loss resulting from estimation error. The latter loss reduces to the posterior variance of the $\log(ED_{95})$ used in the previous section for estimating the appropriate patient-specific strength. We approach the design problem with both nonsequential and sequential techniques and discuss two different computational strategies. One is based on an analytic approximation discussed in Chaloner and Larntz (1989) and implemented in Clyde (1993). The other uses a Monte Carlo–based scheme introduced in Müller and Parmigiani (1995).

The objective of experimentation is to estimate $\log(\lambda)$ using an experimental design, D. D refers to either a collection of design points $\{x_1, \ldots, x_n\}$ or a particular rule on how to determine the design points, which may depend on the data. One component of the utility function is the estimation loss

$$u(\phi, y, D) = -\{\log(\lambda) - E[\log(\lambda) \mid y, D]\}^2 \qquad (2)$$

as in Section 2.3. The second component of the utility function is the sampling cost $C(D, y)$, for which we use $C(D, y) = n(D, y) \cdot c$, where $n(D, y)$ is the number of test strengths associated with the design D and c is a fixed cost per observation.

In this framework, the design problem can be formally described as choosing D to maximize the preposterior expected value of the utility function. We want to select a design that leads to high expected posterior utility, but because we have not observed the data prior to experimentation, we also need to take into account the possible experimental outcomes by averaging over the possible values of Y. The expectation is taken over the posterior distribution of the parameter vector ϕ in the parameter space Φ and the marginal distribution of the data vector y for a new patient in the sample space $\mathcal{Y}$ under design D. Formally, the utility function for a design D is

$$\mathcal{U}(D) = \int_{\mathcal{Y}, \Phi} [u(\phi, y, D) - c \cdot n(D, y)] p_D(\phi \mid y) \, p_D(y)$$

where $p_D(\phi \mid y)$ is the posterior distribution of ϕ for the new patient and $p_D(y)$ is the marginal distribution of Y, given that the design D was used for the experiment. This simplifies to

$$\mathcal{U}(D) = -V(D) - c \cdot \int_{\mathcal{Y}} n(D, y) p_D(y) \qquad (3)$$

where

$$V(D) \equiv \int_{\mathcal{Y}} \mathrm{Var}(\log(\lambda) \mid y, D) p_D(y) \qquad (4)$$

and $\mathrm{Var}(\log(\lambda) \mid y, D)$ is the posterior variance of $\log(\lambda)$ given y and the design D. Elicitation of the trade-off parameter c, which is expressed in the units of the estimation utility, can be complex. General discussion and graphical tools are presented in Parmigiani and Polson (1992) and Verdinelli and Kadane (1992).

Finding optimal designs by maximizing Eq. (3) is difficult computationally because there is no closed-form expression for the posterior variance of the $\log(ED_{95})$ in Eq. (4) and computing the preposterior utility in Eq. (3) requires integrating the posterior variance of the $\log(ED_{95})$ with respect to the marginal distribution of Y. We discuss a Monte Carlo approach for low-dimensional optimization and an analytic approximation for larger nonsequential designs.

3.2 Alternative Criteria

In defibrillator implantation, the goal is to choose the device setting in a way that is beneficial for the patient. Too low a level may result in defibrillation failure and death. On the other hand, higher settings increase the probability of serious myocardial damage, although not necessarily immediate death. Of lesser importance, higher settings also increase the energy consumption of the defibrillator, leading to premature battery failure and early replacement of the device, which requires additional surgery.

Although the squared error loss function on the log scale does reflect the asymmetric nature of the decision problem, the trade-offs between injury versus death cannot be explicitly taken into account. More complex alternatives can be constructed in various ways. For example, one could consider the expected life expectancy of the patient, perhaps adjusting it for quality of life as is often done in cost-utility analysis of medical decision problems (Kamlet, 1991). Maximizing this expectation would lead to the optimal setting. The resulting life expectancy could serve as the design criterion.

A simpler alternative is to focus solely on the first episode of fibrillation after implantation. A setting appropriately balancing the risks of deaths and injury in the first episode is hoped to perform satisfactorily over a number of such episodes. To illustrate the relationship between the simple loss function adopted in this chapter and more realistic ones, we outline this case briefly. We need to model explicitly the probability of death and probability of injury as a function of the strength and directly include the associated costs. These can be elicited from experts and patient, respectively. Let $C_{\mathcal{I}}$ and $C_{\mathcal{F}}$ denote the costs associated with injury and death from failure, respectively. Also let $p(\mathcal{F} \mid x)$ represent the probability of death when the defibrillator fails and $p(\mathcal{I} \mid x)$ represent the probability of injury. The resulting loss function is

$$l(\theta, x) = C_{\mathcal{F}} \, p(\mathcal{F} \mid x) \, \{1 - p(\theta, x)\} + C_{\mathcal{I}} \, p(\mathcal{I} \mid x) \tag{5}$$

where $p(\theta, x)$ models the probability of defibrillation as before. Given the data y from experiment D, we can solve for the shock strength x that minimizes the posterior expected loss,

$$C_{\mathcal{F}} \, p(\mathcal{F} \mid x) \, \{1 - E[p(\theta, x)|y, D]\} + C_{\mathcal{I}} \, p(\mathcal{I} \mid x) \tag{6}$$

The costs for injury could also depend on the strength x. Note that this solution does not actually depend on the parameterization of the model in terms of particular quantiles, such as the ED_{95}.

When Eq. (6) is decreasing in x, the optimum setting is the maximum available shock level. The special case of $p(\mathscr{F} \mid x) = p_0$, independent of x, offers some further insight. Let $f(\theta, x) = \partial p(\theta, x)/\partial x$ and $p'(\mathscr{F} \mid x) = \partial p(\mathscr{F} \mid x)/\partial x$. A necessary condition for a setting x to be optimal is to satisfy

$$p'(\mathscr{F} \mid x) = \frac{p_0 C_{\mathscr{F}}}{C_{\mathscr{I}}} E[f(\theta, x)|y, D] \tag{7}$$

Depending on the ratio of costs, p_0, and on the specific form of p', the solution will be a quantile of the true response function but will not necessarily correspond to the ED_{95}.

For the design problem, $-l(\theta, x)$ evaluated at the optimum setting $x^*(y, D)$, can be used as an alternative criterion to u in Eq. (2). This function is different from the squared error loss in the log scale, although we argued that it may share with it important general features, such as asymmetry. Since the solution to the optimal setting $x^*(y, D)$ may not exist in closed form, this optimal design problem will require an additional optimization step to calculate the utility function.

3.3 A Monte Carlo Strategy for Optimal Design

In theory, Monte Carlo integration can be used to evaluate the design criterion $\mathscr{U}(D)$. A straightforward implementation would evaluate $\mathscr{U}(D_i)$ for each design D_i by simulating a large enough Monte Carlo sample from the joint distribution of ϕ_{J+1} and y and evaluating $\mathscr{U}(D_i)$ by a Monte Carlo average. Unless the Monte Carlo sample is extremely large, the Monte Carlo variability in the estimates of $\mathscr{U}(D_i)$ can cause severe problems with standard optimization programs, such as the simplex algorithm or Newton-Raphson methods, that assume a smooth surface. An algorithm proposed in Müller and Parmigiani (1995) avoids the necessary large Monte Carlo sample by "borrowing strength" from neighboring design points. The method is based on simulating a small number of Monte Carlo experiments under different values of the design variables and smoothing through these simulated points to obtain an expected utility surface, rather than using a pointwise average at each design D_i. The maximum of the surface determines the optimal design.

Formally, the method is described by the following steps:

1. Select designs $D_1, \ldots, D_M$ from the design space, where M is the number of designs that will be evaluated.

2. For each D_i, simulate a pair (ϕ_i, y_i) by generating ϕ_i from the prior distribution for the new patient and generating y_i conditional on ϕ_i from model (1).
3. For each triple (D_i, ϕ_i, y_i) record $n_i = n(D_i, y_i)$ and evaluate $v_i = V(D)$.
4. Fit a surface $\hat{v}(D)$ through the pairs (D_i, v_i), and a surface $\hat{n}(D)$ through the pairs (D_i, n_i), where n_i and v_i are the response variables in the smoothing surface.
5. The expected utility surface is $\hat{U}(D) = -\hat{v}(D) - c\hat{n}(D)$. Find deterministically the extreme point D^* corresponding to the optimal design.

The curve fitting in step 4 replaces the numerical integration for each D_i. In practice, we have used the *loess* function in S-plus (Cleveland *et al.*, 1992), which provides a convenient nonparametric method.

To implement this method we need to determine the prior distribution for a new patient. We discuss this in the next part and then show how to use this method to find an optimal design.

3.4 Prior Distribution for a New Patient

The prior distribution for a prospective patient, based on the current sample, is $p(\phi_{J+1} \mid y_{11}, \ldots, y_{I,J})$. A sample from this distribution can be obtained easily in the MCMC simulation as an additional step by draws from the distribution of ϕ_{J+1} conditional on the currently imputed values of μ and V. Conditional on μ and V, we have that ϕ_{J+1} is $N(\mu, V)$. At the end of each MCMC cycle, we generate ϕ_{J+1} conditional on the current value μ and V which provides a sample from $p(\phi_{J+1} \mid y_{11}, \ldots, y_{I,J})$. We will suppress the dependence of the prior distribution on the previous data to simplify the notation and refer to the prior distribution simply as $p_{\phi_{J+1}}(\phi)$. As before, we excluded the first 1000 samples and used every 10th sample to estimate the distribution. Since there is no analytic expression for the density, we obtained a convenient discrete approximation to this distribution on a 100-by-100 grid of equal-sized cells. The resulting distribution is shown in Figure 3.

With a discretized prior distribution, updating the prior distribution to obtain the posterior distribution for a new patient is straightforward. Let $\mathcal{L}(\phi)$ denote the likelihood function based on the logistic model (1):

$$\mathcal{L}(\phi) = \prod_{i=1}^{I_{j+1}} p(\exp\phi, x_{ij+1})^{y_{ij+1}}(1 - p(\exp\phi, x_{ij+1}))^{1-y_{ij+1}}$$

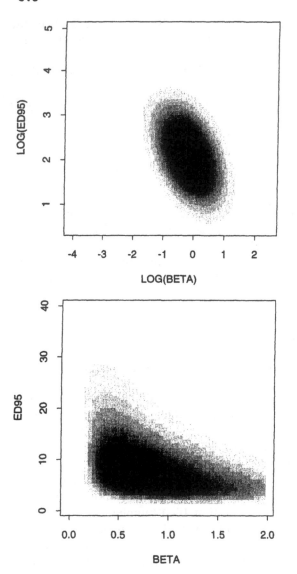

Figure 3 Prior distributions $p_{\phi_{J+1}}(\log(\beta),\ \log(\lambda))$ and $p(\beta,\ \lambda)$ for a new dog obtained from the Markov chain Monte Carlo using the data on the 10 dogs. The triangles represent the posterior estimates for each of the dogs obtained in Section 2.

The posterior distribution for ϕ_{J+1} is

$$p_{\phi_{J+1}}(\phi \mid y) = \frac{\mathscr{L}(\phi)p_{\phi_{J+1}}(\phi)}{\sum\limits_{\phi \in \Phi} \mathscr{L}(\phi)p_{\phi_{J+1}}(\phi)}$$

for ϕ in Φ, the discrete list of points that are the center of each cell that made up the prior distribution $p_{\phi_{J+1}}(\phi)$, and is 0 otherwise. The posterior mean for the $\log(ED_{95})$ is

$$E(\log(\lambda) \mid y) = \sum_{\phi \in \Phi} \log(\lambda)p_{\phi_{J+1}}(\phi \mid y)$$

and the posterior variance is

$$\mathrm{Var}(\log(\lambda) \mid y) = \sum_{\phi \in \Phi} \{\log(\lambda) - E(\log(\lambda)|y)\}^2 p_{\phi_{J+1}}(\phi \mid y)$$

This expression is used in step 3 of the Monte Carlo/smoothing optimization to evaluate $V(D)$. Without any additional experimentation, the prior variance, or the estimation component of the utility function, is .355.

3.5 Equally Spaced Designs

In many bioassay problems, design points are taken to be equally spaced in the log dose scale. To find an optimal design in this class of designs requires finding the initial strength d_0 and the last test strength d_n to maximize the expected utility in Eq. (3). An eight-point design in this class is denoted by $D = \{d_0, d_8\}$. We show how the Monte Carlo strategy can be used to find an optimal design.

We started by randomly generating $M = 2500$ design pairs (d_{0i}, d_{8i}), $i = 1, \ldots, M$ from the design space $\mathscr{D}$. Energy levels range from 0.001 to 32 joules. We used steps 1–5 in Section 3.1 to find the optimal choice of d_0 and d_8. Since the sample size is predetermined, the observation cost is the same for all experiments and can be ignored in calculating the utility. Finding the optimal design then is equivalent to minimizing $\hat{v}(D)$. This surface and the resulting design are shown in Figure 4. The optimal design occurs at $d_0 = 2.87$ and $d_8 = 16.35$ joules, with an expected preposterior variance of 0.09940.

The equally spaced design points in the log scale are not necessarily optimal in the class of non-sequential designs and we could obtain additional gains in efficiency by allowing completely arbitrary spacing of the design points.

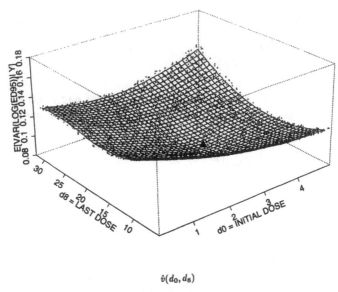

$$\hat{v}(d_0, d_8)$$

Figure 4 Estimated surface for expected posterior variance as a function of the design parameters initial test strength d_0 and last test strength d_8. Fitting the smoothing surface replaces the evaluation of the integral $\int \mathrm{Var}(\log(\lambda) \mid y) \, dP_{d_0, d_8}(\phi, y)$. Without introducing the smoothing surface, this integral would need to be evaluated for many individual design pairs (d_0, d_8) in order to find the optimal choice.

3.6 Arbitrary Spaced Designs Using Approximate Variance

In the case of eight design points, finding the optimal design by the smoothing method is not practically feasible. In Clyde *et al.* (1995), we used an analytic approximation to the design criterion to handle this larger optimization problem. Chaloner and Larntz (1989) used asymptotic approximations based on the expected Fisher information to simplify the design problem for estimating λ with a quadratic loss function. In this particular application, the approximations reduce the number of required integrals from $2n$ to 2. We describe the necessary steps to calculate the approximate design criterion.

The expected Fisher information matrix for ϕ for a single design point x is

$$I(\phi, x) = p(\theta, x)(1 - p(\theta, x))\beta^2 \begin{bmatrix} (x - \mu)^2 & -\lambda(x - \mu) \\ -\lambda(x - \mu) & \lambda^2 \end{bmatrix}$$

For the nonsequential designs, the expected Fisher information matrix is

$$I(\phi, D) = \sum_{i=1}^{n} I(\phi, x_i)$$

Under mild conditions the posterior distribution for ϕ can be approximated by a normal distribution centered at the maximum-likelihood estimate, $\hat{\phi}$, and an approximate variance-covariance matrix for ϕ of

$$\text{Var}(\phi \mid y) \approx (S^{-1} + I(\hat{\phi}, D))^{-1} \qquad (8)$$

where S^{-1} is the prior precision matrix for ϕ. The approximate posterior variance of log λ is given by the (2, 2) element of the approximate covariance matrix in Eq. (8), which will be denoted by $V_{22}(\hat{\phi}, D)$. Since the sample size is fixed, we can ignore the sampling cost in the optimization problem again. Using the approximation in Eq. (4), the approximate expected utility for estimating $\log(\lambda)$ is

$$-\tilde{V}(D) = -\int_{\Phi} V_{22}(\hat{\phi}, D)p(\hat{\phi}) \approx -\sum_{\phi \in \Phi} V_{22}(\phi, D)p_{\phi_{J+1}}(\phi) \qquad (9)$$

where the last approximation uses the prior distribution, $p(\hat{\phi})$, to approximate the predictive distribution of the maximum-likelihood estimate $\hat{\phi}$.

The approximate nonsequential design problem is to find the design D^* that minimizes $\tilde{V}(D)$. This can be found numerically in XLISP-STAT using methods in Clyde (1993) based on standard numerical optimization methods. The approximate posterior variance for this design was 0.0949. In addition, 1000 Monte Carlo samples were used to evaluate the exact posterior variance. The Monte Carlo estimate was 0.094 with a standard deviation of 0.0016. In this example, the normal approximation and the exact calculations are in close agreement. Figure 5 shows the support points of the optimal design for n equal to 8 and, for comparison, the optimal design using eight equally spaced points in the log scale.

3.7 Sequential Designs

In both design and implantation of defibrillators, observations are costly. In implantation, they can also be hazardous. Consequently, it is important to investigate design strategies that stop experimentation early for patients whose ED_{95} can be determined accurately enough after a few observations. Bayesian sequential decision making based on dynamic programming offers a natural and powerful approach to this kind of problems [see DeGroot (1970) and also the introductory chapter of this volume by Berry and Stangl]. At any stage during the experimentation, Bayesian updating allows one to evaluate all information that has been accumulated and to

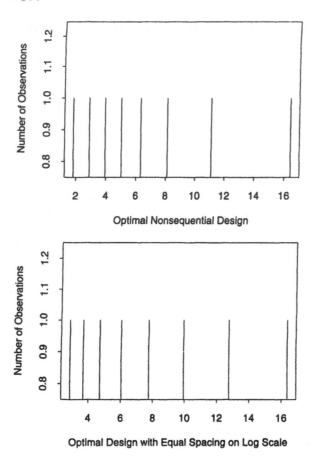

Figure 5 The optimal design with $n = 8$ design points.

decide how to proceed next, by anticipating both future information and one's reaction to it. In particular, one of the options available at every stage is not to take any further observations.

To decide whether or not to take the first observation, and at what dose, requires thinking about what use will be made of the information if it is collected. Some outcomes may lead to stopping, whereas others will require further experimentation, in turn leading to new dilemmas. Options branch out in the future. Postulating that there is a maximum N to the number of observations that can be taken, one can begin at the end and work out the solution backward, one step at a time. This way of proceeding, called backward induction, produces shock levels and stopping strategies that adapt optimally to the accruing information.

More specifically, for any set of shocks $x_1, \ldots, x_{N-1}$ and observed responses $y_1, \ldots, y_{N-1}$, one can derive the optimal last shock level x_N by maximizing

$$\mathcal{U}(x_N) = -V(x_N) - c$$

similarly to what is done in Section 3.1. Using the posterior distribution $\phi \mid y_1, \ldots, y_{N-1}$ as the prior distribution for this last stage incorporates all information available up to this point. If the value of the utility $\mathcal{U}$ at the optimal design is negative, the cost of the new observation outweighs the expected gain and it is best to stop, that is, not to take the Nth observation and proceed immediately with final decision making. Otherwise, it is best to take the Nth observation at an optimal shock level x_N^*. This completely defines the optimal stopping rule, the optimal design point x_N^* $(y_1, \ldots, y_{N-1}, x_1, \ldots, x_{N-1})$, and the resulting expected gain $\mathcal{U}_{N-1}$ $= \max\{0, \mathcal{U}(x^*(y_1, \ldots, y_{N-1}, x_1, \ldots, x_{N-1}))\}$ as a function of the previous design points $x_1, \ldots, x_{N-1}$ and the resulting observations $y_1, \ldots, y_{N-1}$.

One can now move one step back. Fix $x_1, \ldots, x_{N-1}$ and $y_1, \ldots, y_{N-2}$. The criterion for determining the optimal shock x_{N-1} is now

$$\sum_{y_{N-1} = 0,1} [\mathcal{U}_{N-1}(y_{N-1}, x_{N-1}) - c]p_{x_{N-1}}(y_{N-1} \mid y_1, \ldots, y_{N-2})$$

$\mathcal{U}_{N-1}$ also depends on $y_1, \ldots, y_{N-2}$ and $x_1, \ldots, x_{N-2}$, but this has been suppressed in the notation. Again it is best to stop when this is negative.

Iterating this procedure N times defines the optimal shocks and stopping rules at every stage for every possible experimental outcome. Malkin et al. (1993) find sequential designs for fixed sample sizes of 2, 3, and 4, using a design criterion based on minimizing the variance of the ED_{95}.

3.8 Parameterizing an Adaptive Sampling Strategy

Solving optimal sequential design problems can be a difficult computing task. In particular, if updating and optimization at later stages are based on numerical methods of asymptotic approximations, the propagation of approximation error to earlier stages can lead to serious problems. Also, the time required to carry out the computations, or the storage requirements to store the optimal solution, can be too large for use when tests need to be carried out in close sequence and computing resources are not available.

In the defibrillation literature efforts have been concentrated on designing heuristic and easy-to-implement strategies that attempt to mimic the behavior of the optimal sequential solution. Although not optimal in the general class of sequential designs, these designs can be adaptive in that design points do depend on the previous outcomes of the experiments. In particular, early stopping rules can incorporated.

In Clyde et al. (1995), we investigated one such strategy adapted from Bourland et al. (1978): (1) start with an initial dose $d_0 = x_1$; (2) induce fibrillation and observe the response y_1; (3) if $y_1 = 0$ (defibrillation fails), the next dose is increased to $x_2 = x_1(1 + a)$; (4) If $y_1 = 1$ (defibrillation is successful), decrease the dose to $x_2 = x_1(1 - a)$; (5) repeat steps (2) through (4) until either $r = 3$ reversals have occurred or a maximum of $s = 10$ observations have been taken, whichever occurs first. Here a reversal refers to $y_{i+1} = 1 - y_i$. The maximum number of observations s and the number of reversals r are also design parameters, and this framework could be extended by choosing these optimally as well.

We chose the percentage change a and the initial level d_0 to maximize expected utility using the Monte Carlo smoothing approach described in Section 3.1 under three different scenarios. The best sequential design rule for minimizing the preposterior variance ($c = 0$) is $d_0 = 9.77$ and $a = 0.23$. The corresponding approximate posterior variance from the smoothing is 0.092. This design has an expected sample size of 8.6 observations. Using the combined utility surface based on the trade-off parameter $c = 0.02$, the optimal design is $\hat{d}_0 = 9.25$ and $\hat{a} = 0.30$. This design has an expected sample size of 8.3 with an expected posterior variance of 0.095. Introducing observation cost drives the solution toward a much higher value of the percentage change a, to induce the occurrence of reversal and early stopping.

The expected number of observations associated with the adaptive designs is higher than 8, the corresponding value for the nonsequential design. To provide a more direct comparison between the sequential and nonsequential designs, we also considered a class of designs defined by the sequential rule (1) through (4), with the additional restriction that the expected number of design points be constant and equal to 8, the number of points in the nonsequential design. The constrained optimum occurs at $d_0 = 7.96$ and $a = 0.35$, resulting in an expected posterior variance of approximately 0.102. The expected posterior variance for the nonsequential design found in Section 3 is 0.094. The adaptive, up-down strategy is 92% efficient relative to the nonsequential design. All designs provide substantial reductions in variance compared to the prior variance of 0.355.

The design points for the adaptive designs are random. Figure 6 shows possible design points and the proportion of the designs that used the points, based on 1000 simulated experiments. Like the nonsequential designs, the adaptive designs tend to have more support points at the lower strengths, but may have some tests at very high strengths.

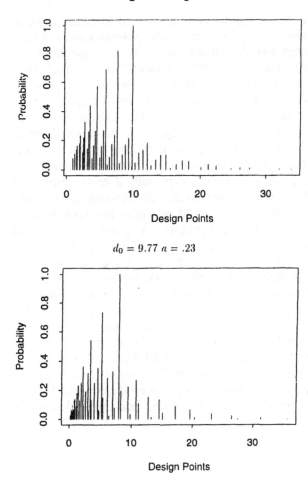

$d_0 = 9.77\ a = .23$

$d_0 = 7.96\ a = .35$

Figure 6 Optimal adaptive designs.

4 DISCUSSION

Design and implantation of heart defibrillators is a growing area of application for statistical design methodologies and one in which there is excellent scope for Bayesian methods. The prospect of incorporating historical data via hierarchical models is very promising. In addition, the high cost of observation and the potential importance of sequential sampling and early stopping make Bayesian methods attractive.

In this chapter, we illustrated the computation of Bayesian designs, focusing on finding optima within restricted sets of strategies. We compared an adaptive strategy based on fixed percentage changes in the energy levels and variable sample size, with a strategy in which all levels are chosen optimally but the sample size is fixed. The comparison of the two classes of strategies is in favor of the second, for the specific prior distributions and loss structure postulated here. The indication that the parameterized sequential strategy is not as efficient as the optimal nonsequential one, or even the design based on equal spacing on the log scale, is of practical interest. However, in other application contexts the parameterized sequential strategy may do better, especially if early stopping is a determinant factor. For a general discussion see Govindarajulu (1988).

Although the adaptive design considered here does incorporate information as it accrues, it is not optimal in the class of sequential designs, and other sequential strategies may improve substantially upon it. Alternative heuristic sampling rules can be constructed based on group-sequential sampling and one-step lookahead procedures. Two-stage sequential designs may be one way to gain in efficiency and maintain relative computational simplicity. For example, the fully optimal sequential design for four design points could be used in the first stage. The 16 possible posterior distributions resulting from this design can be calculated a priori. For each of the possible posterior distributions, one can find the next stage optimal sequential design with four points. Although this requires an intensive computing effort initially, the optimal setting can be determined beforehand for each of the 16^2 possible outcomes, so no real-time calculations have to be performed.

ACKNOWLEDGMENTS

We thank Mark Kroll for supplying the data that were used to construct the prior distribution and Don Berry for many useful comments on an early version of the chapter.

REFERENCES

Albert, J. and Chib, S. (1996). Bayesian probit modeling of binary repeated measures data with application to crossover trials. In *Bayesian Biostatistics*, ed. D. Berry and D. Stangl. Marcel Dekker, New York.

Berger, J. (1985). *Statistical Decision Theory and Bayesian Analysis*, 2nd ed. Springer-Verlag. New York.

Bourland, J. D., Tacker, W. A., Jr., and Geddes, L. A. (1978). Strength-duration curves for trapezoidal waveforms of various tilts for transchest defibrillation in animals. *Medical Instrumentation* 12: 38–41.

Chaloner, K. (1993). A note on optimal Bayesian design for nonlinear problems. *Journal of Statistical Planning and Inference* 37: 229–235.

Chaloner, K. and Larntz, K. (1989). Optimal Bayesian design applied to logistic regression experiments. *Journal of Statistical Planning and Inference* 21: 191–208.

Chaloner, K. and Verdinelli, I. (1994). Bayesian Design: A Review. Technical Report Department of Statistics, Carnegie-Mellon University, Pittsburgh.

Cleveland, W. S., Grosse, E., and Shyu, W. M. (1992). Local regression models. In *Statistical Models in S*, ed. Chambers, J. M. and Hastie, T. J., pp. 309–376. Wadsworth & Brooks/Cole. Pacific Grove, CA.

Clyde, M. (1993). An Object-Oriented System for Bayesian Nonlinear Design Using XLISP-STAT. Technical Report 587, School of Statistics, University of Minnesota, Minneapolis, MN.

Clyde, M., Müller, P., and Parmigiani, G. (1995). Optimal design for heart defibrillators. In *Bayesian Statistics in Science and Engineering: Case Studies II*, ed. Gatsonis, C., Hodges, J. S., Kass, R. E., and Singpurwalla, N. D. Springer-Verlag, New York.

Davy, J. M. (1984). Is there a defibrillation threshold? *Circulation* 70-II: 406.

Davy, J. M., Fain, E. S., Dorian, P., and Winkle, R. A. (1987). The relationship between successful defibrillation and delivered energy in open chest dogs: Reappraisal of the defibrillation concept. *American Heart Journal* 113: 77–84.

DeGroot, M. (1970). *Optimal Statistical Decisions*. McGraw-Hill, New York.

Durham, S. D. and Flournoy, N. (1993). Random walks for quantile estimation. In *Statistical Decision Theory and Related Topics V*, ed. Berger, J. and Gupta, S. Springer-Verlag, New-York. (to appear)

Flournoy, N. (1993). A clinical experiment in bone marrow transplantation: Estimating a percentage point of a quantal response curve. In *Case Studies in Bayesian Statistics*, ed. Gatsonis, C., Hodges, J. S., Kass, R. E., and Singpurwalla, N. D. pp. 324–335. Springer-Verlag, New York.

Freeman, P. R. (1983). Optimal Bayesian sequential estimation of the median effective dose. *Biometrika* 70: 625–632.

Gamerman, D. (1994). Efficient Sampling from the Posterior Distribution in Generalized Linear Mixed Models. Technical Report, Universidade Federal do Rio de Janeiro.

Gatsonis, C. and Greenhouse, J. G. (1992). Bayesian methods for phase I clinical trials. *Statistics in Medicine* 11: 1377–1389.

Gelfand, A. E. and Smith, A. F. M. (1990). Sampling based approaches to calculating marginal densities. *Journal of the American Statistical Association* 85: 398–409.

Gilks, W. R., Clayton, D. G., Spiegelhalter, D. J., Best, N. G., McNeil, A. J., Sharples, L. D., and Kirby, A. J. (1993). Modeling complexity: Applications of Gibbs sampling in medicine. *Journal of the Royal Statistical Society B* 55: 39–52.

Gliner, B. E., Murakawa, Y., and Thakor, N. V. (1990). The defibrillation success rate versus energy relationship: Part I—Curve fitting and the most efficient defibrillation energy. *PACE* 13: 326–338.

Govindarajulu, Z. (1988). *Statistical Techniques in Bioassay*. Karger, Basel.

Kamlet, M. S. (1991). *A Framework for Cost-Utility Analysis of Government Health Care Programs*. U.S. Department of Health and Human Services, Washington, DC.

Malkin, R. A., Pilkington, T. C., Burdick, D. S., Swanson, D. K., Johnson, E. E., and Ideker, R. E. (1993). Estimating the 95% effective defibrillation dose. *IEEE Trans. EMBS* 40(3): 256–265.

Manolis, A. S., Tan-DeGuzman, W., Lee, M. A., Rastegar, H., Haffajee, C. I., Haung, S. K., and Estes, N. A. (1989). Clinical experience in seventy-seven patients with automatic implantable cardioverter defibrillator. *American Heart Journal* 118: 445–450.

McDaniel, W. C. and Schuder, J. C. (1987). The cardiac ventricular defibrillation threshold: Inherent limitations in its application and interpretation. *Medical Instrumentation* 21: 170–176.

Müller, P. and Parmigiani, G. (1995). Optimal design via curve fitting of Monte Carlo experiments. *JASA*, in press.

Müller, P. and Rosner, G. (1994). A Semiparametric Bayesian Population Model with Hierarchical Mixture Priors. ISDS DP 94-17, Duke University, Durham, North Carolina.

O'Quigley, J., Pepe, M., and Fisher, L. (1990). Continual reassessment method: A practical design for phase I clinical trials in cancer. *Biometrics* 46: 33–48.

Parmigiani, G. and Polson, N. G. (1992). Bayesian design for random walk barriers. In *Bayesian Statistics IV*, ed. Bernardo, J. M., Berger, J. O., Dawid, A. P., and Smith, A. F. M., pp. 715–772. Oxford University Press, Oxford.

Smith, A. F. M. and Roberts, G. O. (1993). Bayesian computation via the Gibbs sampler and related Markov chain Monte Carlo methods (with discussion). *Journal of the Royal Statistical Society, Series B* 55: 3–23.

Storer, B. (1989). Design and analysis of phase I clinical trials. *Biometrics* 45: 925–937.

Tsutakawa, R. (1972). Design of an experiment for bioassay. *Journal of the American Statistical Association* 67: 584–590.

Tsutakawa, R. (1980). Selection of dose levels for estimating a percentage point on a logistic quantal response curve. *Applied Statistics* 29: 25–33.

Verdinelli, I. and Kadane, J. (1992). Bayesian designs for maximizing information and outcome *Journal of the American Statistical Association* 86: 510–515.

Wakefield, J. (1994). An expected loss approach to the design of dosage regimens via sampling-based methods. *Statistician* 43: 13–30.

Zacks, S. (1977). Problems and approaches in design of experiments for estimation and testing in non-linear models. In *Multivariate Analysis IV*, ed. Krishnaiah. P. R., pp. 209–223. North-Holland, Amsterdam.

V
MODEL SELECTION

12

Model Selection for Generalized Linear Models via GLIB: Application to Nutrition and Breast Cancer

Adrian E. Raftery   University of Washington, Seattle, Washington
Sylvia Richardson   INSERM, Villejuif, France

ABSTRACT

Epidemiological studies for assessing risk factors often use logistic regression, log-linear models, or other generalized linear models. They involve many decisions, including the choice and coding of risk factors and control variables. It is common practice to select independent variables using a series of significance tests and to choose the way variables are coded subjectively. The overall properties of such a procedure are not well understood, and conditioning on a single model ignores model uncertainty, leading to underestimation of uncertainty about quantities of interest (QUOIs). We describe a Bayesian modeling strategy that formalizes the model selection process and propagates model uncertainty through to inference about QUOIs. Each possible combination of modeling decisions defines a different model, and the models are compared using Bayes factors. Inference about a QUOI is based on an average of its posterior distributions under the individual models, weighted by their posterior model probabilities; the models included in the average are selected by the Occam's Window algorithm. In an initial exploratory phase, the ACE (Alternating Conditional Expectations) algorithm is used to suggest ways to code the variables, but the final coding decisions are based on Bayes factors. The methods can be implemented using GLIB, an S function available free of charge from StatLib. For the special case of logistic regression, the additional S functions ACE.LOGIT and BIC.LOGIT are available. We apply our strategy to an epidemiological study of fat and alcohol consumption as risk factors for breast cancer. In our previous published analysis, the regression model chosen included not only fat and alcohol consumption but also an interaction term between these two variables. Here, however, the Bayes factors favor a simpler and more interpretable model that includes transformed variables but no interaction term.

1   INTRODUCTION

Are fat and alcohol consumption associated with breast cancer? This question has been much discussed in the past 10 years. In this chapter we introduce a new Bayesian modeling strategy for generalized linear models and use it to reanalyze a case-control study in which fat and alcohol consumption and eight other risk factors were measured for 854 women (Richardson *et al.*, 1989). It is usual to analyze such studies by logistic regression (e.g., Breslow and Day, 1980). The analyst must make several decisions, including:

How to code the risk factors? In epidemiological studies, risk factors are often coded in several categories (typically two to five). This has the advantage of allowing easy calculation of relative risks. If the risk factor is not inherently categorical, however, it is somewhat arbitrary and different codings (e.g., a different number of categories, different breakpoints, or coding as a continuous variable) may yield different results.

Which risk factors to include? This is the key question in the study and is analogous to variable selection in regression. It is complicated by the fact that the data at hand do not support the inclusion of some of the "classical" risk factors, for which the evidence comes from previous studies.

Each possible combination of decisions defines a different statistical model for the data, so the number of models initially considered is very large. Even if one considers only two possible ways to code each risk factor, and the possibility of omitting it, there are 3^{10} or about 60,000 possible models. Typically, the analyst selects one of those models by some combination of more or less arbitrary choices (for coding) and significance tests (for risk factor selection). He or she then makes inference about quantities of interest (QUOIs) conditionally on the selected model.

There are three main difficulties with such a strategy. First, the way in which the variables are coded is often somewhat arbitrary, and different choices might well give different results. Second, the basis for risk factor selection strategies based on sequences of significance tests is weak because (a) the properties of the overall strategy (as distinct from the individual tests) are not well understood and (b) the final choice often comes down to a comparison of nonnested models, which cannot easily be done by significance tests. Third, by conditioning on a single model selected out of a large number, the uncertainty associated with the model selection process itself is ignored, and as a result the uncertainty about QUOIs is underestimated. This can lead, for example, to decisions that are riskier than one thinks (Hodges, 1987).

In this chapter we describe a Bayesian modeling strategy that avoids these difficulties. It consists of calculating the posterior probabilities of all the models considered and averaging over them when making inferences about QUOIs, thereby taking account of model uncertainty. Several approximations are suggested to make computation and communication of the results easier. In an initial exploratory phase, the ACE (Alternating Conditional Expectation) algorithm is used to suggest a set of candidate codings for the risk factors. Instead of averaging over all possible models, we average over a reduced set of models given by the Occam's Window algorithm (Madigan and Raftery, 1994).

The result is a practical and formally justifiable modeling strategy that avoids arbitrariness in variable coding and risk factor selection and accounts for model uncertainty. It can be implemented using the GLIB software, available free of charge from StatLib. In the special case of logistic regression, additional software is available to do the full ACE analysis and automatically (if approximately) search through model space, namely the ACE.LOGIT and BIC.LOGIT functions.

In Section 2 we review the Bayesian approach to model selection and model uncertainty using Bayes factors, in the context of generalized linear models. In Section 3 we outline our modeling strategy, and in Section 4 we describe its application to the nutrition and breast cancer study. In Section 5 we give some information about the available software.

2 BAYES FACTORS AND MODEL UNCERTAINTY

In this section, we review the basic ideas of Bayes factors and their use in accounting for model uncertainty. Sections 2.1–2.5 summarize material in Raftery (1988, 1993b) and Kass and Raftery (1995), to which the reader is referred for more information.

2.1 Basic Ideas

We begin with data D assumed to have arisen under one of the two models M_1 and M_2 according to a probability density $P(D \mid M_1)$ or $P(D \mid M_2)$. Given prior probabilities $P(M_1)$ and $P(M_2) = 1 - P(M_1)$, the data produce posterior probabilities $P(M_1 \mid D)$ and $P(M_2 \mid D) = 1 - P(M_1 \mid D)$. Since any prior opinion is transformed to a posterior opinion through considera-tion of the data, the transformation itself represents the evidence provided by the data. In fact, the same transformation is used to obtain the posterior probability, regardless of the prior probability. Once we convert to the odds scale (odds = probability/(1 − probability)), the transformation takes a simple form. From Bayes' theorem we obtain

$$P(M_k \mid D) = \frac{P(D \mid M_k)P(M_k)}{P(D \mid M_1)P(M_1) + P(D \mid M_2)P(M_2)} \qquad (k = 1, 2)$$

so that

$$\frac{P(M_1 \mid D)}{P(M_2 \mid D)} = \frac{P(D \mid M_1)}{P(D \mid M_2)} \frac{P(M_1)}{P(M_2)}$$

and the transformation is simply multiplication by

$$B_{12} = \frac{P(D \mid M_1)}{P(D \mid M_2)} \qquad\qquad\qquad (1)$$

which is the *Bayes factor*. Thus, in words,

Posterior odds = Bayes factor × prior odds

and the Bayes factor is the ratio of the posterior odds of M_1 to its prior odds. When the models M_1 and M_2 are equally probable a priori so that $P(M_1) = P(M_2) = 0.5$, the Bayes factor is equal to the posterior odds in favor of M_1.

In the simplest case, in which the two models are single distributions with no free parameters (the case of "simple versus simple" testing), B_{12} is the likelihood ratio. In other cases, in which there are unknown parameters under either or both of the models, the Bayes factor is still given by Eq. (1) and, in a sense, it continues to have the form of a likeli-hood ratio. Then, however, the densities $P(D \mid M_k)$ $(k = 1, 2)$ are obtained by *integrating* (not maximizing) over the parameter space, so that in Eq. (1),

$$P(D \mid M_k) = \int P(D \mid \theta_k, M_k)P(\theta_k \mid M_k) \, d\theta_k \qquad\qquad (2)$$

where θ_k is the vector of parameters under M_k, $P(\theta_k \mid M_k)$ is its prior density and $P(D \mid \theta_k, M_k)$ is the probability density of D given the value of θ_k, or the likelihood function of θ.

The quantity $P(D \mid M_k)$ given by Eq. (2) is the marginal probability of the data, as it is obtained by integrating the joint density of (D, θ_k) given D over θ_k. It is also sometimes called the marginal likelihood or the integrated likelihood.

One important use of the Bayes factor is as a summary of the evidence for M_1 against M_2 provided by the data. It can be useful to consider twice the logarithm of the Bayes factor, which is on the same scale as the familiar deviance and likelihood ratio test statistics. We use the following rounded scale for interpreting B_{12}, which is based on that of Jeffreys (1961) but is more granular and slightly more conservative than his.

B_{12}	$2 \log B_{12}$	Evidence for M_1
< 1	< 0	Negative (supports M_2)
1 to 3	0 to 2	Not worth more than a bare mention
3 to 20	2 to 6	Positive
20 to 150	6 to 10	Strong
> 150	> 10	Very strong

Kass and Raftery (1995) provide a full review of Bayes factors, to which readers are referred for more information. They discuss issues that we do not dwell on here, including whether it is valid or useful to test point hypotheses or select models at all and the comparison between Bayes factors and non-Bayesian significance testing for nested models.

2.2 Accounting for Model Uncertainty

When more than two models are being considered, the Bayes factors yield posterior probabilities of all the models, as follows. Suppose that $(K + 1)$ models, $M_0, M_1, \ldots, M_K$, are being considered. Each of $M_1, \ldots, M_K$ is compared in turn with M_0, yielding Bayes factors $B_{10}, \ldots, B_{K0}$. Then the posterior probability of M_k is

$$P(M_k \mid D) = \frac{\alpha_k B_{k0}}{\sum\limits_{r=0}^{K} \alpha_r B_{r0}} \tag{3}$$

where $\alpha_k = P(M_k)/P(M_0)$ is the prior odds for M_k against M_0 ($k =$

$0, \ldots, K$); here $B_{00} = \alpha_0 = 1$. For discussion of the prior model probabilities, $P(M_k)$, see Section 2.5.

The posterior model probabilities given by Eq. (3) lead directly to solutions of the prediction, decision-making, and inference problems that take account of model uncertainty. The posterior distribution of a QUOI, Δ, such as a relative risk or the probability that someone who does not have the disease now will get it later, is

$$P(\Delta \mid D) = \sum_{k=0}^{K} P(\Delta \mid D, M_k)P(M_k \mid D) \tag{4}$$

where $P(\Delta \mid D, M_k) = \int P(\Delta \mid D, \theta_k, M_k)P(\theta_k \mid D, M_k)d\theta_k$ (Leamer, 1978, p. 117). In a certain sense, Eq. (4) is guaranteed to give better out-of-sample predictions on average than conditioning on any single model (Madigan and Raftery, 1994). The composite posterior mean and standard deviation (or point estimate and standard error) are given by

$$E[\Delta \mid D] = \sum_{k=0}^{K} \hat{\Delta}_k \, P(M_k \mid D) \tag{5}$$

$$SD[\Delta \mid D]^2 \tag{6}$$

$$= \mathrm{Var}[\Delta \mid D] = \sum_{k=0}^{K} (\mathrm{Var}[\Delta \mid D, M_k] + \hat{\Delta}_k^2) \, P(M_k \mid D) - E[\Delta \mid D]^2$$

where $\hat{\Delta}_k = E[\Delta \mid D, M_k]$ (Leamer, 1978, p. 118). Equations (5) and (6) can be approximated using the output from a standard maximum-likelihood analysis by replacing $\hat{\Delta}_k$ by the MLE and $\mathrm{Var}[\Delta \mid D, M_k]$ by the square of the corresponding standard error (Raftery, 1993a).

This general approach has been used in several previous analyses of medical and epidemiological data. Racine et al. (1986) showed how this method may be used to make inference about a treatment effect in the presence of uncertainty about the existence of a carryover effect. Raftery (1993b) showed that model uncertainty was large and was an important source of uncertainty about the relative risk in a classic study of oral contraceptive use as a risk factor for myocardial infarction. He showed how it could be accounted for using this framework. Madigan and Raftery (1994) did similar analyses of coronary heart disease risk factors and the diagnosis of scrotal swellings. In their examples, they found that out-of-sample predictive performance was better if one took account of model uncertainty than if one conditioned on any single model that might reasonably have been selected.

The posterior probability that a given parameter is nonzero can be calculated from Eq. (4); it is the sum of the posterior probabilities of the models in which the parameter is present. In epidemiology this can be the posterior probability that a particular risk factor is associated with the disease; see Section 4.3 for an example.

The number of models, K, can be so large that it is not practical to compute Eq. (4) directly. Three ways of getting around this have been suggested. One of these, Occam's Window, consists of averaging over a much smaller set of models, chosen by criteria based on standard norms of scientific investigation (Madigan and Raftery, 1994). We use this here and describe it in Section 3.1. A second method, Markov chain Monte Carlo model composition (MC^3), approximates Eq. (4) using a Markov chain that moves through model space (Madigan and York, 1993). The third method, stochastic search variable selection (SSVS), approximates Eq. (4) using a Markov chain that moves through both model space and parameter space (George and McCulloch, 1993).

For recent reviews of Bayesian model selection and model averaging, see Draper (1995), Kass and Raftery (1995), and Raftery (1995).

2.3 Approximating Bayes Factors by the Laplace Method

The Laplace method for integrals (e.g., de Bruijn, 1970, Section 4.4) is based on a Taylor series expansion of the real-valued function $f(u)$ of the p-dimensional vector u and yields the approximation

$$\int e^{f(u)} \, du \approx (2\pi)^{p/2} |A|^{1/2} \exp\{f(u^*)\} \tag{7}$$

where u^* is the value of u at which f attains its maximum and A is minus the inverse Hessian of f evaluated at u^*. When applied to Eq. (2) it yields

$$p(D \mid M_k) \approx (2\pi)^{p_k/2} |\Psi|^{1/2} P(D \mid \bar{\theta}_k, M_k) P(\bar{\theta}_k \mid M_k) \tag{8}$$

where p_k is the dimension of θ_k, θ_k is the posterior mode of θ_k, and Ψ_k is minus the inverse Hessian of $h(\theta_k) = \log\{P(D \mid \theta_k, M_k) P(\theta_k \mid M_k)\}$, evaluated at $\theta_k = \bar{\theta}_k$. Arguments similar to those in the Appendix of Tierney and Kadane (1986) show that in regular statistical models the relative error in Eq. (8), and hence in the resulting approximation to B_{10}, is $O(n^{-1})$.

One can approximate the marginal likelihood $P(D \mid M_k)$ in any regular statistical model using Eq. (8), but this requires the posterior mode $\bar{\theta}_k$ and minus the inverse Hessian Ψ_k, which are not routinely available. Software that calculates the maximum-likelihood estimator (MLE) $\hat{\theta}_k$, the deviance or the likelihood ratio test statistic, and the observed or expected Fisher information matrix, F_k, or its inverse, V_k, is much more widespread and includes, for example, GLIM. Here we describe an approximation proposed by Raftery (1993b) that is based on Eq. (8) but uses only these widely available quantities.

Suppose that the prior distribution of θ_k is such that $E[\theta_k \mid M_k] = \omega_k$ and $\mathrm{Var}[\theta_k \mid M_k] = W_k$. Then approximating $\bar{\theta}_k$ by a single Newton step starting from θ_k and substituting the result into Eq. (8) yields the approximation

$$2 \log B_{10} \approx \chi^2 + (E_1 - E_0) \tag{9}$$

In Eq. (9), $\chi^2 = 2\{l_1(\hat{\theta}_1) - l_0(\hat{\theta}_0)\}$, where $l_k(\theta_k) = \log (P(D \mid \theta_k, M_k))$ is the log-likelihood; χ^2 is the standard likelihood ratio test statistic when M_0 is nested within M_1. Also,

$$\begin{aligned} E_k = {} & 2\lambda_k(\theta_k) + \lambda_k'(\theta_k)^T(F_k + G_k)^{-1}\{2 - F_k(F_k + G_k)^{-1}\}\lambda_k'(\theta_k) \\ & - \log|F_k + G_k| + p_k \log(2\pi) \end{aligned} \tag{10}$$

where $G_k = W_k^{-1}$, $\lambda_k(\theta_k) = \log P(\theta_k \mid M_k)$ is the log-prior density, and $\lambda_k'(\theta_k)$ is the p_k-vector of derivatives of $\lambda_k(\theta_k)$ with respect to the elements of θ_k ($k = 0, 1$).

This approximation is closer to the basic Laplace approximation, Eq. (8), when F_k is the observed rather than the expected Fisher information, so one would expect it also to be generally more accurate in this case. Arguments similar to those of Kass and Vaidyanathan (1992) show that when F_k is the observed Fisher information, the relative error is $O(n^{-1})$, while when F_k is the expected Fisher information, the relative error increases to $O(n^{-1/2})$. When the prior is normal, Eq. (10) becomes

$$E_k = \log|G_k| - (\hat{\theta}_k - \omega_k)^T C_k(\hat{\theta}_k - \omega_k) - \log|F_k + G_k| \tag{11}$$

In Eq. (11), $C_k = G_k\{I - H_k(2 - F_kH_k)G_k\}$, where $H_k = (F_k + G_k)^{-1}$.

2.4 Application to Generalized Linear Models

Raftery (1993b) has shown how the Laplace approximation of Section 2.3 can be used to calculate Bayes factors for generalized linear models using only standard GLIM output or its equivalent.

Suppose that y_i is the dependent variable and that $x_i = (x_{i1}, \ldots, x_{ip})$ is the corresponding vector of independent variables, for $i = 1, \ldots, n$. The model M_1 is defined by specifying $P(y_i \mid x_i, \beta)$ in such a way that $E[y_i \mid x_i] = \mu_i$, $\mathrm{Var}[y_i \mid x_i] = \sigma^2 v(\mu_i)$, and $g(\mu_i) = x_i\beta$, where $\beta = (\beta_1, \ldots, \beta_p)^T$; here g is called the link function. The $n \times p$ matrix with elements x_{ij} is denoted by X, and it is assumed that $x_{i1} = 1$ ($i = 1, \ldots, n$). For the moment we assume that σ^2 is known.

The null model, M_0, is defined by setting $\beta_j = 0$ ($j = 2, \ldots, p$) in M_1. The likelihoods for M_0 and M_1 can be written down explicitly, and so, once the prior has been fully specified, the approximation in Eq. (8) can be computed. However, this approximation is not easy to compute for generalized linear models using readily available software.

By contrast, when applied to generalized linear models, the approximation of Eqs. (9) and (10) is easier to compute. It is an analytic noniterative function of the MLE, the deviance, and the Fisher information matrix and so can be calculated directly from GLIM output or equivalent. If DV_k is the deviance for M_k, then $\chi^2 = (DV_0 - DV_1)/\sigma^2$. The expected Fisher information matrix is $F_1 = \sigma^{-2} X^T W X$, where $W = \text{diag}\{w_1, \ldots, w_n\}$, and $w_i^{-1} = g'(\hat{\mu}_i^{(1)})^2 v(\hat{\mu}_i^{(1)})$ (McCullagh and Nelder, 1989). Here $\hat{\mu}_i^{(1)} = g^{-1}(x_i \hat{\beta}^{(1)})$, where $\hat{\beta}^{(1)}$ is the MLE of β conditional on M_1. Similarly, $F_0 = \sigma^{-2} n g'(\hat{\mu}_i^{(0)})^2 v(\hat{\mu}_i^{(0)})$, where $\hat{\mu}_i^{(0)} = g^{-1}(\hat{\beta}^{(0)})$, $\hat{\beta}^{(0)}$ being the MLE of β_1 conditional on M_0. The observed and expected Fisher information matrices are equal when g is the canonical link function, so the approximations are more accurate in this case. These values of χ^2, $\hat{\beta}^{(0)}$, $\hat{\beta}^{(1)}$, F_0, and F_1 can be substituted directly into Eqs. (9) and (10).

This solution can be extended to situations in which the dispersion parameter, σ^2, is unknown, such as Poisson and binomial models with overdispersion, or models with normal or gamma errors. One may proceed as before with σ^2 replaced by an estimate $\hat{\sigma}^2$, as McCullagh and Nelder (1989) do for estimation. A reasonable estimate would be $\hat{\sigma}^2 = P/(n - p)$, where P is Pearson's goodness-of-fit statistic for the most complex model considered, as advocated by McCullagh and Nelder (1989, pp. 91 and 127).

The method can also be used to compare different link functions. Suppose that we are comparing two models M_1 and M_2, which have the same independent variables X and variance function v but different link functions g_1 and g_2. Then the parameters $\beta^{(1)}$ and $\beta^{(2)}$ under the two models are on different scales and so should have different prior distributions. Thus we calculate $2 \log B_{10}$ and $2 \log B_{20}$ as before, but with different priors obtained separately for each link function, for example, as in Section 2.5. We then compare M_1 and M_2 using the relation $2 \log B_{21} = 2 \log B_{20} - 2 \log B_{10}$.

Finally, the method may be used to compare different error distributions. Consider the comparison of two models, M_1 and M_2, which have the same independent variables X but different variance functions and/or different error distributions; they may also have different link functions. We can continue to use the same general framework because Eq. (8) still gives the marginal likelihood for each model, and Bayes factors and posterior model probabilities are then available from Eqs. (1) and (3) as before.

The parameters $\beta^{(1)}$ and $\beta^{(2)}$ of the two models are on different scales and so should have different prior distributions. The first step is then to specify the prior distribution of β for each model, as in Section 2.5. We may then use the same approximation as before. Equation (9) becomes

$$2 \log B_{21} \approx \chi_{21}^{2*} + (E_2 - E_1) \tag{12}$$

In Eq. (12),

$$\chi_{21}^{2*} = (\sigma_1^{-2} DV_1 - \sigma_2^{-2} DV_2) + 2(l_2^{sat} - l_1^{sat}) \tag{13}$$

where l_k^{sat} is the maximal log-likelihood achievable with the link function and error distribution of model M_k ($k = 1, 2$); this will typically be the log-likelihood under the saturated model. In Eq. (13), σ_k^2 is the dispersion parameter for M_k, which is either known or estimated as above. In Eq. (12), E_k is given as before by Eq. (10).

2.5 A Reference Set of Proper Priors

The prior distributions $P(\theta_k \mid H_k)$ ($k = 1, 2$) are necessary. This may be considered both good and bad. Good, because it is a way of including other information about the values of the parameters. Bad, because these prior densities may be hard to set when there is no such information. Raftery (1993b) considered the situation in which there is little prior information and suggested a reasonable set of prior distributions for this situation. We now briefly review this proposal.

We first consider the case in which $g(\mu) = \mu$ and $v(\mu) = 1$ and the variables have been standardized to have mean 0 and variance 1. In this situation, we denote the parameters by $\gamma = (\gamma_1, \ldots, \gamma_p)$. Here γ_1 is the intercept and $\gamma_2, \ldots, \gamma_p$ are the regression parameters. We assume that the prior distribution of $(\gamma \mid M_1)$ is normal; in fact the results depend rather little on the precise functional form. We also assume that $(\gamma_2, \ldots, \gamma_p)$ are independent a priori; this corresponds to the situation in which the individual variables are of interest in their own right, which is often implicit in the testing situation. We further assume that the prior is *objective* for the testing situation in the sense of Berger and Sellke (1987), that is, symmetric about the null value of γ, namely $(\gamma_1, 0, \ldots, 0)^T$, and nonincreasing as one moves away from the null value.

These assumptions lead to the prior $(\gamma \mid M_1) \sim N(\nu, U)$, where $\nu = (\nu_1, 0, \ldots, 0)$ and $U = \text{diag}\{\psi^2, \phi^2, \ldots, \phi^2\}$. The prior for γ under M_0 is just the conditional prior distribution of γ under M_1 given that $\gamma_2 = \cdots =$

$\gamma_p = 0$, namely $(\gamma_1 \mid M_0) \sim N(\nu_1, \psi^2)$. This is transformed back into a prior on the original parameters β.

This is extended to generalized linear models with other link and variance functions by noting that then estimation is equivalent to weighted least squares with the adjusted dependent variable $z_i = g(\mu_i) + (y_i - \mu_i)g'(\hat{\mu}_i)$ and weights w_i (McCullagh and Nelder, 1989). The prior is the same as before, but in the transformation back to the original β, y is replaced by z and all the summary statistics are weighted.

When several models are considered, it is desirable that the priors be consistent with each other in the sense that if M_2 is defined by setting restrictions $\rho(\beta) = 0$ on the parameters of M_1, then $P(\beta \mid M_1) = P(\beta \mid M_2, \rho(\beta) = 0)$. A reasonable way to ensure this is to obtain a prior for the largest model as above and then derive the priors for other models by conditioning on the constraints that define them.

The prior distribution has three user-specified parameters: ν_1, ψ, and ϕ. Bayes factors tend to be insensitive to ν_1 and ψ; see Kass and Raftery (1995) for an explanation based on Kass and Vaidyanathan (1992). Raftery (1993b) found $\nu_1 = 1$ and $\psi = 1$ to be reasonable values. Bayes factors are more sensitive to ϕ, however, with larger values of ϕ tending to favor simpler models. Raftery (1993b) defined a reasonable range of values of ϕ by requiring that the prior not contribute much evidence in favor of either model, whether the models being compared are nested or not. These requirements are to some extent in conflict: for nonnested models it implies that ϕ be large, and for nested models it implies that ϕ not be too large. Balancing these two desiderata in a certain sense gives $\phi = e^{1/2} = 1.65$, and requiring that the priors not contribute evidence "worth more than a bare mention" beyond what is unavoidable leads to the range $1 \le \phi \le 5$. The resulting priors are then transformed back to the original scale for the variables; results for other choices of $g(\mu)$ and $v(\mu)$ are obtained by weighting the cases appropriately.

The result is a *reference set of proper priors* for generalized linear models. Although they are mildly data dependent, they do have properties that one would associate with genuine subjective data-independent priors that represent a small amount of prior information.

2.6 Prior Model Probabilities

When there is little prior information about the relative plausibility of the models considered, taking them all to be equally likely a priori is a reasonable "neutral" choice. When the number of models is small or moderate, this is intuitively appealing and understandable. When the number of models is very large, however, one could worry about whether this choice might have unintended perverse consequences, as has happened before with other "uniform" priors. In our experience with very large model spaces (up to 10^{12} models) involving several kinds of model and about 20 data sets, we have found no such perverse effects (Madigan *et al.*, 1994; Raftery *et al.*, 1993; Madigan and Raftery, 1994). In addition, we have found inference using Occam's Window (Section 3.1) to be quite robust to moderately large changes in prior model probabilities (e.g., halving or doubling the prior odds).

If it is available, prior information can easily be taken into account by adjusting the prior model probabilities. In variable selection problems, prior information often takes the form of prior evidence for the inclusion of a variable, rather than for an individual model. Suppose that this is the only kind of prior information available and that model M_k is specified by a vector $(\delta_{k1}, \ldots, \delta_{kp})$, where $\delta_{ki} = 1$ if the ith variable is included and 0 if not. Then, if π_i is the prior probability that the ith variable has an effect, and if the prior information about different variables is approximately independent, it is reasonable to specify

$$P(M_k) \propto \prod_{i=1}^{p} [\pi_i^{\delta_{ki}} (1 - \pi_i)^{(1 - \delta_{ki})}] \qquad (14)$$

In epidemiology, prior information often takes the form of strong evidence from previous studies for the inclusion of a particular risk factor. If the previous studies combined are more informative about the risk factor than the present one, then it is a reasonable approximation to set the corresponding $\pi_i = 1$ in Eq. (14) and hence to consider only models that include that risk factor. This approximation is reasonable in the sense that the true prior probability π_i is so close to 1 that the posterior probability would also be close to 1, almost regardless of the data at hand. Thus setting $\pi_i = 1$ yields a posterior distribution of QUOIs close to what would result from using the true π_i. This provides a formal rationale for the common practice of "controlling" for particular independent variables even when the data at hand provide little evidence for their inclusion in the model.

In our experience with moderate to large data sets, we have found it sufficient to restrict attention to prior model probabilities of the form of Eq. (14) with $\pi_i = 0, \frac{1}{2}$, or 1. This might not be enough, however, for small data sets such as small clinical trials, where more careful assessment and elicitation of prior model probabilities might be needed.

3 MODELING STRATEGY

We now describe our modeling strategy. We start by reviewing Occam's Window, which embodies our criteria for selecting a set of models. We then describe how we use ACE to select an initial candidate set of codings for each risk factor. Finally, we give the iterative algorithm used for combining these two steps.

3.1 Occam's Window

This algorithm was proposed by Madigan and Raftery (1994) and selects a subset of the models initially considered. It involves averaging over a much smaller set of models than in Eq. (4), thereby making it easier to compute and to communicate model uncertainty.

They argued that if a model is far less likely a posteriori than the most likely model, then it has been discredited and should no longer be considered. Thus models not belonging to

$$\mathcal{A}' = \left\{ M_k : \frac{P(M_k \mid D)}{\max_l \{ P(M_l \mid D) \}} \geq c \right\}$$

should be excluded from Eq. (4), where c is a fairly small number ($\ll 1$) chosen by the data analyst. By analogy with the common 5% significance level used in frequentist tests, we have used $c = 0.05$. Appealing to Occam's razor, they also exclude from Eq. (4) any model that receives less support from the data than a simpler model that is nested within it, namely those belonging to

$$\mathcal{B} = \left\{ M_k : \exists M_l \in \mathcal{A}, M_l \subset M_k, \frac{P(M_l \mid D)}{P(M_k \mid D)} > 1 \right\}$$

Then Eq. (4) is replaced by

$$P(\Delta \mid D) = \frac{\sum_{M_k \in \mathcal{A}} P(\Delta \mid M_k, D) P(D \mid M_k) P(M_k)}{\sum_{M_k \in \mathcal{A}} P(D \mid M_k) P(M_k)}$$

where $\mathcal{A} = \mathcal{A}' \backslash \mathcal{B}$.

Typically the number of terms in Eq. (4) is reduced to 25 or less and often to as few as one or two, even when the number of models initially considered is very large. This procedure mimics the evolutionary process of model selection that is typical of science. The final solution is fairly robust to the initial class considered, in the sense that most initial classes that contain $\mathcal{A}$ give the same result.

When the total number of models initially considered, K, is relatively small, the models in $\mathcal{A}$ can be found directly by considering each model in turn. When K is very large, however, this may not be feasible, and Madigan and Raftery (1994) proposed a tree-based algorithm (the Up-Down algorithm) that is computationally feasible in this case. It has been successfully applied to situations in which there are up to 10^{12} models.

An alternative approach when K is very large is to apply the leaps-and-bounds algorithm of Furnival and Wilson (1974) to an approximating weighted linear regression problem to reduce the number of models and then find the models in $\mathcal{A}$ directly from within this reduced set. The BIC.LOGIT software described in Section 5.3 does this, but using only the BIC approximation to the Bayes factors, which is somewhat cruder than the approximations used here.

3.2   Selecting Transformations with ACE

The ACE algorithm (Breiman and Friedman, 1985) is a way of finding nonparametric transformations of the independent and dependent variables in linear regression such that the relationship is as linear as possible. See DeVeaux (1989) for an exposition.

This can be extended to generalized linear models by noting that estimation there is equivalent to weighted least squares with the adjusted dependent variable z_i and the weights w_i given in Section 2.5. The z_i and the w_i involve the conditional expectations μ_i, and the method can be implemented by initially estimating the μ_i from the full model with all risk factors, assuming linearity. For logistic regression, $z = \text{logit}(\mu) + (y - \mu)/\{\mu(1 - \mu)\}$ and $w = \mu(1 - \mu)$, while for Poisson log-linear models, $z = \log \mu + (y - \mu)/\mu$ and $w = \mu$ (dropping the subscript i's). The transformation of the dependent variable must always be specified to be linear. We implemented ACE using the S-PLUS function ace. The ACE.LOGIT function described in Section 5.2 performs this full procedure for the special case of logistic regression.

We use the ACE output to *suggest* parametric transformations of the variables, if possible, ones that are substantively interpretable; we do *not* use the nonparametric transformations estimated by ACE in their raw form. The transformations suggested often have the form of a threshold effect, with no change in the expected value of the dependent variable below or beyond a certain threshold value. This kind of transformation often fits the data better *and* is more interpretable than the commonly recommended addition of a quadratic term to deal with nonlinearity. Sometimes, the ACE output suggests that the variable be categorized and how this should be done. ACE has also been used in this way by Raftery *et al.* (1995a,b).

Often it is reasonable to assume that the relationship between disease and risk factor is monotonic, and imposing this constraint can be very helpful in obtaining useful output from ACE. We generally make this assumption unless the relationship is clearly nonmonotonic. To detect strong nonmonotonicities, we run ACE a first time without the monotonicity constraints and then a second time with them.

Generalized additive models provide an approach to modeling in the presence of nonlinearity that is similar in concept to ACE and is conceived as a direct generalization of generalized linear models (Hastie and Tibshirani, 1990). However, the software to estimate them (in S-PLUS) does not yet allow for monotonicity constraints, so we have found ACE more useful.

3.3 An Iterative Strategy

The methods we have described are linked in the following iterative algorithm:

1. Run ACE with the adjusted dependent variable z_i and weights w_i, specifying the transformation of the dependent variable to be linear, but without monotonicity constraints. Identify independent variables whose estimated ACE transformations are strongly nonmonotonic.
2. Run ACE again, this time with the monotonicity constraints for all the independent variables except those for which strongly nonmonotonic relationships were identified in step 1.
3. Using the ACE output, identify candidate parametric transformations of the independent variables. Possibilities include power and log transformations, threshold and piecewise linear functions, categorical (i.e. piecewise constant) transformations, and the addition of quadratic and higher order polynomial terms.
4. Using the "preferred" codings of the independent variables from step 3, run GLIB using as models all relevant subsets of the independent variables.
5. For each of the most likely models a posteriori from step 4, calculate Bayes factors for it against perturbed versions of itself in

which other candidate transformations are used, for each independent variable in turn. If any of the other transformations is preferred, use it and return to step 4.
6. Apply the rules of Section 3.1 to the models from step 4 so as to
find the models in Occam's Window. Run GLIB again using only
those models. Use the results to make inference about QUOIs or
to report the main conclusions from the analysis and the remaining
uncertainties.

We have found that ACE is highly effective at suggesting good transformations and that the iteration between steps 4 and 5 is often not needed.
If the number of models is too large, it will not be possible to execute
step 4 directly, and it will have to be replaced by a version of the Up-
Down algorithm of Madigan and Raftery (1994). Software to do this for
generalized linear models has not yet been produced, but it can sometimes
be done manually; see Raftery (1993) for an example. The BIC.LOGIT
software described in Section 5.3 does it automatically, but rather approximately, for the special case of logistic regression.

4 AN EPIDEMIOLOGICAL APPLICATION: NUTRITION
AND BREAST CANCER

This section illustrates the use of Bayes factors in an epidemiological
study. We start by a brief description of the study.

4.1 Description of Study

The association of dietary factors and breast cancer (BC) has been widely
discussed during the past decade and interest has focused on the role
played by fat and alcohol intake. Epidemiological studies have given inconsistent results with respect to these risk factors, even though alcohol
has been found to be positively associated with BC in a majority of studies
(Howe et al., 1991; Van der Brandt et al., 1993).
The part played by dietary risk factors was tested in a case-control
study that took place in Montpellier (France) between 1983 and 1987.
Cases were women in the age group 21–66 with histologically confirmed
primary carcinoma of the breast who were hospitalized in a cancer institute and had not previously undergone any therapy. Controls in the same
age group were all the women admitted for the first time to one of three
wards, of which one was neurological, one was neurosurgical in a nearby
hospital, and one housed all those hospitalized for general surgery in a

large clinic. All these women came for a first diagnosis and hence were not currently treated for chronic diseases. These wards were chosen because the pathologies they treat are not usually related to nutritional factors. We excluded from the controls women admitted for neoplasic diseases. Cardiovascular diseases were also excluded from the hospital wards because in neurology they are overrepresented. The cases and controls were of similar ages and lived in the same area. A dietary history questionnaire administered by interview was used to measure the intake of total fat and its constituents, other nutrients, and alcohol consumption.

In the original analysis of this study, a positive association was found with alcohol (Richardson *et al.*, 1989) and fat intake (Richardson *et al.*, 1991), but the evidence for the role of fat intake was not overwhelming. Classical risk factors for BC were also analyzed, in particular those related to reproductive history (Ségala *et al.*, 1991). We shall focus our reanalysis on the association with alcohol and fat intake.

The sample consists of 854 women (379 cases and 475 controls) for which all the risk factors were recorded. Simple summary statistics, highlighting comparisons between cases and controls, are shown in Table 1 and Figure 1. For the classical risk factors, the differences between cases and controls relate mainly to the age of the end of schooling, family history of BC, and past history of benign breast disease. A shift in the distribution of alcohol consumption and a smaller shift in the lower end of the distribution of fat consumption can also be seen between cases and controls.

4.2 Choice of Transformation of Risk Factors

The risk factors considered in this analysis are the classical risk factors for BC and in addition alcohol consumption and fat intake (total fat or saturated fat). There is strong prior evidence from the literature for a link between BC and classical risk factors such as age, menopausal status, age at menarche, parity, family history of BC, history of benign breast disease, educational level characterized by age at the end of schooling, and Quetelet's body mass index (weight in kg/height^2 in meters). It is thus necessary to control for the effect of these risk factors in any analyses of the role of alcohol and fat intake. This will be done by including classical risk factors in all the subsequent models even though in our study sample only some of these risk factors are clearly different between cases and controls. Nevertheless, there is no consensus on whether some of these risk factors should be treated as continuous or categorical variables and, if categorical, on the class limits to be used. A similar question arises for alcohol and fat intake.

Table 1 Comparison of Cases and Controls for the Dichotomous Variables
Considered in the Epidemiological Study

		Cases		Controls		Odds ratio[a]	95% CI
		N	%	N	%		
Menopausal	No	197	52.0	275	57.9	1	
status	Yes	182	48.0	200	42.1	0.93	[0.65–1.34]
Family	No	355	93.7	463	97.5	1	
history of BC	Yes	24	6.3	12	2.5	2.61	[1.29–5.29]
Past history	No	325	85.7	442	93.1	1	
of benign breast disease	Yes	54	14.3	33	6.9	2.23	[1.41–3.51]
Consumption	No	145	38.3	272	57.3	1	
of alcohol	Yes	234	61.7	203	42.7	2.16	[1.64–2.85]

[a] The odds ratios are age adjusted.

As outlined in Section 3.2, we investigated suitable transformations
of all these variables by an ACE analysis first with and then without
monotonicity constraints. The variables were included in the ACE fit without any categorization. Only one of total fat or saturated fat was present
in the model at once; otherwise all variables were included.

The plots of the transformed variables against the original ones are
shown in Figure 2. The only risk factor that clearly exhibits a linear pattern
is age, so age will not be transformed and will be entered in all the models
as a continuous variable. For parity, a dichotomy is apparent and parity
is transformed into a categorical variable: three or less children versus
four or more children.

For age at menarche, age at the end of schooling, and total and saturated fat intakes, simple linear transformations with thresholds seem to
be suggested; that is, the transformed variable X^* is equal to the original
variable X when $X \leq a$ and X^* is equal to a when $X > a$. For age at
menarche, $a = 11.5$ years was chosen; for age at the end of schooling,
$a = 16$ years; for total fat, $a = 800$ g per week; for saturated fat, $a = 250$
g per week. For Quetelet's index, two thresholds are indicated with X^*
$= X$ when $X \leq 20$, $X^* = 20$ when $20 \leq X \leq 25.5$, $X^* = 25.5$ when $X >$
25.5. Finally, an interesting phenomenon appears when looking at alcohol
consumption. The relationship seems fairly linear among the drinkers but
there is a noticeable jump between the initial point representing the group

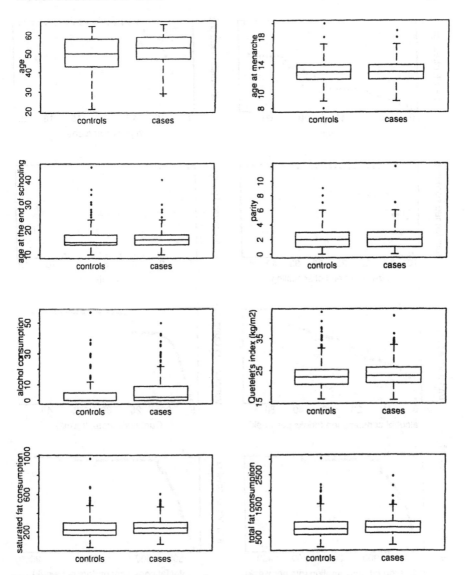

Figure 1 Box plots of risk factors by case/control status.

of nondrinkers (about half the sample) and the low drinkers. Hence, apart from treating alcohol consumption linearly, a dichotomous variable distinguishing drinkers from nondrinkers will also be considered.

Some of the thresholds found above have an epidemiological interpretation. Early menarche is usually considered a detrimental risk factor,

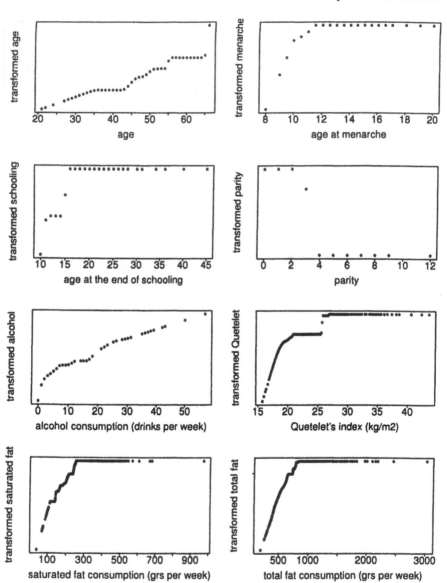

Figure 2   ACE transformations of the risk factors.

whereas high parity would be protective. Here it is found that the risk seems to vary exclusively among women with very early menarche. The plot for parity highlights a break when women have four or more children, a fact that had been noticed in previous analyses of these data. Similarly,

BC risk is known to increase with higher socioeconomic status, and women prolonging their secondary education after age 16 (which is a natural stopping point in the French educational system) would tend to have different careers than those less well educated. The two thresholds found for Quetelet's index identify a group of slim women ($Q \leq 20$) and a group of corpulent women ($Q \geq 25$). For the nutritional variables, it is important to keep in mind that the thresholds do not necessary relate to an underlying biological phenomenon but could be connected to measurement error. For example, the distinction between drinkers and nondrinkers is less subject to measurement error than the quantitative measurement of alcohol consumption.

4.3 Choice of Independent Variables

From now on, in line with our prior knowledge, all the models considered include the classical risk factors transformed as suggested by the ACE plots (see Section 2.6).

Our first step is to study the effect of alcohol and fat intake on the risk of BC, using the transformed variables. Alcohol intake is thus characterized by two variables: "alc," which is equal to the number of alcoholic drinks per week (the alcohol content—calculated using volume and average percent pure alcohol—is nearly identical for each type of drink), and "alc0," which dichotomizes drinkers and nondrinkers. Fat intake is represented by "tfat" and "tsat," the transformed threshold variables measuring consumption of these nutrients per week (in grams).

The dichotomous inclusion of each of these four variables leads to 16 models, which were compared using GLIB (Table 2). Let us first look at the column $\phi = 1.65$, which is the recommended value for the prior variance parameter of the regression parameters (see Section 2.5). The highest value for $2 \log B_{10}$ is obtained for model 10, M_{10}. This value will be used as a "yardstick," and our first selection rule is to exclude all models that predict the data far less well than the best model, M_{10}. Applying the first rule of the Occam's Window strategy of Section 3.1, we excluded M_k for which $P(M_k|D)/P(M_{10}|D) \leq 0.05$, that is, for which the difference between $2 \log B_{10}$ for the best model M_{10} and $2 \log B_{10}$ for model M_k was greater than 6. This excluded models 1, 2, 4, 5, 11, 14, 15, and 16.

Our second rule of model selection is to compare nested models and to exclude a model involving more parameters if a simpler model has a higher posterior probability. We note that model 9 is nested in model 12 and has a higher posterior probability, so model 12 is excluded. By similar reasoning, model 13 is excluded by reference to model 10 and 6 by reference to 3. We are thus left with five models: 3, 7, 8, 9, and 10.

Table 2 Selection of Independent Variables[a]

Model	alc	alc0	tfat	tsat	$\phi = 1.00$	$\phi = 1.65$	$\phi = 5.00$	Deviance	df
						2 log B_{10}			
1	0	0	0	0	0.0	0.0	0.0	1107.3	844
2	1	0	0	0	14.6	13.6	11.4	1086.2	843
3	0	1	0	0	18.9	17.9	15.7	1081.8	843
4	0	0	1	0	6.1	5.1	2.9	1094.6	843
5	0	0	0	1	5.9	4.9	2.7	1094.7	843
6	1	1	0	0	16.8	14.8	10.3	1077.6	842
7	1	0	1	0	17.3	15.3	10.9	1076.8	842
8	1	0	0	1	18.3	16.4	11.9	1075.7	842
9	0	1	1	0	21.7	19.7	15.3	1072.4	842
10	0	1	0	1	22.5	20.5	16.1	1071.6	842
11	0	0	1	1	1.4	−0.6	−5.0	1093.7	842
12	1	1	1	0	18.7	15.7	9.1	1069.0	841
13	1	1	0	1	19.8	16.9	10.2	1067.8	841
14	1	0	1	1	13.2	10.2	3.5	1075.5	841
15	0	1	1	1	17.4	14.4	7.8	1071.3	841
16	1	1	1	1	14.6	10.6	1.8	1067.6	840

[a] All models include the classical risk factors (age, menopausal status, age at menarche, parity, familial BC, history of benign breast disease, age at the end of schooling, and Quetelet's index).

This model selection is robust to changes in the value of ϕ, the parameter characterizing the prior variance of the regression parameters. The same five models remain whether we apply our selection rule for $\phi = 1$, $\phi = 5$, or anything in between. The values of the Bayes factor are indeed sensitive to the choice of ϕ but the relative rankings of the models are comparable across ϕ. There is some evidence for a small amount of overdispersion, with an estimated overdispersion parameter of about $\bar{\sigma} = 1.13$. Taking account of this would make relatively little difference to the results and we have not done so. However, it can be done using our approach; see Section 2.4.

Note that classical model comparisons between nested models using deviance differences would lead to somewhat different conclusions. For example, if we compare models 3 and 10, which are nested, the difference in deviance is 10.2 for 1 df, which classically would give an overwhelming preference for model 10; that is to say that saturated fat has an effect above that of alcohol. On the other hand the posterior probabilities that can be calculated from the Bayes factor indicate that model 10 is about 3.5 times more probable than model 3. So from a Bayesian point of view, the evidence for an effect of saturated fat, while positive, is not over-whelming.

Our next step was to rerun GLIB, including only the five selected models, to find their posterior probabilities (Table 3). Note that these posterior probabilities allow us to compare nonnested models. Overall we see that the models including alc0 have higher posterior probabilities. Some influence of the chosen values for ϕ is also apparent, with higher values of ϕ favoring smaller models, particularly model 3, which has fewer parameters. This is as expected; see Section 2.5.

The posterior probabilities that the regression coefficient is different from zero for each of the four variables included in the five models, which is also an output of GLIB, are given in Table 4. These probabilities are not much influenced by the values of ϕ. All five retained models include either alc0 or alc as a variable, so that the evidence for an association between BC and alcohol consumption is very strong. The evidence for the part played by fat consumption (either total fat or saturated fat only) is less strong. There are odds of only $7:1$ ($\phi = 1.65$) supporting it. We can also compare the qualitative and quantitative effects of alcohol: there are odds of $9:1$ in favor of the main effect of alcohol being of the dichoto-mous type. On the other hand, there is only weak evidence for the effect of fat consumption to be restricted to saturated fat, as the odds are only roughly $1.5:1$ in favor of saturated fat. This is an area where substantial uncertainty remains.

4.4 Confirmatory Analysis of the Choice of Transformations

The appropriateness of the choice of transformations for total fat and saturated fat can be confirmed by comparing the chosen models to similar

Table 3   Posterior Model Probabilities for the Five Models Selected by Occam's Window

| | | Posterior probabilities (%) | | |
Model	Independent variable	$\phi = 1.00$	$\phi = 1.65$	$\phi = 5.00$
3	alc0	8	13	31
7	alc, tfat	4	4	3
8	alc, tsat	6	6	5
9	alc0, tfat	33	31	25
10	alc0, tsat	49	46	37

[a] All models include the classical risk factors (see Table 1).

Table 4 Posterior Probabilities for the Inclusion of Each
Independent Variable

Independent variable	Posterior probabilities (%)		
	$\phi = 1.00$	$\phi = 1.65$	$\phi = 5.00$
alc	10	9	7
alc0	90	91	93
tfat	37	35	27
tsat	55	52	41

models in which the variables have their original quantitative values. This
was done systematically by including variables without thresholds in
models 7 to 10 (Table 5). Furthermore, a variable alc3 discretizing alcohol
consumption into three classes (0, 1–7, > 7) was also created and compared to the dichotomized alc0 variable. The results in Table 5 indicate
without exception that the models including the transformed variables
with thresholds have substantially higher posterior probabilities. The variable alc3 does not perform well either.

Thus the coding chosen on the basis of ACE as in Section 4.2 is
adequate. If it had not been, we would have iterated as in Section 3.3,
changing the codings and redoing the independent variable selection.

Table 5 Confirmatory Analysis of the Choice of
Transformations of Independent Variables via
Bayes Factors ($\phi = 1.65$)

Model	Independent variables	$2\log B_{k0}$
3	alc0	17.9
7	alc, tfat	15.3
	alc, fat	8.4
8	alc, tsat	16.4
	alc, sat	11.1
9	alc0, tfat	19.7
	alc0, fat	12.4
10	alc0, tsat	20.5
	alc0, sat	14.8
	alc3, tsat	14.4

4.5 Model Uncertainty

In our example the regression coefficients estimated under the different models do not vary much (Table 6). Furthermore, they are similar to the maximum-likelihood estimators (results not shown). For example the coefficient for the variable alc0 given by GLIB ($\phi = 1.65$) varies between 0.73 (model 3), 0.69 (model 9), and 0.70 (model 10) and is equal to the corresponding maximum-likelihood estimate up to two decimal places. This lack of variation of coefficients across models is not a general occurrence. When some but not all the models considered include interaction terms, the regression coefficients estimated can be substantially different, as was the case in the study of oral contraceptive use and myocardial infarction analyzed in Raftery (1993b).

In this study, the main consequences of model uncertainty are thus concentrated on evaluating the risk for particular individuals. For example, for a women having three alcoholic drinks per week, models 7 and 8 estimate an odds ratio of 1.1 whereas models 3, 9, and 10 lead to an odds ratio of 2.0. Overall, an odds ratio of 1.9 for three drinks per week would be estimated if the odds ratio for each model were weighted by the corresponding model probability. On the other hand, for a women having 18 drinks per week, the odds ratio for alcohol consumption would be equal under all the models.

There is a strong correlation between overall fat intake and saturated fat intake ($r = 0.89$), with the intakes usually related by a ratio of 3.2. This is well reflected by the ratio of the corresponding regression coefficients. So for women whose consumption of saturated fat is about a third of their overall fat consumption, odds ratio estimates given by models 7 to 10 will be comparable. On the other hand, there are a few women consuming a much higher proportion of saturated fat, such as patient 41,

Table 6 Posterior Means of the Regression
Coefficients Under the Five Models ($\phi = 1.65$)

Model	alc0	alc	tfat	tsat
	\multicolumn{4}{Variable}
3	.734	—	—	—
7	—	.038	.0017	—
8	—	.039	—	.0052
9	.691	—	.0017	—
10	.704	—	—	.0051

who consumed 390 g per week of saturated fat and 645 g per week of total fat. For these women, the odds ratio given by model 9 is 3.0 while that given by model 10 is 3.6. Note that this discrepancy would have been much bigger if the threshold of 250 had not been used for saturated fat.

4.6 Comments

In the original analysis of this study, an elevation of BC risk associated with alcohol consumption, categorized in either five classes (0, 1–2, 3–9, 10–17, > 17), or three classes (0, 1–7, > 7), had been found. The effect of fat consumption was not apparent unless a negative interaction (significant at the 5% level) was introduced in the model. It had also been noted that the odds ratio (1.8) corresponding to the lowest class of alcohol consumption (one or two drinks per week) was higher than would be expected if the risk increased linearly with consumption of alcohol.

As in all observational studies, great care has to be taken to address all the possibilities of biases before drawing any causal inference (Kleinbaum *et al.*, 1982). In case-control studies, selection bias (which takes multiple forms such as referral bias and bias linked to the choice of diseases for which the controls were hospitalized) is an important issue that was carefully investigated in our previous work. Another concern is the validity of the dietary instrument used. Indeed, dietary intake assessment is notoriously subject to measurement error, leading generally to risk estimates that are biased conservatively. Despite these inferential limitations, case-control studies have substantially contributed to advancing the epidemiological knowledge on diet and cancer (Hill *et al.*, 1994).

In the new analysis that we present here, thresholds were strongly indicated for the effect of fat consumption. With the introduction of these thresholds, there is substantial but not overwhelming evidence for an effect of fat consumption, whether total or restricted to saturated fat. No interaction comes into play. This was checked by calculating Bayes factors for models including interactions between alcohol and fat consumption. In essence, the negative interactions, which were somewhat unnaturally introduced in the previous analyses, are superceded by the use of variables with thresholds.

The new analysis has also yielded thought-provoking results about alcohol consumption. Indeed, there is clear evidence that models including the simple dichotomy, drinker versus nondrinker, are preferred. From a biological point of view, this is a surprising result because none of the several biological hypotheses that have been put forward to explain the role of ethanol in mammary carcinogenesis could account for such a threshold. Hence we would favor either or both of two other explanations of this result. The first is that the notorious difficulties in measuring alcohol consumption have resulted in imprecise measurements of the quantity of alcohol consumed to such a degree that the dichotomy, drinker versus nondrinker, better explains the variation in BC risk in our data. Measurement error in the assessment of the consumption of alcohol, fat, and other nutrients is indeed an issue here. The second possible explanation is that potential unobserved lifestyle factors could partly confound the association with alcohol.

In conclusion, by quantifying the model uncertainties well, areas where future research is needed are more clearly identified. There is a need to improve the assessment of alcohol and fat consumption, and dietary history questionnaires could usefully be complemented by other measuring instruments. It was not our purpose to assess in detail the specific role that could be played by different components of fat intake. Substantial model uncertainty remains concerning a specific role of saturated fat intake, and populations in which there is less correlation between total fat and saturated fat intakes should be sought.

5   SOFTWARE

5.1   GLIB: Generalized Linear Bayesian Modeling

GLIB is an S-PLUS function for implementing the methods that we have described. It can be obtained free of charge from StatLib by sending the e-mail message "send glib from S" to *statlib@stat.cmu.edu*. Here are some of its features:

> For each model considered, GLIB returns the posterior model probability and the Bayes factor for it against the null model (i.e., the one with no independent variables).
> For each independent variable, GLIB returns the posterior probability that the corresponding regression parameter is nonzero (obtained by summing across models), and its composite posterior mean and standard deviation given that it is nonzero, from Eqs. (5) and (6).
> GLIB uses the reference proper priors of Section 2.5, and the user can specify a set of values of the prior dispersion parameter ϕ. The Bayesian results are returned for each value of ϕ, thus assessing sensitivity to the prior. The default values are $\phi = 1.00$, 1.65, 5.00, that is, the lower bound, the recommended value, and the upper bound from Section 2.5. The numerical values of the prior

mean and variance matrix used for each model and each value of ϕ are output on request by specifying **priorvar** = **T**.

The user can specify the prior model probabilities via the input **pmw**. Actually, these are prior model *weights* and do not have to sum to one; GLIB renormalizes them automatically. This is convenient if the program is being rerun after removing some of the models initially considered. By default, all models are assumed equally likely a priori.

GLIB returns standard frequentist GLIM results (deviances, MLEs, standard errors, and so on), as well as their Bayesian analogues. This allows an arbitrary number of generalized linear models to be estimated with a single command, which can be useful even if the Bayesian aspects are not of primary interest.

The models to be fit are specified by the input design matrix **x** and the input matrix **models**, which specifies the subsets to be considered. Columns 2–5 of Table 1 constitute an example of the input **models** matrix.

5.2 ACE.LOGIT: ACE Analysis for Logistic Regression

ACE.LOGIT is an S-PLUS function that carries out the ACE analysis described in Section 3.2 for the special case of logistic regression. It automatically produces graphs such as those in Figure 2. It can be obtained free of charge from StatLib by sending the message "send ace.logit from S" to *statlib@stat.cmu.edu*.

5.3 BIC.LOGIT: Approximate Bayesian Model Selection for Logistic Regression

BIC.LOGIT is an S-PLUS function that can be obtained free of charge by sending the e-mail message "send bicreg from S" to *statlib@-stat.cmu.edu*. It implements the Occam's Window algorithm for logistic regression automatically but in an approximate way using the somewhat crude BIC approximation of Eq. (15) below.

For a given dependent variable and set of candidate independent variables, it finds the models in Occam's Window and their posterior probabilities, and for each independent variable it finds $Pr[\beta_j \neq 0|D]$ and the posterior mean and standard deviation.

It exploits the fact that at the MLE, logistic regression is approximately a weighted least-squares problem with an adjusted dependent variable (McCullagh and Nelder, 1989). To reduce the set of models to a manageable number, it replaces the logistic regression problem by the equivalent weighted least-squares problem. It then uses the leaps and bounds algorithm of Furnival and Wilson (1974) to identify a reduced set of good models. Finally, it calculates BIC exactly for the remaining models and finds those that lie in Occam's Window. When there are more than 30 variables, it first uses backward elimination to reduce the initial set of variables to 30. For more details see Raftery (1995).

6 DISCUSSION

We have described a Bayesian model-building strategy for generalized linear models that avoids the difficulties of commonly used ad hoc strategies, namely arbitrariness in the coding of risk factors, lack of knowledge of the overall properties of model selection strategies based on significance tests, and failure to account for model uncertainty. It delivers a parsimonious set of models together with their posterior probabilities, thereby facilitating effective communication of model uncertainty. As a result, the conclusions from the analysis are clear, as are the remaining uncertainties. This helps to define future research priorities.

In our strategy, model selection is based on Bayes factors rather than on significance tests. This has been argued to be a more appropriate form of inference (Edwards *et al.*, 1963; Berger and Delampady, 1987), and it tends to give more reasonable results for large samples (e.g., Raftery, 1986), as well as allowing easy comparison of nonnested models (Kass and Raftery, 1995). The methods can be implemented using the GLIB software available free of charge from StatLib. This software automatically assesses the sensitivity of the results to the prior dispersion.

The strategy yields the posterior probability of inclusion for each risk factor considered and thus answers what is often the main question asked in epidemiological studies, namely, how likely is it that there is an association between risk factor and disease? It also provides a formal justification for including "classical" risk factors even when the data at hand provide little support for their inclusion. The justification is that the evidence for their inclusion comes from prior studies and is reflected in low prior model probabilities for models that exclude them.

These points are well illustrated by the epidemiological application of Section 4. Previous analyses using a more standard strategy had led to somewhat ambiguous and counterintuitive conclusions, particularly concerning the joint role of alcohol and fat consumption (see Section 4.6). Our analysis, by contrast, revealed positive (but not decisive) evidence for an association with fat consumption. It is also clear that the data do not allow us to determine whether the effect is due to total fat or just to saturated fat. Our analysis also showed that alcohol consumption is a risk factor for breast cancer but that the finding of a dose-effect relationship might be hampered by measurement error. Thus future research that replicates this study should focus on better measurement of alcohol and fat consumptions and on finding populations in which it is possible to distinguish between the effects of different types of fat consumption.

Accurate Bayesian analysis is possible for generalized linear models because a good approximation is available for the Bayes factors via the Laplace method. That may not be the case for other biostatistical models, although approximations of similar quality are available for some hierarchical models used in health services research: beta-binomial models (Kahn and Raftery, 1995) and Poisson-gamma models (Rosenkranz, 1992). For other biostatistical models for which such approximations are not yet available, it is often possible to use the BIC or Schwarz approximation (Schwarz, 1978), namely

$$\log P(D \mid M_k) \sim \log P(D \mid \hat{\theta}_k, M_k) - \tfrac{1}{2} p_k \log n \tag{15}$$

On the face of it, this is a crude approximation, but it has been observed to be surprisingly accurate. Kass and Wasserman (1992) have pointed out that it is, in fact, quite an accurate approximation for a particular, reasonable, prior. This opens up a wide class of models for the kind of model selection strategy and model uncertainty analysis outlined in this chapter, as discussed by Raftery (1995) in general and by Raftery, Madigan and Volinsky (1995) for survival analysis. It is the basis of the BIC.LOGIT software described in Section 5.3.

The general approach outlined in this chapter can be used to choose variable codings and transformations, as well as which variables to include, as we have described and illustrated. It could also be extended to take account of uncertainty due to this choice (such as uncertainty about the threshold value in a threshold transformation), and this would be a worthwhile topic for future research.

ACKNOWLEDGMENTS

Raftery's research was supported by ONR contract N-00014-91-J-1074, by the Ministère de la Recherche et de l'Espace, Paris, by the Université de Paris VI, and by INRIA, Rocquencourt, France. Raftery thanks the latter two institutions, Paul Deheuvels, and Gilles Celeux for hearty hospitality during his Paris sabbatical in which part of this chapter was written. The authors are grateful to Christine Monfort for excellent research assistance and to Don Berry, Norman Breslow, Michel Chavance, Mariette

Gerber, Jennifer Hoeting, David Madigan, and Dalene Stangl for helpful discussions and comments on an earlier version of this chapter.

REFERENCES

Berger, J. O. and Delampady, M. (1987). Testing precise hypotheses (with discussion). *Statistical Science* 3: 317–352.

Berger, J. O. and Sellke, T. (1987). Testing a point null hypothesis: The irreconcilability of P values and evidence (with discussion). *Journal of the American Statistical Association* 82: 112–122.

Breiman, L. and Friedman, J. H. (1985). Estimating optimal transformations for multiple regression and correlation (with discussion). *Journal of the American Statistical Association* 80: 580–619.

Breslow, N. E. and Day, N. E. (1980). *Statistical Methods in Cancer Research*. Vol. I, *The Analysis of Case-Control Studies*. International Agency for Research on Cancer, Lyon, France.

de Bruijn, N. G. (1970). *Asymptotic Methods in Analysis*. North-Holland, Amsterdam.

DeVeaux, R. D. (1989). Finding transformations for regression using the ACE algorithm. *Sociological Methods and Research* 18: 327–359.

Draper, D. (1995). Assessment and propatation of model uncertainty (with discussion). *Journal of the Royal Statistical Society, Series B* 57, in press.

Edwards, W., Lindman, H. and Savage, L. J. (1963). Bayesian statistical inference for psychological research. *Psychological Review* 70: 193–242.

Furnival, G. M. and Wilson, R. W., Jr. (1974). Regression by leaps and bounds. *Technometrics* 16: 499–511.

George, E. I. and McCulloch, R. E. (1993). Variable selection via Gibbs sampling. *Journal of the American Statistical Association* 88: 881–889.

Hastie, T. and Tibshirani, R. (1990). *Generalized Additive Models*. Chapman & Hall, London.

Hill, M. J., Giacosa, A., and Caygill, C. P. J. (1994). *Epidemiology of Diet and Cancer*. Ellis Horwood, Chichester.

Hodges, J. S. (1987). Uncertainty, policy analysis and statistics. *Statistical Science* 2: 259–291.

Howe, G., Rohan, T., Decarli, A., *et al.* (1991). The association between alcohol and breast cancer risk: Evidence from the combined analysis of six dietary case-control studies. *International Journal of Cancer* 47: 707–710.

Jeffreys, H. (1961). *Theory of Probability*, 3rd ed. Oxford, Clarendon Press.

Kahn, M. J. and Raftery, A. E. (1995). Discharge rates of medicare stroke patients to skilled nursing facilities: Bayesian logistic regression with unobserved heterogeneity. *Journal of the American Statistical Association* 90, in press.

Kass, R. E. and Raftery, A. E. (1995) Bayes factors. *Journal of the American Statistical Association* 90, 773–795.

Kass, R. E. and Vaidyanathan, S. (1992). Approximate Bayes factors and orthogo-

352

Raftery and Richardson

nal parameters, with application to testing equality of two Binomial proportions. *Journal of the Royal Statistical Society, Series B* 54: 129–144.

Kass, R. E. and Wasserman, L. (1992). The Surprising Accuracy of the Schwarz Criterion as an Approximation to The log Bayes Factor. Technical Report, Department of Statistics, Carnegie-Mellon University, Pittsburgh.

Kleinbaum, D. G., Kupper, L. L., and Morgenstern, H. (1982). *Epidemiologic Research: Principles and Quantitative Methods*. Lifetime Learning Publications, Belmont, CA.

Leamer, E. E. (1978). *Specification Searches*. Wiley, New York.

Madigan, D. and Raftery, A. E. (1994). Model selection and accounting for model uncertainty in graphical models using Occam's window. *Journal of the American Statistical Association* 89: 1535–1546.

Madigan, D. and York, J. (1993) Bayesian graphical models for discrete data. *International Statistical Review* 63: 215–232.

Madigan, D., Raftery, A. E., York, J. C., Bradshaw, J. M., and Almond, R. G. (1994) Strategies for graphical model selection. In *Selecting Models from Data: AI and Statistics IV*, ed. Cheeseman, P. and Oldford, R. W., pp. 91–100. Springer-Verlag, New York.

McCullagh, P. and Nelder, J. A. (1989). *Generalized Linear Models*, 2nd ed. Chapman & Hall, London.

Racine, A., Grieve, A. P., Fluhler, H., and Smith, A. F. M. (1986). Bayesian methods in practice: experiences in the pharmaceutical industry (with discussion). *Applied Statistics* 35: 93–150.

Raftery, A. E. (1986). Choosing models for cross-classifications. *American Sociological Review* 51: 145–146.

Raftery, A. E. (1988). Approximate Bayes Factors for Generalized Linear Models. Technical Report No. 121, Department of Statistics, University of Washington.

Raftery, A. E. (1993a). Bayesian model selection in structural equation models. In *Testing Structural Equation Models*, ed. Bollen, K. A. and Long, J. S., pp. 167–180. Sage, Beverly Hills, CA.

Raftery, A. E. (1993b). Approximate Bayes Factors and Accounting for Model Uncertainty in Generalized Linear Models. Technical Report No. 255, Department of Statistics, University of Washington.

Raftery, A. E. (1995). Bayesian model selection in social research (with discussion). In *Sociological Methodology 1995*, ed. Marsden, P. V., in press, Oxford, Blackwell.

Raftery, A. E., Lewis, S. M., and Aghajanian, A. (1995a). Demand or ideation? Evidence from the Iranian martial fertility decline. *Demography*, 32: 159–182.

Raftery, A. E., Lewis, S. M., Aghajanian, A., and Kahn, M. J. (1995b). Event History Modeling of World Fertility Survey Data. *Mathematical Population Studies*, in press.

Raftery, A. E., Madigan, D. M., and Hoeting, J. (1993). Model Selection and Accounting for Model Uncertainty in Linear Regression Models. Technical Report No. 262, Department of Statistics, University of Washington.

Raftery, A. E., Madigan, D. M., and Volinsky, C. T. (1995). Accounting for model uncertainty in survival analysis improves predictive performance. In *Bayesian Statistics 5*, ed. Bernardo, J. M., Berger, J. O., Dawid, A. P., and Smith, A. F. M. Oxford University Press, Oxford, in press.

Richardson, S., de Vincenzi, I., Gerber, M., and Pujol, H. (1989). Alcohol consumption in a case-control study of breast cancer in southern France. *International Journal of Cancer* 44: 84–89.

Richardson, S., Gerber, M., and Cénée S. (1991). The role of fat, animal protein and some vitamin consumption in breast cancer: A case-control study in Southern France. *International Journal of Cancer* 48: 1–9.

Rosenkranz, S. (1992). The Bayes Factor for Model Evaluation in a Hierarchical Poisson Model for Area Counts. Ph.D. dissertation, Department of Biostatistics, University of Washington,

Schwarz, G. (1978). Estimating the dimension of a model. *Annals of Statistics* 6: 461–464.

Ségala, C., Gerber, M., and Richardson S. (1991). The pattern of risk factors for breast cancer in a southern France population: Interest for a stratified analysis by age at diagnosis. *British Journal of Cancer* 64: 919–925.

Tierney, L. and Kadane, J. B. (1986). Accurate approximations for posterior moments and marginal densities. *Journal of the American Statistical Association* 81: 82–86.

Van der Brandt, P. A., Van't Veer, P., Goldbohm, R. A., *et al.* (1993). A prospective cohort study on dietary fat and the risk of postmenopausal breast cancer. *Cancer Research* 53: 75–82.

VI
HIERARCHICAL MODELS

13
Bayesian Analysis of Population Pharmacokinetic and Instantaneous Pharmacodynamic Relationships

Amy Racine-Poon Ciba-Geigy, Basel, Switzerland
Jon Wakefield Imperial College, London, England

ABSTRACT

A typical population pharmacokinetic data set consists of measured drug concentrations with associated sampling times and known dose histories along with patient characteristics (covariates) such as age, weight, and gender. A typical population pharmacodynamic data set consists of measured responses, dose histories, and covariates. Models for such data attempt to identify sources of variability in the observed data and are increasingly being seen as an important aid in drug development. In particular, the identification of subgroups of patients (for example, the elderly) who exhibit altered behavior is important for the selection of optimal doses. In this chapter we stress the importance of predictive distributions for concentrations/responses. Such distributions can be used for dose selection and for model validation both within and between studies. Here we consider data from four phase I studies collected after administration of the drug recombinant hirudin. We use the data from two of these studies to construct the population model and then validate the model using the data from the remaining two studies. The studies were used to investigate the absorption characteristics of the drug and the effect of age, repeated

dosing, and steady-state dosing on the pharmacokinetic/pharmacody-
namic relationship.

1 INTRODUCTION

Pharmacokinetic (PK) models describe the time course of a drug and its
metabolites after its introduction into the body. Pharmacodynamic (PD)
models describe how some response measure changes as a function of
dose/concentration. It is commonly believed that the actual concentration
of a drug in blood plasma is a much better predictor of the efficacy or
tolerability of the drug than the administered dose. When the kinetic profile or the kinetic/dynamic profile is well understood, the dose schedule
required to achieve a certain desirable pharmacodynamic level for an individual may be achieved.

PK models often arise from simple compartmental systems in which
the body is modeled as a system of homogeneous compartments within
each of which the kinetics of the drug are assumed to be similar (Gibaldi
and Perrier, 1982). Such a system leads to a model function predicting drug
concentrations that is of nonlinear form. These compartmental models are
great simplifications of the many complex processes that affect a drug
molecule, but certain parameters of the model such as the volume (blood
space required to convert total amount of drug to concentration), the
clearance (defined as the volume of plasma cleared of drug in unit time),
and the rate of absorption, do have interpretations.

A typical population PK data set consists of measured drug concentrations with associated sampling times and known dose histories along with
patient characteristics (covariates) such as age, weight, and gender. A
typical population PD data set consists of measured responses, dose histories, and covariates. The analysis of such data sets has several aims that
can be summarized as attempting to explain the variation that is observed
in measured concentrations/responses. The classification of this variability into intra- and interindividual error is of particular importance. It is
therefore of interest not only to assess the average kinetic profile and the
average dynamic profile but also the between-individual variation of these
profiles.

The detection of subgroups of patients who exhibit altered behavior
is of great interest. This variability is often quantified by covariate models
for PK or PD parameters. For example, it may be that an individual's
clearance is a function of that individual's weight. Such characterization
allows a prior distribution for a new individual's parameters to be determined based on the individual's own covariates. This is important for

two reasons. First, when a dose size is to be decided on for a particular population, for example, the elderly, predictive distributions for response can be examined under different regimens and these, when combined with an appropriate therapeutic end point (and perhaps a loss function reflecting the relative losses associated with under- and overdosing) can be used to select a particular dose. With a substance that shows a small between-individual variation in kinetics and dynamics we would expect to be able to achieve a single "optimal" dosing scheme for most patients. This knowledge may be valuable in selecting doses for phase II and phase III trials. It is also necessary to have such knowledge for monitoring the concentrations in randomized-to-concentration trials.

Second, in cases in which the therapeutic range of desired responses is narrow, individuals may undergo on-line therapeutic monitoring. In such situations only limited concentration/response data are available and these data alone are not sufficient to provide accurate parameter estimates and hence predictive distributions for concentration. The individual data combined with population information can, however, be used to provide reliable estimates of the kinetic/dynamic parameters. Such estimates allow predictive concentrations and hence responses under different dosage regimens to be determined and adjusted as more information becomes available. This procedure is known as Bayesian individualization.

Throughout this chapter we emphasize the importance of predictive distributions. Predictive distributions are useful for a number of reasons. Obtaining predictive distributions for concentrations/responses under different dosage regimens allows an optimal regimen to be determined once the distribution is compared to the desirable range for concentrations/responses. Predictive distributions can also be used for model validation either within a particular study or between different studies. In the former case predictive distributions for particular concentrations/responses may be obtained with those observations deleted. The validation of the model between studies is of great importance because one would hope that the deterministic model that is appropriate for one study is valid for others also; if not, one loses predictive power and it is then not clear how to obtain a predictive distribution for a new individual. For the application considered in this chapter we had four studies with which we could construct and validate our model.

Population PK/PD models fit naturally into a hierarchical framework. At the first stage of the hierarchy the individual's concentration/response profiles are modeled, conditional on the values of the individual's parameters. At the second stage of the hierarchy the distribution from which the individual's parameters arise is specified. Such a model contains a large number of parameters and this, combined with the nonlinearity at the first stage, makes inference difficult whether the approach is classical or Bayesian. We discuss the benefits of a Bayesian approach in Section 2 and describe a general Bayesian hierarchical model appropriate for PK/PD data. We also describe the specific elements of the model for the data that we shall be analyzing in this chapter.

The aforementioned nonlinearity means that the integrations required for a Bayesian analysis cannot be carried out analytically, and the large dimensionality of the parameter space precludes the use of numerical integration or Laplacian methods. We will show, however, that a Markov chain Monte Carlo strategy is particularly convenient given the conditional structure of the hierarchical model. There are many advantages to such an approach that produces *samples* from the posterior distribution. For example, samples from the posterior distributions of functions of interest are easily recovered, as are predictive distributions for parameters/responses.

An important requirement for population pharmacokinetic analyses is the ability to address the robustness of the inferences to distributional assumptions. Regulatory authorities are particularly interested in this aspect. With a Markov chain Monte Carlo approach it is straightforward to examine different distributional assumptions. For example, the extension to *t*-distributions at the first or the second stage is straightforward and provides analyses that are robust to both outlying concentration measurements and outlying individuals.

We shall look at the combination of data from different studies. This aspect is important, since typically numerous studies are carried out during a drug's development. For example, different studies may address the linearity of the kinetic model, the effect of coadministration, and the constancy of kinetic parameters under different administration routes/dosing strategies and in different patient populations. A rational method for combining such studies is essential.

In this chapter we base our description of population pharmacokinetic/pharmacodynamic models on a particular group of studies for the drug recombinant hirudin (rec-hirudin). Rec-hirudin has properties similar to those of the leech *Hirudo medicinalis*, a naturally occurring anticoagulant. It is an antithrombotic agent that is a highly selective, potent, and almost irreversible inhibitor of human thrombin. One instance of its use is following an operation, and its goal is to to prevent blood clots from forming. A series of phase I studies was carried out in order to understand the tolerability and the pharmacokinetic profiles following single and repeated dosing and with different routes of administration in different groups of healthy volunteers. In addition, a measure of the clotting time, the activated partial thrombin time (APTT), was determined from blood samples in an attempt to understand the possible influence of dosing schedule and route of administration on this pharmacodynamic (response) measurement. Note that the APTT can be measured from blood samples quickly, whereas the concentration takes a considerably longer time to determine. Hence for future patients it will be possible to obtain on-line APTT measurements but not concentration measurements.

We briefly describe the possible routes of administration that we consider. An *intravenous bolus* is an instantaneous introduction of drug directly into the blood stream via an injection, an *intravenous infusion* is a constant introduction of drug directly into the blood stream over some specified period, and a *subcutaneous dose* is an injection beneath the skin and is, consequently, not directly into the blood stream.

In our example, four studies will be used. Two of the studies will be used for developing the model and obtaining parameter estimates; the other two studies will be used for validation of the model assumptions. One of the goals of these evaluations is the investigation of the possibility of recommending a dose for twice-daily subcutaneous application such that the average APTT at steady state is 1.5 to 2 times the baseline APTT value. We now describe the studies we will analyze.

1.1 Study 1 (Learning Study)

In addition to gaining an understanding of the PK and PK/PD profiles, the studies were carried out to investigate the absorption characteristics for subcutaneous administration. Two different routes of single application, intravenous and subcutaneous, were used for administration to 16 young healthy male volunteers in a two-period randomized crossover open (that is, not "blind") study. Each of eight volunteers received, on separate days, doses of 0.3 mg/kg body weight, one as an intravenous bolus and the other subcutaneously, while the remaining eight volunteers each received 0.5 mg/kg in both forms. Notice, therefore, that the doses depend on the weight of the individual. On each of the days of administration, blood samples were taken during a 24-hour period following administration. The plasma concentration of rec-hirudin of each sample was determined along with the clotting measure APTT. The observed data for subject 1 are graphically illustrated in Fig. 1. The solid lines in this figure represent drug concentrations and the dotted lines responses. Squares represent the intravenous data, triangles the subcutaneous data. Notice that the maximum concentration achieved with the subcutaneous dose is far lower than with the intravenous dose, although the concentrations following the subcutaneous dose do not fall so quickly. This is explained by the lower rate of absorption into the blood stream.

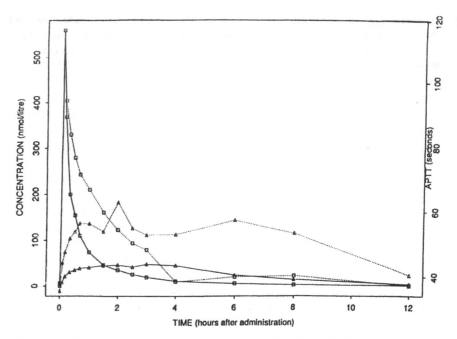

Figure 1 Observed concentration and response data for individual 1 of study 1. Solid lines denote drug concentrations and dotted lines the APTT response. Squares represent the intravenous data, triangles the subcutaneous data. The concentration and response were also measured at 24 hours, but these points have been omitted for clarity.

1.2 Study 2 (Learning Study)

The purpose of this study was to investigate the possible modification of the PK profile and the PK/PD relationship by the covariate age. The study was carried out with 12 elderly healthy volunteers who had the normal range of renal function. Each of the volunteers received a single subcutaneous application. Six of the volunteers received a dose of 0.3 mg/kg, and the remaining volunteers received 0.5 mg/kg. Blood samples were taken for a 24-hour period after application. The plasma concentration of rechirudin and the APTT were measured.

1.3 Study 3 (Validation Study)

The purpose of this study was to investigate the PK and PK/PD relationship after repeated subcutaneous dosing. Eight healthy volunteers re-

Table 1 Summary of the Four Studies that We Analyze[a]

Study	Type of administration	Patient population	No. of volunteers/ samples
1	IV bolus and SC single doses	Healthy volunteers	16/16 and 16
2	SC single dose	Elderly volunteers	12/16
3	Repeated SC dosing	Elderly volunteers	8/27
4	IV infusion	Healthy volunteers	8/18

[a] Intravenous and subcutaneous doses are denoted IV and SC, respectively. The final column gives the number of volunteers and the number of samples collected per volunteer. For example, in study 1 each of the volunteers had 16 blood samples for each of the administration routes.

ceived subcutaneous dosing twice daily for six consecutive days. Four of the volunteers received 0.3 mg/kg every 12 hours, whereas the remaining four received 0.5 mg/kg every 12 hours. Two blood samples were taken in each administration period, one immediately before the administration (trough) and one 3 hours after administration (peak). On the seventh day following the morning dose three samples were collected.

1.4 Study 4 (Validation Study)

The purpose of this study was to investigate the PK and PK/PD relationship at equilibrium. If an intravenous infusion is continued indefinitely, a constant concentration/time profile results as the amount of drug entering the body is equal to the amount being eliminated from the body. In practice, equilibrium is approximately reached after some finite time that depends on the elimination rate of the drug and the infusion rate. Eight young male volunteers received a constant infusion over 72 hours, four at a rate of 0.2 mg/kg per hour and four at 0.3 mg/kg per hour. Blood samples were taken during the infusion period and for 12 hours after the infusion was completed. These four studies are summarized in Table 1.

We now define a general Bayesian hierarchical model for PK/PD data.

2 POPULATION PHARMACOKINETIC/PHARMACODYNAMIC MODELS

Many examples of population PK analyses can be found in the literature. For a review see Steimer *et al.* (1994). For a discussion of methodology

refer to Racine-Poon and Smith (1990). Population PK/PD analyses are
now becoming more common; for example, see Aarons *et al.* (1991), Holford and Peace (1992a,b), and the reviews of Girard *et al.* (1990) and
Steimer *et al.* (1993).

We first define notation for the situation in which we simply have
data for a single individual before making the extension to a population.
Let $Y(t)$ and $Z(t)$ denote random variables that describe concentration and
response at time t, respectively, and θ and β denote kinetic and dynamic
parameters. For the studies considered in this chapter $Y(t)$ represents the
concentration of rec-hirudin at time t given a particular dose history and
$Z(t)$ represents the APTT at time t. In what follows we suppress the dependence on t. We first discuss the form of the likelihood function for θ, β.
We assume that the response is related to the parameters of the PK model
in some way and so is dependent on θ as well as β. Specifically, we shall
assume that the plasma concentration at time t is modeled by $\eta(\theta, t)$ and
the response is modeled by $\delta(\beta, \theta, t)$. Typically η will take the form of a
sum of exponentials (three exponentials in our example) that describes
the *disposition* (that is, the distribution throughout the body and the elimination) convolved with another function that depends on the route of
administration. The response $\delta(\beta, \theta, t)$ can take one of a variety of forms
depending on the context; here we use a sigmoid Emax model, which we
describe below.

2.1   Kinetic Models for Rec-Hirudin

Intravenous Model

Based on information from kinetic specialists, the kinetics profile after a
single intravenous dose of size D can be described by a sum of three
exponential terms,

$$\eta(\theta, t) = \text{observed concentration}$$
$$= D \times [A_1 \exp(-\alpha_1 t) + A_2 \exp(-\alpha_2 t) + A_3 \exp(-\alpha_3 t)]$$

To have identifiability, the constraint $\alpha_1 > \alpha_2 > \alpha_3 > 0$ is applied. To
ensure that positive concentrations result, each of A_1, A_2, A_3 is assumed
to be positive. Here the concentration in an individual is assumed to be
proportional to the dose in the individual with the constant of proportionality being independent of dose; that is, the parameters of the model for a
particular individual are independent of dose. This assumption was validated in previous studies in which different doses were administered to
the same volunteers in a randomized order.

Subcutaneous Model

A first-order absorption model was assumed: the concentration after a single subcutaneous dose is given by

$\eta(\theta, t)$ = observed concentration

$$= F_a D\, k_a \left[A_1 \frac{\exp(-\alpha_1 t)}{(k_a - \alpha_1)} + A_2 \frac{\exp(-\alpha_2 t)}{(k_a - \alpha_2)} \right.$$

$$\left. + A_3 \frac{\exp(-\alpha_3 t)}{(k_a - \alpha_3)} - B \exp(-k_a t) \right]$$

where $0 < F_a < 1$ is the fraction of the dose that is absorbed, k_a is the positive absorption rate constant and

$$B = \frac{A_1}{(k_a - \alpha_1)} + \frac{A_2}{(k_a - \alpha_2)} + \frac{A_3}{(k_a - \alpha_3)}$$

Note that F_a can be estimated only when both intravenous and bolus doses are given to the same individual. We further assume that the kinetics are linear, which means that the profile from repeated doses can be obtained from the sum of the single-dose profiles using the superposition principle (see Gibaldi and Perrier, 1982). This assumption was later validated using study 3.

We parameterize in terms of

$$\theta = (\text{logit } F_a \log k_a, \log A_1, \log A_2, \log A_3, \log(\alpha_1 - \alpha_2),$$

$$\log(\alpha_2 - \alpha_3), \log \alpha_3)$$

2.2 Kinetic/Dynamic Model for Rec-Hirudin

The observed APTT measurements were plotted against the concentration for each volunteer and each application formulation. Figure 2 shows such a plot for individual 1 of study 1. The consecutive observations were joined to identify whether hysterious loops were evident (Holford and Sheiner, 1981, 1982) and to identify the possible functional form between the observed APTT and observed concentration.

If a hysterious loop was evident, the assumption of an instantaneous relationship between concentration and dynamic effect would have been unreasonable. In this case one possibility would be the incorporation of an effect compartment (Segre, 1968; Sheiner et al., 1979; Holford and Sheiner, 1981, 1982). No such loop was detected in our data; we therefore assumed that the concentration and effect relationship was instantaneous. In addition, the concentration-effect relationship was assumed to be iden-

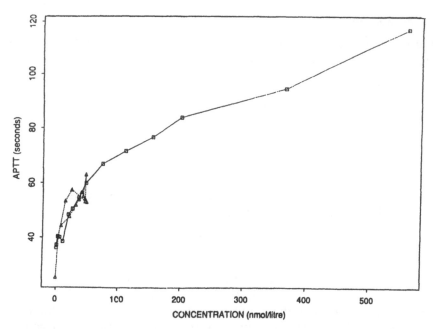

Figure 2 Observed concentration versus observed APTT for individual 1 of study 1. Squares represent the intravenous data, triangles the subcutaneous data. The solid line joins up the intravenous data, the dotted line the subcutaneous data.

tical for both routes of administration. This is evident from the observed data in study 1. As already stated, the expected effect of rec-hirudin is inhibition of human thrombin, which is measured by the reciprocal of the APTT. The relationship between the actual inhibition effect $\delta(\beta, \theta, t)$ and the actual concentration $\eta(\theta, t)$ is given by the inhibition sigmoid Emax model (Holford and Sheiner, 1981, 1982):

$$\delta(\beta, \theta, t) = \text{APTT}_0^{-1} \left[1 - \frac{\eta(\theta, t)^\omega}{(\eta(\theta, t)^\omega + \text{IC}_{50}^\omega)} \right]$$

where $\text{APTT}_0 > 0$ is the baseline APTT (that is, the APTT when no drug is present) and $\text{IC}_{50} > 0$ is a parameter that represents the concentration which would be required to produce 50% inhibition. Let z denote the observed inhibition. Maximum-likelihood estimates of Hill's coefficient ω were obtained for the 16 individuals of study 1 (with additive independent errors) and were found to be close to 0.5, and so ω was fixed at 0.5. The adequacy of this choice of Hill's coefficient in general will be validated in studies 3 and 4.

We parameterize the PK/PD parameters as $\beta = (\log \text{APTT}_0, \log \text{IC}_{50})$. Ignoring for the moment the dependence on variance terms, the likelihood for a particular individual is given by

$$l(\theta, \beta) = f(\mathbf{y}, \mathbf{z} \mid \theta, \beta) = f(\mathbf{y} \mid \theta, \mathbf{z}, \beta)f(\mathbf{z} \mid \theta, \beta)$$
$$= f(\mathbf{y} \mid \theta)f(\mathbf{z} \mid \theta, \beta)$$

since $\mathbf{y}$ and $\mathbf{z}$ and $\mathbf{y}$ and β are independent conditional on θ. Here $f(\cdot \mid \cdot)$ is a generic notation for a density function.

2.3 Modeling the Individual Measurement Errors

The Concentration Measurement Error

The measurement error for the concentration is chosen to be additive normal on a log scale. The precisions of the measurement errors of the intravenous and subcutaneous formulations are allowed to differ. Based on the information available in our analytic laboratory for method validation of the analytic procedure, the measurement error appeared to be proportional to the actual concentration; the lognormal error assumption is therefore reasonable in this regard. However, the so-called measurement error in the individual concentration is a combination of the error in the analytic determination of the concentration and the lack of fit of the model. The lack of fit of the intravenous model is expected to be smaller than that of the subcutaneous model due to the additional approximation of first-order absorption kinetics. We therefore have

$$\log y_{ijk} = \log \eta(\theta, t_{ijk}) + \epsilon_{ijk}$$

where i and j index individuals and sampling times within individuals, respectively, and ϵ_{ijk} is $N(0, \tau_k^{-1})$, $k = 1, 2$, with $k = 1$ representing the intravenous experiment and $k = 2$ representing the subcutaneous experiment.

The Response Measurement Error

The residual diagnostics from the maximum-likelihood fits to estimate ω for study 1 appeared consistent with a lognormal error structure, because the residual error appeared to be proportional to that of the magnitude of the reciprocal APTT:

$$\log z_{ijk} = \log \delta(\beta, \theta, t_{ijk}) + \epsilon_{ijk}$$

where i, j, and k index individuals, sampling times within individuals, and experiments, respectively, and ϵ_{ijk} $(k = 3)$ is $N(0, \tau_3^{-1})$.

For the moment, for notational simplicity, we consider a single study. Suppose that we have I individuals for whom PK/PD data have been collected. Denote these data by $\mathbf{y}_i = (y_{i1}, \ldots, y_{in_i})$ and $\mathbf{z}_i = (z_{i1}, \ldots, z_{im_i})$, for $i = 1, \ldots, I$. This notation makes explicit the fact that there may be missing PK or PD data at some time points. In fact, the PK and PD data can be collected at different time points. Let t_{ij} denote the times at which data were collected; that is, for each $y_{ij}, j = 1, \ldots, n_i$ and/or $z_{ij}, j = 1, \ldots, m_i$ there is an associated time point t_{ij}. Finally, let θ_i and β_i denote the PK and PD parameters for individual i. The population approach assumes that, possibly after conditioning on covariates, that is, for different patient subgroups, the PK and PD parameters are drawn from some distribution. Here we shall, for simplicity, assume no covariate relationships. Furthermore, we shall assume that the PK and PD parameter vectors are drawn from independent distributions, that is

$$\theta_i \sim p_\theta(. \,|\, \phi_\theta) \quad \text{and} \quad \beta_i \sim p_\beta(. \,|\, \phi_\beta)$$

where ϕ_θ, ϕ_β represent population PK and PD parameters, respectively. In particular $p_\theta(. \,|\, \phi_\theta)$ and $p_\beta(. \,|\, \phi_\beta)$ may represent multivariate normal or Student t-distributions, in which case $\phi_\theta = (\mu_\theta, \Sigma_\theta)$ and $\phi_\beta = (\mu_\beta, \Sigma_\beta)$ where μ represents a population mean parameter and Σ a population scale matrix.

2.4 Modeling the Between-Individual Variation for Rec-Hirudin

To be able to accommodate possible outlying individuals in the PK parameter, a Student t-distribution with 4 degrees of freedom was chosen to describe the variation between individuals (Wakefield et al., 1994), that is,

$$\theta_i \sim t_4(\phi_\theta)$$

where the ϕ_θ contain the location vector and the dispersion matrix of the seven- or eight-dimensional t density, depending on whether the experiment is intravenous or subcutaneous. There is some arbitrariness in this choice of degrees of freedom, and other low values could equally have been chosen. A bivariate normal distribution was chosen to model the PD parameters:

$$\beta_i \sim N(\phi_\phi)$$

where the ϕ_β contain the mean vector and the variance/covariance matrix of the bivariate normal density.

2.5 Covariate Model for Rec-Hirudin

One of the goals of study 2 was to investigate the influence of age on the PK and PK/PD profiles. Study 1 and study 2 were evaluated together using the foregoing between-individuals variation model. Based on study 2 alone, the parameters F_a, A_1, A_2, and A_3 are identifiable only up to a positive constant. It would therefore not be possible to estimate the data of study 2 alone without an assumption about the size of F_a (say). Also, the study 2 data were insufficient to identify all three phases of distribution and elimination together. The posterior median for each individual's PK parameters $(k_a, A_1, A_2, A_3, \alpha_1, \alpha_2, \alpha_3)$ and PD parameters (AP-TT_0, IC_{50}) were plotted with the age group identified. The medians were obtained from an analysis utilizing Markov chain Monte Carlo methodology, which we describe in Section 3. Figures 3 and 4 give illustrations. We plot on a logarithmic scale because the logarithms of the parameters were assumed to arise from normal/Student's t-distributions.

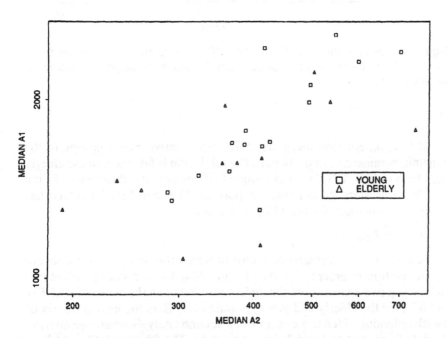

Figure 3   Posterior medians of parameter A_1 versus posterior medians of parameter A_2 for the young (squares) and the elderly (triangles) volunteers of studies 1 and 2.

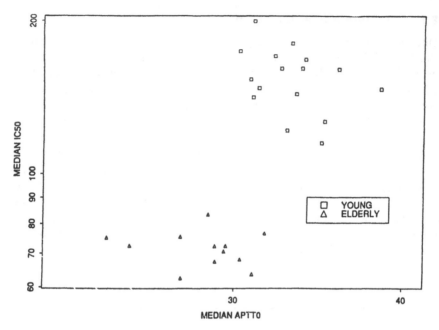

Figure 4 Posterior medians of parameter IC_{50} versus posterior medians of parameter $APTT_0$ for the young (squares) and the elderly (triangles) volunteers of studies 1 and 2.

It is apparent that there are differences between age groups in the dynamic parameters. In particular, the inhibition is far lower in the elderly and the baseline APTT is also reduced. However, the differences in the kinetic parameters do not seem important. We therefore introduce age group as a covariate for the PD parameters

$$\beta_i \sim N(X_i\mu_\beta, \Sigma_\beta)$$

where μ_β is a four-dimensional vector in which the first two elements are the population intercepts for $APTT_0$ and IC_{50} for the young volunteers and the last two elements are the differences between the means for $APTT_0$ and IC_{50} for the elderly and young populations. X_i is the design matrix of the ith individual. Hence we carry out a second analysis where we analyse the data from studies 1 and 2 simultaneously. The 5th percentile, median, and 95th percentile of the population PK and PK/PD parameters are listed in Table 2. These values were also obtained from an analysis utilizing Markov chain Monte Carlo methodology. Note the differences in the PD

Table 2 Posterior Summaries for Population PK and PK/PD Location
Parameters and Error Precisions for Studies 1 and 2[a]

Parameter	5th percentile	Median	95th percentile
k_a (h^{-1})	0.276	0.301	0.327
F_a	0.991	0.992	0.993
A_1 (nmol/L mg^{-1})	1278	1500	1795
A_2 (nmol/L mg^{-1})	324	364	415
A_3 (nmol/L mg^{-1})	13.2	17.1	21.7
$\alpha_1 - \alpha_2$ (h^{-1})	2.82	3.32	3.97
$\alpha_2 - \alpha_3$ (h^{-1})	0.337	0.383	0.428
α_3 (h^{-1})	0.116	0.136	0.157
IC_{50} (nmol/L) (young)	127	144	163
IC_{50} (nmol/L) (elderly)	60.1	72.3	87.2
$APTT_0$ (sec) (young)	31.1	32.8	34.4
$APTT_0$ (sec) (elderly)	26.8	28.5	30.3
τ_1 (intravenous)	21.3	25.9	31.2
τ_2 (subcutaneous)	14.7	17.0	19.7
τ_3 (APTT)	170	191	216

[a] The location parameters are the medians of the appropriate population distribution. The values tabulated are the medians of the posterior distributions of these parameters along with the 5 and 95% points of this posterior.

parameters for the two populations and also the decrease in concentration precision in the subcutaneous study (τ_2) compared with concentration precision in the intravenous study (τ_1).

Graphical examination of the fits reveals that the PK model appears to be adequate. The median of the posterior distribution of the PK/PD parameters for each individual was then used to obtain fitted curves. These fitted curves for subject 1 of study 1 are illustrated graphically in Figures 5 and 6. Examination of the fitted curves shows that we can identify the three phases of distribution and elimination for the intravenous data, but this is not so obvious for the fitted curve for the subcutaneous data.

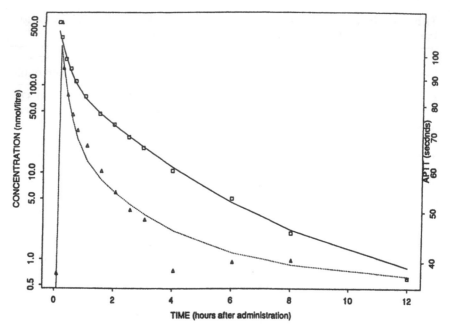

Figure 5   Fitted versus predicted for intravenous data of individual 1 of study 1. Squares represent observed concentrations, triangles observed responses. We plot the logarithms of the concentrations and responses because we have assumed additive normal errors on this scale.

The discussion above describes a two-stage hierarchical model; the Bayesian approach that we use completes this hierarchy with a third stage comprising priors for ϕ_θ and ϕ_β. Such a hierarchical model seems appropriate for data of these type, since it explicitly quantifies the variability that is observed both *within* a particular individual and *between* different individuals. The first stage of the model addresses the former source of variability and the second stage addresses the latter.

We advocate a Bayesian approach for the following reasons.

1.   The PK models used are invariably nonlinear in the parameters, and the PD models are sometimes nonlinear also. Consequently, classical inference depends on asymptotic arguments for interval estimates and hypothesis testing. For example, likelihood ratio tests are used to test whether covariates are significant. It is difficult to check whether such arguments are valid for a given data

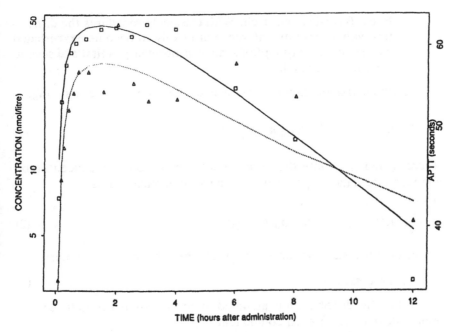

Figure 6 Fitted versus predicted for subcutaneous data of individual 1 of study 1. Squares represent observed concentrations, triangles observed responses. We plot the logarithms of the concentrations and responses because we have assumed additive normal errors on this scale.

set. Furthermore, a second level of approximation is often incorporated when two-stage models are analyzed via first-order or linearization methods (Beal and Sheiner, 1982; Lindstrom and Bates, 1990), since only an approximate likelihood is obtained. This is the approach adopted in the commonly used package NONMEM (Beal and Sheiner, 1982).

2. One of the principal aims of a population PK/PD analysis is prediction, and as we shall illustrate shortly, the Bayesian approach is well suited to this requirement.

3. The third-stage prior may be used to include information from different or previous studies. Later we illustrate how an informative third-stage prior based on a previous study may also be used to assess sensitivity.

4. It is of great importance to test the robustness of inferences to modeling assumptions. This is straightforward with a sampling-

based Bayesian approach. With classical procedures the difficulties with asymptotic inference are likely to become exacerbated when common assumptions, such as normality at first and second stages, are relaxed.

To summarize the hierarchical structure we have at the first stage

$$\prod_{i=1}^{I} f(\mathbf{y}_i \mid \theta_i, \tau_\theta) \times f(\mathbf{z}_i \mid \theta_i, \beta_i, \tau_\beta) \tag{1}$$

where τ_θ and τ_β denote the intraindividual precisions for concentration and response data, respectively. At the second stage we have

$$\prod_{i=1}^{I} p_\theta(\theta_i \mid \phi_\theta), \qquad \prod_{i=1}^{I} p_\beta(\beta_i \mid \phi_\beta) \tag{2}$$

and at the third stage we have a prior for the remaining parameters:

$$p(\phi_\theta, \phi_\beta, \tau_\theta, \tau_\beta) \tag{3}$$

Generally, this distribution is assumed to be the product of independent prior distributions for each component.

3 IMPLEMENTATION

In this section we briefly describe how to perform the calculations required to carry out a Bayesian analysis. Inference with a Bayesian approach centers on the posterior distribution of the unknown parameters of the model. We ignore for the moment the assumption of the second-stage t distribution. In our case, writing $\theta = (\theta_1, \ldots, \theta_I)$ and $\beta = (\beta_1, \ldots, \beta_I)$ we have, taking the product of Eqs. (1), (2), and (3),

$$p(\theta, \tau_\theta, \beta, \tau_\beta, \phi_\theta, \phi_\beta \mid y, z) \propto \prod_{i=1}^{I} f(\mathbf{y}_i \mid \theta_i, \tau_\theta) \, f(\mathbf{z}_i \mid \theta_i, \beta_i, \tau_\beta)$$

$$\times \prod_{i=1}^{I} p_\theta(\theta_i \mid \phi_\theta) p_\beta(\beta_i \mid \phi_\beta) \tag{4}$$

$$\times p(\phi_\theta, \phi_\beta, \tau_\theta, \tau_\beta)$$

If we are interested in marginal distributions, these are obtained by integrating out all other parameters. Moment summaries are found via integration over marginal distributions. Unfortunately, for population PK/PD models the nonlinearity at the first stage and the unknown variances do not allow these integrations to be carried out analytically. Numerical integration or Laplacian methods are also out of the question because of the large dimensionality of the parameter space.

A great deal of attention has been paid to Markov chain Monte Carlo methods in Bayesian inference (Geman and Geman, 1984; Gelfand *et al.*, 1990; Tierney, 1994; Smith and Roberts, 1993; Gilks *et al.*, 1993). The idea behind such methods is to construct a Markov chain whose stationary distribution is the required posterior distribution, in our case Eq. (4). Let ψ denote all of the unknowns of the model. Starting from an initial point $\psi^{(0)}$ we simulate from the Markov chain and so obtain a sequence $\psi^{(t)}$, $t = 1, 2, \ldots$. Under mild regularity conditions, as $t \rightarrow \infty$ the random variable $\psi^{(t)}$ tends in distribution to a realization from the posterior distribution of ψ.

In our context a Gibbs sampling approach would require samples from the following conditional distributions:

$$p(\theta_i \mid \theta_j, j \neq i, \tau_\theta, \beta, \tau_\beta, \phi_\theta, \phi_\beta, \mathbf{y}, \mathbf{z})$$

$$p(\beta_i \mid \beta_j, j \neq i, \tau_\theta, \theta, \tau_\beta, \phi_\theta, \phi_\beta, \mathbf{y}, \mathbf{z})$$

$$p(\tau_\theta \mid \theta, \beta, \tau_\beta, \phi_\theta, \phi_\beta, \mathbf{y}, \mathbf{z})$$

$$p(\tau_\beta \mid \theta, \tau_\theta, \beta, \phi_\theta, \phi_\beta, \mathbf{y}, \mathbf{z})$$

$$p(\phi_\theta \mid \theta, \tau_\theta, \beta, \tau_\beta, \phi_\beta, \mathbf{y}, \mathbf{z})$$

$$p(\phi_\beta \mid \theta, \tau_\theta, \beta, \tau_\beta, \phi_\theta, \mathbf{y}, \mathbf{z})$$

For the model that we have described each of these distributions is straightforward to sample (Wakefield *et al.*, 1994), apart from the first two when nonlinear models are used for the PK and PD relationships. For such distributions it is possible to use a Metropolis-Hastings algorithm (see Bennett *et al.*, 1995). For the analyses in this chapter we used a Metropolis-Hastings algorithm centered at the current point and with variance/covariance matrix based on the information at the approximate posterior mode. The extension to t-distributions at the second stage is straightforward; for details see Wakefield *et al.* (1994).

It is easy to make inferences using the realized points from the Markov chain. If we wish to estimate moments, for example, we simply work out the ergodic average of the appropriate quantity.

4 VALIDATION

In order to validate the appropriateness of the PK/PD model and the distributional assumptions chosen for studies 1 and 2 we did the following. We

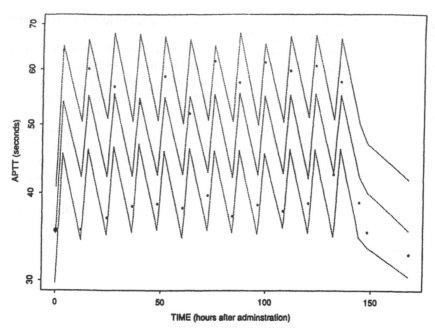

Figure 7 Median prediction along with 5% and 95% predictive intervals for individual 1 of study 3. The dots represent the observed data. The prediction is based on the baseline APTT measurement only and population information. The baseline APTT point is enclosed in a square.

first obtained the posterior distribution of the population PK/PD parameters using studies 1 and 2. We then took one, two, or three observed APTT values from each of the individuals of studies 3 and 4 and then attempted to predict each of these individual's remaining responses using these data and the aforementioned population information.

4.1 Implementation of the Validation

Let z_p be the available APTT values of a future patient to be used for prediction for his or her complete APTT profile. Let (θ_p, β_p) be the parameters of the future individual. The posterior distribution for this individual's parameters is given by

$$p(\theta_p, \beta_p \mid z_p, y, z) \propto p(z_p \mid \theta_p, \beta_p) \times p(\theta_p, \beta_p \mid y, z)$$

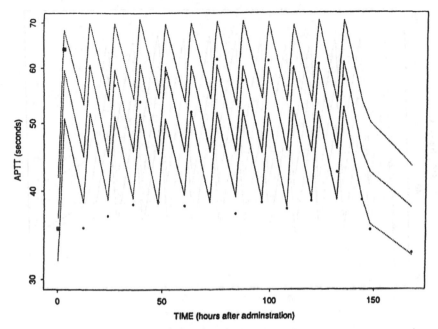

Figure 8 Median prediction along with 5% and 95% predictive intervals for individual 1 of study 3. The dots represent the observed data. The prediction is based on baseline and 3-hour APTT measurements and population information. These points are enclosed in squares.

where y and z denote the *learning data*, that is, the data available from studies 1 and 2. The second term on the right-hand side is the predictive distribution for a new individual's parameters and can be viewed as the prior distribution for the future individual. To obtain samples from the density $p(\theta_p, \beta_p \mid z_p, y, z)$, 10000 predictive samples were generated by retaining samples of the population parameters from studies 1 and 2 and then generating θ_p, β_p samples from the appropriate normal/Student's t-distributions. That is, we sample from the elderly population distribution for study 3 and the young population distribution for study 4. For each of these samples the likelihood of the observed data was evaluated. These values were then normalized and 1000 samples were resampled using these importance weights (Rubin, 1988; Smith and Gelfand, 1992). Hence we obtain an approximate sample from the posterior of the individual. The

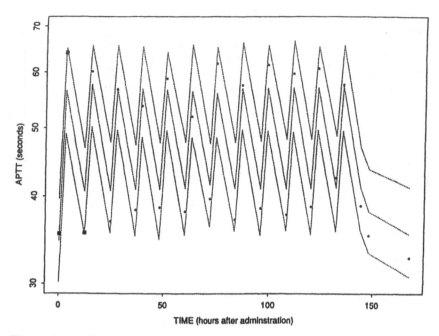

Figure 9 Median prediction along with 5% and 95% predictive intervals for individual 1 of study 3. The dots represent the observed data. The prediction is based on baseline and 3- and 12-hour APTT measurements and population information. These points are enclosed in squares.

predictive APTT profile of this individual can then be calculated. These profiles were obtained by evaluating $\delta(\beta, \theta, t)$ at each time point and then ordering these samples. In fact, a separate set of 1000 samples was obtained for each time point. Hence we are evaluating the *expected* APTT at each of the time points. The measurement error could be incorporated but we choose to plot on a logarithmic scale, which means that the additional measurement error would merely widen the intervals symmetrically.

For each of the subjects in study 3, three sets of predictions were produced. The first of these predictions was based on the baseline APTT measurement of the subject alone. The second set of predictions was based on the baseline and 3-hour APTT measurements, and the third set was based on baseline and 3- and 12-hour APTT measurements. For each of the subjects the median prediction was produced, along with the 5% and

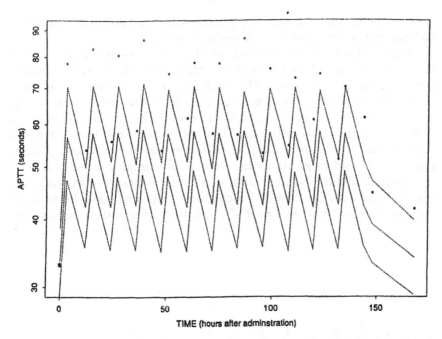

Figure 10 Median prediction along with 5% and 95% predictive intervals for individual 5 of study 3. The dots represent the observed data. The prediction is based on the baseline APTT measurement only and population information. The baseline APTT point is enclosed in a square.

95% prediction envelopes. We discuss two particular subjects, numbers 1 and 5. Figures 7–9 show the predictions for subject 1, who received doses of 0.3 mg/kg. The predictions of Figure 8 appear worse than those of Figure 7 even though an additional data point is available. Careful inspection of Figure 7 indicates that this subject's 3- and 12-hour measurements are at the extremes of the predictive intervals. This indicates that this subject has a high rate of absorption (resulting in a high 3-hour measurement) and also a fast elimination rate (resulting in a low 12-hour measurement). When the 3-hour measurement is taken into account, an attempt is made to fit the high 3-hour measurement and this results in the predictive intervals being shifted upward. Consequently all of the trough measurements are now missed. Using the 3- and 12-hour measurements along with the baseline, we obtain a considerably improved set of predic-

tions. Note that the variability in the predictive distributions across administration periods reflects the resampling variability.

Figures 10–12 show a second set of predictions for subject 5, who
received doses of 0.5 mg/kg. Here the predictions based on the baseline
measurement only are poor. The peak and trough values are all much
higher than predicted. However, including the 3-hour measurement (Figure 11) greatly improves the predictions. The addition of the 12-hour point
(Figure 12) makes little difference. The predictions here do not take into
account the natural day-to-day variation within an individual patient.
Hence one would expect to underestimate the variation in the profiles.

For each of the subjects in study 4, a set of predictions was produced,
based on the individual's APTT measured at baseline and with an additional measurement taken 1 hour after the beginning of the infusion. The

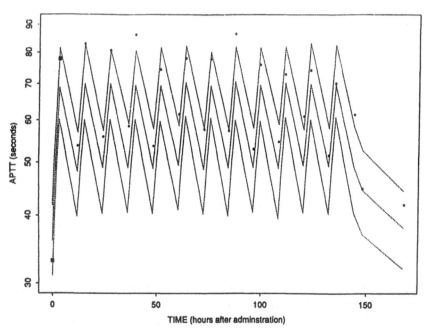

Figure 11   Median prediction along with 5% and 95% predictive intervals for
individual 1 of study 3. The dots represent the observed data. The prediction is
based on baseline and 3-hour APTT measurements and population information.
These points are enclosed in squares.

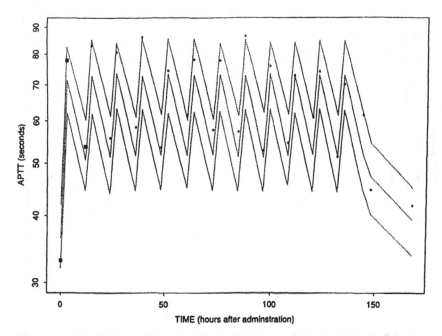

Figure 12 Median prediction along with 5% and 95% predictive intervals for individual 1 of study 3. The dots represent the observed data. The prediction is based on baseline and 3- and 12-hour APTT measurements and population information. These points are enclosed in squares.

median prediction of individual 3 along with the 5% and the 95% prediction envelope is shown in Figure 13. The prediction captures the observed APTT reasonably well. We therefore conclude that the model chosen for the PK/PD modeling is reasonable.

We chose to use the sampling importance resampling technique for the prediction. An alternative method is to rerun the Gibbs sampler adding the observations of the new individual. However, the resampling method is much more efficient. Another possibility is to use a rejection algorithm to obtain points that are from the exact posterior distribution of the individual (as described in Wakefield, 1994), but the high dimensionality of the parameter space would result in extremely low acceptance probabilities and hence low efficiency.

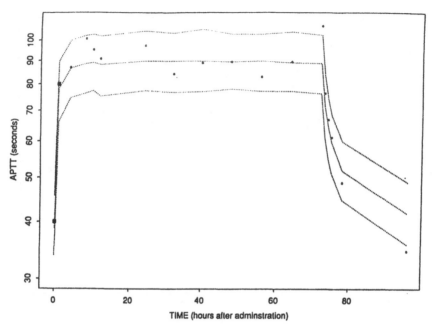

Figure 13 Median prediction along with 5% and 95% predictive intervals for individual 3 of study 4. The dots represent the observed data. The prediction is based on baseline and 1-hour APTT measurements and population information. These points are enclosed in squares.

5 DOSE RECOMMENDATION

One of the primary aims of these studies was to see whether it was possible to identify three different doses of twice daily subcutaneous applications for a future phase II study of elderly patients. The goal of a phase II dose-finding study is to illustrate a dose-response relationship using a chosen efficacy outcome measure within a well-tolerated dose range. A good choice of doses should therefore be able to illustrate the differences in response between a number of the chosen doses while also showing a benefit over placebo or positive control (that is, a competitor treatment). Based on safety considerations, the dose range is restricted to be below 20 mg twice daily. Hence, assuming that the steady-state APTT is a reasonable surrogate measure of both the efficacy of the drug and the drug's safety in the patient population, we would like to assess the possibility of using 10, 15, and 20 mg twice daily in a subcutaneous application. The aim is to maintain the ratio of average steady-state to baseline APTT within 1.5–2.0. To investigate this, 1000 future individuals' PK/PD parameters were generated from the predictive distribution of the elderly patients. The predictive steady-state plasma profiles corresponding to elderly individuals with body weights 50, 70, and 90 kg after administration of the three doses were then calculated. The corresponding average APTT to baseline ratios were then calculated. Figure 14 shows box plots of the average steady-state to baseline ratios at the three weights and under the three doses. It appears that these doses would be able to illustrate the dose-response relationship without compromising the safety requirements. To maintain the average steady-state APTT to baseline ratio in the range 1.5–2, 15 and 20 mg are reasonable for the elderly population of weights 50 and 70 kg.

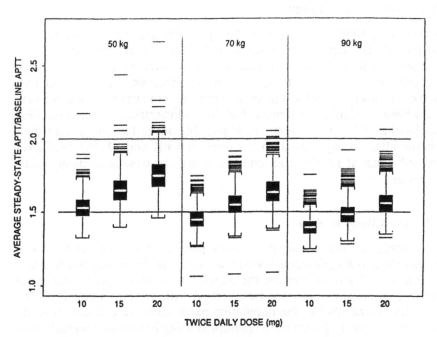

Figure 14 Box plots of predictive ratios of steady-state to baseline APTT under different doses for elderly individuals of weight 50, 70, and 90 kg.

6 DISCUSSION

6.1 Convergence

To assess convergence of the Gibbs sampler, two series with very different starting values were run. The parameters that were used for monitoring the convergence were the intraindividual PK and PK/PD parameters τ_1, τ_2, and τ_3, the population mean IC_{50} parameter for the elderly population, and the average expected steady-state concentration λ mg/kg dose of the twice-daily subcutaneous study. The three precision parameters are global indicators of how well all of the parameters fit the data. The IC_{50} and λ parameters were chosen as they are the two parameters that are directly related to the dose required to produce the required inhibition effect. The series with bad starting values took about 30,000 iterations to mix with the series with good starting values. This is not an uncommonly large number for a model of this type. The nonlinearity of the model leads to a large increase in convergence time over the linear hierarchical model.

6.2 Predictive Probability

We have demonstrated that predictive probabilities can be used not only to predict the required dose for an individual patient or patient group but also for model validation and diagnostic purposes. If the predicted concentration ranges fail to agree with the observed data, the model assumptions may be incorrect or some important covariate relationship may have been overlooked. For individual dose adjustments all that is required is a baseline APTT measurement and possibly one or two additional values shortly after dosing. Based on the predicted PK and PK/PD profile for the individual at steady state, one can adjust the dose accordingly when necessary.

6.3 Sensitivity to Prior

To investigate prior sensitivity, a prior partially based on the results of Wakefield and Racine-Poon (1995) was used for the population location parameters. The prior used for the population location parameters of the original analysis was independent normal with variances equal to 100. Samples of size 5000 for the population parameters were drawn from the original analysis and 1000 were resampled using importance weights proportional to the ratio of the new to the old prior (Smith and Gelfand, 1992). Summaries of the population parameters and the precisions based on both priors are shown in Tables 3 and 4.

Table 3 Sensitivity of Posterior Summaries to Two Priors for Population PK and PK/PD Location Parameters[a]

Parameter	Prior 5th, 50th, and 95th percentiles	Diffuse 5th, 50th, and 95th percentiles	Informative 5th, 50th, and 95th percentiles
$\log k_a$	(−1.18, −0.79, −0.39)	(−1.29, −1.20, −1.13)	(−1.20, −1.12, −1.05)
$\text{logit } F_a$	(2.59, 3.74, 4.89)	(4.8, 4.87, 4.97)	(4.67, 4.70, 4.74)
$\log A_1$	(6.89, 7.30, 7.71)	(7.16, 7.32, 7.50)	(7.19, 7.32, 7.46)
$\log A_2$	(5.49, 5.90, 6.31)	(5.76, 5.86, 5.94)	(5.88, 5.97, 6.01)
$\log A_3$	(2.39, 2.80, 3.21)	(2.54, 2.79, 3.10)	(2.85, 3.01, 3.11)
$\log(\alpha_1 - \alpha_2)$	(1.19, 1.60, 2.01)	(1.01, 1.17, 1.35)	(1.17, 1.26, 1.38)
$\log(\alpha_2 - \alpha_3)$	(−0.91, 0.0, 0.91)	(−1.11, −0.987, −0.878)	(−0.954, −0.886, −0.784)
$\log \alpha_3$	(−2.11, −1.70, −1.29)	(−2.18, −2.01, −1.85)	(−2.10, −1.99, −1.80)
$\log \text{APTT}_0$	(2.68, 3.50, 4.32)	(3.44, 3.49, 3.54)	(3.44, 3.49, 3.53)
$\log \text{IC}_{50}$	(1.36, 3.01, 4.65)	(4.84, 4.96, 5.09)	(4.86, 4.98, 5.06)
$\Delta\log \text{APTT}_0$	(−1.64, 0.0, 1.64)	(−0.219, −0.139, −0.058)	(−0.220, −0.139, −0.052)
$\Delta\log \text{IC}_{50}$	(−1.64, 0.0, 1.64)	(−0.931, −0.705, −0.482)	(−0.822, −0.688, −0.409)

[a] The first column gives the percentiles of the informative prior based partly on the analysis of Wakefield and Racine-Poon (1995) on an independent study. The second and third columns give the posterior percentiles using a diffuse prior and the informative prior, respectively.

Table 4 Sensitivity of Posterior Summaries for Population PK and PK/PD Scale Parameters and Precisions[a]

Parameter	Diffuse 5th, 50th, and 95th percentiles	Informative 5th, 50th, and 95th percentiles
τ_1	(19.7, 25.4, 31.0)	(21.0, 25.6, 29.7)
τ_2	(14.7, 17.1, 19.9)	(15.6, 17.4, 19.4)
τ_3	(167.6, 190.2, 214.9)	(165.3, 184.5, 213.8)
var(log k_a)	(0.020, 0.037, 0.077)	(0.020, 0.034, 0.062)
var(logit F_a)	(0.004, 0.016, 0.049)	(0.008, 0.014, 0.032)
var(log A_1)	(0.062, 0.125, 0.264)	(0.069, 0.139, 0.228)
var(log A_2)	(0.015, 0.034, 0.101)	(0.019, 0.034, 0.073)
var(log A_3)	(0.117, 0.263, 1.03)	(0.132, 0.260, 0.400)
var(log($\alpha_1 - \alpha_2$))	(0.071, 0.138, 0.297)	(0.072, 0.140, 0.268)
var(log($\alpha_2 - \alpha_3$))	(0.020, 0.046, 0.107)	(0.032, 0.049, 0.088)
var(log α_3)	(0.059, 0.119, 0.339)	(0.063, 0.115, 0.165)
var(log $APTT_0$)	(0.009, 0.014, 0.023)	(0.008, 0.014, 0.020)
var(log IC_{50})	(0.026, 0.053, 0.109)	(0.030, 0.050, 0.098)

[a] The second and third columns give the posterior percentiles using a diffuse prior and the informative prior based on Wakefield and Racine-Poon (1995), respectively.

The medians and 5th and 95th percentiles of the population locations for all PK and PK/PD parameters are given in Table 3 along with the 5th and 95th percentiles of the informative prior. Table 4 gives the medians and 5th and 95th percentiles of the population scale parameters along with the intraindividual precision parameters. Comparison of the values for the two priors shows them to be quite similar.

6.4 Importance Sampling

The importance sampling technique for prediction and prior sensitivity has the advantage of being efficient. However, if the individual is an outlier (relative to the individuals on whom the prediction was based) or the new prior is quite different from the original prior, then the approximation to the density of interest will be poor. It is therefore of great importance not to rely simply on the summary statistics based on the importance sampling for the conclusion; the importance weights also have to be inspected. If these weights concentrate on few values, the approximation to the predictive or posterior density is likely to be poor. In these cases, one should rerun the analysis or use other simulation techniques.

6.5 Modeling Kinetic and Dynamic Relationships Simultaneously

Using the Bayesian approach, the PK and PK/PD relationship can be modeled simultaneously; that is, the PK parameter information can be used to update the PD parameter information and vice versa. There is a direct link between the two sets of parameters. The most naive approach would be to ignore the uncertainty in the PK measurements and simply use the observed concentrations. A more refined approach is to estimate the individuals' PK parameters based on the concentration information alone and then on the basis of these PK parameters obtain "fitted" concentrations. The population PD parameter estimation will then be based on the observed responses and on these fitted concentrations. This approach has the disadvantage that the variation in the PD parameters will be underestimated, because the variation in the fitted concentrations is not acknowledged. Also, the PD information is not used to update the PK information. Finally, since the model fitting is divided into two parts, prediction based on PD values alone is possible only via many approximations. The model that we have described allows the effect of not only dose size but also route of administration to be investigated.

The analysis was carried out on an IBM RX6000 workstation, and the computing time was about 24 hours.

ACKNOWLEDGMENTS

The second author carried out this work while visiting the Biometrics Group of Ciba-Geigy AG, Basel. Both authors have had helpful discussions with Drs P. Lloyd, P. Graf, and D. Gygax of Bioanalytics and Pharmacokinetics in Ciba-Geigy. The analytical work and the pharmacokinetics evaluation were carried out by Dr. G. Lefevre in the bioanalytical unit in Ciba-Geigy, Paris. The authors have benefited from discussions with Dr. L. Aarons of the Department of Pharmacy, Manchester University.

REFERENCES

Aarons, L., Mandema, J. W., and Danhof, M. (1991). A population analysis of the pharmacokinetics and pharmacodynamics of midazolam in the rat. *Journal of Pharmacokinetics and Biopharmaceutics* 5: 485–496.

Beal, S. L. and Sheiner, L. B. (1982). Estimating population kinetics. *CRC Critical Reviews in Biomedical Engineering* 8: 195–222.

Bennett, J. E., Racine-Poon, A., and Wakefield, J. C. Markov chain Monte Carlo for nonlinear hierarchical models. In *Markov Chain Monte Carlo in Practice*, eds. Gilks, W. R., Richardson, S., and Spiegelhalter, D. J. Chapman and Hall, New York, 1995.

Gelfand, A. E., Hills, S. E., Racine-Poon, A., and Smith, A. F. M. (1990). Illustration of Bayesian inference in normal data models using Gibbs sampling. *Journal of the American Statistical Association* 85: 972–985.

Geman, S. and Geman, D. (1984). Stochastic relaxation, Gibbs distributions and the Bayesian restoration of images. *IEEE Transactions on Pattern Analysis and Machine Intelligence* 6: 721–741.

Gibaldi, M. and Perrier, D. (1982). *Drugs and the Pharmaceutical Sciences*, Vol. 15, *Pharmacokinetics*, 2nd ed. Marcel Dekker, New York.

Gilks, W. R., Clayton, D. G., Spiegelhalter, D. J., Best, N. G., McNeil, A. J., Sharples, L. D., and Kirby, A. J. (1993). Modelling complexity: Applications of Gibbs sampling in Medicine. *Journal of the Royal Statistical Society Ser. B* 55: 39–52.

Girard, P., Nony, P., and Boissel, J. P. (1990). The place of simultaneous pharmacokinetic pharmacodynamic modelling in drug development: Trends and perspectives. *Fundamentals of Clinical Pharmacology* 4(2): 103s–115s.

Holford, N. H. G. and Peace, K. E. (1992a). Methodologic aspects of a population pharmacodynamic model for cognitive effects in Alzheimer patients treated with tacrine. *Proceedings of the National Academy of Sciences USA* 89: 11466–11470.

Holford, N. H. G. and Peace, K. E. (1992b). Results and validation of a population pharmacodynamic model for cognitive effects in Alzheimer patients treated with tacrine. *Proceedings of the National Academy of Sciences USA* 89: 11471–11475.

Holford, N. H. G. and Sheiner, L. B. (1981). Understanding the dose-effect relationship: Clinical application of PK-PD models. *Clinical Pharmacokinetics* 6: 429–453.

Holford, N. H. G. and Sheiner, L. B. (1982). Kinetics of pharmacological response. *Pharmacology and Therapeutics* 16: 143–166.

Lindstrom, M. J. and Bates, D. M. (1990). Nonlinear mixed effects models for repeated measures data. *Biometrics* 46: 673–687.

Racine-Poon, A. and Smith, A. F. M. (1990). Population models. In *Statistical Methodology in the Pharmaceutical Sciences*, ed. Berry, D. Marcel Dekker, New York.

Rubin, D. R. (1988). Using the SIR algorithm to simulate posterior distributions. In *Bayesian Statistics 3*, ed. Bernardo J. E. *et al*), pp. 395–402. Oxford University Press, New York.

Segre, G. (1968). Kinetics of interaction between drugs and biological systems. *Il Farmaco* 23: 907–9

Sheiner, L. B., Stanski, D. R., Vozeh, S., Miller, R. D., and Ham, J. (1979). Simultaneous modeling of pharmacokinetics and pharmacodynamics: Application to d-tubocurarine. *Clinical Pharmacology and Therapeutics* 29: 358–371.

Smith, A. F. M. and Gelfand, A. E. (1992). Bayesian statistics without tears: A sampling resampling perspective. *The American Statistician* 46: 84–88.

Smith, A. F. M. and Roberts, G. O. (1993). Bayesian computation via the Gibbs sampler and related Markov chain Monte Carlo. *Journal of the Royal Statistical Society Ser. B* 55: 3–23.

Steimer, J.-L., Ebelin, M.-E., and Van Bree, J. (1993). Pharmacokinetic and pharmacodynamic data and models in clinical trials. *European Journal of Drug Metabolism and Pharmacokinetics* 18: 61–76.

Steimer, J.-L., Vozeh, S., Racine, A., Holford, N. G., and O'Neill, R. (1994). The population approach: Rationale, methods and applications in clinical pharmacology and drug development. *Handbook of Experimental Pharmacology*, Volume 110, *Pharmacokinetics of drugs*, eds. Welling, P. and Balant, L. Springer Verlag, New York, 1995.

Tierney, L. (1994). Markov Chains for Exploring Posterior Distributions. *Annals of Statistics* 22, 1701–1762.

Wakefield, J. C. (1995). An expected loss approach to the design of dosage regimens via sampling-based methods. *The Statistician* 43: 13–29.

Wakefield, J. C. and Racine-Poon, A. An application of Bayesian population pharmacokinetic/pharmacodynamic models to dose recommendation. *Statistics in Medicine*, 14, 971–986, 1995.

Wakefield, J. C., Smith, A. F. M., Racine-Poon, A., and Gelfand, A. E. (1994). Bayesian analysis of linear and nonlinear population models using the Gibbs sampler. *Applied Statistics* 43, 201–221.

14

Bayesian and Frequentist Analyses of an In Vivo Experiment in Tumor Hemodynamics

Russell D. Wolfinger   SAS Institute, Inc. Cary, North Carolina
Gary L. Rosner   Duke University Medical Center, Durham, North Carolina

ABSTRACT

This chapter considers the analysis of experimental data consisting of multiple measurements taken at various locations in an animal. A traditional mixed linear model is proposed and fitted to the data in order to draw statistical inferences about tumor hemodynamics. Assuming normality of the data, frequentist and Bayesian approaches are compared, the former employing the method of residual/restricted maximum likelihood, and the latter a straightforward simulation from the joint posterior density function formed from standard noninformative priors. Although more computationally intensive, the Bayesian approach provides more information and makes inference more readily attainable than traditional frequentist statistical methods.

1 INTRODUCTION

1.1 Background and Design of Experiments

Hydralazine, an antihypertensive drug, has been investigated to determine its ability to reduce blood flow within tumors. The motivation for these experiments is the theory that manipulating blood flow within a tumor might enhance the effectiveness of certain treatment modalities. For example, hyperthermia treats cancerous tumors by heating them, but heat transfer by flowing blood is the single most important factor limiting the ability to achieve therapeutic temperatures clinically. If one could slow blood flow within the tumor, greater efficacy might be achieved.

Although empirical evidence exists that hydralazine tends to decrease blood flow selectively in tumors compared with surrounding normal tissue, the physiological mechanism behind this phenomenon remains unknown. One theory attributes this to a "vascular steal" phenomenon, namely, that relatively greater arteriolar dilation occurs in normal tissue relative to tumor tissue, "stealing" blood away from flowing through the tumor.

Earlier experiments were carried out in 20 rats to determine the effect of hydralazine on arterioles in normal and tumor tissue. Diameters were measured before and after hydralazine administration in 15 animals having tumor implants and 5 without. Further, measurements were made in normal arterioles for 8 of the 15 tumor-bearing animals and in tumor-feeding arterioles in the remaining 7. The investigators did not find a significant difference in arteriole dilation as a function of location after adjusting for pretreatment diameters, casting doubt in their minds that the vascular steal phenomenon accounts for differential blood flow.

We analyze data arising from experiments evaluating hemodynamics in venules (small veins) in 23 tumor-bearing rats (for details see Dewhirst et al., 1994). In each animal, measurements were made in vessels located in only one of three locations (namely, in the tumor center, the peripheral boundary of the tumor, or in adjacent normal tissue) after the implanted tumor had grown for 9 or 14 days. Baseline measurements were taken before administering the drug hydralazine and again between 20 and 45 minutes later. Table 1 shows the number of animals in each group studied, along with the number of venules studied per animal, and Figure 1 is a plot of the data.

Table 1 Number of Animals (Vessels) Studied Under Each Experimental Condition

	Tumor center	Tumor periphery	Normal
9-day	5 (57)	5 (16)	3 (34)
14-day	4 (52)	4 (22)	2 (39)

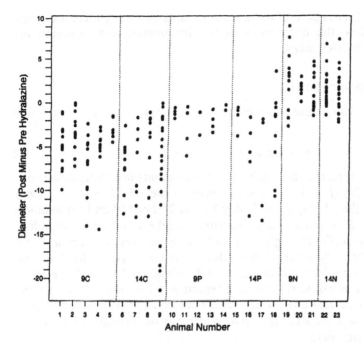

Figure 1 Diameter differences by animal grouped by age of tumor in days (9 or 14) and location (C = center, P = periphery, and N = normal).

We analyze the change in diameters seen in these vessels after the animals were given the drug. If the vessels in the tumor constrict but those in normal tissue do not, this may explain the blood flow reduction seen in tumors after hydralazine administration. That is, the drug causes increased flow resistance in tumor vessels, reducing blood flow, rather than dilating arterioles in normal tissues and, thereby, stealing blood destined for the tumor.

1.2 Analysis Strategy

The standard statistical approach to investigating an experiment of this sort is as follows:

1. Set forth a model and assumptions that reasonably describe the data.
2. Fit the model.
3. Use the fitted model to draw relevant inferences.

We now discuss each of these steps in terms of the hydralazine example, highlighting the differences in the frequentist and Bayesian approaches to all three steps.

Model and Assumptions

We use a two-way model with nested (hierarchical) errors and a baseline covariate:

$$y_{ijkl} = m + a_i + b_j + cx_{ijkl} + u_{ijk} + e_{ijkl} \tag{1}$$

Here y_{ijkl} represents the difference (post- minus pre-hydralazine) observed in the lth ($l = 1, \ldots, n_{ijk}$) vessel in the kth ($k = 1, \ldots, n_{ij}$) animal within the jth day [$j = 1$ (day 9) and 2 (day 14)] and ith location [$i = 1$ (center), 2 (periphery), and 3 (normal)], and x_{ijkl} is the corresponding baseline diameter. On the right-hand side, m is an overall mean, a_i is the ith location effect, b_j is the jth day effect, c is the slope for the baseline diameter, u_{ijk} is the deviation due to the kth animal, and e_{ijkl} is the residual error. It is also possible to add an interaction term, ab_{ij}, to the model, but for theoretical and analytical reasons we considered it unnecessary. Although simple in nature, this model addresses the fundamental aspects of the hydralazine data.

All of the parameters on the right-hand side of the model except x_{ijkl} are unobserved, and a typical frequentist analysis would label m, a_i, b_j, and c as fixed-effects parameters and u_{ijk} and e_{ijkl} random-effects parameters, resulting in a traditional mixed model. The fixed-effects parameters m, a_i, b_j, and c are regarded simply as constants to be estimated, whereas the random-effects parameters u_{ijk} and e_{ijkl} are assumed to be independent normal random variables with zero means and variances σ_u^2 and σ_e^2, respectively. The variance components σ_u^2 and σ_e^2 then become additional parameters to be estimated from the data. See Searle *et al.* (1992) for a thorough discussion and history of this model.

The distinction between random and fixed effects in this context is as follows. The two random effects correspond to the animals and the venules nested within animals, and both can be viewed as samples from theoretical populations about which we wish to draw inferences. The fixed effects, on the other hand, correspond to specific experimental conditions upon which we wish our inferences to be conditioned.

A Bayesian regards all parameters as "random" in the sense that all uncertainty about them should be described in terms of probability distributions. A joint prior distribution needs to be specified for the parameters, and this, multiplied by the likelihood function, is the (unnormalized) joint posterior density. The final joint posterior distribution and its marginals are the basis for all Bayesian inference. For simplicity and comparability with the frequentist approach, we make use of noninformative priors (see Box and Tiao, 1973). As a rule for forming them, we retain the aforementioned frequentist setup of the model in terms of fixed and random effects. Schervish (1992) calls these the "fixed selection" and "random selection" cases. We then place a flat prior on the fixed-effects parameters and Jeffreys' priors on the variance components.

Model Fitting

Frequentists commonly assume the random effects are normally distributed and construct the likelihood or residual likelihood function accordingly. This function is subsequently maximized over all unknown parameters to obtain their estimates (Harville, 1977). The estimates of the fixed effects are typically in estimated generalized-least-squares form, and the random effects can also be estimated using empirical Bayes methods, as in Laird and Ware (1982) and Strenio et al. (1983).

In contrast, Bayesians require the joint posterior density of the parameters and its marginals, but even for simple unbalanced mixed models such as Eq. (1), its computation involves high-dimensional integrals. Not until recent advances in computing hardware and algorithms has such an analysis been feasible for the everyday analyst. Section 2 describes a straightforward simulation approach to the problem.

Inference

The frequentist uses sampling distributions of the parameter estimates to make probability statements defined across hypothetical repetitions of the experiment, producing such quantities as p-values, t- and F-tests, and confidence intervals for relevant linear combinations. When estimates of the variance components are employed in generalized-least-squares formulas, care must be taken to adjust for the uncertainty resulting from estimating them. This is accomplished via degrees-of-freedom adjustments to the t- and F-distributions; however, for unbalanced data sets such as ours, these adjustments usually result in only approximations to the true sampling distributions of the relevant statistics.

On the other hand, the Bayesian joint posterior distribution and its marginals inherently account for all uncertainty in the model, including that associated with unknown variance components. Bayesian inference uses quantities such as joint and marginal means, medians, modes, quantiles, and highest posterior density regions. Hypothesis testing is also possible through the use of one-dimensional summary parameters and Bayes factors (Berger, 1985).

After covering some details of Bayesian estimation in Section 2, we present both frequentist and Bayesian inferences for the blood vessel example in Section 3. These developments show that the Bayesian approach not only is implementable but also able to provide insights superior to those gained by typical frequentist statistics.

2 BAYESIAN ANALYSIS DETAILS

2.1 Background

Following initial work by Hill (1965), only a few textbooks describe a Bayesian approach to variance components; among them are Zellner (1971), Box and Tiao (1973), Broemeling (1985), and Searle et al. (1992). These books take an analytic approach to the problem and derive the expressions describing the joint and marginal posterior distributions of the mean parameters and variance components. Even in the balanced case where $n_{ijk} = n$ and $n_{ij} = n_1$ for all i, j, k, the formulas become rather cumbersome, and approximations are required for several useful functions of the parameters. For unbalanced situations like the data analyzed here, the analysis is, to quote Searle et al. (1992, p. 108), "very intractable."

Instead of trying to delve into complicated mathematical expressions, our strategy is to generate a pseudorandom sample of say, 10,000 observations, from the joint posterior distribution of the parameters (Smith and Gelfand, 1992). This sample can then be input to a graphical statistical package for exploratory data analysis in order to perform Bayesian inference. Analyzing any one parameter from this sample is equivalent to looking at its marginal posterior distribution because of the implicit Monte Carlo integration carried out in the sampling process.

For example, Figure 2 displays a distribution analysis from Release 6.10 of SAS/INSIGHT Software (SAS Institute Inc., 1990) using 10,000 observations simulated from the posterior distribution of the least-squares mean of the center tumor location [see Eq. (2) in Section 3.1 for a definition]. Pictured are a histogram, smooth kernel density estimate, moments, and quantiles, all of which enable a thorough Bayesian analysis of this difference. For instance, assuming the posterior density is symmetric, the 2.5th and 97.5th quantiles represent a 95% highest posterior density region, the Bayesian analogue of a confidence interval.

In addition to basic summary statistics from the posterior distributions, predictive densities and Bayes factors can be computed from the sample. Bivariate marginal information can be displayed through bivariate

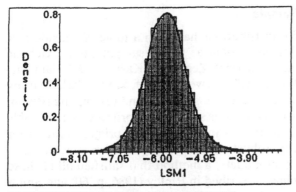

Moments			
N	10000.0	Sum Wgts	10000.00
Mean	−5.78	Sum	−57831.03
Std Dev	0.51	Variance	0.26
Skewness	0.07	Kurtosis	0.63
USS	337062	CSS	2618.67
CV	−8.85	Std Mean	5.118E−03

Quantiles			
100% Max	−3.12	99.0%	−4.50
75% Q3	−5.47	97.5%	−4.74
50% Med	−5.78	95.0%	−4.94
25% Q1	−6.11	90.0%	−5.15
0% Min	−8.04	10.0%	−6.41
Range	4.91	5.0%	−6.62
Q3−Q1	0.64	2.5%	−6.80
Mode	−8.04	1.0%	−7.03

Figure 2   Posterior density of the least-squares mean of the center tumor location.

density estimation and three-dimensional graphs. A further attractive feature of the entire approach is that posterior summaries can be computed
for any function of the original parameters: one just constructs a new
variable by evaluating the desired function at each of the 10,000 sample
points and then analyzes this variable graphically.

Some accuracy is sacrificed because of the pseudorandomness and
simulation error, but these are usually negligible in practice or can be
made so by increasing the Monte Carlo sample size. The primary question
thus becomes "How do we generate the sample from the joint posterior?"

2.2 Generating the Sample

Simulation from posterior distributions has been a topic of considerable recent research, involving areas such as Gibbs sampling (e.g., Gelfand and Smith, 1990; Gelfand *et al.*, 1990; Zeger and Karim, 1991), data augmentation (Tanner and Wong, 1987; Geweke, 1992), Monte Carlo Markov chains (Tierney, 1994), and importance sampling and resampling (Rubin, 1987). A large percentage of the various sampling algorithms are applicable to our simple variance components model, and our primary criteria for choosing one are simplicity, robustness, efficiency, and implementability.

For variance component models, a method that satisfied all of these criteria is rejection sampling, described in Ripley (1986, p. 60) and Smith and Gelfand (1992). The basic idea is to generate a pseudorandom observation from a convenient distribution (chosen to be as close as possible to the posterior) and then retain that observation in the final sample with probability proportional to the ratio of the two densities.

For our variance components model, we form the joint posterior density (up to a proportionality constant) from the normal likelihood and the reference priors discussed briefly in the next subsection. The fixed-effects parameters can be analytically integrated out of this joint posterior, leaving the marginal posterior density of the variance components. After an orthogonalizing transformation, this marginal density can be closely approximated by a product of inverted gamma densities, which form the convenient rejection sampling distribution. Then, after generating a pseudorandom value for each of the variance components, the values for the fixed-effects parameters are simulated from a multivariate normal distribution, which is their posterior conditional distribution given the variance components.

2.3 Priors

We employ the conventional Jeffreys' reference prior in order to provide a readily implemented, unified method for Bayesian inference in mixed models. This improper prior is locally uniform in the fixed-effects parameters and the logarithms of the variance components and is described in Zellner (1971), Box and Tiao (1973), and Broemeling (1985). If in Eq. (1) $n_{ijk} = K$ for all i, j, and k (that is, the same number of vessels are measured within each animal), then the Jeffreys' prior is

$$p(m, a_1, a_2, a_3, b_1, b_2, c, \sigma_u^2, \sigma_e^2) \propto \sigma_e^{-2}(\sigma_e^2 + K\sigma_u^2)^{-1}$$

where $\propto$ denotes equivalence up to a proportionality constant. The hydralazine data are not balanced in this way, but the corresponding prior is similar in that it varies in proportion to σ_u^{-2} and σ_e^{-4}.

Jeffreys' prior is certainly not intended as a replacement for real prior information, and, in fact, incorporating informative priors into the rejection sampling algorithm is straightforward. Jeffreys' prior lets the data "speak for themselves" in the sense that the prior has little influence on the posterior, thus making inferences comparable with those from classical frequentist techniques.

3 ANALYSES

3.1 Frequentist Analysis

We first carry out a typical frequentist analysis. The appropriate code for the SAS MIXED procedure (SAS Institute Inc., 1992) is as follows:

```
proc mixed;
    class loc day animal;
    model deldiam = loc day prediam / s cl;
    random animal (loc*day);
    lsmeans loc / diff cl;
run;
```

Since LOC, DAY, and ANIMAL are all categorical variables and not continuous ones, they are placed on the CLASS statement. The MODEL statement specifies LOC, DAY, and PREDIAM as fixed effects, corresponding to a_i, b_j, and c, respectively, in Eq. (1). The S option requests that the estimated generalized least-squares solution for them be printed, and the CL option requests 95% confidence limits.

ANIMAL (LOC*DAY) is specified as a random effect on the RANDOM statement, indicating animals nested within each combination of location and day. A residual error, corresponding to e_{ijkl}, is also included in the model by default, and REML is the default estimation method for σ_u^2 and σ_e^2.

The LSMEANS statement requests least-squares means (estimators of the adjusted marginal means that would be expected had the design been balanced) for locations. In terms of the model parameters, the least-squares mean for location i is an estimate of

$$l_i = m + a_i + \frac{b_1 + b_2}{2} + c\bar{x} \tag{2}$$

where $\bar{x}$ is the overall mean of the baseline diameters. The parameter l_i can be viewed as the population marginal mean for location i adjusted to the mean of the covariate. The DIFF option estimates the differences between the $l_i s$, and the CL option requests 95% confidence limits for both the $l_i s$ and their differences.

The output resulting from this code applied to the hydralazine data is summarized in Table 2. The REML estimates of σ_u^2 and σ_e^2 are 1.05 and 10.90, respectively, and their asymptotic standard errors are from the inverse of the Hessian matrix (the observed Fisher information). The p-values and confidence limits for σ_u^2 and σ_e^2 are computed from the normal distribution, explaining the negative lower limit for σ_u^2.

The REML optimization required three Newton-Raphson iterations, with the value of the maximized residual log likelihood being -581.4. From this one can compute several information criteria (Bozdogan, 1987), the most popular ones being AIC$= -583.4$ and BIC$= -586.8$, in larger-is-better form.

Table 2 also lists the estimated generalized least-squares estimates of m, a_1, a_2, a_3, b_1, b_2, and c, along with their approximate standard errors. The associated p-values and 95% confidence intervals are based on the t-distribution. Because of linear dependencies, the estimates of a_3 and b_2

Table 2 Parameter Estimates and Associated Statistics for Model (1) Fitted to the Hydralazine Data

Parameter	Estimate	Standard error	Estimate/ std. error	p-value	95% confidence interval	
σ_u^2	1.05	0.79	1.33	0.1847	-0.50	2.59
σ_e^2	10.90	1.10	9.93	0.0000	8.75	13.05
m	3.56	0.90	4.01	0.0008	1.70	5.41
a_1	-7.26	0.77	-9.43	0.0000	-8.88	-5.65
a_2	-4.40	0.89	-4.93	0.0000	-6.27	-2.54
a_3	0.00					
b_1	0.66	0.68	1.02	0.3205	-0.72	2.10
b_2	0.00					
c	-0.06	0.01	-5.07	0.0000	-0.09	-0.04
l_1	-5.79	0.47	-12.25	0.0000	-6.78	-4.80
l_2	-2.93	0.65	-4.51	0.0000	-4.29	-1.57
l_3	1.48	0.61	2.42	0.0163	0.20	2.75
$l_1 - l_2$	-2.86	0.80	-3.56	0.0005	-4.54	-1.18
$l_1 - l_3$	-7.26	0.77	-9.43	0.0000	-8.88	-5.65
$l_2 - l_3$	-4.40	0.89	-4.93	0.0000	-6.27	-2.54

are set to 0. Table 2 concludes with the output from the LSMEANS statement, showing highly significant differences in the locations.

Consistent with the significant location differences is the type III F-statistic for locations, which equals 44.46 with 2 and 19 degrees of freedom. The resulting p-value is less than 0.00005. On the other hand, the F-statistic for days equals 1.04 with 1 and 19 degrees of freedom, and the p-value is 0.3205.

3.2 Bayesian Analysis

To generate a sample from the posterior distribution of the mixed-model parameters using PROC MIXED, the following lines must be added to the previous code:

```
prior jeffreys / nsample=1e4 seed=21346 rbound=1.34;
make 'sample' out=sample noprint;
```

The PRIOR statement is an undocumented enhancement to Release 6.10 of PROC MIXED. JEFFREYS indicates Jeffreys' prior, and the only current alternative is FLAT. The NSAMPLE=option specifies the number of posterior samples (1000 by default) and SEED=indicates the random number seed (clock time by default).

The RBOUND=option specifies an upper bound for the rejection sampling ratio. Without this option, PROC MIXED attempts to estimate this data-dependent quantity with a simplex algorithm before carrying out the sampling; however, sometimes the simplex method fails to obtain the correct value. PROC MIXED prints a warning message if the bounding value is violated during the sampling process, and if this violation occurs, the entire job should be rerun using the higher bound.

The MAKE statement creates an SAS data set containing the posterior sample and the NOPRINT option suppresses its listing.

All calculations for the hydralazine data set required approximately 1 minute 30 seconds on an HP 9000/720 workstation running HP-UX 9.01 with 32 mb. The 10,000-observation data set was then input to SAS/INSIGHT Software (SAS Institute Inc., 1990) to perform Bayesian inference.

Pictured in Figure 2 is the posterior density of the least-squares mean for the center location. For comparison, the frequentist statistics are the ones just referred to in Table 2. For example, the frequentist point estimate is -5.79, while the estimated posterior mean is -5.78. Also provided in the Bayesian analysis are estimates of the posterior median and mode, -5.78 and -5.77, respectively. The 95% confidence interval is $(-6.78, -4.80)$, while the 95% highest posterior density region is slightly wider and asymmetric: $(-6.80, -4.74)$. Both analysis methods thus lead to nearly identical conclusions about the center location.

However, to emphasize a key advantage of the Bayesian method, suppose the quantity of interest is actually the ratio of the magnitude of the normal-tissue-location least-squares mean to that of the center-tumor location, which in terms of the parameters is

$$\frac{|l_3|}{|l_1|} = \frac{|m + a_3 + (b_1 + b_2)/2 + c\bar{x}|}{|m + a_1 + (b_1 + b_2)/2 + c\bar{x}|}$$

A frequentist would typically make some distributional approximation for their joint distribution and then marginalize, often making use of the delta method approximation (Billingsley, 1986). This task is tedious for this particular example because the covariance between the two estimates is not directly available in the PROC MIXED output. On the other hand, a Bayesian need only create a new variable in the already-formed sample from the posterior and produce an estimate of the posterior distribution, which is precisely what is done in Figure 3.

A conclusion from Figure 3 is that the probability is more than 0.95 that the adjusted normal-tissue venule magnitudes are less than half the size of those in the center of the tumor. An analogous statement may not have been as easily determined by a frequentist analysis. Furthermore, this type of Bayesian analysis is just as simple for any well-defined function of the parameters, whereas a frequentist would need to redo the analytical work (derivatives, etc.) for each new function.

The posterior density of the coefficient c of the baseline diameter covariate is plotted in Figure 4. The Bayesian and frequentist results for c are very similar, with both the PROC MIXED estimate and posterior mean equaling -0.062 and the standard deviations equaling 0.012 and 0.013, respectively. Both methods thus concur that the larger vessels change gradually less from baseline than do the smaller ones. An additional analysis (not shown) revealed that separate baseline coefficients for each location are unnecessary.

Next we consider the posterior distributions of σ_u^2 and σ_e^2. Figure 5 displays their marginal densities. Typical frequentist information is found in the first two lines of Table 2, and, because based on asymptotics, is much less informative than Figure 5. Figure 6 displays the joint density of the two variance components computed using bivariate kernel density estimation. The joint mode is (0.99, 10.87), and the lower bound of 0 has a significant impact on the density in the σ_u^2 axis. Such graphics enable the Bayesian to make joint inferences about σ_u^2 and σ_e^2. For example, the

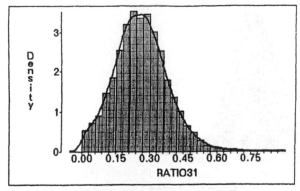

Moments			
N	10000.0	Sum Wgts	10000.00
Mean	0.26	Sum	2606.87
Std Dev	0.12	Variance	0.01
Skewness	0.36	Kurtosis	0.36
USS	813.52	CSS	133.94
CV	44.40	Std Mean	1.157E−03

Quantiles			
100% Max	0.86	99.0%	0.57
75% Q3	0.33	97.5%	0.51
50% Med	0.26	95.0%	0.46
25% Q1	0.18	90.0%	0.41
0% Min	3.98E−04	10.0%	0.11
Range	0.85	5.0%	0.08
Q3−Q1	0.15	2.5%	0.04
Mode	3.98E−04	1.0%	0.02

Figure 3 Posterior density of the ratio of the absolute value of the normal-tissue-location least-squares mean to that of the center-tumor location.

outermost contour in Figure 6 represents the 99% highest posterior density region, an analogue of the frequentist confidence ellipsoid.

We now return to consideration of the extent of the differences of the blood vessel locations. The F-statistic for locations and its associated p-value are widely used statistics summarizing the fixed-effect differences. However, as discussed by Schervish (1992), the classical F-statistic fails to separate the true effect size from the sample size. In other words,

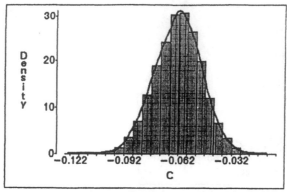

Moments			
N	10000.000	Sum Wgts	10000.000
Mean	−0.062	Sum	−619.732
Std Dev	0.013	Variance	1.6609E−04
Skewness	−0.026	Kurtosis	−0.065
USS	40.068	CSS	1.661
CV	−20.796	Std Mean	1.2888E−04

Quantiles			
100% Max	−0.017	99.0%	−0.032
75% Q3	−0.053	97.5%	−0.037
50% Med	−0.062	95.0%	−0.041
25% Q1	−0.071	90.0%	−0.046
0% Min	−0.118	10.0%	−0.079
Range	0.101	5.0%	−0.084
Q3−Q1	0.017	2.5%	−0.087
Mode	−0.118	1.0%	−0.092

Figure 4 Posterior density of the baseline diameter coefficient.

classical statistics become more significant with larger sample sizes, regardless of the differences in the true parameters. How shall Bayesian inference proceed along these lines?

One possibility is to consider directly the trivariate marginal posterior distribution of the differences, $a_1 - a_2$, $a_1 - a_3$, and $a_2 - a_3$, which are the parameters estimated by the differences of least-squares means in the bottom of Table 2. One could study the individual marginals and also create some bivariate plots like Figure 6.

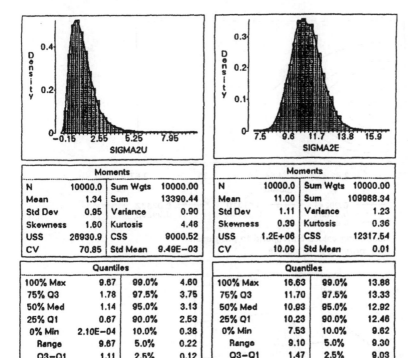

Moments					Moments			
N	10000.0	Sum Wgts	10000.00		N	10000.0	Sum Wgts	10000.00
Mean	1.34	Sum	13390.44		Mean	11.00	Sum	109968.34
Std Dev	0.95	Variance	0.90		Std Dev	1.11	Variance	1.23
Skewness	1.60	Kurtosis	4.48		Skewness	0.39	Kurtosis	0.36
USS	26930.9	CSS	9000.52		USS	1.2E+06	CSS	12317.54
CV	70.85	Std Mean	9.49E-03		CV	10.09	Std Mean	0.01

Quantiles					Quantiles			
100% Max	9.67	99.0%	4.60		100% Max	16.63	99.0%	13.88
75% Q3	1.78	97.5%	3.75		75% Q3	11.70	97.5%	13.33
50% Med	1.14	95.0%	3.13		50% Med	10.93	95.0%	12.92
25% Q1	0.67	90.0%	2.53		25% Q1	10.23	90.0%	12.46
0% Min	2.10E-04	10.0%	0.36		0% Min	7.53	10.0%	9.62
Range	9.67	5.0%	0.22		Range	9.10	5.0%	9.30
Q3-Q1	1.11	2.5%	0.12		Q3-Q1	1.47	2.5%	9.03
Mode	2.10E-04	1.0%	0.07		Mode	7.53	1.0%	8.71

Figure 5 Marginal posterior densities of σ_u^2 and σ_e^2.

Schervish (1992) recommends defining a one-dimensional parameter that summarizes the differences in the trivariate distribution and directly measures the magnitude of the hypothesis tested by the frequentist F-statistic. Different possibilities exist, and the one we wish to consider is described by Ghosh in the discussion of Schervish (1992). Defining β to be the vector of fixed-effects parameters, that is,

$$\beta = (m, a_1, a_2, a_3, b_1, b_2, c)'$$

and

$$L = \begin{bmatrix} 0 & 1 & 0 & -1 & 0 & 0 & 0 \\ 0 & 0 & 1 & -1 & 0 & 0 & 0 \end{bmatrix}$$

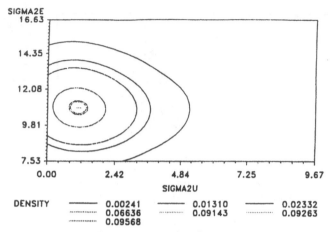

DENSITY   ——— 0.00241   ——— 0.01310   ——— 0.02332
             ·········· 0.06636   ·········· 0.09143   ·········· 0.09263
             ·········· 0.09568

Ascending values correspond to progression
from outer to inner curves.

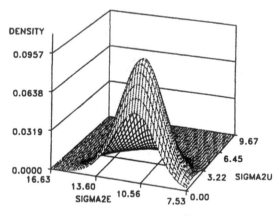

Figure 6   Joint posterior density of σ_u^2 and σ_e^2.

the summary parameter is defined as

$$\rho^* = \frac{(L\beta)'(LL')^{-1}L\beta}{\text{rank}(L)}$$

The construction of ρ^* is similar to that of the standard F-statistic, except that the overall variance matrix does not enter the central quadratic form. As shown by Schervish (1992), $2\rho^*$ can be interpreted as the average squared deviation (root-mean-squared error) of the a_i. Figure 7 details the posterior density of $(2\rho^*)^{1/2}$. From it, one can make statements such as

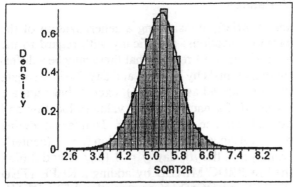

Moments			
N	10000.0	Sum Wgts	10000.00
Mean	5.22	Sum	52190.80
Std Dev	0.60	Variance	0.36
Skewness	0.04	Kurtosis	0.64
USS	275958	CSS	3570.57
CV	11.45	Std Mean	5.976E−03

Quantiles			
100% Max	8.51	99.0%	6.69
75% Q3	5.59	97.5%	6.42
50% Med	5.22	95.0%	6.19
25% Q1	4.84	90.0%	5.95
0% Min	2.76	10.0%	4.47
Range	5.74	5.0%	4.24
Q3−Q1	0.75	2.5%	4.03
Mode	2.76	1.0%	3.76

Figure 7   Posterior density of $(2\rho^*)^{1/2}$ for locations.

"There is approximately a 1% chance that the average deviation of the locations is less than 3.76." These types of inferences can be very useful to experimenters, often more so than the standard p-value explanation.

Both frequentist and Bayesian approaches produce strong evidence that hydralazine leads to constriction of venules in tumor but not in normal tissue, with significantly more constriction occurring in the center of the tumors than in their periphery. These results, along with earlier experiments showing little difference between tumor and normal arteriole dilation, argue against the vascular steal phenomenon in this setting.

3.3 Heterogeneous Variances

We conclude this chapter by briefly considering a generalization of the model analyzed in the previous sections, specifically with regard to the residual variance parameter σ_e^2. Figure 1 reveals that there may be a different degree of variability within animals by location and day. To investigate this possibility, we fit the same model as in Eq. (1), except that separate e_{ij} variance components were fit for each different value of location and day instead of a single residual variance parameter, σ_e^2. To this end, define σ_{eij}^2 to be the residual error's variance for the ith location [$i = 1$ (center), 2 (periphery), and 3 (normal)] and the jth day [$j = 1$ (day 9) and 2 (day 14)]. (This is accomplished in PROC MIXED by adding a REPEATED statement, specifying GROUP = LOC*DAY as an option.) The REML parameter estimates from this analysis are presented in Table 3.

Table 3 reveals a significant degree of heterogeneity, especially across days within the tumor locations. The value for the maximized residual log likelihood is -554.6, with AIC$=-561.6$ and BIC$=-573.4$. Comparing with the homogeneous values of AIC$=-583.4$ and BIC$=-586.8$, both information criteria prefer this heterogeneous model to the homogeneous one.

In spite of the heterogeneity, the estimates of the least-squares means remain relatively unchanged, and, therefore, the same conclusions regarding vascular steal apply. The effect of this heterogeneity on the estimated generalized-least-squares estimates of the fixed effects is to weight all of

Table 3 Parameter Estimates and Approximate Standard Errors for Model (1) with Heterogeneous Residual Variances Fitted to the Hydralazine Data

Parameter	Estimate	Standard error
σ_u^2	0.99	0.61
σ_{e11}^2	6.76	1.36
σ_{e12}^2	23.85	4.78
σ_{e21}^2	1.89	0.76
σ_{e22}^2	16.41	5.15
σ_{e31}^2	5.73	1.47
σ_{e32}^2	4.78	1.11
l_1	-5.79	0.49
l_2	-2.89	0.52
l_3	1.50	0.53

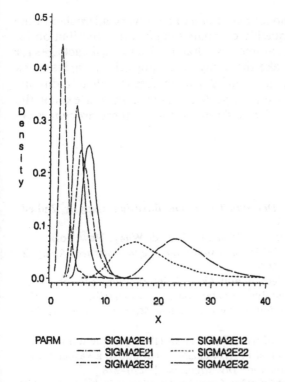

Figure 8 Marginal posterior densities of σ_{e11}^2, σ_{e12}^2, σ_{e21}^2, σ_{e22}^2, σ_{e31}^2, and σ_{e32}^2.

the data from day 9 as well as the normal tissue data from day 14 more heavily than the rest.

Figure 8 displays the marginal posterior densities of the six parameters, computed using kernel density estimation on 10,000 observations from our rejection sampling algorithm. Heterogeneity in both location and dispersion is apparent here as well.

4 CONCLUSION

We have set forth a straightforward and flexible sampling-based Bayesian inference procedure for a simple variance component model and compared the results with a traditional frequentist analysis. The procedure generates a pseudorandom independent sample from the joint posterior distribution of the mixed-model parameters and is very adaptable in the sense that the analyst can quickly compute the posterior distribution for arbitrary functions of these parameters. Graphical statistical packages for exploratory data analysis make the study of the posterior samples quick and extensive, often yielding more informative inferences than traditional frequentist summaries. It is our hope that statisticians will exploit the availability of graphical statistical software and apply these procedures in Bayesian analyses.

REFERENCES

Berger, J. O. (1985). *Statistical Decision Theory and Bayesian Analysis*, 2nd ed. Springer-Verlag, New York.

Billingsley, P. (1986). *Probability and Measure*, 2nd ed. Wiley, New York.

Box, G. E. P. and Tiao, G. C. (1973). *Bayesian Inference in Statistical Analysis*. Addison-Wesley, Reading, MA.

Bozdogan, H. (1987). Model selection and Akaike's information criterion (AIC): The general theory and its analytical extensions. *Psychometrika* 52: 345–370.

Broemeling, L. D. (1985). *Bayesian Analysis of Linear Models*. Marcel Dekker, New York.

Dewhirst, M. W., Madwed, D., Meyer, R. E., *et al.* (1994). Reduction in tumor blood flow in skin flap tumor after hydralazine is not due to a vascular steal phenomenon. *Radiation Oncology Investigations* 1: 270–278.

Gelfand, A. E. and Smith, A. F. M. (1990). Sampling-based approaches to calculating marginal densities. *Journal of the American Statistical Association* 85: 398–409.

Gelfand, A. E., Hills, S. E., Racine-Poon, A., and Smith, A. F. M. (1990). Illustration of Bayesian inference in normal data models using Gibbs sampling. *Journal of the American Statistical Association* 85: 972–985.

Geweke, J. (1992). Evaluating the accuracy of sampling-based approaches to the calculation of posterior moments. In *Bayesian Statistics 4*, ed. Bernardo, J. M., Berger, J. O., Dawid, A. P., and Smith, A. F. M., pp. 169–193. Oxford University Press, New York (with discussion).

Harville, D. A. (1977). Maximum likelihood approaches to variance component estimation and to related problems. *Journal of the American Statistical Association* 72: 320–338.

Hill, B. M. (1965). Inference about variance components in the one-way model. *Journal of the American Statistical Association* 60: 806–825.

Laird, N. and Ware, J. H. (1982). Random-effects models for longitudinal data. *Biometrics* 38: 963–974.

Ripley, B. D. (1987). *Stochastic Simulation*. Wiley, New York.

Rubin, D. B. (1987). The SIR algorithm—a discussion of Tanner and Wong's: The calculation of posterior distributions by data augmentation. *Journal of the American Statistical Association* 82: 543–546.

SAS Institute Inc. (1990). *SAS/INSIGHT User's Guide, Version 6, First Edition*. SAS Institute Inc., Cary, NC.

SAS Institute Inc. (1992). SAS Technical Report P-229. *SAS/STAT Software: Changes and Enhancements, Release 6.07*. Chapter 16: The MIXED Procedure. SAS Institute Inc., Cary, NC.

Schervish, M. J. (1992). Bayesian analysis of linear models. In *Bayesian Statistics 4*, ed. Bernardo, J. M., Berger, J. O., Dawid, A. P., and Smith, A. F. M., pp. 419–434. Oxford University Press, New York (with discussion).

Searle, S. R., Casella, G., and McCulloch, C. E. (1992). *Variance Components*. Wiley, New York.

Smith, A. F. M. and Gelfand, A. E. (1992). Bayesian statistics without tears: A sampling-resampling perspective. *American Statistician* 46: 84–88.

Strenio, J. F., Weisberg, H. I., and Bryk, A. S. (1983). Empirical Bayes estimation of individual growth-curve parameters and their relationship to covariates. *Biometrics* 39: 71–86.

Tanner, M. A. and Wong, W.-H. (1987). The calculation of posterior distributions by data augmentation. *Journal of the American Statistical Association* 82: 528–550.

Tierney, L. (1994). Markov Chains for Exploring Posterior Distributions (with discussion). *Annals of Statistics* 22: 1701–1762.

Zeger, S. L. and Karim, M. R. (1991). Generalized linear models with random effects; a Gibbs sampling approach. *Journal of the American Statistical Association* 86: 79–86.

Zellner, A. (1971). *An Introduction to Bayesian Inference in Econometrics*. Wiley, New York.

15

Bayesian Meta-Analysis of Randomized Trials Using Graphical Models and BUGS

Teresa C. Smith and David J. Spiegelhalter   Medical Research Council Biostatistics Unit, Institute of Public Health, Cambridge, England
Mahesh K. B. Parmar   Medical Research Council Cancer Trials Office, Cambridge, England

ABSTRACT

In this chapter we describe how hierarchical random-effects models can be applied to meta-analysis using a fully Bayesian approach. Using a meta-analysis of randomized trials of selective decontamination of the digestive tract as an example, inferences are made using Gibbs sampling via BUGS, a freely available software package. We illustrate the usefulness of graphical modeling techniques for expressing the conditional independence assumptions of the parameters in the model and show how specification of the model in BUGS leads naturally from the graph formulation. Problems with using a standard noninformative prior distribution for the population variance are discussed and suitable alternative prior distributions are derived and compared.

1   INTRODUCTION

Hierarchical random-effects models are becoming increasingly popular for tackling problems involving complex structured data. In this chapter we describe how meta-analysis fits naturally into the hierarchical framework. Using an example of a meta-analysis of randomized trials of selective decontamination of the digestive tract [1], we illustrate how a Bayesian analysis can be carried out. Due to recent developments in Markov chain Monte Carlo [2], implementing such models within a fully Bayesian framework is now a viable option, without recourse to approximations in which uncertainty about certain parameters is ignored.

Inferences on the parameters of interest are based on the posterior marginal distributions, which are obtained using the simulation technique known as Gibbs sampling. This is implemented using BUGS, a software package available free of charge, which provides a means of analyzing complex models essentially by describing their structure and automatically deriving the expressions necessary for the Gibbs sampling. In addition, we show how graphical modeling techniques may be used to express the conditional independence assumptions of the parameters in the model: not only can this provide valuable insight, but BUGS also exploits the resulting factorization of the full distribution of all the parameters and data as a product of simple conditional distributions.

Focus then centers on the choice of prior distribution and most importantly on the problems with using a standard noninformative prior for the population (or between-study) variance in a random-effects meta-analysis. Techniques for deriving suitable alternative priors based on three different approaches are also explored and their consequences compared.

2 THE PROBLEM

Preventing infections in intensive care units is a major area of concern. However, controversy still surrounds the best way to avoid them. One suggested strategy involves selectively decontaminating the digestive tract to prevent carriage of potentially pathogenic microorganisms from the oropharynx, stomach, and gut. An international collaborative group investigated the clinical benefits of selective decontamination of the digestive tract by carrying out a meta-analysis of 22 randomized trials [1]. In each trial, patients in intensive care units were randomized to either a treatment or a control group, where treatment consisted of different combinations of oral nonabsorbable antibiotics, with some studies in addition including a systemic component of the treatment. Patients in the control groups were given no treatment. For each trial the number who developed respiratory tract infections in the treatment and the control groups were recorded (Table 1).

The collaborative group analyzed the data using the classical Mantel-Haenszel-Peto method [3] to obtain estimates of both the individual treatment effects in each study and the pooled effect. The pooled-effect estimate is based on assuming a common effect across all studies.

Table 1 Respiratory Tract Infections in Control and Treatment Groups of 22 Trials, with Individual and Pooled Estimates of Odds Ratios (95% Confidence Intervals) Obtained Using Mantel-Haenszel-Peto Method

Study	Infections/total Control	Treated	Odds ratio (95% confidence interval)
1	25/54	7/47	0.24 (0.10, 0.55)
2	24/41	4/38	0.13 (0.05, 0.32)
3	37/95	20/96	0.42 (0.23, 0.78)
4	11/17	1/14	0.10 (0.02, 0.40)
5	26/49	10/48	0.25 (0.11, 0.58)
6	13/84	2/101	0.17 (0.06, 0.48)
7	38/170	12/161	0.31 (0.17, 0.57)
8	29/60	1/28	0.14 (0.05, 0.36)
9	9/20	1/19	0.13 (0.03, 0.54)
10	44/47	22/49	0.11 (0.04, 0.25)
11	30/160	25/162	0.79 (0.44, 1.41)
12	40/185	31/200	0.67 (0.40, 1.12)
13	10/41	9/39	0.93 (0.33, 2.59)
14	40/185	22/193	0.48 (0.28, 0.82)
15	4/46	0/45	0.13 (0.02, 0.95)
16	60/140	31/131	0.42 (0.26, 0.70)
17	12/75	4/75	0.33 (0.12, 0.92)
18	42/225	31/220	0.72 (0.43, 1.18)
19	26/57	7/55	0.21 (0.09, 0.47)
20	17/92	3/91	0.21 (0.08, 0.54)
21	23/23	14/25	0.09 (0.02, 0.33)
22	6/68	3/65	0.52 (0.13, 1.99)
Pooled			0.36 (0.31, 0.43)

3 A FULL BAYESIAN RANDOM-EFFECTS MODEL

A random-effects model, unlike the fixed-effect method, adopts a probability model for individual study effects whose joint distribution is assumed not to depend on the order in which the studies are placed. In other words, we have no prior reason for thinking any particular study is different from another, and hence we are formally expressing a belief in *similarity*, as opposed to the extremes for *equivalence* or *complete independence*. This formal assumption, known as exchangeability, is mathematically equivalent to assuming these effects are randomly drawn from some population.

Let r_i^C denote the number of patients in the control group with infections in the ith study, arising from n_i^C patients randomized to the control group, each assumed to have probability of p_i^C of developing an infection. Adopting equivalent notation for the treatment group, the full model can be written

$$r_i^C \sim \text{binomial}(p_i^C, n_i^C)$$
$$r_i^T \sim \text{binomial}(p_i^T, n_i^T)$$
$$\text{logit}(p_i^C) = \mu_i - \delta_i/2$$
$$\text{logit}(p_i^T) = \mu_i + \delta_i/2$$
$$\delta_i \sim \text{normal}(d, \sigma^2)$$

where $\text{logit}(p) = \log_e[p/(1 - p)]$. The estimates of primary interest are the study-level treatment effects $\delta_i = \text{logit}(p_i^T) - \text{logit}(p_i^C)$, which is the log(odds ratio) in the ith study, and the population, or pooled, treatment effect d. $\mu_i = [\text{logit}(p_i^T) + \text{logit}(p_i^C)]/2$ may be considered the "average" (on the logit scale) infection rate in the ith trial.

Standard "empirical Bayes" methods [4] make inferences conditional on estimates of d and σ^2 that are obtained using a moment-matching procedure. This, however, ignores the uncertainty in the estimates of σ^2 when making inferences about d, and in σ^2 and d when estimating the precision of estimates of the individual trial effects, and this neglect may have a considerable impact if there are only a few studies.

A fully Bayesian analysis allows for this uncertainty by placing prior distributions on the unknown parameters, μ's, σ^2, and d. These priors will, however, generally be noninformative. Inferences about the parameters of interest (d, the population effect, and δ_i's, the individual trial effects) can then be made from the joint distribution formed by the prior and the likelihood by integrating out the unknown parameters (see Section 4). We may also be interested in predicting the true effect in a new study, as this in effect produces a prior distribution that may be used, for example, in assessing sample size for a further confirmatory trial.

The model can be expressed in the form of a graph [5] in which the nodes in the graph denote the data and parameters of the model (Figure 1). The idea of such a representation is to display qualitative aspects of the model without requiring algebraic formulas and hence to call attention to the essential assumptions. Constants fixed by the design of the study are denoted as double-edged rectangles (n_i^C, n_i^T), observed variables as

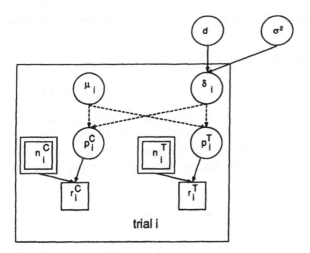

Figure 1 Graphical model for random-effects meta-analysis.

single-edged rectangles (r_i^C, r_i^T), and unobserved variables as circles (p_i^C, p_i^T, μ_i, δ_i, d, σ^2).

The graph is directed and acyclic, and the arrows drawn between nodes indicate the conditional independence assumptions of the model. Direct influences on a node are known as its "parents", using an obvious genetic analogy, so, for example, d and σ^2 are the parents of each δ_i. The directed links express the assumption that given its parent nodes, denoted pa(v), each node v is independent of all other nodes except descendants of v. For example, from Figure 1, it can be seen that conditional on knowing p_i^T (the true treatment response rate in the ith trial), our beliefs in r_i^T would be independent of the population parameters (d and σ^2) and data for the other trials. The links may either indicate a stochastic dependence (solid arrow), for example, $r_i^C \sim$ binomial(p_i^C, n_i^C), or a logical function (dashed arrow) such as logit(p_i^C) = $\mu_i - \delta_i/2$. Due to the conditional independence assumptions as expressed in the graph, the full joint probability distribution of all the quantities V can be specified as a simple factorization of the conditional parent-child distributions $p(v \mid \text{pa}(v))$ [6]. So

$$p(V) = \prod_{v \in V} p(v \mid \text{pa}(v)) \qquad (1)$$

For the model described above, this joint distribution (ignoring the constants $\underline{n}^C$, $\underline{n}^T$, and using the fact that $p_i{}^C$, $p_i{}^T$ can be expressed in terms of μ_i, δ_i) takes the form

$$p(\underline{r}^C, \underline{r}^T, \underline{\mu}, \underline{\delta}, d, \sigma^2) \propto \prod_i [p(r_i{}^C \mid \mu_i, \delta_i) \, p(r_i{}^T \mid \mu_i, \delta_i) \, p(\mu_i) \tag{2}$$

$$p(\delta_i \mid d, \sigma^2)] \, p(d) \, p(\sigma^2)$$

A technical problem is to obtain the appropriate posterior marginal distributions for parameters of interest, conditional on having observed the data $\underline{r}^C$, $\underline{r}^T$. For example, inferences on the population effect d should be based on $p(d \mid \underline{r}^C, \underline{r}^T)$, which by Bayes' theorem is proportional to $p(\underline{r}^C, \underline{r}^T \mid d) \, p(d)$, so that

$$p(d \mid \underline{r}^C, \underline{r}^T) \propto \int p(\underline{r}^C, \underline{r}^T, \underline{\mu}, \underline{\delta}, d, \sigma^2) \, d\underline{\mu} \, d\underline{\delta} \, d\sigma^2$$

We thus need to integrate the joint distribution (2) over $\underline{\mu}$, $\underline{\delta}$, and σ^2. The integrand does not have a closed-form solution, even though it is composed of a product of terms each of which has a simple form. Some form of approximation or simulation technique is necessary for this and other inferences based on integrating out parameters from the full joint distribution. Developments in computer-intensive methodology have established what are known as Markov chain Monte Carlo methods as a practical proposition, a particular form of which is known as Gibbs sampling [7].

The graph is thus translated into a full probability model, and in the next section we now show how this model description directly forms the basis of the computational method.

4 INFERENCE USING GIBBS SAMPLING AND BUGS

A simple, although computationally demanding procedure for the numerical integration of complex functions, Gibbs sampling has come from its origins in statistical mechanics through image processing to play a major role in modern statistics. Initial values are given to all unknown quantities, which include all parameters, missing data, latent variables, and so on. Samples are then successively drawn from the conditional distribution of each variable in turn, given the current value of all the other variables, both observed data and unknown parameters set at their temporary values. It can be shown that under broad conditions *eventually* one will be sampling from the correct posterior distributions of the unknown parameters. There is a large literature on this topic: both methodological [2] and applications [8].

For any node v it is therefore necessary to sample from $p(v \mid V\backslash v)$, the full conditional distribution given all other nodes $V\backslash v$. However, the factorization (1) of the joint distribution expressed by the graphical model can be exploited to obtain,

$$p(v \mid V\backslash v) \propto \text{terms in } p(V) \text{ containing } v$$

$$= p(v \mid \text{pa}(v)) \prod_{v \in \text{pa}(w)} p(w \mid \text{pa}(w))$$

The full conditional distribution of any node v therefore depends only on a prior component $p(v \mid \text{pa}(v))$ and likelihood components arising from each of its children w.

The Gibbs sampling was implemented using the BUGS software [9, 10]. The BUGS language allows the model to be specified in much the same way as it was represented by the graphical model. The specification file has two sections. The first contains a declaration of all the nodes in the graph and the names of the files containing the data and the initial values, and the second is a list of the dependence relations expressed by the graph. This is shown below. The "average" infection rate for the ith trial, μ_i, and the overall treatment effect, d, were assigned normal(0, 4) and normal(0, 10) prior distributions, respectively. We note that BUGS parameterizes the normal distribution in terms of the precision rather than the variance in order to provide a conjugate prior analysis, such that "tau" represents $\tau = 1/\sigma^2$. The gamma(0.001, 0.001) prior distribution given to τ is a noninformative prior that is approximately equivalent to $p(\sigma^2) \propto 1/\sigma^2$. Suitable priors for τ are discussed in more detail in Sections 5 and 6.

```
for (i in 1 : Num) {
        rt[i] ~ dbin(pt[i], nt[i]);
        rc[i] ~ dbin(pc[i], nc[i]);
    logit(pc[i]) < - mu[i] - (delta[i]/2);
    logit(pt[i]) < - mu[i] + (delta[i]/2);
        delta[i] ~ dnorm(d, tau);
        mu[i] ~ dnorm(0.0, 0.25);
}
        d ~ dnorm(0.0, 0.1);
        tau ~ dgamma(1.0E-3, 1.0E-3);
    sigma < - 1/sqrt(tau);
delta.new ~ dnorm(d, tau);
```

The syntax of the language should be largely self-explanatory and essentially consists of using the graph to express the joint distribution (2) as concisely as possible. Two forms of relation are shown: $\sim$ translates to "is distributed as" and $< -$ represents "is equal to." It is important to understand that the language is *declarative* (describing a model), rather than *procedural* (specifying a sequence of steps as in a standard computer program). For example, the lines may be given in any order. (Details of how to obtain the program are given at the end of this chapter.)

From this model specification BUGS works out the parents and children of each node, constructs an internal representation of the graph, derives the necessary full conditional distributions, and carries out the Gibbs sampling. In fairly simple problems of this type a "burn-in" of, say, 500 iterations is generally sufficient to reach convergence and then summary statistics such as means and standard deviations of generated parameters values are monitored over, say, 1000 further iterations: formal techniques are available for checking convergence [11] and here a method described by Geweke [12] was adopted.

The full Bayesian random-effects model specified above gives a mean estimate for d of -1.39 (95% probability interval -1.82, -1.01). Figure 2 shows the individual study estimates for the random-effects model are drawn toward a central overall effect and have smaller intervals than the fixed-effects (Mantel-Haenszel-Peto) estimates. The pooled effect, however, has a wider interval. This is the expected pattern, due to some of the *within*-study variability in the fixed-effects analysis being accounted for as *between*-study variability in the random-effects model.

To obtain an estimate of the overall or pooled effect in a meta-analysis, the individual study estimates are combined in a way that enables some studies to contribute more to the pooled estimate than others. The amount that a study contributes to the pooled estimate is determined by its *weight*. In a fixed-effects analysis the studies are given weights that are inversely proportional to the variance of the estimated study effect, so

$$\hat{d} = \frac{\sum_{j=1}^{k} \hat{\delta}_j w_j}{\sum_{j=1}^{k} w_j}, \qquad w_j = \frac{1}{\text{Var}(\hat{\delta}_j)}$$

and hence each weight w_j is proportional to study size.

In a random-effects analysis, however, the between-study variability σ^2, as well as the within-study variability, is taken into account in the weighting. If there is little heterogeneity between studies, then σ^2 will be small and the fixed-effects and random-effects estimates will be very similar. However, as the heterogeneity increases we will find that the weighting in the random-effects analysis will be less influenced by the within-study variance and become dominated by the between-study variance.

In the preceding example we see that random-effects analysis estimates a stronger relationship between infections and selective decontamination upon pooling. The random-effects analysis is placing less weight

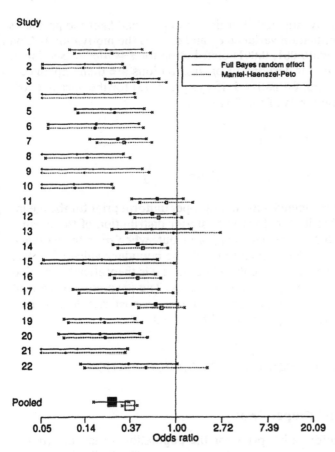

Figure 2 Estimated odd ratios for Bayesian random-effects model and Mantel-Haenszel-Peto method (area of mark is proportional to the sample size of the study).

on larger studies (with small variances) than the fixed-effect method due to the presence of heterogeneity, and hence this shift in estimated pooled log odds ratios is a result of the larger studies tending to have less strong treatment effects.

5 ISSUES WITH IMPROPER PRIORS IN HIERARCHICAL MODELS

It is important that appropriate consideration is given to the choice of prior distributions and that the effect they have on estimates of the parameters of interest is investigated. In this section we highlight the problems of priors for the population variance σ^2 and discuss the motivation behind the use of the gamma(0.001, 0.001) distribution for $\tau = 1/\sigma^2$ in Section 4. Alternative methods for deriving prior distributions are then explained.

The standard noninformative Jeffreys prior [13] for the variance, σ^2, of a normal distribution is of the form

$$p(\sigma^2) \propto \frac{1}{\sigma^2}$$

or equivalently $\quad p(\tau) \propto 1/\tau, \quad$ where $\tau = \dfrac{1}{\sigma^2}$

which leads, when combined with an improper uniform prior for the mean, to the classical t-distribution for the posterior distribution of the mean.

However, as DuMouchel [14, 15] points out, when σ^2 is the variance of a random effect in a hierarchical model, the boundary value $\sigma^2 = 0$ is supported by a nonnegligible likelihood because it is theoretically possible that there are no trial-specific random effects. The asymptote at zero of $p(\sigma^2) \propto 1/\sigma^2$ is then sufficient to lead to an *improper* posterior distribution, and so suitable alternative priors should be used.

6 ALTERNATIVE PROPER PRIORS FOR POPULATION VARIANCE

6.1 Prior 1: "Just" Proper Prior

The standard noninformative prior for the population variance, $p(\sigma^2) \propto 1/\sigma^2$, is formally equivalent to an inverse gamma(0, 0) distribution. A "just" proper approximation to this is an inverse gamma(0.001, 0.001) distribution for σ^2 (Figure 3), or equivalently a gamma(0.001, 0.001) distribution for $\tau = 1/\sigma^2$. This was the prior used in Section 4.

6.2 Prior 2: Proper Prior by Introspection

By using knowledge of the particular context of the problem being analyzed a reasonable prior may be derived based on judgments about the likely size of between-study variability.

First, suppose that before looking at the data we consider it is plausible to observe one order of magnitude spread in odds ratios between the studies, so that the ratio of the maximum odds ratio to the minimum odds ratio could be 10. Converting to a log scale, this can be interpreted as having a prior belief that 95% of studies (contained in a range $\pm 1.96\sigma$)

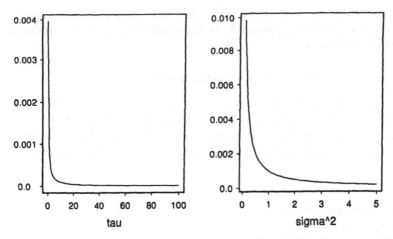

Figure 3 Inverse gamma (0.001, 0.001) distribution for σ^2.

cover log odds ratios in a range of log 10 = 2.3, and hence that a reasonable estimate of σ^2 is $(2.3/(2*1.96))^2 = 0.34$. Suppose in addition that we believe it would be very unlikely to observe two orders of magnitude difference (i.e., a hundredfold difference) between odds ratios in a meta-analysis. The ratio of the maximum odds ratio to the minimum odds ratio would then be 100 and this would lead to a "high" value of σ^2 of 1.37. A gamma(3, 1)

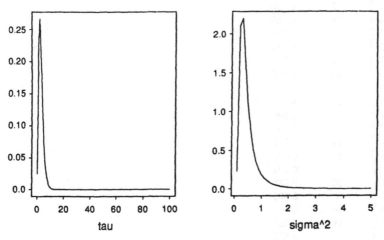

Figure 4 Inverse gamma (3, 1) distribution for σ^2.

distribution has a mean of 1/0.33 and a 96% probability of exceeding 1/1.37, and hence an inverse gamma(3, 1) distribution for σ^2 could represent these beliefs (Figure 4).

6.3 Prior 3: Proper Prior by Empirical Methods

From a large number of meta-analyses, each providing an estimate of the between-study variability, a prior for σ^2 can be obtained empirically. Parmar *et al.* [16] reviewed 30 meta-analyses of trials in a number of areas (Table 2) and the resulting estimates of σ^2 were used to find an inverse

Table 2   Estimates of $\hat{\sigma}^2$ from 30 Meta-analyses [16]

	No. of trials	$\hat{\sigma}^2$
1	21	0.044
2	5	0.055
3	21	0.000
4	64	0.145
5	7	0.918
6	8	0.005
7	21	0.000
8	26	0.000
9	4	0.000
10	6	0.063
11	13	0.111
12	13	0.012
13	7	0.023
14	10	0.109
15	7	0.011
16	11	0.477
17	6	1.297
18	19	0.056
19	11	1.352
20	21	0.000
21	31	0.000
22	28	0.000
23	31	0.016
24	9	0.000
25	23	0.004
26	20	0.076
27	8	0.000
28	10	0.064
29	14	0.327
30	8	0.533

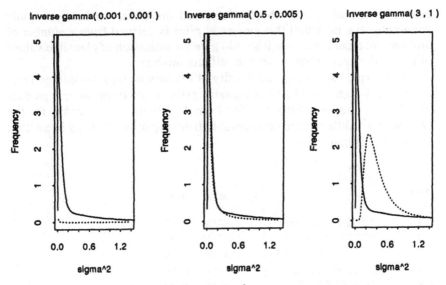

Figure 5 Kernel estimate of density for $\hat{\sigma}^2$ from meta-analyses of cancer trials (solid line = kernel estimate, dotted line = inverse gamma density).

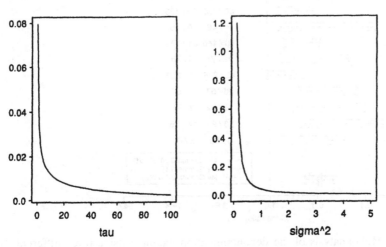

Figure 6 Inverse gamma (0.5, 0.005) distribution for σ^2.

gamma distribution that "best" described the observed between-study variability. We note that this empirical prior is derived from a number of different medical areas, but it should give an indication of plausible values of population variances in random-effects analyses.

The kernel estimate of the density of the sample was considered (Figure 5), and it suggested that a reasonable prior would be an inverse gamma (0.5, 0.005), shown in Figure 6. The kernel estimate of the density for the sample of $\hat{\sigma}^2$s differs quite considerably from the priors derived in Section 6.1 and 6.2 (Figure 5).

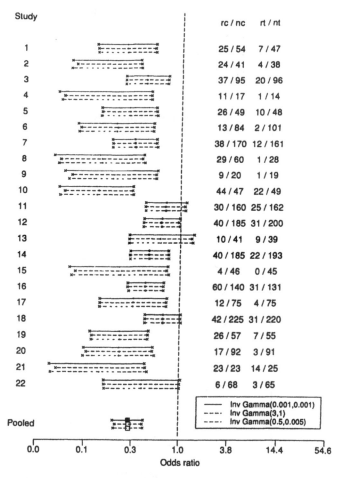

Figure 7 Meta-analysis of the decontamination example using three different priors for σ^2. Posterior mean and 95% credible intervals obtained from Gibbs sampling.

7 COMPARISON OF ALTERNATIVE PRIORS

The three different priors gave almost identical results when applied to the meta-analysis of the selective decontamination example (Figure 7). The only difference was that the estimated credible intervals from the introspective inverse gamma(3, 1) were slightly narrower compared with those from the other two priors (particularly for the smaller study estimates). This is due to the just proper prior and the data-driven prior both favoring small values of between-study variability σ^2, and therefore more of the residual variability is accounted for as within-study variability.

Although the three approaches led to different inverse gamma priors, there was actually little practical difference between the distributions. This suggests that the just proper prior may well be a reasonable prior for σ^2, as it provides reasonable support to a wide range of plausible values of σ^2.

8 DISCUSSION

The selective decontamination of the digestive tract meta-analysis was used here to highlight a number of problems that frequently occur in meta-analysis. In particular, any variability between studies should be taken into account in the analysis. The fixed-effect analysis estimated a 64% reduction in infections with selective decontamination compared with the controls (odds ratio 0.36, 95% confidence interval 0.31, 0.43). When the random-effects model is used, however, a stronger relationship is found with an estimated 75% reduction in infections (odds ratio 0.25, 95% probability interval 0.16, 0.36). Because the random-effects analysis acknowledges the presence of heterogeneity, the uncertainty of estimating a pooled effect when individual study estimates vary greatly is reflected in its much wider 95% interval.

When heterogeneity among the studies in a meta-analysis is apparent, possible differences between the studies that could be instrumental in causing the heterogeneity could be adjusted for. In the decontamination example the impression gained from Figure 2 is that the larger studies have the less extreme results. In the Bayesian random-effects model we could easily include the log of the study size ($n_i = n_i^T + n_i^C$) as a covariate so that

$$\text{logit}(p_i^T) - \text{logit}(p_i^C) = \delta_i + \beta(\log n_i - \log n.)$$

where $\log n. = (1/I) \sum_i \log n_i$. The estimated overall treatment effect then has the slightly odd interpretation as the effect expected in an "average" size trial.

Other factors that could explain the difference in the effects of the studies include the variations in patient mix both within and between the studies. For example, the percentages of trauma, surgical, and medical patients varied widely. Differences in study design (whether they were double blind or not and variations in diagnostic measures) and treatment regimens could also have been important contributors. Heterogeneity should be explained, if possible, and not simply accommodated within a random-effects analysis.

Although the fully Bayesian model described here enables adjustments that can account for some of the between-study variability, interpretation requires care. In the example above it may be felt that the larger studies carry more credibility and we might expect carefully controlled studies to tend to show smaller effects. The larger centers are also more likely to stay up with recent developments and use more standard definitions of disease and treatment. The more extreme effects being observed in smaller studies might also point to publication bias, as small studies with small observed effects may not have been significant enough to be published.

The three population variance priors compared in this chapter, although themselves specific to the particular model, were derived using approaches that can be used to develop priors in many other areas.

This chapter illustrates just one application of BUGS. The software is capable of handling a wide range of problems, including hierarchical random effects and measurement error in generalized linear models, latent variable and mixture models, and various forms of missing and censored data. It also provides a mechanism for attaching models of different types within a single structure: essentially the graphical formalism permits models of arbitrary complexity. Naturally, for complex models issues of convergence of the sampled values becomes crucial, and a range of diagnostics are possible. BUGS is freely available, with full documentation, by anonymous ftp from the second author (e-mail: bugs@mrc-bsu.cam.ac.uk).

ACKNOWLEDGMENT

The BUGS project is supported by a grant from the UK Economic and Social Science Research Council's initiative for the Analysis of Large and Complex Datasets.

REFERENCES

1. Selective Decontamination of the Digestive Tract Trialists' Collaborative Group. (1993). Meta-analysis of randomised controlled trials of selective decontamination of the digestive tract. *British Medical Journal* 307: 525–532.

2. Gelfand, A. E. and Smith, A. F. M. (1990). Sampling-based approaches to calculating marginal densities. *Journal of the American Statistical Association* 85: 398–409.
3. Antiplatelet Trialists' Collaboration. (1988). Secondary prevention of vascular disease by prolonged antiplatelet treatments. *British Medical Journal* 296: 320–331.
4. DerSimonian, R. and Laird, N. (1986). Meta-analysis in clinical trials. *Controlled Clinical Trials* 7: 177–188.
5. Whittaker, J. (1990). *Graphical Models in Applied Multivariate Analysis.* John Wiley & Sons, Chichester, UK.
6. Lauritzen, S. L., Dawid, A. P., Larsen, B. N., and Leimer, H.-G. (1990). Independence properties of directed Markov fields. *Networks* 20: 491–505.
7. Geman, S. and Geman, D. (1984). Stochastic relaxation, Gibbs distributions, and the Bayesian restoration of images. *IEEE Transactions on Pattern Analysis and Machine Intelligence* 6: 721–741.
8. Gilks, W. R., Clayton, D. G., Spiegelhalter, D. J., Best, N. G., McNeil, A. J., Sharples, L. D., and Kirby, A. J. (1993). Modelling complexity: applications of Gibbs sampling in medicine (with discussion). *Journal of the Royal Statistical Society Series B* 55: 39–102.
9. Thomas, A., Spiegelhalter, D. J., and Gilks, W. R. (1992). BUGS: A program to perform Bayesian inference using Gibbs sampling. In *Bayesian Statistics 4*, ed. Bernardo, J. M., Berger, J. O., Dawid, A. P., and Smith, A. F. M. pp. 837–842. Clarendon Press, Oxford.
10. Gilks, W. R., Thomas, A., and Spiegelhalter, D. J. (1994). A language and program for complex Bayesian modelling. *The Statistician*, 43: 169–178.
11. Gelman, A. and Rubin, D. B. (1992). Inference from iterative simulation using multiple sequences. *Statistical Science* 7: 457–472.
12. Geweke, J. (1992). Evaluating the accuracy of sampling-based approaches to the calculation of posterior moments. In *Bayesian Statistics 4*, ed. Bernardo, J. M., Berger, J. O., Dawid, A. P., and Smith, A. F. M., pp. 169–194. Clarendon Press, Oxford.
13. Box, G. E. P. and Tiao, G. C. (1973). *Bayesian Inference in Statistical Analysis.* Addison-Wesley, Reading, MA.
14. DuMouchel, W. H. (1990). Bayesian meta-analysis. In *Statistical Methodology in the Pharmaceutical Sciences*, ed. Berry, D. A., 509–529. Marcel Dekker, New York.
15. DuMouchel, W. and Waternaux, C. (1992). Hierarchical models for combining information and for meta-analyses (discussion). In *Bayesian Statistics 4*, ed. Bernardo, J. M., Berger, J. O., Dawid, A. P., and Smith, A. F. M., pp. 338–341. Clarendon Press, Oxford.
16. Parmar, M. H., Hughes, M., and Freedman, L. S. (1995). Heterogeneity and small trials in meta-analysis. In preparation.

16
Hierarchical Analysis of Continuous-Time Survival Models

Dalene K. Stangl   Duke University, Durham, North Carolina

ABSTRACT

This chapter considers the problem of pooling information across study sites in a multicenter clinical trial designed to assess the prophylactic effects of imipramine hydrochloride in preventing the recurrence of affective disorders. Assuming exponentially distributed survival times, two- and three-stage Bayesian hierarchical models are used to model the heterogeneity between study sites. Advantages and disadvantages of each approach are discussed. Laplace approximations are used to calculate posterior distributions. Results show clear quantitative, but not qualitative, interaction between centers and treatment effects.

1 INTRODUCTION

In multicenter clinical trials there is often considerable heterogeneity among centers not only in terms of patient populations but also in terms of protocol interpretation and adherence. Because of these differences, inter- and intracenter analysis can be a rich source of information about for whom and how treatments are working. At the same time, incorporating treatment center as a covariate can be difficult because of the increased complexity of the statistical model and small sample sizes within centers. Fleiss (1986) discusses the controversy over how these data are analyzed and points out the paucity of available statistical methodology. He posits that pooling, in the sense of averaging within-clinic differences, is almost always justified and that pooling, in the sense of throwing all the data together, is only rarely justified.

Numerous tests for heterogeneity in treatment response across subgroups have been derived. The simplest and most commonly used is based on comparing odds ratios across m 2 × 2 tables (see Mantel and Haenszel, 1959). Paul and Donner (1989) conducted a simulation study to compare the performance of nine such procedures. Albert (1987) used a hierarchical model to compare odds ratios from m 2 × 2 tables. In the first stage, the distribution of the data in each group was binomial, and in the second stage the log odds of each group was distributed independently and identically (i.i.d.) $N(\mu, \sigma^2)$. He then described the behavior of the posterior modes of the log odds ratios in terms of shrinkage toward the model of homogeneity. Using a similar model, Skene and Wakefield (1990) examined whether the relative efficacy of two treatments varied among m study centers. Both Albert and Skene consider only a binary response variable, and neither considers censored observations.

The purpose of this chapter is to apply analogous hierarchical methodology to survival models with a continuous-time response variable and censoring. The chapter begins by introducing the Collaborative Study of Long-Term Maintenance Drug Therapy in Recurrent Affective Illness, a multicenter clinical trial funded by the National Institute for Mental Health's Pharmacologic Research Branch (Prien et al., 1984). A general discussion of Bayesian hierarchical models and numerical methods useful in solving integration difficulties that arise in such models follows. Finally, the proposed methodologies are carried out on the data from the National Institute of Mental Health (NIMH) collaborative maintenance therapy study.

2 EXAMPLE: RECURRENCE OF DEPRESSION

Psychiatric research has shown that depression is a recurrent illness (see Nystrom, 1979; Zis and Goodwin, 1979). Most patients who suffer from an initial episode of depression will recover but then go on to suffer one or more recurrences. The efficacy of medications in preventing the occurrence of new episodes of illness has been studied far less adequately than treatment directed at the acute episode. To fill this gap, the National Institute of Mental Health funded a multicenter clinical trial designed to determine the comparative efficacies of lithium carbonate, imipramine hydrochloride, and a combination of lithium and imipramine in preventing the recurrence of unipolar and bipolar affective disorders. As presented in Figure 1, the design of the study consisted of two phases: a preliminary phase and a maintenance phase. The purpose of the preliminary phase was to control the index episode, stabilize the patient's clinical condition, and establish stable maintenance dose levels of lithium plus imipramine in preparation for the maintenance phase. Upon meeting specified entrance criteria that ensured that the patient was in an acute episode, patients received the treatment of choice of the psychiatrist responsible for preliminary phase care. After acute symptoms were controlled, the patient received a combination of both lithium and imipramine. Once the patient remained on predetermined medication dosages and met specified "wellness" criteria for two consecutive months, the patient entered the maintenance phase of the study. The maintenance phase was the major experimental phase of the study and involved a 2-year double-blind comparison of the treatments involved. One hundred and fifty patients were eligible for the maintenance phase of the study and were randomized into a group that remained on imipramine or a group that was withdrawn from imipramine. After randomization, patients were followed for up to 2 years, the end of the study, or until they experienced a recurrence of depression. The response variable of interest is time between randomization and the first recurrence of a depressive episode. Five centers participated in the

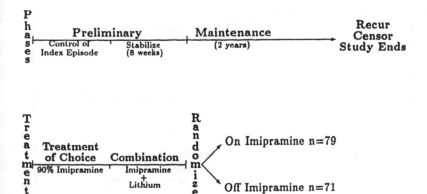

Figure 1 Study design.

project. The original analysis is presented in Prien *et al.* (1984) and several design issues are explored in Greenhouse *et al.* (1991).

The Kaplan-Meier curves for all centers combined are presented in Figure 2. This figure shows that the separation in the two survival curves occurs in the first 8 to 10 weeks. There are many recurrences in the off-imipramine group during this time. After this time the survival curves appear to drop at approximately the same rate. Figure 3 shows the Kaplan-Meier curves for the two centers with the largest sample sizes. It is seen from this figure that the centers differ in several ways. For center D, the cumulative survival at 52 weeks was 0.69 ± 0.10 for the on-imipramine group and 0.57 ± 0.10 for the off-imipramine group. This yields a difference between treatments of 0.12 ± 0.14. For center E, the cumulative survival at 52 weeks was 0.42 ± 0.13 for the on-imipramine group and 0.08 ± 0.08 for the off-imipramine group. This yields a difference between treatments of 0.34 ± 0.15. A more dramatic difference between the centers is seen in the first few weeks. At 10 weeks we see very little difference

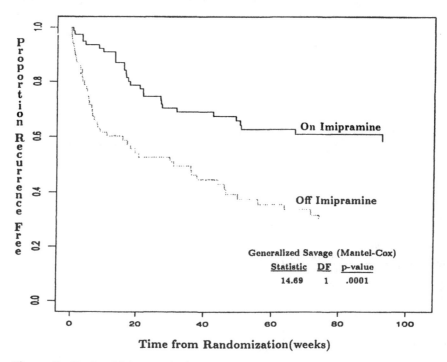

Figure 2 Kaplan-Meier survival curves.

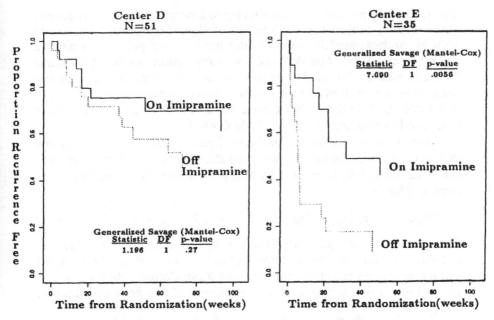

Figure 3 Kaplan-Meier survival curves for centers D and E.

between the treatment groups at center D. The cumulative survival of the on- and off-imipramine groups, respectively, are 0.92 ± 0.05 and 0.84 ± 0.07. The difference between treatments is 0.08 ± 0.08. However, at 10 weeks there is more than a 50% difference between the treatment groups at center E. The cumulative survival at 10 weeks for the on-imipramine group at center E is 0.83 ± 0.09 and for the off-imipramine group is 0.29 ± 0.11. This yields a treatment difference of 0.54 ± 0.14. In summary, this figure suggests differences in treatment response between centers D and E. These plots show that study center differences warrant further attention. The goal is to develop general methods both to assess the variability between centers in terms of the relative treatment effect and to get estimates of the survival distribution for each treatment group within each center as well as for the population in general.

3 HIERARCHICAL MODELS

Modeling subgroup heterogeneity can be complicated. Often this complexity can be reduced with a sequence of relatively simple models placed in a hierarchy. Such is the case when trying to incorporate the heterogeneity between study centers into the analysis of clinical trial data. One way to view hierarchical models is to consider not only the patients as being a sample from a population but also the study centers as being a sample from a population. The goal is to estimate parameters specific to particular study centers as well as population parameters. Comparison of the posterior distributions of the study center parameters allows assessment of the degree of heterogeneity between study centers.

Suppose observations are gathered from m study centers. Assume each of the n_i observations, t_{ik}, observed at center i is distributed independently and identically from a family of distributions $f(\cdot)$ indexed by parameters θ_i. That is,

Stage I: $\quad t_{ik} \mid \theta_i \sim f(t_{ik} \mid \theta_i), \qquad i = 1,2,\ldots m, k = 1,\ldots n_i$

At the second stage a joint distribution is specified for the θ_i's. This second-stage distribution can be thought of in two ways. It can be assumed that the m centers are also a sample from a population with a particular distribution, or the second stage can be thought of as a prior distribution.

Stage II: $\quad \theta_1, \theta_2, \ldots, \theta_m \mid \lambda \sim g(\theta_1, \theta_2, \ldots, \theta_m \mid \lambda)$

If a priori it is believed that the effectiveness of each treatment will be the same across centers, and the joint density is invariant to any permutation of the subscripts, then the θ_i can be modeled as exchangeable. Otherwise, appropriate adjustments must be made in the joint distribution. Before proceeding it must be decided how to deal with the second-stage hyperparameters. Two alternatives include directly specifying a third-stage distribution for λ or estimating λ from the data. The former alternative will henceforth be referred to as a three-stage Bayesian model and the latter alternative the two-stage or empirical Bayes model. A third stage would simply be

Stage III: $\quad \lambda \mid \gamma \sim h(\lambda \mid \gamma)$

If little or nothing is known about the third-stage distribution, diffuse distributions, i.e., distributions with large variances, can be used. Proper posteriors typically require that the diffuse distribution be restricted to a compact space. In contrast to the parametric empirical Bayes methodology that assumes no uncertainty about the λ's, the use of diffuse priors can incorporate as much uncertainty as desired. Excellent introductions to the parametric empirical Bayes methodology are provided by Deeley and Lindley (1981), Morris (1983), and Casella (1985).

Reasons for choosing one of these methodologies over the other depend on the goal of the analysis and resources. The three-stage model requires more extensive modeling of prior information via the third stage. The parametric empirical Bayes methodology requires only specification of the form of the second-stage distribution. Second-stage parameters are estimated from the data. Unless noninformative priors are used, the three-stage model requires specification of both the form of the distribution and the distribution parameters. While this may be difficult, the payoff is posterior distributions for the population (second-stage) parameters. These distributions are not available in the empirical Bayes methodology. A disadvantage of the empirical Bayes methodology is that posterior variances are underestimated. By using the maximum-likelihood estimates of the second-stage parameters, we ignore the uncertainty in their estimation. The result is underestimated variances in the posterior distributions. Kass and Steffey (1990) derive a second-order approximation to the true posterior variance. This approximation can be used to determine the degree to which the posterior variance is underestimated.

4 NUMERICAL METHODS

The calculation of posterior distributions requires multidimensional integration. In hierarchical models the dimension of the integrals increases quickly, and unless conjugate distributions are used, these integrations can be quite difficult. Because of this, approximation methods are helpful. One such method is the Laplace method adapted by Tierney and Kadane (1986) and exemplified in hierarchical models by Kass and Steffey (1989). The Laplace method is suited for integration problems in which the integrand is smooth and has a single dominant peak in a region that is roughly known. If h is a smooth function of a k-dimensional parameter θ with $-h$ having a maximum at $\hat{\theta}$, Laplace's method approximates an integral of the form

$$I = \int f(\theta) \exp(-nh(\theta)) \, d\theta$$

by expanding h and f about $\hat{\theta}$. The result is

$$\hat{I} = \left(\frac{2\pi}{n}\right)^{k/2} \det(\Sigma)^{1/2} f(\hat{\theta}) \exp(-nh(\hat{\theta}))$$

where $\Sigma^{-1} = D^2 h(\hat{\theta})$, the Hessian of h at $\hat{\theta}$. Often, Bayesian computations require computation of a ratio of such integrals. In these cases the Laplace method is applied to the numerator and denominator. For more detailed discussion, readers are referred to Tierney and Kadane (1986) and Tierney *et al.* (1989).

Another method for approximating posterior distribution is through substitution sampling. This approach was developed and illustrated in Gelfand and Smith (1990) and Gelfand *et al.* (1990) and has become known as the Gibbs sample algorithm. This methodology can be used when the joint density of a collection of random variables can be uniquely determined by the full conditional densities. For examples, suppose the joint density $f(\theta_1, \theta_2, \theta_3)$ is unknown but that the three full conditional posterior distributions $f_1(\theta_1 \mid \theta_2, \theta_3, y)$, $f_2(\theta_2 \mid \theta_1, \theta_3, y)$ and $f_3(\theta_3 \mid \theta_1, \theta_2, y)$ are known. By iteratively sampling from the conditional posterior distributions a sample from the joint posterior $f(\theta_1, \theta_2, \theta_3) \mid y)$ is obtained. Through Monte Carlo integration this sample is used to estimate the posterior marginal distributions.

5 PARAMETRIC EMPIRICAL BAYES

The initial analysis of the data from the Maintenance Therapy Trial was carried out using an empiricial Bayes methodology. For each treatment group (on-imipramine and off-imipramine) the first-stage distribution is modeled by exponential survival times.
Stage I:

$$f(t_{ijk} \mid \theta_{ij}) = \theta_{ij} e^{-\theta_{ij} t_{ijk}}, \qquad \underbrace{i = 1, \ldots, 5}_{\text{center}}, \underbrace{j = 1, 2}_{\text{treatment}}, \underbrace{k = 1, \ldots, n_{ij}}_{\text{subjects}}$$

In the second stage the θ_{ij} within treatments are modeled as independent and identically distributed gamma random variables.
Stage II:

$$g(\theta_{1j}, \theta_{2j}, \theta_{3j}, \theta_{4j}, \theta_{5j} \mid \alpha_j, \beta_j) = \prod_{i=1}^{5} \frac{\beta_j^{\alpha_j}}{\Gamma(\alpha_j)} \theta_{ij}^{\alpha_j - 1} e^{-\beta_j \theta_{ij}}$$

Due to the conjugacy of this distribution the posterior distribution of θ_{ij} is gamma$(\alpha_j + n_{ij}, \beta_j + n_{ij} \bar{t}_{ij.})$ with density

$$f(\theta_{ij} \mid t, \alpha_j, \beta_j) = \frac{(\beta_j + n_{ij} \bar{t}_{ij.})^{\alpha_j + n_{ij}}}{\Gamma(\alpha_j + n_{ij})} \theta_{ij}^{\alpha_j + n_{ij} - 1} e^{-\theta_{ij}(\beta_j + n_{ij} \bar{t}_{ij.})}$$

To estimate α_j and β_j the likelihood is written in terms of α_j and β_j as

$$L(\mathbf{t} \mid \alpha_j, \beta_j) = \prod_{i=1}^{5} \int_{\theta_{ij}} \left[\prod_{k=1}^{n^{ij}} f(t_{ijk} \mid \theta_{ij}, \alpha_j, \beta_j) \right] f(\theta_{ij} \mid \alpha_j, \beta_j) \, d\theta_{ij}$$

$$= \prod_{i=1}^{5} \frac{\beta_j^{\alpha_j}}{\Gamma(\alpha_j)} \frac{\Gamma(\alpha_j + n_{ij})}{(\beta_j + n_{ij}t_{ij}.)^{\alpha_j + n_{ij}}}$$

Using numerical techniques one can find the joint estimates of α_j and β_j that maximize this likelihood. Upon substituting these estimates, $\hat{\alpha}_j$ and $\hat{\beta}_j$, into the conditional distribution of θ_{ij} we obtain a gamma distribution with mean

$$\tilde{\theta}_{ij} = \frac{\hat{\beta}_j}{\hat{\beta}_j + \sum\limits_{k=1}^{n^{ij}} t_{ijk}} \frac{\hat{\alpha}_j}{\hat{\beta}_j} + \frac{\sum\limits_{k=1}^{n^{ij}} t_{ijk}}{\hat{\beta}_j + \sum\limits_{k=1}^{n^{ij}} t_{ijk}} \hat{\theta}_{ij}$$

where $\hat{\alpha}_j/\hat{\beta}_j$ is the second-stage mean evaluated at the maximum-likelihood estimates of α_j and β_j, and $\hat{\theta}_{ij} = n_{ij}/\sum_{k=1}^{n_{ij}} t_{ijk}$ is the maximum-likelihood estimate of the hazard rate using only observations from center i. Hence θ_{ij} is a weighted average of a "pooled" hazard rate and the ith center's hazard rate. Note that $\hat{\alpha}_j$ and $\hat{\beta}_j$ are estimated using all the data. The weights depend on β_j and $\sum_{k=1}^{n^{ij}} t_{ij}$. As $\sum_{k=1}^{n^{ij}} t_{ij}$ increases more weight is put on the individual center's hazard rate.

Analogous results are derived when allowing for censored observations. Survival times again are exponential, but $\theta_{ij} = \sum_{k=1}^{n^i} \delta_{ijk}/\sum_{k=1}^{n^i} t_{ijk}$, where δ_{ijk} is an indicator of whether the kth observation from center i is a failure or censored. That is, $\delta_{ijk} = 1$ if the observation fails, and $\delta_{ijk} = 0$ if the observation is censored. Again using a gamma prior and the maximum-likelihood estimates $\hat{\alpha}_j$ and $\hat{\beta}_j$, the posterior of θ_{ij} is gamma with parameters $\hat{\alpha}_j + \sum_{k=1}^{n_{ij}} \delta_{ijk}$ and $\hat{\beta}_j + \sum_{k=1}^{n_{ij}} t_{ijk}$. Hence the posterior mean is $\bar{\theta} = (\hat{\alpha}_j + \sum_{k=1}^{n_{ij}} \delta_{ijk})/(\hat{\beta}_j + \sum_{k=1}^{n_{ij}} t_{ijk})$.

The results of the empirical Bayes analysis of the Maintenance Therapy Trial are presented in Table 1 and in Figures 4 and 5. Table 1 displays the individual center recurrence rates (column 5), the pooled recurrence rate (column 6), and the expectation of the individual center posterior recurrence rates (column 7). The shrinkage toward the pooled rate is clearly shown. Looking at the on-imipramine group, we see that center

Table 1 Prior and Posterior Expectations for the Center Hazard Rates

Center	n_{ij}	$\sum_{k=1}^{n_{ij}} t_{ijk}$	$\sum_{k=1}^{n_{ij}} \delta_{ijk}$	$\theta_{ij} = \dfrac{\sum_{k=1}^{n_{ij}} \delta_{ijk}}{\sum_{k=1}^{n_{ij}} t_{ijk}}$	$\dfrac{\hat{\alpha}_j}{\hat{\beta}_j}$	$E\left(\theta_{ij}\mid t, \hat{\alpha}_j, \hat{\beta}_j\right)$
On Imipramine						
A	15	1205	3	0.002	.006	0.005
B	15	770	6	0.007	.006	0.006
C	5	332	3	0.009	.006	0.006
D	26	1503	8	0.005	.006	0.006
E	18	846	9	0.010	.006	0.007
Off Imipramine						
A	11	499	6	0.012	.025	0.014
B	10	391	6	0.015	.025	0.017
C	8	177	7	0.039	.025	0.035
D	25	1367	12	0.008	.025	0.009
E	17	240	15	0.062	.025	0.052

A and center D have recurrence rates less than the pooled rate of .006. The posterior recurrence rates for these two centers are pulled up toward the pooled rate. For centers B, C, and E, which have recurrence rates less than the pooled rate, the posterior rates are pulled down toward the pooled rate. Comparing the variability in recurrence rates in column 5 to the variability in recurrence rates in column 7, we see that the heterogeneity between centers has been reduced. This effect is seen again in the off-imipramine group; however, more variability remains between the posterior recurrence rates for this group.

Figure 4 shows the posterior distributions for the individual center hazard rates of each group. The first plot displays the posterior distributions of the θ_{ij} for the on-imipramine group, and the second plot shows these distributions for the patients taken off imipramine. In the upper right-hand corner of each plot the 95% highest posterior density (HPD) region and the posterior mode of each density are presented. For patients remaining on imipramine the mode for each center's hazard is between .0049 and .0072. There is much similarity between these distributions. On the other hand, clear differences are seen between the posterior densities of the hazard rates for the off-imipramine group. These densities show

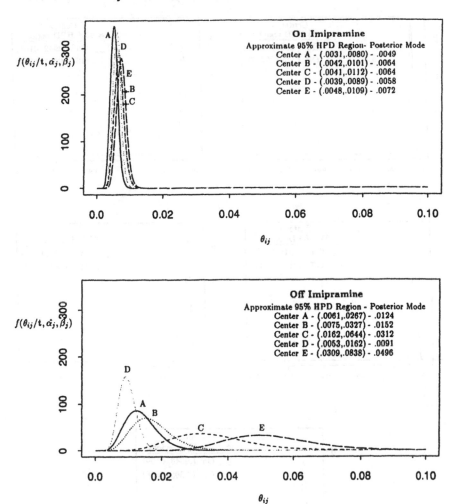

Figure 4 A comparison of posterior densities of θ_{ij} between centers.

much more variability within centers as well as much more heterogeneity between the centers. The posterior modes range from .0091 for center D to .0496 for center E. The posterior distributions for centers D and E are quite distinct. It is also of interest to note that the off-imipramine group at center D appears to have done as well as the on-imipramine group at centers C and E.

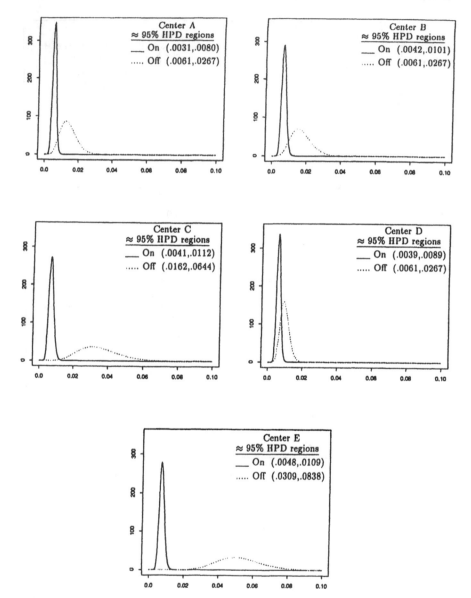

Figure 5 A comparison of posterior densities of θ_{i1} and θ_{i2} within centers.

As previously mentioned, one problem with the empirical Bayes methodology is that posterior variances are underestimated. To examine the impact of underestimating the posterior variances, posterior distributions were recomputed in the following manner. The parameters for the posterior gamma distributions were altered in such a way that the posterior modes were left unchanged, but posterior variances were increased to include the uncertainty in estimating the prior parameters. That is, the equations

$$m = \frac{\phi_j - 1}{\psi_j} \tag{1}$$

$$v = \frac{\phi_j}{\psi_j^2} \tag{2}$$

where m is the mode of the empirical Bayes posterior gamma distribution and v is the approximation to the variance presented by Kass and Steffey (1989), were solved for ϕ_j and ψ_j, the parameters of the revised posterior gamma distribution. Because the revised posterior distributions were only slightly different from the original, they are not presented here. The same conclusions are reached. Increasing the variances for the on-imipramine group makes the conclusion of no difference between centers even more clear. Likewise, increasing the variances for the off-imipramine group did not alter the conclusion of clear differences between the centers.

Figure 5 shows a comparison of the on- and off-imipramine groups within study centers. Across all centers we see that the posterior density for the on-imipramine groups is consistently to the left of that for the off-imipramine group, reflecting longer survival times for patients remaining on imipramine, but the degree of difference varies. At center D there is much overlap between the two groups, whereas at center E there is very little. Hence, although there appears to be no qualitative interaction in the relative efficacy of the two treatments across centers, there does appear to be a quantitative one. That is, across centers imipramine appears to prolong the time to recurrence; however, the degree to which it prolongs recurrence varies.

We can get a more direct assessment of the heterogeneity in relative treatment effect between centers if we examine the posterior distribution of the logarithm of the ratio of the hazard rates for the two groups, $\zeta_i = \log(\theta_{i1}/\theta_{i2})$. That is, θ_{i1} is the hazard rate for patients on imipramine, and θ_{i2} is the hazard rate for patients taken off imipramine. Since the posterior distributions of θ_{i1} and θ_{i2} are independent gamma distributions, we can write the joint distribution as

$$f(\theta_{i1}, \theta_{i2} \mid t, \alpha_j, \beta_j)$$

$$= \prod_{j=1}^{2} \frac{\left(\beta_j + \sum_{k=1}^{n_{ij}} t_{ijk}\right)^{\alpha_j + \delta_{ij}}}{\Gamma(\alpha_j + \delta_{ij})} \theta_{ij}^{\alpha_j + \delta_{ij} - 1} \exp\left[-\theta_{ij}\left(\beta_j + \sum_{k=1}^{n_{ij}} t_{ijk}\right)\right]$$

Reparameterizing to $\zeta_i = \log(\theta_{i1}/\theta_{i2})$ and $\eta_i = \log(\theta_{i2})$, finding the joint distribution of ζ_i and η_i, and integrating out η_i leaves the posterior distribution of ζ_i:

$$f(\zeta_i \mid t, \alpha_j, \beta_j) = \frac{\Gamma(\alpha_1 + \alpha_2 + \delta_{i1} + \delta_{i2})}{\Gamma(\alpha_1 + \delta_{i1})\Gamma(\alpha_2 + \delta_{i2})}$$

$$\frac{\left(\beta_1 + \sum_{k=1}^{n_{i1}} t_{i1k}\right)^{\alpha_1 + \delta_{i1}} \left(\beta_2 + \sum_{k=1}^{n_{i2}} t_{i2k}\right)^{\alpha_2 + \delta_{i2}}}{\left[e^{\zeta_i}\left(\beta_1 + \sum_{k=1}^{n_{i1}} t_{i1k}\right) + \left(\beta_2 + \sum_{k=1}^{n_{i2}} t_{i2k}\right)\right]^{\alpha_1 + \alpha_2 + \delta_{i1} + \delta_{i2}}} e^{\zeta_i(\alpha_1 + \delta_{i1})}$$

Figure 6 displays the posterior distributions of the ζ_i for the five centers. Here again we see the heterogeneity in the relative treatment effects. While all centers show a smaller hazard rate for the on-imipramine group (i.e., most of the posterior probability is on $\zeta_i < 0$), center E with the leftmost posterior distribution shows the largest treatment difference, and center D with the rightmost posterior distribution shows the smallest treat-

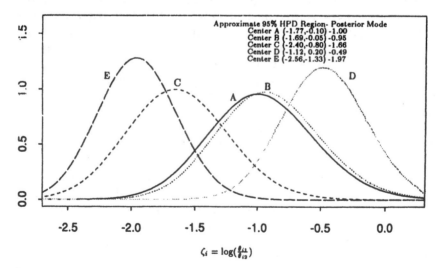

Figure 6 A comparison of posterior densities of relative treatment effect across centers.

ment difference. Using the posterior modes to estimate the relative treatment effects, we see that recurrence times for patients on imipramine are almost 6 times longer than for patients taken off imipramine at center E but only about 1.5 times longer at center D. The 95% highest posterior density regions for centers D and E do not overlap. Using the value of ζ_i at which the posterior distributions for centers D and E meet, $\zeta_i = -1.2$, we have $P(\zeta_E > -1.2) \approx .01$ and $P(\zeta_D < -1.2) \approx .01$. From this we would conclude that there are clear differences between the study centers in terms of the relative treatment effect.

6 THREE-STAGE MODEL

Because the relative treatment effect is of primary interest, the three-stage model will be parameterized in a way that allows direct assessment of the heterogeneity in the relative treatment effect. Hence focus will be placed on the posterior distribution of $\zeta_i = \log(\theta_{i1}/\theta_{i2})$. Let n_{ij} be the number of patients recruited at center i, assigned treatment j. Let δ_{ij} be the number of patients at center i experiencing a recurrence on treatment j and $\bar{t}_{ij}$ be the mean of the observation times at center i for treatment j. Again it will be assumed that recurrences within each treatment-center combination follow independent exponential distributions with parameters θ_{ij}. The logarithm of the relative hazard rates will again be denoted by $\zeta_i = \log(\theta_{i1}/\theta_{i2})$ and the logarithm of the hazard rate for the second treatment by $\eta_i = \log(\theta_{i2})$. It will be assumed that the ζ_i follow independent normal distributions each with a mean μ_ζ and standard deviation σ_ζ. The mean μ_ζ can be interpreted as the population's mean, and σ_ζ is a measure of the heterogeneity between centers. To obtain posterior distributions for the parameters of interest, prior distributions for μ_ζ, μ_η, σ_ζ, and σ_η are needed. Constant priors will be used for μ_ζ and μ_η and normal priors will be used for $\log \sigma_\zeta$, and $\log \sigma_\eta$. Before explaining the choice of these priors, the three-stage hierarchical model can be summarized as follows:
Stage I:

$$f(t_{ijk} \mid \theta_{ij}) = \theta_{ij} e^{-\theta_{ij} t_{ijk}}, \qquad \underbrace{i = 1, \ldots, 5}_{\text{center}} \underbrace{j = 1, 2,}_{\text{treatment}} \underbrace{k = 1, \ldots, n_{ij}}_{\text{subjects}}$$

This results in the following marginal density in terms of ζ_i and η_i:

$$f(t \mid \zeta, \eta) = \prod_{i=1}^{5} (e^{\zeta_i + \eta_i})^{\delta_{i1}} e^{-e^{(\zeta_i + \eta_i)} n_{i1} \bar{t}_{i1}} (e^{\eta_i})^{\delta_{i2}} e^{-e^{\eta_i} n_{i2} \bar{t}_{i2}}$$

$$= \prod_{i=1}^{5} (e^{\zeta_i})^{\delta_{i1}} (e^{\eta_i})^{\delta_{i1} + \delta_{i2}} e^{-e^{\eta_i} (e^{\zeta_i} n_{i1} \bar{t}_{i1} + n_{i2} \bar{t}_{i2})}$$

Stage II:

$$g(\zeta_1, \zeta_2, \zeta_3, \zeta_4, \zeta_5 \mid \mu_\zeta, \sigma_\zeta) = \prod_{i=1}^{5} \frac{1}{\sqrt{2\pi}\sigma_\zeta} \exp\left[-\frac{1}{2}\left(\frac{\zeta_i - \mu_\zeta}{\sigma_\zeta}\right)^2\right]$$

$$g(\eta_1, \eta_2, \eta_3, \eta_4, \eta_5 \mid \mu_\eta, \sigma_\eta) = \prod_{i=1}^{5} \frac{1}{\sqrt{2\pi}\sigma_\eta} \exp\left[-\frac{1}{2}\left(\frac{\eta_i - \mu_\eta}{\sigma_\eta}\right)^2\right]$$

Stage III:

$$\mu_\zeta \sim \text{constant}$$
$$\mu_\eta \sim \text{constant}$$
$$\log(\sigma_\zeta) \sim N(a, b)$$
$$\log(\sigma_\eta) \sim N(-.22, 1)$$

Although the priors on μ_ζ, μ_η, and σ_η, are important, it is the prior on σ_ζ that reflects prior beliefs about the heterogeneity in the relative treatment effects across centers. For this reason, results using two prior distributions on σ_ζ are presented. The first prior reflects a prior belief of only small differences in relative treatment effects across centers, and the second reflects a prior belief of large differences in relative treatment effect across centers. Reparameterization to $\log(\sigma_\zeta)$ allows the use of a normal prior distribution. Two normal priors that represent these divergent prior beliefs about center heterogeneity are $\log(\sigma_\zeta) \sim N(-1.61, .5)$ and $\log(\sigma_\zeta) \sim N(0, .5)$. Since $e^{-1.61} = .20$, the first prior places high probability on small standard deviations for the relative treatment effect. Likewise, since $e^0 = 1$, the second prior places high probability on large standard deviations. Our prior uncertainty on the amount of heterogeneity is reflected in the variance of the distribution of σ_ζ. The larger the variance, the more uncertain we are about how much heterogeneity will occur. For this example, a variance of .25 will be used for each prior distribution representing moderate uncertainty about the degree of heterogeneity among centers. In conjunction with both normal priors for $\log(\sigma_\zeta)$, a constant or flat prior will be used for the means, μ_ζ and μ_η. Finally, a normal distribution will be used for the prior of $\log(\sigma_\eta)$. To represent an expectation of moderate heterogeneity in the effect of being taken off imipramine, a mean of $-.22$ will be used, and to represent a large uncertainty about the degree of heterogeneity a variance of 1 will be used.

In the three-stage model, μ_ζ and σ_ζ are the mean and standard deviation of the common sampling distribution of the ζ_i. Hence the posterior distribution of μ_ζ provides a population estimate of the relative efficacy of two treatments, and the posterior distribution of σ_ζ or equivalently the posterior distribution of $\log(\sigma_\zeta)$ provides a direct measure of the extent of heterogeneity in relative efficacy of the treatments across the study centers. Posteriors of $\log(\sigma_\zeta)$ concentrated at values less than -1 correspond to a small standard deviation for the sampling distribution of the ζ_i and hence suggest that the relative efficacy of the treatments is similar across centers. Posteriors of $\log(\sigma_\zeta)$ concentrated at values greater than -1 correspond to a large standard deviation for the sampling distribution of the ζ_i and hence suggest heterogeneity in relative treatment effect across centers.

Figure 7 compares the posterior distributions of the relative treatment effect using two stage III priors on $\log(\sigma_\zeta)$ and the empirical Bayes methodology. Because nonconjugate priors were used, posterior distributions did not have a simple closed form. Hence posterior distributions were approximated using the Laplace approximation discussed earlier. Column 1 shows the posterior distributions of ζ_i, μ_ζ, and $\log(\sigma_\zeta)$ when the $N(-1.61, .5)$ prior was used for $\log(\sigma_\zeta)$. The second column shows the same distributions when the $N(0, .5)$ prior was used. Beginning with the first column of plots, we see that the posterior of $\log(\sigma_\zeta)$ is not much different from the prior. A large mass of both falls on values less than -1, indicating only small differences between the centers. The mode of both is approximately -1.61, corresponding to a standard deviation in the relative treatment effect across centers of $.2$. In the first plot in column 1, we see this similarity between centers. Here the posterior distributions of the relative treatment effects at each center show striking homogeneity. The distribution for each ζ_i has been shrunk very close to that of the population mean. The posterior distribution for the population mean, μ_ζ, is shown in the second plot in column 1. It is centered about -1.2 and has a standard deviation of $.25$. Using the posterior mean of -1.2 as a population estimate of the relative efficacy, this indicates that patients on imipramine recurred at about one-third the rate of the patients taken off imipramine. In other words, time to recurrence was more than three times longer for patients remaining on imipramine.

Moving to the second column, we see a different result. Again we begin by comparing the prior and posterior distributions of $\log(\sigma_\zeta)$. This time the posterior is shifted to the left of the prior distribution. This means that the data provided evidence of smaller differences in the relative treatment effect across centers than were anticipated by the prior. Even so, the mass of this posterior distribution lies almost entirely to the right of -1, indicating center differences. Comparing the posterior distributions of the relative treatment effect for each center reflects these differences. Much less shrinkage toward the population mean is seen; however, the posteriors distributions for even the two most discrepant centers, D and

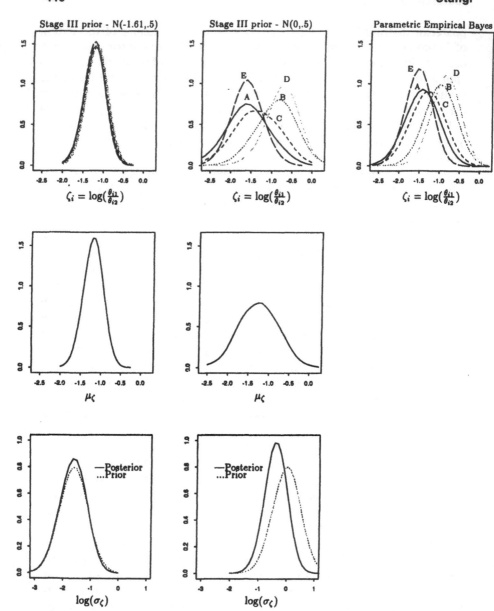

Figure 7 A comparison of posterior densities.

E, still show considerable overlap. If we look at the value of ζ for which the posterior distributions of centers D and E meet, $\zeta = -1.16$, we see that $P(\zeta_E > -1.16 \mid t) \approx .17$, and $P(\zeta_D < -1.16 \mid t) \approx .11$. The heterogeneity is also reflected in the posterior distribution of the population mean. The posterior distribution of μ_ζ now has a larger standard deviation. Although the mean of μ_ζ is again -1.2, the standard deviation is .39. So while we would still estimate that recurrence times for patients on imipramine were three times longer, the uncertainty associated with this estimate is greater.

Comparing the prior and posterior distributions of $\log(\sigma_\zeta)$, we see that the posterior distribution is not far from the prior distribution for either of the two prior distributions. Using the first prior, most of the mass of the posterior distribution is placed on values of $\log(\sigma_\zeta)$ less than -1, while for the second prior most of the mass of the posterior distributions is still on values of $\log(\sigma_\zeta)$ greater than -1. This means that the data provide relatively little information to change our opinions from what we believed a priori. This is consistent with conclusions reached by Goel and DeGroot (1981). They found that for many widely used information measures, the information about the hyperparameters decreases as one moves to higher levels away from the data. This result highlights the importance of carefully specifying prior distributions and of thoroughly investigating the sensitivity of inferences on the priors.

The plot in the third column displays the results of an empirical Bayes analysis. Here second-stage parameters were estimated from the data. The degree of heterogeneity falls between those found using the two prior distributions. If we look at the value of ζ_i for which the posterior distributions of centers D and E meet, $\zeta = -1.2$, we see that $P(\zeta_E > -1.2 \mid t) \approx .17$ and $P(\zeta_D < -1.2 \mid t) \approx .17$. This result is quite similar to that found with the second prior for $\log(\sigma_\zeta)$. It is of interest to note that it is quite different from the empirical Bayes analysis of the previous section; however, there were differences in the modeling that account for this. In the parametric empirical Bayes methodology in the previous section, center hazard rates for each treatment were shrunk toward their respective pooled estimates, whereas in this section the logarithms of the relative hazard rates were shrunk toward a pooled estimate.

Calculations using Laplace approximations were done in the XLISP-STAT environment of Tierney (1990). The calculations required five commands: **defun, bayes-model, moments, margin1, and plot-lines. Defun** is used to define the likelihood and prior distributions. **Bayes-model** gives first-order approximations to the posterior means and standard deviations of the parameters. **Moments** provides the second-order moment approximations. **Margin1** provides approximated density values for a sequence of points, and **plot-lines** produces plots of the posterior distributions. For this chapter, output from **margin1** was transferred to S (Becker *et al.*, 1988) as it allowed more flexible plotting configurations.

XLISP-STAT is public-domain software and can be obtained from the StatLib archive at Carnegie-Mellon University. Instructions for doing so may be obtained by sending an electronic mail message to *statlib@-temper.stat.cmu.edu*. The message should say:

> **send index from xlispstat**

There are no restrictions on copying the software.

7 DISCUSSION

This chapter explores the heterogeneity between centers in a multicenter clinical trial. The trial was designed to assess the prophylactic effects of imipramine in preventing the recurrence of depression. Results showed a clear quantitative but not qualitative interaction between centers and treatment effects. That is, the direction of the treatment effect did not vary between centers, but the magnitude of the effect varied a great deal. Some centers showed a greater than threefold decrease in recurrence rates with the prolonged use of imipramine, whereas others showed only small to modest gains. In the typical definition of qualitative interaction it is only the direction of the effects that matters. From a decision theoretic perspective the differences in magnitude of effect should also be considered. The underlying criterion would be whether optimal treatment decisions vary across the centers. Depending on the balance between improvement in health and quality of life and the monetary costs and side effects of taking imipramine, small to modest increases may not justify prolonged drug usage. Stangl (1995) uses this data set and hierarchical Bayesian models to show how to incorporate a Bayesian analysis into a decision theoretic framework and shows how to explore the sensitivity of decisions to the choice of utility function.

Two methodologies, empirical Bayes and three-stage Bayesian models, were presented as alternatives for modeling study center heterogeneity. The former has the advantage that only the parametric form of the prior distribution must be specified. Values of the hyperparameters are estimated from the data. The disadvantages of this method are that it does not incorporate available knowledge about the hyperparameters, it underestimates posterior variances, and it does not provide posterior distributions for the population treatment effect and variance of treatment effect between centers. These disadvantages are overcome by the three-stage Bayesian models. Both methodologies can be tailored to a wide variety of problems. The methods are adaptable to Weibull, lognormal, or other commonly used survival distributions. Stangl (1991, 1995) uses this example, but also fits exponential mixture and exponential change-point models to the data to show the sensitivity of the results not only to the prior distributions but also to the likelihood.

In the example presented treatment effects were assumed independent and identically distributed. This assumption requires careful consideration. If centers can be grouped according to patient characteristics, protocol variations, or any other variable that may be pertinent to treatment effect, then these similarities and differences should be modeled. For example, in the three-stage Bayesian model, if a priori some centers were expected to be similar and others different this could be modeled at stage II using a multivariate normal distribution with a nondiagonal covariance matrix. Caution is warranted, however, because each nonzero entry in this matrix results in another parameter about which the data may provide little information.

In this chapter, Bayesian hierarchical models were used to model center differences; however, this is but one example of their use. Any grouping of patients that is believed to affect the treatment response can be modeled in this way. Similarly, meta-analyses in which data sets from several studies are combined can be analyzed using these techniques.

REFERENCES

Albert, J. H. (1987). Bayesian estimation of odds ratios under prior hypotheses of independence and exchangeability. *J. Statist. Comput. Simul.* 27: 251–268.

Becker, R. A., Chambers, J. M., and Wilks, A. R. (1988). *The New S Language, A Programming Environment for Data Analysis and Graphics.* Wadsworth and Brook, Pacific Grove, CA.

Casella, G. (1985). An introduction to empirical Bayes Data Analysis. *The American Statistician* 39(2): 83–87.

Deely, J. J. and Lindley, D. V. (1981). Bayes empirical Bayes. *Journal of the American Statistical Association* 76(376): 833–841.

Fleiss, J. (1986). Analysis of data from multicenter clinical trials. *Controlled Clinical Trials* 7: 267–275.

Gelfand, A. E., Hills, S. E., Racine-Poon, A. and Smith A. F. M. (1990). Illustration of Bayesian Inference in normal data models using Gibbs Sampling. *Journal of the American Statistical Association* 85(412): 972–985.

Gelfand, A. E. and Smith, A. F. M. (1990). Sampling based approaches to calculation of marginal densities. *Journal of the American Statistical Association* 85: 398–409.

Goel, P. K. and DeGroot, M. H. (1981). Information about hyperparameters in hierarchical models. *Journal of the American Statistical Association* 76: 140–147.

Greenhouse, J., Stangl, D., Kupfer, D., and Prien, R. (1991). Methodological issues in maintenance therapy clinical trials. *Archives of General Psychiatry* 48: 313–318.

Kaplan, E. L. and Meier, P. (1958). Nonparametric estimation from incomplete observations. *Journal of the American Statistical Association* 53: 457–481.

Kass, R. and Steffey, D. (1989). Approximate Bayesian inference in conditionally independent hierarchical models. *Journal of the American Statistical Association* 84: 717–726.

Mantel, N. and Haenszel, W. (1959). Statistical aspects of the analysis of data from retrospective studies of disease. *J. Natl. Cancer Inst.* 22: 719–748.

Morris, C. (1983). Parametric empirical Bayes inference: Theory and applications. *Journal of the American Statistical Association* 78(381): 47–65.

Nystrom S. (1979). Depressions: Factors related to ten-year prognosis. *Acta Psychiatr. Scand.* 60: 225–238.

Paul, S. R. and Donner, A. (1989). A Comparison of tests of homogeneity of odds ratios in k 2 × 2 tables. *Statistics in Medicine* 8: 1455–1468.

Prien, R. F., Kupfer, D. J., Mansky, P. A., Small, J. G., Tuason, V. B., Voss, C. B., and Johnson, W. E. (1984). Drug therapy in prevention of recurrences in unipolar and bipolar affective disorders. *Archives of General Psychiatry* 41: 1096–1104.

Skene, A. M. and Wakefield, J. C. (1990). Hierarchical models for multicentre binary response studies. *Statistics in Medicine* 9: 919–929.

Stangl, D. (1991). Modeling Heterogeneity in Multi-center Clinical Trials using Bayesian Hierarchical Survival Models. Ph.D. dissertation, Carnegie-Mellon University, Pittsburgh.

Stangl, D. (1995). Prediction and decision making using Bayesian hierarchical survival models. *Statistics in Medicine* 14(20): 2173–2190.

Stangl, D. and Greenhouse, J. (1995). Assessing Placebo Response Using Bayesian Hierarchical Survival Models. Technical Report No. 95-01, Institute of Statistics and Decision Sciences, Duke University, Durham, NC.

Tierney, L. (1990). *LISP-STAT, an Object-Oriented Environment for Statistical Computing and Dynamic Graphics.* Wiley, New York.

Tierney, L. and Kadane, J. B. (1986). Accurate approximations for posterior moments and marginal densities. *Journal of the American Statistical Association* 81: 82–86.

Tierney, L., Kass, R., and Kadane, J. B. (1989). Fully exponential Laplace approximations to expectations and variances of nonpositive functions. *Journal of the American Statistical Association* 84: 710–716.

Zis A. P. and Goodwin, F. K. (1979). Major affective disorder as a recurrent illness: A critical review. *Arch Gen Psychiatry* 36: 835–839.

17

Hierarchical Bayesian Linear Models for Assessing the Effect of Extreme Cold Weather on Schizophrenic Births

William DuMouchel and Christine Waternaux   Columbia University, New York, New York
Dennis Kinney   McLean Hospital, Belmont, Massachusetts

ABSTRACT

The hierarchical Bayesian method combines and summarizes the results from multiple sites using a weighted regression analysis in which the unit of observation is the site and the covariates are characteristics of the sites. There are two sources of random error in this model: the usual within-site sampling error and an additional random effect due to unpredictable differences among sites. The hierarchical model allows sites with small sample sizes to "borrow strength" from the others, to the extent that the between-site variance is estimated to be small. A special graph, called a *trace plot*, displays the posterior distribution of the among-site standard deviation and its effect on the other parameter estimates. This chapter combines and reanalyzes data on schizophrenic births collected across 17 states of the United States. We found additional evidence that severe weather may be associated with increased risk of schizophrenia and confirmed the existence of a dose-response curve, where the risk for schizophrenia increases as severity of winter increases.

1 INTRODUCTION

The hierarchical Bayesian method combines and summarizes the results from multiple sites (or studies) using weighted regression methods in which the unit of observation is the site and the covariates, or independent variables, are characteristics of the sites. There are two sources of random error in this model: the usual within-site sampling error and an additional random effect due to unpredictable differences among sites. Inevitably, results from different sites have some degree of heterogeneity. The hierarchical Bayes methods permit analyses of multisite data that are more ambitious than a recital of the estimation results at each site and more sophisticated than a pooled analysis that assumes that the only source of random error is sampling variation within each site.

This chapter discusses hierarchical Bayesian linear models (HBLM) for combining data on schizophrenic births collected across 17 states of the United States. Season of birth is among the most widely reported risk factors for schizophrenia. Most of the several dozen studies around the world have reported that birth in the winter and early spring months of the year is associated with significantly increased risk of schizophrenia. These findings, however, remain controversial (*Schizophrenia Bulletin*, 1990). Severe weather is the most obvious variable that varies markedly across different seasons and is a plausible candidate for producing disruptive effects on early brain development, because embryological studies of laboratory animals indicate that extreme temperatures can produce birth defects.

Torrey *et al.* (1977) studied the birth months of persons later diagnosed as schizophrenic. They collected data from state mental hospitals on over 50,000 schizophrenic patients born between 1920 and 1955. The controls were counts of general births in the same states for the same years. Table 1 shows the number of schizophrenic births in the coldest months (A) and in the mildest months (B) and the number of general births in the coldest months (C) and in the mildest months (D) for 17 states. Weather extremes were assessed based on wind-chill factor. To identify the coldest months, we used published meteorological records (Conway and Liston, 1990) that provide historical average temperatures and wind speeds by month for major cities. Each monthly wind speed–temperature pair was converted to a wind-chill factor. The coldest month was used to rank order states. Further details are given in Kinney *et al.* (1993). Table 1 shows states in decreasing order of weather severity in winter.

We chose $y_i = \log_e$ of the odds ratio of state i to be the summary statistics of interest characterizing the association of season of birth to the odds of a schizophrenic birth across states. (Because schizophrenic births are quite rare, the odds ratio is practically equal to the risk ratio.) The associated standard error is computed as $s_i = [1/A_i + 1/B_i]^{1/2}$. The variation in the s_i reflects the quite variable sample size from each state. In Table 1, the columns labeled "lnOR" and "se.lnOR" contain the values of y_i and s_i, respectively. Figure 1 shows 95% confidence intervals, computed as $y_i \pm 2s_i$, for each state's log odds ratio. We assume that y_i estimates θ_i, the state-specific parameter for the association between schizophrenia and season of birth. The odds ratios seem to be larger for the more northern states, but the confidence ranges are wide.

We have many reasons to suspect state-to-state variation in odds ratios—reporting methods, socioeconomic status (SES), population differences, and weather extreme differences. The primary hypothesis is that risk for schizophrenia increases as the severity of weather in the month of birth increases. Thus we seek a "dose-response" curve. The pattern of weather severity by month varies from state to state, so we did not assume that the same calendar months should be treated as "cold" or "mild" across states. Tables for average humidity, minimum and maximum temperature, and wind speed were used to assign heat-discomfort and wind-chill values for each month in a given state. The "mild" months

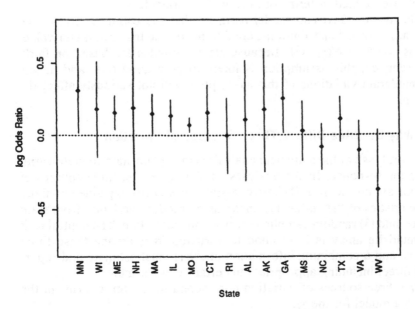

Figure 1 Nonsimultaneous 95% confidence intervals for the odds ratio from each state.

were not always the warmest months, because we wanted to avoid con-
taminating the months used as baseline with months of extreme heat,
which might also have an adverse effect on the fetus.

2 THE MODEL

The hierarchical Bayesian approach is distinguished by the construction
and use of a formal statistical model at two levels. At the first level, a
parametric model is set up for each of the individual sites, in which a
likelihood function relates the distribution of the sample statistics to one
or more unknown parameters characterizing that site—in this case, θ_i,
the true log odds ratio for the ith state. At the second level, a parametric
statistical model is constructed to relate the parameters from the separate
sites to each other. This makes sense in a Bayesian formulation, as in the
Bayesian approach parameters are not just fixed unknown constants; they
are also thought of as random variables having distributions of their own.
The computations involved in combining the two levels of the model are
derived by the application of Bayes' formula for conditional probability.
The Bayesian analysis provides a posterior distribution for each of the
study-specific parameters, thus providing satisfying information if the
models are defined in terms of meaningful parameters.

Our formal Bayesian model assumes that the estimates from each
study have normal distributions, conditional on the true parameter value.
That is, $y_i \mid \theta_i \sim N(\theta_i, s_i^2)$. Because state estimates are based on fairly
large samples, this assumption is likely to be a good one, and in any
case moderate violations of this assumption will not seriously affect the
analysis.

2.1 Why Do Sites Differ? Three Sources of Variation

The central task of the statistical analysis is to explain state-to-state variation in the response. In the analyses combining the y_i, three sources of
variation are considered: (1) between-state differences explained by fixed
characteristics of the states, (2) unexplained random variation from state
to state, and (3) random sampling error within each state. The central task
of a multisite analysis is to allocate variation in y_i among these three
sources, and a combining-information analysis must estimate and report
on all three, including appropriate uncertainty assessments.

The three sources of variation correspond to the three terms in the
following model for the y_i:

$$y_i = x_i\beta + \delta_i + \epsilon_i \tag{1a}$$

$$\theta_i = x_i\beta + \delta_i \tag{1b}$$

$$\delta_i \sim N(0, \tau^2) \tag{1c}$$

$$\epsilon_i \sim N(0, s_i^2) \tag{1d}$$

The term $x_i\beta$ in Eq. (1a) represents source 1 above, the predictable part of the variation in y. Here x_i denotes the row vector $(1, x_{i1}, \ldots, x_{iJ})$, where the J covariates x_{ij} ($j = 1, \ldots, J$) are used to explain state differences and the elements of the parameter column vector $\beta = (\beta_0, \beta_1, \ldots, \beta_J)^t$ are coefficients that need to be estimated. For this problem we considered two versions of this model. In one, there are no covariates, and the term $x_i\beta$ in Eq. (1a) is replaced by the single parameter μ. Thus, the first model assumes that there are no systematic variations in the seasonality of schizophrenic births from state to state. In the second model, two covariates are used, both intended to describe the severity of winter versus mild months for the state. The variable "Winter" in Table 1 is a ranking of the severity of the weather in the winter months in each state, with Minnesota ranked most severe as "9," Rhode Island ranked "1" as the least severe among the nonsouthern states, and all eight of the southern states given the rank "0." The variable "South" in Table 1 is

Table 1 Number of Schizophrenic Births in the Coldest Months (A), and in the Mildest Months (B), and Number of General Births in the Coldest Months (C) and in the Mildest Months (D) for Each of 17 States

	A	B	C*	D*	OR	lnOR	se.lnOR	Winter	South
MN	161	63	3357	1791	1.363	0.310	0.149	9	0
WI	76	68	1843	1975	1.198	0.181	0.167	8	0
ME	625	541	1072	1084	1.168	0.155	0.059	7	0
NH	27	24	273	293	1.207	0.188	0.281	6	0
MA	428	387	2418	2531	1.158	0.147	0.070	5	0
IL	656	620	4110	4440	1.143	0.134	0.056	4	0
MO	2583	4779	3388	6723	1.073	0.070	0.024	3	0
CT	227	202	956	992	1.166	0.154	0.097	2	0
RI	116	122	394	414	0.999	-0.001	0.130	1	0
AL	104	29	5213	1615	1.111	0.105	0.210	0	1
AR	108	27	2946	879	1.193	0.176	0.215	0	1
GA	152	120	3456	3513	1.288	0.253	0.122	0	1
MS	142	276	3085	6157	1.027	0.027	0.103	0	1
NC	307	349	2591	2720	0.923	-0.080	0.078	0	1
TX	833	204	9709	2649	1.114	0.108	0.078	0	1
VA	191	219	1914	1982	0.903	-0.102	0.099	0	1
WV	40	61	1080	1139	0.692	-0.368	0.203	0	1

*Columns for general births are in 100s of births

a dummy variable identifying the southern states. The second model has three parameters involving fixed effects:

$$x_i\beta = \beta_0 + \beta_1 \text{ Winter} + \beta_2 \text{ South}$$

The use of the South dummy variable reflects the fact that it was impossible to rank accurately the severity of winters in the southern states. The second term in Eq. (1a), δ_i, represents source 2 listed above, unexplained random variation from state to state. This term is interpreted in light of Eqs. (1b) and (1c). The equation $\theta_i = x_i\beta + \delta_i$ shows that δ_i denotes a random component of the "true values" θ_i and should not be thought of as measurement error. The specification (1c) denotes that each δ_i has a normal distribution with mean 0 and standard deviation τ. (Therefore, the model predicts that about two-thirds of the δ_i are within the range $\pm\tau$ and about 95% of studies have $\delta_i = \theta_i - x_i\beta$ within the range $\pm 2\tau$.) The size of τ measures how much unobserved state-specific factors like population variations, choice of hospital, uncontrolled conditions, and so forth contribute to unpredictable deviations from the prediction $x_i\beta$. These deviations are unpredictable because no x variables have been identified that isolate them, and they are not reduced in size or importance as the sample sizes at each site increase. If and when such an x variable is identified, putting it into the $x_i\beta$ term would tend to reduce the value of τ in the revised model and correspondingly reduce the role of the δ_i.

The third term in Eq. (1a) is ϵ_i, which is the effect of sampling error within the ith site. The specification (1d) states that ϵ_i has a normal distribution with standard deviation s_i. This is usually approximately true, because the statistic y_i is an estimate based on all the data from site i, and thus the central limit theorem applies in our example.

Because the values of s_i are assumed known, the allocation of variation in the y_i is achieved by estimating the unknown parameters in the other two terms, namely by estimating β and τ. If τ were known, this estimation problem would be nothing more than a weighted regression problem in which the variance of each observation has an additive formula

$$\text{Var}(y_i) = \tau^2 + s_i^2 \tag{2}$$

Since in our example (and in most others) τ is not known a priori, there is an extra level of analysis involving inference about τ, as discussed below.

2.2 The Hierarchical Bayesian Linear Model

The mixed model (including both fixed and random effects) defined by Eqs. (1) is not explicitly Bayesian. A Bayesian approach to this problem has the advantage of being able to incorporate prior distributions for β and τ into the analysis and to properly incorporate uncertainty about τ. The Bayesian version of the model (1) expresses each source of variation as a conditional distribution:

$$y_i \mid \theta_i \sim N(\theta_i, s_i^2) \tag{3a}$$

$$\theta_i \mid (\beta, \tau) \sim N(x_i\beta, \tau^2) \tag{3b}$$

$$\beta_j \sim N(b_j, d_j^2) \tag{4a}$$

$$\tau \sim \pi(\tau) \tag{4b}$$

The two parts of Eq. (3) are equivalent to the four parts of Eq. (1), and the two parts of Eq. (4) define prior distributions for β and τ. The θ_i, $i = 1, \ldots, K$ in Eq. (3b) are assumed to be independent conditional on β and τ, and the β_j, $j = 1, \ldots, J$ in Eq. (4a) are assumed to be independent of each other and of τ. In Eqs. (4) it is assumed that the coefficient means b_j and standard deviations d_j, and the prior probability density for τ, $\pi(\tau)$, are provided as part of the specification of the prior distributions.

Note that this specification emphasizes the θ_i rather than the δ_i, which is often appropriate because in many applications the θ_i from some or all of the studies are of greater interest than either the δ_i or β. Using standard Bayesian calculations, the posterior expectation of θ_i, conditional on β and τ, is

$$E[\theta_i \mid y_i, \beta, \tau] = x_i\beta + \frac{(y_i - x_i\beta)\tau^2}{\tau^2 + s_i^2} \tag{5}$$

Formula (5) is the famous "shrinkage" formula for estimating θ_i. It shows that the naive estimate of θ_i, namely y_i, is "shrunk" toward the value of $x_i\beta$ by the factor $\tau^2/(\tau^2 + s_i^2)$. The word shrinkage is used because the dispersion of the posterior expectations in Eq. (5) is less than that of the y_i. If τ is small compared to s_i there is a lot of shrinkage and we say that y_i is "borrowing strength" from the rest of the y's (which were used to estimate β and τ). On the other hand, if τ is large there is hardly any shrinkage or strength worth borrowing.

Based on the definitions in Eqs. (3), the value of $x_i\beta$ is often referred to as the "prior mean," and the shrinkage estimate (5) is often referred to as the "posterior mean." Some practical calculations of the posterior mean of θ_i may just plug in estimates of β and τ into Eq. (5)—this is the empirical Bayes estimate—but the fully Bayesian calculation requires the computation of $\pi(\tau \mid y)$, the posterior distribution of τ, and $\beta(\tau)$, the estimate of β given τ. Then the posterior expectation of θ_i is the average of shrinkage estimates:

$$E[\theta_i \mid \mathbf{y}] = \int \pi(\tau \mid \mathbf{y}) \left[x_i\beta(\tau) + \frac{(y_i - x_i\beta(\tau))\tau^2}{(\tau^2 + s_i^2)} \right] d\tau \tag{6}$$

2.3 Prior Distributions

The Bayesian specification is completed by adding the prior distributions for β and τ, as in Eqs. (4).

In the absence of more specific information, we use the *diffuse prior* distribution for β, defined as Eq. (4a) with arbitrary b_j and $d_j = \infty$. This diffuse prior distribution is equivalent to assuming that the prior density of β is constant over the whole multidimensional range of β, also called a *flat prior*. Examples of informative prior distributions for β are given in DuMouchel and Harris (1983) and DuMouchel (1990); only flat priors for β are used in this chapter. We encourage the use of a *proper prior* distribution for τ, namely $\pi(\tau)$ must satisfy the requirements of a true probability distribution: $\pi(\tau) \geq 0$, $\int \pi(\tau)\, d\tau = 1$. Here we assume a reference prior distribution for τ, namely

$$\pi(\tau) = \frac{s_0}{(s_0 + \tau)^2} \tag{7}$$

where

$$s_0^2 = \frac{K}{\sum s_i^{-2}} \tag{8}$$

The prior in Eq. (7) has median equal to s_0, where s_0^2 is the harmonic mean of the K sampling variances s_i^2, and is extremely highly dispersed, since the expectations of both τ and $1/\tau$ are infinite. The interpretation of this prior distribution is aided by considering the implied prior distribution of the multiplier from Eq. (5) of a y having a typical standard error s_0, namely $Z_0 = \tau^2/(\tau^2 + s_0^2)$, which in the actuarial literature (see, e.g., Venter, 1990) is called the *credibility* of y. The prior distribution of Z_0 over the unit interval is symmetric about 1/2, having quartiles (0.1, 0.5, 0.9), so that this prior is very "open-minded" about how much shrinkage is appropriate.

3 RESULTS

First, the posterior distribution of τ and its interpretation are presented, as well as how estimates of β and of the θ_i depend on τ. The posterior distributions of the other parameters, which are formed by averaging with respect to the distribution of τ, are discussed later.

3.1 Conditioning on τ

From the model (3), if τ is known, the variables y_i are independently normally distributed with mean $x_i\beta$ and variance $\tau^2 + s_i^2$. If β has a flat prior distribution, it can be shown that the posterior expectation of β, conditional on τ and on $y = (y_1, \ldots, y_K)^t$, is the coefficient vector of a weighted regression of y on X, where the weights are inversely proportional to $\tau^2 + s_i^2$. That is,

$$\beta^*(\tau) \equiv E[\beta \mid y, \tau] = [X^t W(\tau) X]^{-1} X^t W(\tau) y \qquad (9)$$

where

$$W(\tau) = \text{diag}[(\tau^2 + s_i^2)^{-1}] \qquad (10)$$

Also, the expectation of each θ_i, given y and τ, is the shrinkage estimate, an average of $\mu^*(\tau)$ and y_i:

$$\theta_i^*(\tau) \equiv E[\theta_i \mid y, \tau] = x_i\beta^*(\tau) + \frac{[y_i - x_i\beta^*(\tau)]\,\tau^2}{(\tau^2 + s_i^2)} \qquad (11)$$

We use the $\theta_i^*(\tau)$ as estimates of the θ_i for each state. These estimates make use of the information in all the states given the model. Note that they also depend on τ, which is still unknown. Figure 2 displays a plot of these quantities, versus τ, for the schizophrenia data given in Table 1, using the model involving the two covariates. This plot is called a *trace plot* of the hierarchical analysis. The curves in Figure 2, scaled by the right axis of the figure, show the estimates $\theta_i^*(\tau)$ versus τ. When τ is small, say $\tau < 0.01$, these estimates fit the model $\beta_0 + \beta_1$ Winter $+ \beta_2$ South very well, which leads to the appearance on Figure 2 of the nonsouthern states (labeled A–I on the figure) being equally spaced and the southern states (labeled J–Q) being shrunk to a single point. However, as τ becomes larger, the estimates for the states begin to diverge from this pattern, with GA and WV (labeled L and Q) diverging the most. At the extreme right of Figure 2, the estimates of θ_i are the same as y_i.

Finally, by combining the assumption $y_i \sim N(x_i\beta, \tau^2 + s_i^2)$ with the flat prior for β and the prior distribution $\pi(\tau)$ for τ, the posterior distribution for τ, $\pi(\tau \mid y)$, is computed as in DuMouchel and Harris (1983), namely

$$\pi(\tau \mid y) \propto \pi(\tau) \mid W(\tau) \mid^{.5} \mid X^t W(\tau) X \mid^{-.5} \exp\{-y^t S(\tau) y/2\} \qquad (12)$$

where $S(\tau) = W(\tau) - W(\tau)X[X^t W(\tau)X]^{-1}X^t W(\tau)$, and where $|\cdot|$ denotes determinant.

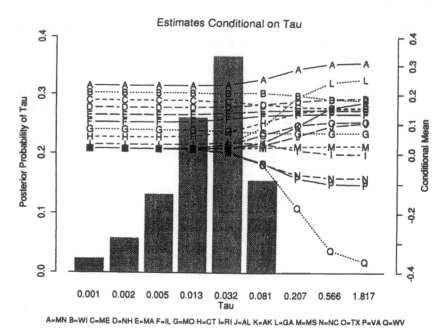

A=MN B=WI C=ME D=NH E=MA F=IL G=MO H=CT I=RI J=AL K=AK L=GA M=MS N=NC O=TX P=VA Q=WV

Figure 2 Trace plot for the schizophrenia HBLM analysis.

The histogram in Figure 2, scaled by the left axis of the plot, shows $\pi(\tau \mid y)$ for nine values of τ. The τ values are chosen automatically, after a computer exploration of $\pi(\tau \mid y)$, so that the posterior distribution is well represented by the nine-point distribution depicted in the histogram. Note that the values of τ are not equally spaced; instead they are closer to, but not exactly, a geometric series, as if the horizontal axis had a log scale. (The nine points are approximate Gauss-Hermite integration points with respect to the posterior distribution of log τ.) The most likely value of τ is about 0.032, but the probability of all values $0.004 < \tau < 0.01$ is high, and even the values $\tau = 0.001$ and $\tau = 0.2$ show up as low-probability values on the histogram in Figure 2. For comparison, Table 1 shows that s_i varies from 0.024 to 0.281 [with median 0.103, and, from Eq. (8), $s_0 = 0.070$], so that most sampling standard errors are probably larger than τ, by a factor of 2 or 3.

The overall estimates of the θ_i are weighted averages of the values graphed in Figure 2, where the weights are equal to the heights of the histogram bars, as in formula (6).

3.2 Regression Coefficients

Figure 3 shows the printed output of an HBLM regression program written for the S-PLUS statistical computing environment. The first line of Figure 3 shows the calling sequence for the program, which identifies the variables that are response, covariates, and standard errors, respectively. The table labeled "Coefficients" in the output displays the estimated regression function, with posterior mean coefficients (standard deviations):

$$\ln(OR) = 0.014 + 0.024 \text{ Winter} + 0.003 \text{ South} + \delta + \epsilon$$
$$(0.014) \qquad\qquad (0.076)$$

The rightmost column of the Coefficients table shows that the posterior probability $P(\beta_{\text{winter}} > 0 \mid y) = 0.958$, while the coefficient for South

```
Call:  hblm(lnOR ~ Winter + South, s.e. = se.lnOR)

DF for Resids= 14       RSS = 14.525

Coefficients:
             Mean   S.D.  Prob > 0
(Intercept) 0.014  0.064  0.584
     Winter 0.024  0.014  0.958
      South 0.003  0.076  0.523

         RSS Estimate of Tau =  0.017
         Prior Median of Tau =  0.07
 exp(post.mode(log(Tau))) =  0.032    (s.d. = 0.029 )
      Posterior Mean of Tau =  0.029    (s.d. = 0.026 )

    Shrinkage of Means:

        Y  Prior Mn (Y-Prior)/SE Post.Mn Post.SD Prob > 0
MN  0.310  0.230    0.537      0.234  0.076  0.999
WI  0.181  0.206   -0.149      0.205  0.067  0.997
ME  0.155  0.182   -0.456      0.176  0.046  1.000
NH  0.188  0.158    0.107      0.158  0.052  0.995
MA  0.147  0.134    0.187      0.135  0.037  0.999
IL  0.134  0.110    0.431      0.114  0.033  0.999
MO  0.070  0.086   -0.661      0.077  0.022  1.000
CT  0.154  0.062    0.950      0.071  0.049  0.952
RI -0.001  0.038   -0.299      0.035  0.057  0.750
AL  0.105  0.017    0.419      0.020  0.055  0.651
AR  0.176  0.017    0.739      0.022  0.055  0.661
GA  0.253  0.017    1.934      0.035  0.057  0.734
MS  0.027  0.017    0.096      0.018  0.049  0.649
NC -0.080  0.017   -1.245      0.003  0.049  0.547
TX  0.108  0.017    1.165      0.030  0.049  0.730
VA -0.102  0.017   -1.203      0.005  0.051  0.562
WV -0.368  0.017   -1.897      0.005  0.058  0.569
```

Figure 3 HBLM regression output for the schizophrenia data using the covariates "Winter" and "South."

has mean and median almost exactly 0, implying that a Winter score of 0 for the southern states is appropriate and that there is strong evidence of an increasing relationship between excess of schizophrenic births in cold versus mild months with severity of winter. The estimated coefficients predict a ratio of odds ratios between Minnesota and the southern states of $\exp\{9(0.024) - 0.003\} = e^{0.213} = 1.24$, that is, a 24% increase in the incidence of schizophrenic births in the most unfavorable months and locations.

The next block of output below the coefficients in Figure 3 shows the estimation results concerning τ. The "RSS Estimate" ($\tau_{RSS} = 0.017$) is a method-of-moments estimate derived in DuMouchel and Harris (1983), which is a generalization of the estimate of DerSimonian and Laird (1986) that allows for the effects of covariates. The "Prior Median" (= 0.070) is computed using Eq. (8), while the posterior mean and standard deviation

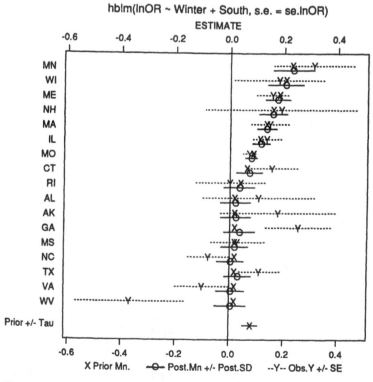

Figure 4 Summary plot for the HBLM schizophrenia data analysis.

of τ are $\tau^* = 0.029 \pm 0.026$, respectively. It is interesting to compare these results with the results from fitting the simple model with no covariates. In that case, $\mu^* = 0.086(0.024)$, $\tau_{RSS} = 0.051$, and $\tau^* = 0.038(0.032)$. Dropping the influential weather covariates results, as expected, in an increase of "unexplained" state-to-state variation, as measured by τ.

3.3 Summary Plot

Figure 4 compares classical estimation of each θ_i with its posterior distribution. For each state, the interval $y_i \pm s_i$ is displayed as a "Y" and a dotted error bar, while the posterior mean $\theta_i^* \pm$ posterior standard deviation is displayed as an "O" and a solid error bar. The shrinkage center, $x_i\beta$, is displayed as an "X" for each state. It can be seen from Figure 4 that the posterior error bars are much shorter than the classical error bars, showing the benefits of borrowing strength. In only two of the states, GA and WV, are the y_i much more than one standard error away from the prior mean denoted by "X". The bottom block of output of Figure 3 shows detailed calculation of the shrinkage factors for each state. The rightmost column of the Shrinkage of Means table contains $P(\theta_i > 0 \mid y)$ and shows that there is extremely strong evidence ($P > 0.99$) for seasonal variation in schizophrenia in the seven states with most extreme winter, fairly strong evidence ($P = 0.95$) for CT, and not much evidence ($P \leq 0.75$) for the other nine states.

4 DISCUSSION

We have used a hierarchical Bayesian linear model to fit multisite data on birth months of schizophrenics. The random-effects model is formally identical to that used in Bayesian meta-analyses (DuMouchel, 1990, 1995). HBLM methods go beyond simply summarizing the results of multiple studies (or multisite data) in several respects. First, they allow use of mixed models (having both covariates and a random effect) to model variation among sites. As seen in the present example, judicious choice of covariates quantifying site characteristics (here severity of winter) helped reduce the "unobservable" component of variance. Second, HBLM methods permit an examination of each site separately. We were able to establish that Rhode Island, a northern state in the study that has comparatively mild winters, tends to behave more like the southern states.

Our reanalysis of the Torrey *et al.* (1977) data is consistent with the hypothesis that severe weather may be associated with increased risk of schizophrenia. We found odds ratios for the risk of births of schizophrenics significantly in excess of 1.1 for winter months in some states. Our model shows the existence of a dose-response curve where the risk for schizophrenia increases as severity of winter increases; for the seven states with most extreme winters, proportionally more schizophrenics than controls were born in the winter months. However, because we had to use aggregate data, individual variations are not taken into account. For instance, yearly fluctuations in monthly averages were not taken into account, and we use the largest city in each state as a proxy to judge the weather of the whole state. In addition, other important psychiatric differences between northern and southern states, besides the coldness of their winters, could possibly account for these results.

The use of plots is an important aspect of a multisite analysis. The side-by-side 95% confidence intervals of the results from each site, as in Figure 1, provide an immediate and model-free comparison of the within-site and between-site variation. In our experience, the trace plot, as in Figure 2, is a very informative tool for understanding the estimation process that goes on in hierarchical modeling. It illustrates the dependence of $\theta_i^*(\tau)$ on τ (the shrinkage idea), and the superposition of the posterior distribution of τ on the trace plot, which will be used to "integrate out τ," further helps to demystify the black box aspect of models having hyperpriors. Finally, the summary plot, as in Figure 4, allows a before-and-after comparison of how the inferences at each site have been affected by the use of the HBLM methodology. The analyses and plots presented in this chapter were produced using programs written in the S-PLUS programming language and are available from the StatLib online library at Carnegie-Mellon University.

ACKNOWLEDGMENT

This work was supported in part by grants from the National Institute of Mental Health, MH15758, and by a grant from the John D. and Catherine T. MacArthur Research Network on the Psychobiology of Depression 85–191.

REFERENCES

Conway, M. and Liston, L. (1990). *The Weather Handbook*, Random House, New York.
DerSimonian, R. and Laird, N. (1986). Meta-analysis in clinical trials. *Controlled Clinical Trials* 7: 177–188.

DuMouchel, W. (1990). Bayesian meta-analysis. In *Statistical Methods for Pharmacology*, ed. Berry, D., pp. 509–529. Marcel Dekker, New York.

DuMouchel, W. (1995). Meta-analysis for dose-response models. *Statistics in Medicine*, 14: 679–685.

DuMouchel, W. and Harris, J. (1983). Bayes methods for combining the results of cancer studies in humans and other species (with discussion). *Journal of the American Statistical Association* 78: 293–315.

Kinney, D. K., Waternaux, C., Spivak, C., LeBlanc, C., and Vernooy, A. (1993). Schizophrenia risk predicted by meteorologic extremes near birth. *Schizophrenia Research* 9(2,3): 135.

Schizophrenia Bulletin. (1990). Entire issue 16(1) devoted to the possible effects of season of birth on development of schizophrenia.

Torrey, E. F., Torrey, B. B., and Peterson, M. R. (1977). Seasonality of schizophrenic births in the United States. *Archives of General Psychiatry* 34: 1065–1070.

Venter, G. G. (1990). Credibility. In *Foundations of Casualty Actuarial Science*, Chapter 7. Casualty Actuarial Society, New York.

18
Fitting and Checking a Two-Level Poisson Model: Modeling Patient Mortality Rates in Heart Transplant Patients

Cindy L. Christiansen   Harvard Medical School and Harvard Pilgrim Health Care, Boston, Massachusetts

Carl N. Morris   Harvard University, Cambridge, Massachusetts

ABSTRACT

The three-step iterative process of model specification, model fitting, and model checking is relevant to most applied statistical work. We demonstrate these steps using heart transplant mortality data and a two-level Poisson model. Assuming the level-one model to be correctly specified, we focus on graphical and statistical methods for checking the structural, level-two, model. Using the three-step process, a violation of the exchangeability assumption is noted and a correction is proposed. This work primarily provides a tutorial on the modeling process for hierarchical models. Second, it provides insight and a structured process for specifying and checking future medical profiling models for transplant programs when more complete data become available.

1 BACKGROUND AND INTRODUCTION

The supply of transplantable organs has remained relatively constant over the past few years, while the number of patients waiting for organs and the number of transplant programs continue to increase. For example, at the end of 1989 there were 1320 patients on the United States heart transplant waiting list and about 130 heart transplant programs. Five years later, in 1994, these numbers increased to 2900 and 165, respectively [United Network for Organ Sharing (UNOS), 1994; UNOS, 1993]. Health care policymakers need accurate estimates of transplant surgery outcomes to make optimal decisions regarding organ allocation, to evaluate or develop standards of care, and to regulate the number and location of transplant programs. UNOS analyzes and reports program-specific graft and survival rates [Health Resources and Services Administration (HRSA), 1992]. The U.S. government and many patient advocacy groups support the release of transplant data and analyses to assist patients, families, and physicians in their decision making. The practice of "profiling" health care providers (the quantification and reporting of patient outcomes by individual physicians, hospitals, or programs) in all areas of the health care system is expected to continue (Kassirer, 1994; Kolata, 1994). Accurate medical profiling depends on having complete and correct data and on making estimates supported by appropriate models and statistical tools. This chapter focuses on developing well-specified models for profiling analyses.

Hosenpud *et al.* (1994) analyze all U.S. heart transplant surgeries performed between October 1987 and December 1991 to determine how heart transplant program size affects patient mortality. They use generalized additive models (Hastie and Tibshirani, 1986) and locally weighted regressions to show that programs averaging eight or fewer treated patients annually have higher mortality risks than those averaging nine or more. This result has policy implications for Medicare rules regarding reimbursement eligibility, because Medicare requires that heart transplants are provided by programs that do 12 or more transplants per year, partly seeking to optimize transplant surgery outcomes.

We study the 1-month mortality rates of heart transplant patients, using the UNOS data (HRSA, 1992), for transplants done between October 1987 and December 1989. We show that a hierarchical model describes the data better than does a standard log-linear regression model. We find that the small programs, those averaging four or fewer transplants per year, have higher 1-month mortality rates than the larger ones and are not adequately described by the model. This collateral result, like that of Hosenpud *et al.* (1994), also suggests that the current Medicare reimbursement requirement might be reset below 12 patients per year without increasing mortality rates.

This chapter is a tutorial on hierarchical model fitting and model checking for the "level-two" distribution [Eq. (5)] assuming the "level-one" distribution [Eq. (4)] to be specified correctly. We extend the patient-level statistical model UNOS employs (HRSA, 1992) to account for the hierarchical structure of the data, and we demonstrate data discovery methods and model-checking techniques for the level-two specification. We arrive at our preferred model for these data through a three-step iterative process of model specification, model fitting, and model checking. Actual profiling results for the transplant programs, i.e., an inferential and comparative study of program-level performance outcomes, are not the focus here, although they are discussed briefly in the concluding remarks. It is good statistical practice to sequence a profile analysis in this way, i.e., understand the problem, investigate the data, specify, fit, and check the model, and finally obtain the profile results.

Section 2 introduces the data and the notion of "exposure," which is computed from patient-level data. Appendix A.1 provides additional information for the exposure calculations. Section 3 describes the general two-level Poisson model applied in this chapter to the heart transplant program data listed in Appendix A.2. Appendix A.3 gives details about the authors' publicly available S-PLUS (StatSci, 1991) program that implements fitting the data to the model. Section 4 offers some informal techniques useful for preliminary thinking about hierarchical models, e.g., data sequencing, histograms, and checking for extra variability. The first hierarchical model is fit in Section 5.1. There, a violation of the level-two gamma distribution assumption is identified. The results in Section 5.2 suggest strongly that the expected mortality rates, which are commonly proposed and used as exposure, do not serve that function well for the smallest 37 transplant programs. Section 5.3 gives the final model for mortality at the 94 largest heart transplant programs. In these programs, $n_i \geq 10$ patients during the 27-month period and exposures seem to serve their function adequately. Section 6 concludes with cautionary and encouraging notes for checking and using hierarchical models in health service evaluations.

2 DATA

A public-use data file provided by UNOS includes all heart transplant surgeries performed in the United States between October 1, 1987 and December 31, 1989 (27 months). Encrypted patient and hospital identities in the database offer anonymity, yet make patient- and hospital-level analyses possible. The file contains no hospital-level covariates, except for the number of heart transplant patients and patient-level data that can be aggregated to the hospital level.

2.1 Patient-Level Data

The unadjusted mortality rate (death within 30 days of the transplant surgery) for the 3646 heart transplant patients in this data set is 8.6%. The variables we use to predict deaths are given in Eq. (1), together with their estimated logistic regression coefficients. UNOS's health status variable indicates whether the patient was working, homebound (the omitted category), hospitalized, in intensive care, or on life support prior to the transplant surgery. (The health status variable in this public-use data set has received the criticism that it does not capture all of the patients' comorbid conditions prior to transplant surgeries.)

$$
\begin{aligned}
\text{logit}(\hat{p}_{ij}) \equiv\ & \text{logit(estimated probability of death for patient } j \text{ in} \\
& \text{hospital } i \text{ within 30 days of transplant)} \\
=\ & -2.93 + 2.01(\text{congenital heart disease})_{ij} \\
& + 1.29(\text{repeat transplant})_{ij} + .305(\text{female})_{ij} \\
& - 2.05(\text{working})_{ij} - .0693(\text{hospitalized})_{ij} \\
& + .0976(\text{in intensive care})_{ij} \\
& + .803(\text{on life support})_{ij} \\
& + .0283(\text{black})_{ij} - .264(\text{other race})_{ij} \\
& + .0202(\text{age}_{ij} - 45) + .000715(\text{age}_{ij} - 45)^2 \\
& + .0350(\text{age}_{ij} - 45)(\text{congenital heart disease})_{ij}
\end{aligned}
\tag{1}
$$

In Eq. (1), $\hat{p}_{ij} \equiv$ estimated probability of death within 30 days after surgery for patient j in transplant program i; $j = 1, \ldots, n_i \equiv$ number of patients in transplant program i and $i = 1, \ldots, k \equiv$ number of transplant programs = 131. The double subscript coding ij is needed only because our interest lies in studying the hospitals, but the covariates actually do not depend on the program i. If our interests were at the patient level, a better notation might have been to let $j = 1, \ldots, 3646 =$ number of patients. More information on the patient-level analysis is given in Appendix A.1.

2.2 Hospital-Level Data

The hospital-level data (observation number i, $e_i \equiv$ exposure, $z_i \equiv$ number of deaths, and $n_i \equiv$ number of patients for $i = 1, \ldots, 131$) appear in Table A of Appendix A.2 (the first four columns). The data are sorted on the number of patients receiving heart transplants, n_i, during the 27-month period. A secondary sort is on e_i, which is defined in Eq. (2). The 131 × 4 data matrix is available through StatLib. The next six columns of Table

A contain computed values for use in Section 4, and the final column is discussed in Section 6.

The estimated expected number of deaths for the 131 heart transplant programs,

$$e_i \equiv \sum_{j=1}^{n_i} \hat{p}_{ij}, \qquad i = 1, \ldots, k = 131 \tag{2}$$

are calculated from Eq. (1). These expected values differ slightly from the UNOS values because UNOS uses a patient-level logistic model different from Eq. (1) in the HRSA (1992) report. Throughout the rest of this work, e_i is used as a known value.

In Section 3, we propose that e_i represents "exposure"; that is, transplant program i was exposed to e_i deaths given that they treated n_i patients. Exposure is a function of the patient case mix described by the variables included in Eq. (1). A hospital-level index of case-mix severity is the average of the estimated probabilities within a transplant program,

$$\bar{e}_i = \sum_{j=1}^{n_i} \frac{\hat{p}_{ij}}{n_i} = \frac{e_i}{n_i} \tag{3}$$

A commonly used profiling technique tests for differences $y_i - \bar{e}_i$ in the observed death rates $y_i \equiv z_i/n_i$ (where z_i is the number of deaths at transplant program i) and the expected death rates $\bar{e}_i$, using asymptotically normal distribution assumptions. For these data, that method identifies three programs that have significantly better mortality rates than their expected rates and 13 programs that have significantly worse mortality rates than their expected rates (all with .025 probability in each tail). Assuming the 131 standardized differences are identically and normally distributed, one would expect about 3.3 differences to be significant in each direction at random. Because there are 13 large differences, this suggests either that there are some programs with unexplained high mortality rates (possibly due to unmeasured patient-level variables or to poor program performance) or that the model assumptions used are not appropriate for the data.

Correct model specification is crucial before concluding that the unexplained high rates are due to program performance. It is unlikely that the mortality rates of programs having a small number of patients are normally distributed. We propose and check a different model for these data by assuming that the number of deaths in the transplant programs can be modeled as Poisson-distributed random variables, given that the programs are exposed to e_i expected deaths and given the Poisson parameters λ_i that govern the rates of death. (If all of the 131 hospitals treated many patients, for example, 500 or more patients each, we would expect our profile analysis results to be similar to the results from the standard analysis.)

3 A POISSON REGRESSION INTERACTIVE MULTILEVEL MODEL (PRIMM)

In this work, one of our interests is to investigate the functional relationship between the observed number of deaths and the expected number of deaths at the programs and thus to understand the appropriateness of using the expected number of deaths as an exposure measurement in a profile analysis. Possible transplant profiling analyses involve binomial and Poisson distributions where the random variable is the number of postoperative deaths (within 30 days of surgery), which occur randomly and independently (Johnson and Kotz, 1969). Here we assume z_i, the number of deaths observed within 30 days of surgery for program i, follows a Poisson distribution with mean $e_i \lambda_i$, where e_i represents the program's exposure to death [Eq. (2)] and λ_i governs the program's rate of death. This work employs the hierarchical Poisson model of Christiansen and Morris (1994a). The S-PLUS program PRIMM described in Christiansen and Morris (1994b) and used in these analyses is available through Statlib. (See Appendix A.3.)

Poisson distributions are the limiting form of binomial distributions when the number of independent Bernoulli trials is large and the probability of an occurrence is small. A binomial distribution for z_i, representing the number of deaths out of n_i patients, might be used to analyze the true proportion of deaths in transplant program i. A regression then could account for aggregate differences in the programs' case-mix severity levels $\{e_i\}$. The binomial model assumes that the patients within programs are identically distributed, which is not true. Thus, the sum of patient death indicators would not have a binomial distribution. All profile analyses should include an investigation of model assumption violations and their effect on the profile results.

The two-level Poisson model contains two mathematically equivalent joint distributions, one for the model in its "descriptive" form and the other for the model in its "inferential" form. Equations (4), (5), and (6) comprise the descriptive model. The observed data, $z_i \equiv$ number of deaths at transplant program i, are modeled as Poisson random variables in level one [Eq. (4)]. The individual rate parameters, λ_i, are assumed to follow gamma distributions in level two [Eq. (5)]. The distribution for the hyperparameter vector, $\alpha = (\zeta, \beta)$, is given in Eq. (6). Technically, e_i serves as a legitimate exposure if it enters into the model through Eq. (4) but not Eq. (5); that is, it captures all the patient-level information although not necessarily all the program information. The level-two rates λ_i are exchangeable (i.e., identically distributed because the λ_i's are independent) if and only if $E(\lambda_i \mid \alpha) = \mu_i$ does not depend on i. Sections 4 and 5 describe techniques that help to check this exchangeability assumption.

A note on notation: Throughout, square brackets indicate the means and the variances of the distributions, as noted in Eqs. (4), (5), (9), and (12). The gamma density for $U \sim \text{Gam}(\zeta, 1)/b$ is $(bu)^{\zeta} \exp(-bu)/u\Gamma(\zeta)$ for $u > 0$, and in square bracket notation $U \sim [\zeta/b, \zeta/b^2]$

Level one:

$$z_i \mid \lambda_i \sim \text{Pois}(e_i\lambda_i) = \text{Pois}[e_i\lambda_i, e_i\lambda_i], \text{ independently,} \tag{4}$$

$$i = 1, \ldots, k = 131 \text{ transplant programs}$$

Level two:

$$\lambda_i \mid \alpha \sim \mu_i \left\{ \frac{\text{Gam}(\zeta, 1)}{\zeta} \right\} = \text{Gam}\left[\mu_i, \frac{\mu_i^2}{\zeta} \right], \text{ independently,} \tag{5}$$

$$\log(\mu_i) = x_i'\beta$$

Distribution for the hyperparameter vector:

$$h(\zeta, \beta) = \frac{z_0}{(z_0 + \zeta)^2} \quad \text{(with respect to } d\zeta \, d\beta)$$

$$-\infty < \beta_j < \infty, j = 0, \ldots, \tag{6}$$

$$r = \text{number of regression coefficients};$$

$$\zeta > 0; \text{ on } \Re^{r+1}, \text{ independently}$$

The choice of an improper prior distribution for β is standard, made here to provide good frequentist properties to the resulting rules. The distribution for ζ in Eq. (6) is specified by choosing a value z_0 and assuming that

$$B_0 \equiv \frac{\zeta}{\zeta + z_0} \sim \text{Uniform}(0, 1) \tag{7}$$

Thus, ζ has a proper prior distribution, with median z_0. For the analyses in this work, we use the PRIMM default values for $z_0 = e_0 m_0$, where $e_0 = \min(e_i)$ and $m_0 = \bar{y} \equiv \text{average}(z_i/e_i)$. These values are chosen to be conservative and to prevent overshrinking. Although ζ is unknown, one can compare the estimated value ζ with z_0 to check this point informally. We do this throughout our iterative model-fitting process in Section 5.

For example, for the first model specification (see Section 5.1, model Reg0.k131), z_0 is .08 and ζ is 5.30, which is much larger, implying that the default choice of z_0 is conservative in this case. We encourage readers to try nondefault z_0 values in their exploration of these data and the PRIMM program.

For the distribution of the observed data z_i given the hyperparameter $\alpha = (\zeta, \beta)$, define the shrinkage factor

$$B_i \equiv \frac{\zeta}{\zeta + e_i\mu_i} \tag{8}$$

Equations (9), (12), and (6) define the inferential model derived from Eqs. (4)–(6). Note that the joint densities on $(z, \lambda, \zeta, \beta)$ are the same when defined by either the three parts of the descriptive model or the three parts of the inferential model.

Level one:

$$z_i \mid (\zeta, \beta) \sim \text{NB}\,(\zeta, (1 - B_i))$$

$$= \text{NB}\left[e_i\mu_i, \frac{e_i\mu_i}{B_i}\right]$$

$$= \text{NB}\left[e_i\mu_i, e_i\mu_i + \frac{(e_i\mu_i)^2}{\zeta}\right], \tag{9}$$

$$i = 1, \ldots, k, \text{ independently, } z_i$$

$$= 0, 1, 2, \ldots, \zeta > 0, 0 < B_i < 1$$

The independent marginal densities in Eq. (9) are negative binomial (NB) with probability mass functions

$$f(z_i \mid \zeta, \beta) = \frac{\Gamma(z_i + \zeta)}{\Gamma(\zeta)z_i!}\,(1 - B_i)^{z_i}B_i^{\zeta} \tag{10}$$

In the inferential model, the level two distributions of the Poisson rate parameters $\{\lambda_i\}$ given the data and hyperparameters (ζ, β) are

$$g(\lambda_i \mid \text{data}, \zeta, \beta) = \frac{\{\lambda_i(e_i + \zeta/\mu_i)\}^{z_i + \zeta}\,\exp(-\lambda_i(e_i + \zeta/\mu_i))}{\Gamma(z_i + \zeta)\lambda_i}, \tag{11}$$

independently, $i = 1, \ldots, k$

and are gamma distributions as defined in Eqs. (12)–(14).

Level two:

$$\lambda_i \mid (\text{data}, \zeta, \beta) \sim \text{Gam}\left(z_i + \zeta, e_i + \frac{\zeta}{\mu_i}\right) = \text{Gam}[\lambda_i^*, (\sigma_i^*)^2], \qquad (12)$$

$$i = 1, \ldots, k, \text{ independently}$$

where we define the posterior means

$$\lambda_i^* \equiv E(\lambda_i \mid \text{data}, \zeta, \beta) = (1 - B_i)y_i + B_i\mu_i \qquad (13)$$

and the posterior variances

$$(\sigma_i^*)^2 \equiv \text{Var}(\lambda_i \mid \text{data}, \zeta, \beta) = \lambda_i^* \frac{(1 - B_i)}{e_i} \qquad (14)$$

Formulas (12)–(14) provide intermediate steps for the analysis of this Poisson model, but they cannot be used directly because the hyperparameters ζ and β are unknown. Christiansen and Morris (1994a) analyze this model in a way that accounts for the uncertainty in ζ and β, by using accurate approximations to the unconditional distribution of the $\{\lambda_i\}$, given all of the observations $\{z_i\}$. This is the basis of the computations made in Section 5.

4 INFORMAL TECHNIQUES

Several informal statistical techniques, described below, can help with our initial data exploration to understand better the role of exposure and the need for hierarchical modeling. From these informal looks at the data, we will learn:

Section 4.1: There might be problems with (1) using $\{e_i\}$ as exposures in our specification of the level one model and (2) assuming exchangeability in our initial specification of the level two model. We notice this by looking at sorted data, cumulative sums (z_i^*, e_i^*), and at the ratios of cumulative sums (z_i^*/e_i^*) in Table A. The table suggests that the observed-to-expected death ratios are not proportional across the programs. The smaller programs seem to be treating slightly more difficult cases, on average, than the larger programs treat. We detect this by considering the ratios of the exposures to the number of patients, e_i/n_i, and by computing weighted averages of groups of these ratios using cumulative sums when the data have been sorted by n_i.

Section 4.2: When $\log(e_i)$ is introduced as an explanatory variable for the analysis of all 131 programs (as a check for $\{e_i\}$'s role as exposure and as one check for nonexchangeability), the smaller programs may be highly influential. A histogram of $\log(e_i)$, which has left skewness, indicates this, and another histogram shows this will not be a problem if the small programs are removed.

Section 4.3: There is more variability than Poisson variability in the data, yet some variability can be explained by regression. Figure 2 will help us to see this.

By first exploring the data using these informal statistical techniques, we obtain important insights for the two-level model specification, model fitting, and model checking iterations that are demonstrated in Section 5. Note that none of the techniques in this section requires using a hierarchical model.

4.1 Sorting the Data

As displayed, the data in Appendix A.2 are sorted according to increasing n_i, the number of patients receiving heart transplants in each program during the 27-month period. Note that n_i varies from 1 to 152 patients, a wide range. Listing data in sorted form often shows useful patterns. For example, we see from Table A that as the transplant programs increase in size (increasing n_i) e_i tends to become smaller than the observed number of deaths, z_i. To clarify this point, we computed and listed in Table A z_i^*, e_i^*, and z_i^*/e_i^*, where z_i^* is the cumulative sum of the z_i's from observation 1 through observation i and e_i^* is the cumulative sum of the e_j's from observation 1 through observation i. These values suggest possible problems with using e_i as exposure and with the exchangeability assumption in Eq. (5) for a model with no covariates. A glance down the z_i^*, the e_i^*, and the z_i^*/e_i^* columns of Table A shows that the observed deaths are approximately twice the predicted deaths for transplant programs with fewer than 10 patients. This indicates that the observed-to-expected death ratios are not proportional across the programs. For $n_i \geq 10$, the z_i^* and e_i^* values begin to converge, and if we remove the 37 smallest programs, the new ratios z_i^*/e_i^* (not shown) for the remaining 94 transplant programs hover around 1.0 with a maximum of 1.24.

Although the patient-level model of Eq. (1) includes a five-category health indicator, many researchers in the field of transplantation question this covariate's ability to capture the degree of patient illness and the comorbid conditions the patients present. Therefore, such a large value of the average ratio might result if the transplant programs with few patients treat the sicker patients but the patient-level health status variable captures this poorly. To investigate this informally, we explore the ratio $\bar{e}_i$ of the expected number of deaths to the number of patients, and the cumulative sums e_i^* and n_i^* in Table A. The $\{\bar{e}_i\}$ range from .036 to .462 with 10 of the 37 smallest programs having $\bar{e}_i$ values greater than .091, the third quartile value of $\{\bar{e}_i\}$. The weighted mean of $\{\bar{e}_i\}$, weighted by the number of patients for $i = 1, \ldots, 131$, is .085. For the 37 small transplant programs the weighted mean is .097, whereas for the 31 large transplant programs, the weighted mean is .086. These weighted sums can be computed easily using the columns e_i^* and n_i^* and the formula $(e_p^* - e_q^*)/(n_p^* - n_q^*)$ which gives the weighted mean for the $i = q + 1, \ldots, p$ programs (here, $q = 0, \ldots, k$; $p = 1, \ldots, k$; $p > q$; e_0^* and n_0^* are defined to be zero). The ratio $.097/.086 = 1.13$ indicates that the 37 small programs have a case mix that is 13% more difficult than in the remaining 94 programs, on average. From this it seems possible that the small transplant programs have a disproportionate number of sicker patients if exposures are capturing the case mix adequately. However, this does not explain fully the finding that the small programs have about twice as many observed deaths as expected.

4.2 Explanatory Variables and Their Histograms

When using inferential methods, as in Section 5, care must be taken when skewed variables are used as regressors, particularly when the true functional form of the variable is unknown, because the values in the long tail might be more influential than others due to high leverage values. Histograms of explanatory variables (and functions of the variables) help to detect high-leverage observations.

We recommend including the covariate $\log(e_i)$ in the final model as a check for the appropriateness of using $\{e_i\}$ as exposures and, when no other regressors are available, as a check for the exchangeability assumption in Eq. (5). The functional form $\log(e_i)$ is used here because with the log link (5) exposures enter into the mean $E(z_i) = e_i\mu_i = \exp(\beta_0)e_i^{1+\beta_1}$ of the negative binomial distribution (9) in this form with β_1 assumed to be zero if $\{e_i\}$ is acting as exposure and if the unobserved rates $\{\lambda_i\}$ are exchangeable.

The histogram of $\log(e_i)$, Figure 1, using all 131 programs, is left skewed. Therefore, when $\log(e_i)$ is included in the model to check the assumption that $\{e_i\}$ serves as exposures and as a check for exchangeability, the observations at the small programs may be highly influential, because the small hospitals will have high leverages. This means that the estimate of the regression coefficient for $\log(e_i)$ might be unduly influenced by the outcomes at the 37 smaller transplant programs, which we noted in Section 4.1 look systematically different from the outcomes at the 94 larger programs. The histogram in Figure 1 of $\log(e_i)$ for the 94

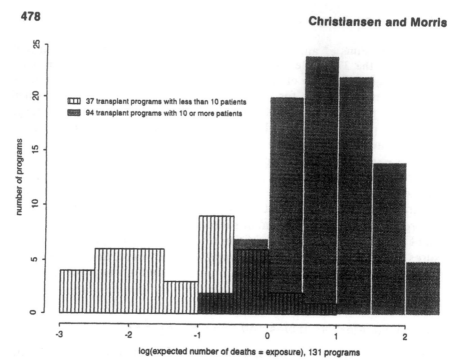

Figure 1 The histogram of $\log(e_i)$ includes 131 heart transplant programs. It is important to consider the distributions of covariates in a regression because observations in the long tail of a skewed distribution could have unduly high leverage when the true functional form of a covariate is unknown.

largest transplant programs ($n_i \geq 10$) looks symmetric and therefore we expect no problems with high leverages if only these programs are included in the analysis.

Histograms of the standardized exposure, $\bar{e}_i$, and $\log(\bar{e}_i)$ for the 131 programs have long right tails. These standardized exposures could be used as a covariate and should be investigated further. In particular, some of the smaller hospitals have the large values of $\log(\bar{e}_i)$, so several of them will have high leverage in the regression that includes this covariate.

4.3 Other Graphical Exploration

Two-level hierarchical models capture the within-program variability, Eq. (4), and the between-program variability, Eq. (5). Figure 2 depicts these two levels of variation for the 94 hospitals in the largest three quartiles

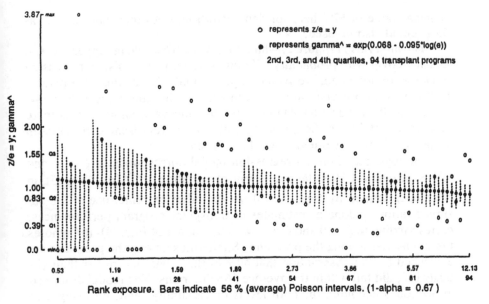

Figure 2   Using a two-level model to analyze these data seems necessary based on the information in this figure. See Section 4.3 and Appendix A.4 for more information.

$(n_i \geq 10)$. The horizontal axis of Figure 2 is the rank of the predicted number of deaths, e_i (also listed on this axis are some of the e_i values). We use the rank of the exposures instead of the exposures to make better use of the graphing space and to avoid overlays. The vertical axis is scaled for the ratio $y_i \equiv z_i/e_i$, $i = 38, \ldots, 131$, of the observed deaths to the expected deaths, shown by open circles, and for the estimate of $\gamma_i = E(y_i)$ $= \exp(\beta_0 + \beta_1 \log(e_i))$, $\hat{\gamma}_i = \exp(.068 - .095\log(e_i))$, shown by solid circles. This estimate of γ_i does not assume a hierarchical model. Instead, the coefficients .068 and $-.095$ are obtained from a standard Poisson log-linear regression model (McCullagh and Nelder, 1989) that can be calculated in S-PLUS using the command "glm" or "glim".

The vertical dashed lines in Figure 2 are intervals that correspond nominally to the middle two-thirds of the 94 Poisson distributions, assuming the z_i's are independent Poisson random variables with means $e_i\hat{\gamma}_i$. Because the z_i's are discrete, some arbitrary rule is necessary for determining the end points of these intervals. We use the rule outlined in Appendix A.4, which forces the interval to cover the value $\hat{\gamma}_i$. We chose the nominal value of 67% here for demonstration, but other values, such as 95%, could also be used.

The intervals in Figure 2, which follow the 67% rule of Appendix A.4, actually cover with probability 56% on average. If the Poisson regression model were adequate, we would expect about 52.7 of the 94 programs (56%) to have observed ratios z_i/e_i contained within these intervals. However, only 38 of the 94 (40%) fall within the intervals, implying that there is more variability in the observed data than is represented by Poisson variation. Thus, a two-level model is needed.

The log-linear Poisson regression model corresponds to $\zeta = \infty$ and CV = 0, where CV is the coefficient of variation $1/\sqrt{\zeta}$ from Eq. (5), in the model defined by Eqs. (4)–(6). At the other extreme, use of the observed ratios y_i alone as estimates of transplant program performances, corresponding to $\zeta = 0$ and CV = ∞ in the model of Eqs. (4)–(6), ignores the similarities among the programs. Similarities among the programs are evidenced by the observation that 40% of the programs, rather than virtually 0%, did fall within the average 56% intervals. The model described in Section 3 accounts for both levels of variability, and, by using it to analyze the transplant program mortality rates, we estimate $\hat{\zeta} = 6.3$ and CV = .40 in our final model (Reg1.k94 from Section 5.3).

5 ITERATIVELY SPECIFYING, FITTING, AND CHECKING MODELS

Good statistical practice for analyses involves a sequence of:

1. Model specification
2. Model fit
3. Model checking

This is done iteratively, in that whenever model checking fails, one attempts to respecify and refit the model until the new round of checks are satisfactory. This is a familiar process, for example, with regression models. We emphasize that the same process is appropriate for checking the more complex specification of a multilevel model.

The new feature in checking the Poisson two-level model involves testing the assumption that the level two distributions for the Poisson rates λ_i follow the gamma distributions specified in Eq. (5). We do this by using the marginal distributions of Eq. (9) for the deaths z_i, $i = 1, \ldots, 131$. Assuming the level one distributions in Eq. (4) are valid, then the negative binomial distributions (9) hold if and only if the gamma model in Eq. (5)

is true. Thus, we only need to test whether the $\{z_i\}$ follow the distributions presented in Eq. (9).

The conditional independence assumption for the z_i's in Eq. (4) follows because, given $\{\lambda_i\}$, the deaths in one transplant program do not affect those in others. (Technically, we also should check that deaths within a transplant program are independent. While this is plausible, we have no means for verifying it at this time.) Analysis of the individual observations within each transplant program would also provide an empirical basis for validating the level one model. Such methods for checking Eq. (4) are well developed and well known and do not involve the features of hierarchical models. (See, for example, Weisberg, 1985; McCullagh and Nelder, 1989.) For the remainder of this chapter, we accept the model of Eq. (4) and focus on fitting and checking the two-level model defined by Eqs. (4) and (5).

For the e_i legitimately to be called exposures, the mean μ_i of Eq. (5) must not depend on e_i. The initial model then will fit all 131 programs and assume that $\log(\mu_i) = \beta_0$ is unknown, but not dependent on e_i. In this case, with no other hospital-level covariates available, the e_i's are exposures if and only if the λ_i's in Eq. (5) are exchangeable, meaning that because they are independent, the λ_i's are identically distributed. (Aggregate patient-level data could be used as hospital-level data. Results from using aggregated data require careful interpretation. See Section 6 for further information.)

5.1 Specifying, Fitting, and Checking Model Reg0.k131

Our first iteration involves fitting all $k = 131$ transplant programs to the exchangeable model of Eq. (5), which we label Reg0.k131 (meaning the regression model with $r = 1$, i.e., no covariates, and with $k = 131$ transplant programs). Table 1 reports results for this model and for several other two-level Poisson models that follow the structure in Section 3.

The PRIMM program outputs the estimated hyperparameters, their standard errors and correlations, the shrinkage values, and the posterior means and variances, among other items (see an example of the output in Figure 6 and additional information on the program in Appendix A.3.) In all models reported here, z_0 is the default computer value, listed in Table 1, which is always less than 1 in these cases and always much less than ζ. As long as ζ is estimated to be larger than z_0, this choice of prior information is conservative, because large values of ζ indicate very little level two heterogeneity; see Eq. (5).

For model Reg0.k131, the chi-square "2*adjpost.ratio" is 189.1 on 129 degrees of freedom, which is much too large to justify the fitting of

Table 1   Estimation Summaries for Three Models

Model	z_0	2*adjpost.ratio	k-(r+1)	$\hat{\beta}_0$	$\hat{\beta}_1$ (s.e.)	$\hat{\zeta}$	log($\hat{\zeta}$) (s.e.)	ave. $\hat{B}_i$	max. $\hat{B}_i$	min. $\hat{B}_i$
k=131										
Reg0.k131	.08	189.1	129	.05	-	5.30	1.67 (.42)	.73	.99	.29
Reg1.k131	.08	177.7	128	.31	-.27 (.08)	5.83	1.76 (.43)	.74	.97	.40
k=37										
Reg0.k37	.13	43.2	35	.76	-	2.54	.93 (.86)	.76	.95	.38
Reg1.k37	.13	42.3	34	.57	-.19 (.25)	2.43	.89 (.85)	.74	.92	.42
k=94										
Reg0.k94	.53	136.1	92	-.05	-	6.37	1.85 (.48)	.71	.93	.35
Reg1.k94	.53	134.3	91	.11	-.13 (.11)	6.29	1.84 (.47)	.71	.90	.39

Model	ave. y	max.y	min.y	ave. $\hat{\mu}_0$	max. $\hat{\mu}_0$	min. $\hat{\mu}_0$	ave. $\hat{\lambda}_i$	max. $\hat{\lambda}_i$	min. $\hat{\lambda}_i$
k=131									
Reg0.k131	1.38	15.15	0.00	1.05	1.05	1.05	1.05	1.68	.49
Reg1.k131	1.38	15.15	0.00	1.35	3.05	.71	1.35	3.31	.46
k=37									
Reg0.k37	2.35	15.15	0.00	2.18	2.18	2.18	2.17	3.90	1.32
Reg1.k37	2.35	15.15	0.00	2.56	3.82	1.80	2.53	4.68	1.24
k=94									
Reg0.k94	1.00	3.87	0.00	.96	.96	.96	.96	1.42	.51
Reg1.k94	1.00	3.87	0.00	1.00	1.24	.81	1.00	1.52	.49

Reg0 means no explanatory variable is used in the model; Reg1 means one explanatory variable (e_i) is included. The notation "k131" means that the number of programs is 131, similarly for k37 and k94.

one constant as the common death rate for all programs. Equivalently, this may be thought of as rejecting a test that ζ is infinite. In fact, the fitted value for ζ is 5.3. Coupled with the estimate $\hat{\beta}_0 = .05$ the fitted model, from Eq. (5) is

$$\lambda_i \mid \underset{\sim}{\alpha} \sim 1.05 \frac{\text{Gam}(5.3, 1)}{5.3}, \qquad i = 1, \ldots, k = 131 \qquad (15)$$

In this model λ_i is approximately distributed χ^2_{11}, standardized to have deaths exceeding expected deaths (exposure) by 5%.

Other information provided by the PRIMM printout includes the shrinkages $\{\hat{B}_i\}$ for each program, which range between .29 for the larger programs and .99 for those that treated just one patient. These values average to .73, meaning that the estimated ratios $\{y_i\}$ are shrunk, on average, 73% of the way to the fitted value $\hat{\mu}_i = 1.05$. Although the maximum and the minimum y_i values range from 0 to 15.15, these estimates contain excessive variability, and the PRIMM estimates range only from .49 (program 122) to 1.68 (program 28). This is an example of regression to the mean.

Having fit our initial model, Reg0.k131, we proceed with graphical and formal statistical checking procedures. Figure 3 graphs the Pearson residuals, computed from Eq. (9).

$$r_i = \frac{z_i - e_i\mu_i}{\sqrt{e_i\mu_i/B_i}} \tag{16}$$

with fitted values $\hat{\zeta} = 5.3$ and $\hat{\mu}_i = 1.05$ for these data. (The PRIMM program gives these residuals as optional output.) For the horizontal axis we use the observation number which, except for ties, ranks the transplant programs by the number of patients, n_i. We use this scale because it provides the best use of graph space and minimizes the amount of over-

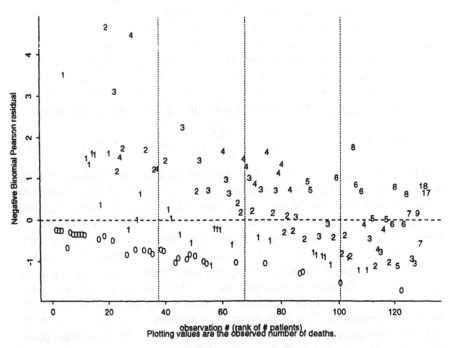

Figure 3 The negative binomial Pearson residuals are plotted against the rank of the number of patients n_i. This plot shows problems with the model specification and the exchangeability assumption, particularly for the subgroup of the 37 smallest programs.

plotting. The numbers plotted are the actual number of observed deaths, z_i.

Readers are cautioned that the distributions involved are negative binomial and that because ζ is not large, the distributions of the residuals will be long tailed and skewed positively. Thus, Figure 3 looks different from standard residual plots for normal data. Negative binomial residual plots will have many points near but below the expected value, zero, and fewer above but some rather far above zero. One also expects to see that the average values and standard deviations of the points in every vertical section (chosen here as the quartiles based on n_i) are near zero and one, respectively, given the correct specification of the level two model. We actually see a higher mean and a larger variance for the small transplant programs on the left of Figure 3 than for the large ones on the right. This suggests, at minimum, that the number of patients, n_i, which is highly correlated with exposure, e_i (the correlation is .99) explains something not captured in the initial model Reg0.k131. The downsloping line pattern in the plotted z_i values is explained at the end of this section.

To illustrate more precisely the heterogeneity of Figure 3, we separate the transplant programs into four roughly equal-size groups, based on the quartiles of n_i. The first quartile contains the 37 smallest programs, those that treated fewer than 10 patients. The next three quartiles include those with $10 \leq n_i \leq 18$, $19 \leq n_i \leq 40$, and $41 \leq n_i \leq 152$. In Table 2, model Reg0.k131, the average residual, $\bar{r}_j$, appears for each quartile, $j = 1, 2, 3, 4$, and the standard deviation s_j also is listed. The chi-square tests,

$$\sqrt{M_j - 1}\, \bar{r}_j \sim N(0, 1) \quad \text{and} \quad \sqrt{\frac{M_j - 1}{2}}\, \log(s_j^2) \sim N(0,1) \tag{17}$$

test the $\bar{r}_j$ and the s_j^2, $j = 1, 2, 3, 4$, against their nominal values of zero and one, respectively. The mean of the 37 residuals of the transplant programs with less than 10 patients is .653, which is 3.9 nominal standard deviations away from the nominal value zero. This is a very bad fit, as Figure 3 suggests visually. The standard deviation for these 37 programs, $s_1 = 1.473$, from Table 2, is 3.3 standard errors away from the nominal value of one (logarithmic scale, using the delta method, and a chi-square approximation for these nonnormal data), which strongly suggests extra variation. The other three quartiles fit better using this model, although the largest programs (fourth quartile, $j = 4$) show some tendency to be different from the others. It seems, from this fit, that the small transplant programs either do not fit the same model as the larger ones or perhaps that a covariate must be introduced to improve the model fit. Either of these involves a new specification, and thus we iterate to a new sequence of model specification, model fitting, and model checking.

5.2 Specifying, Fitting, and Checking Model Reg1.k131

The next model, Reg1.k131, includes an explanatory variable in the form of $x_i = \log(e_i)$. We alternatively might also have included a function of n_i, the number of patients, as a second covariate. However, n_i is highly collinear with e_i and so that provides little help. This Poisson model is "nonexchangeable" if the regression coefficient on x_i differs from 0 because the distribution (5) is not exchangeable when μ_i is not constant.

The covariate e_i is considered for two reasons. First, if e_i truly acts as exposure, then it should not have further predictive power. Fitting a nonexchangeable model (5) that includes $\log(e_i)$ as a covariate means that the negative binomial distributions in Eq. (9) have expected values, assuming the log link,

$$E\left(\frac{z_i}{e_i}\right) \equiv \mu_i = \exp(\beta_0 + \beta_1 x_i) = \exp(\beta_0)e_i^{\beta_1} \tag{18}$$

Thus, $e_i \dagger \equiv e_i^{(1 + \beta_1)}$ would be a more appropriate version of exposure if β_1 is not equal to zero, implying that e_i is not a valid exposure. Second, if $\log(e_i)$ predicts, then the model (5) is not exchangeable. This is true for any covariate that has predictive power. Using $\log(e_i)$ as a covariate and testing that $\beta_1 = 0$ gives us one way of checking that $\{e_i\}$ is acting as true exposures and that the exchangeability assumption holds for the gamma distributions, Eq. (5).

Another covariate choice might have been a function of the standardized exposures $\{\bar{e}_i\}$, $\bar{e}_i \equiv e_i/n_i$ in Eq. (3). The reader is encouraged to investigate further the use of this and other functional forms of $\{e_i\}$ and $\{n_i\}$ in the regression model and as exposures.

The exposures vary dramatically from .057 to 12.1 and have a long right tail; their logarithms vary from -3.2 to 2.2. We have seen already (see Figure 1) that the histogram of the logarithm of the 131 exposures shows a long left tail. We prefer a transformation of the explanatory variable that reduces the leverage of a few outlying points. For now, we refit the data using $\log(e_i)$ knowing that the programs with small e_i's might have high leverages in this model specification.

The results, corresponding to Reg1.k131, are reported in Table 1. Note that the introduction of the explanatory variable has acted to provide a slightly better fit, and correspondingly slightly larger shrinkages B_i, but not nearly enough to adopt the regression curve. This is again indicated by the 2*adjpost.ratio calculation. From Table 1, $\hat{\beta}_1 = -.27$, differs from zero by more than three standard errors. Thus, in this model, e_i seems not to be a valid measure of exposure if required to govern the full range of programs, but values near e_i^{73} would serve better as exposures. Note that $\hat{\beta}_0 = .31$ provides a mean of $\hat{\mu}_i = \exp(.31) = 1.4$ deaths per expected death for a program with expected death $e_i = 1$. One expects transplant programs serving about 12 patients to have exposures in this range.

The residuals from this Reg1.k131 model are shown in Figure 4. Although this is an improvement over Figure 3, there still is the appearance of a "smile" for the mean, rather than a constant mean, and of larger variability at the two end points. The calculations of the mean and the standard deviation of the residuals in four subsets of the data, $\bar{r}_j$ and s_j, $j = 1, 2, 3, 4$, in Table 2 under Reg1.k131 bear this out. Although the small transplant programs in the first quartile, $j = 1$, seem to fit much better now (e.g., $\bar{r}_1$ is only 1.23 nominal standard errors removed from zero), the fit remains unsatisfactory. Quartiles 2–4 show a monotone increasing pattern in the mean. The first quartile denies this monotonicity. These 37 programs account for only 4.8% of all patients served, but because these points have high leverages due to their outlying values for $\log(e_i)$, they seem to have distorted the linear fit for the remaining pro-

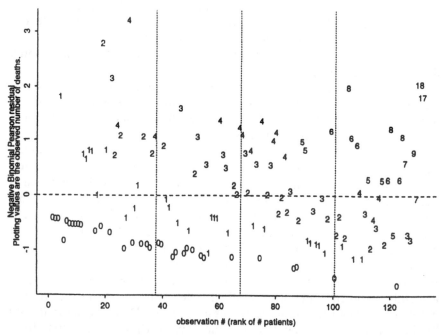

Figure 4 Although less severe than in Figure 3, this residual plot for the nonexchangeable model Reg1.k131 shows problems with model fit. The plotted values are, again, the observed death counts.

grams. Also, the standard deviation, s_1, of these points is still large and still not in accord with the values of s_2, s_3, and s_4 for the other three quartiles.

5.3 Specifying, Fitting, and Checking Model Reg1.k94

We propose next to investigate the fit of our model to the 94 transplant programs that served 10 or more patients (quartiles 2, 3, and 4). Not only does this seem justified on the basis of the analyses thus far, but other studies commonly omit programs with fewer than 10 patients because they are viewed as too small to analyze. Note here that the model described in Section 3 allows any program, no matter how small, to be included for analytical reasons. Estimates for programs with even one patient can be made, by borrowing strength from all the others. Indeed, one even can estimate the effectiveness of programs that have served no patients with this model, assuming only that it fits Eqs. (4)–(6). Eliminating these 37 programs (28% of the 131 programs) accounts for only 175 patients (4.8% of the 3646 total patients) and 35 of the 312 deaths (11.2% of the deaths).

The third iteration uses the same model specification as Reg1.k131 but restricts the model to the 94 transplant programs that served at least 10 patients. This model, again with the $x_i = \log(e_i)$, is denoted Reg1.k94 with results reported in Table 1. Despite the good fit, the average shrinkage (71%) is diminished because shrinkage is less for observations with more data (larger exposures) as these observations have. The new $\beta_1 = -.13$ in this model is only slightly more than one standard error from zero, so that, by common inference standards, e_i would not be rejected as a valid exposure. Further, $\beta_0 = .11$, meaning that transplant programs with $e_i = 1$ (roughly, the smallest programs in this analysis of the 94 programs) are estimated to have ratios exceeding exposure by only $1 - \exp(.11) = 12\%$. The convolution parameter $\zeta = 6.29$, making the rate λ_i in Eq. (5) an approximate scaled chi-square with 13 degrees of freedom and with mean $\hat{\mu}_i = 1.12/e_i^{-.13}$.

Shrinkage in this model can be expressed as

$$\hat{B}_i = \frac{6.37}{6.37 + 1.12e_i^{.87}} \tag{19}$$

using the values from Table 1. Large shrinkage occurs when a program has a small exposure to death, that is, small e_i. Of course, e_i is correlated positively with n_i, so in general, shrinkage is large for the small transplant programs.

The residual plot, Figure 5, appears satisfactory. The means in all three quartiles are only a fraction of a standard deviation away from the

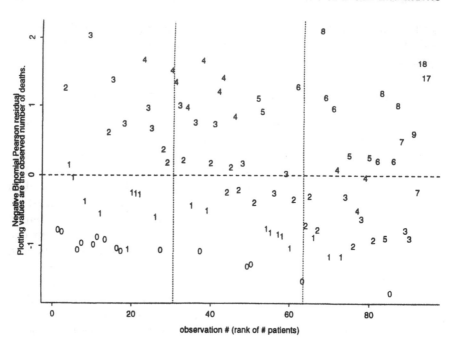

Figure 5 The horizontal axis in this residual plot for our final model Reg1.k94 is the rank of e_i. (Note that Figures 3 and 4 use the rank of n_i as the horizontal axis.) The downsloping linear patterns in observations with the same z_i can be explained by the residual formula (16) and the simplified form (20) for the case of $z_i = 0$.

nominal value zero; see Reg1.k94, Table 2. The three standard deviations, s_2, s_3, and s_4, are similar to each other although all are less than one. For reference, we include the residuals of the 37 observations excluded from this analysis. They were not used to estimate the hyperparameter $\alpha = (\zeta, \beta)$, and they fit the assumptions poorly. For example, the average residual for that quartile is 2.78 standard errors above zero.

To conclude, we adopt this final model, Reg1.k94, as the best we can do for these data. This choice helps to remind us that there is still some slight pattern, so we will continue to investigate the validity of the exposures used as we look forward to enlarging the data set with more years of information and as we consider possible missing components that need to be represented in computing the exposure from Eq. (1). The reader might prefer, given $\hat{\beta}_1$ does not differ significantly from zero, to use the

Table 2   Residuals

Model	$j = 1,2,3,4$	#programs $= M_j$	$\bar{r} \equiv$ mean	$s \equiv$ st.dev.	(max-min)	$(M_j - 1)\,\bar{r}_j^2 \sim \chi^2_1$	$\dfrac{\{M_j-1\}\{\log(s_j^2)\}^2}{2}$ $\sim \chi^2_1$
Reg0.k131	all	131	.118	1.130	6.36	1.34^2	1.97^2
	1st quartile*	37	.653	1.473	5.51	3.92^2	3.29^2
	2nd quartile	30	-.006	.969	3.34	$-.03^2$	$.24^2$
	3rd quartile	33	.006	.883	3.16	$.04^2$	$.99^2$
	4th quartile	31	-.283	.799	3.45	-1.55^2	1.74^2
Reg1.k131	all	131	.001	.952	4.86	$.01^2$	$.80^2$
	1st quartile	37	.204	1.079	4.19	1.23^2	$.65^2$
	2nd quartile	30	-.205	.858	2.74	-1.11^2	1.17^2
	3rd quartile	33	-.053	.847	2.86	$-.30^2$	1.33^2
	4th quartile	31	.015	.977	3.69	$.08^2$	$.18^2$
Reg1.k94	1st, 2nd, 3rd	94	-.006	.929	3.79	$-.06^2$	1.00^2
	1st quartile	37	.464	1.290	4.88	2.78^2	2.16^2
	2nd quartile	30	-.045	.955	3.12	$-.24^2$	$.35^2$
	3rd quartile	33	.062	.915	3.17	$.35^2$	$.72^2$
	4th quartile	31	-.041	.946	3.79	$-.22^2$	$.43^2$

* Quartiles are based on number of patients, n_j.
1st quartile: $1 \le n_j \le 9$ 2nd quartile: $10 \le n_j \le 18$ 3rd quartile: $19 \le n_j \le 40$ 4th quartile: $41 \le n_j \le 152$.
The first quartile for Reg1.k94 is in italics to distinguish it from the other quartiles. The first quartile data are not part of the fitted model Reg1.k94, but the model was used to calculate their residuals.

model Reg0.k94, accepting e_i as a legitimate "exposure" and so return to an exchangeable model for the 94 transplant programs with n_i greater than or equal to 10. Reg0.k94 would be preferred over Reg1.k94 if ecological correlation (discussed briefly in Section 6, item 2) is a problem because to account for ecological correlation, the regression coefficients must be specified rather than estimated. We include the Reg0.k94 values in Table 1 for reference, but because, from additional analyses, it appears that ecological correlation is not a problem for these data, we prefer to use Reg1.k94.

Readers may wonder about the patterns represented by the z_i values 0–18 in Figure 5 and also in Figures 3 and 4. The horizontal axis in Figure 5 involves the rank of e_i rather than the rank of n_i (as in Figures 3 and 4) to further show the pattern that transplant programs with $z_i = 0$ (the bottom curve with zeros) produce a downsloping curve in their residuals. There is a similar pattern for $z_i = 1, 2, 3$, etc. Graphs like this, although they look strange, are common when working with discrete data. In this case, for example, the curve for $z_i = 0$ is given by

$$r_i = \frac{-e_i\mu_i}{\sqrt{e\mu_i/B_i}} = -\sqrt{\frac{e_i\mu_i}{1 + (e_i\mu_i/\zeta)}} \tag{20}$$

As the subscript i increases, this is a monotonic decreasing convex function of $e_i\mu_i$ and therefore, because μ_i is fairly constant, it is a monotone decreasing convex function of e_i. Each larger value of z_i principally acts to cause an upward shift of these curves that are no longer monotone. Because of the high correlation between e_i and n_i, Figures 3 and 4 have similar patterns.

6 CONCLUSION

With three iterations of model specification, model fitting, and model checking, we draw the following conclusions regarding the heart transplant data just analyzed. First, small transplant programs have 30-day mortality ratios that are different from those of larger programs. This is partly because the value used for exposure, the predicted number of deaths, is not a true exposure for the set of 131 programs. A better exposure would be e_i^{73}, based on the regression coefficient from model Reg1.k131. This result also indicates a violation of exchangeability. To fix that, we performed a subset analysis on the 94 largest programs. After our analyses of these data, we prefer using model Reg1.k94. This estimates exposure as e_i^{87}, but it is not significantly different from e_i^1.

Second, residual plots, although different for discrete data models, still help to check model fit. Simple graphical methods, such as histograms, are valuable when considering functions of independent variables for model specification. Third, sorting the data helps one to understand it better, and simple statistics, such as means, standard deviations, ranges, and cumulative sums allow the data to "speak" more clearly to the researcher. Fourth, extra variation in the data, as shown in Figure 2, is handled best through hierarchical modeling techniques.

Several topics are not fully addressed in this work. Comparisons of the results presented here with other statistical models, e.g., a generalized linear overdispersed Poisson model, and other statistical methods, e.g., Monte Carlo sampling, are left to the reader.

The results of our analyses on the smaller 37 programs are included in Table 1. Does the model in Section 3 fit these data? Are the unobserved ratios, λ_i, exchangeable? What conclusions can be made about these 37 programs?

Although this chapter focuses on fitting and checking a two-level model for transplant data, a profile of the programs' mortality rates would be the final goal of the analyses. Readers may want to pursue this for themselves. To complete a profile analysis for the 94 heart transplant programs, one would work with the distributions fitted to Eqs. (12)–(14) and the estimated values of their means and their variances. From the partial output in Figure 6, observations 4 and 93 have similar "lam^" (posterior mean) estimates, 1.211 and 1.228, respectively. What statements can be made about the performance of these programs based on their estimated means and standard deviations if one approximates their true ratios of observed to expected deaths as having these gamma distributions?

$$\lambda_4 \sim \frac{.4892^2}{1.211} \text{ Gam} \left(\frac{1.211^2}{.4892^2}, 1 \right)$$

$$\lambda_{93} \sim \frac{.2667^2}{1.228} \text{ Gam} \left(\frac{1.228^2}{.2667^2}, 1 \right) \tag{21}$$

If programs $i = 38, \ldots, 131$ perform transplant surgeries on n_i "average" patients (45-year-old white males without congenital heart disease who were home-bound prior to transplant surgery and who were receiving their first transplant), the model Reg94.r1 can be used to estimate the program's performance for such individuals. The estimated expected value, denoted $\hat{\lambda}_i^{**}$, based on the approximations to Eqs. (10)–(12), are listed in the final column of Table A. The 10 programs with the smallest estimated expected relative death rates, $\hat{\lambda}_i^{**}$, are observations {122, 100, 127, 121, 110, 107, 113, 126, 118, 86}. A standard profile analysis, using p-values calculated from the statistic $(z_i - e_i)/\sqrt{e_i}$ and (falsely) assuming normally distributed data, ranks these programs as (1, 2, 7, 3, 5, 4, 6, 13, 11, 8), where 1 = the smallest ratio, etc.

In concluding, we offer three cautionary notes that arise in profiling applications. The first involves biases that can be caused by using grouped data to draw inferences. The second warns that the ecological regression problem could cause biases if one does not control for hospital effects in the level one regression. The third is that not using key data that distinguish among different patient severity levels can lead to unfair comparisons in a hospital profile analysis.

1. Inferences using aggregate data, here patient-level data grouped within programs, require special care. In our data, at the individual level the "repeat transplant" indicator variable sharply re-

$header:

```
                                                                                    [,1]
[1,] "                                         ************* Model Reg1.k94 *************"
[2,] "                                                       Tue Oct 18 18:04:01 EDT 1994"
[3,] "                                              Hierarchical Poisson Regression Model II"
[4,] "                              Copyright Oct. 1992 C. L. Christiansen & C. N. Morris, primm (v2)"
[5,] "                                  NOTE:  IN THIS VERSION (v2), r = # REGRESSION COEFFICIENTS"
```

$lrt:

k	94.0000000
saturated	-225.3326274
max(log.adjpost)	-292.4793758
2*adjpost.ratio	134.2934967
k-(# regr.coef)-1	91.0000000

$e.mu0:

	e0	munot0	e0*munot0
prior	0.532	0.9992	0.5316

$hyperparameter:

	estimate	se	est/se
beta^0	0.07177	0.1235	0.5812
beta^1	-0.13270	0.1106	-1.2010
zeta	6.29100	2.9380	2.1410
log(zeta)	1.83900	0.4670	3.9380

$correlation:

	beta^0	beta^1	log(zeta)
beta^0	1.00000	-0.78890	-0.04893
beta^1	-0.78890	1.00000	0.06296
log(zeta)	-0.04893	0.06296	1.00000

$output:

	obs#	y	munot^	B^	lam^	se(lam^)	lam^0.025	lam^0.975
1	38	0.0000	1.2440	0.9048	1.1260	0.5098	0.3589	2.323
2	39	0.0000	1.2270	0.8978	1.1010	0.4937	0.3563	2.259
3	40	2.9760	1.2000	0.8863	1.4020	0.5401	0.5519	2.642
4	41	1.3850	1.1870	0.8801	1.2110	0.4892	0.4504	2.342
5	42	1.1060	1.1480	0.8584	1.1420	0.4521	0.4354	2.184
6	43	0.0000	1.0970	0.8227	0.9023	0.3809	0.3171	1.788
7	44	0.0000	1.1400	0.8532	0.9724	0.4172	0.3345	1.945
8	45	0.7117	1.0770	0.8061	1.0060	0.3862	0.3973	1.892
9	46	3.8660	1.1750	0.8734	1.5150	0.5552	0.6303	2.782
10	47	0.0000	1.1290	0.8462	0.9553	0.4080	0.3306	1.906
85	122	0.0000	0.8901	0.5497	0.4893	0.2205	0.1572	1.007
86	123	0.9927	0.8853	0.5404	0.9347	0.2720	0.4800	1.538
87	124	1.4370	0.8945	0.5581	1.1340	0.3120	0.6074	1.822
88	125	1.1420	0.8838	0.5373	1.0030	0.2822	0.5286	1.627
89	126	0.4801	0.8816	0.5331	0.6942	0.2359	0.3117	1.227
90	127	0.4284	0.8693	0.5082	0.6525	0.2240	0.2901	1.159
91	128	1.1460	0.8572	0.4832	1.0070	0.2657	0.5551	1.590
92	129	0.7312	0.8369	0.4399	0.7777	0.2200	0.4080	1.265
93	130	1.4940	0.8144	0.3906	1.2280	0.2667	0.7624	1.804
94	131	1.4010	0.8138	0.3892	1.1730	0.2581	0.7226	1.730

Figure 6 The S-PLUS statement primm(e, z, log(e) − 2.379, title = "Model Reg1.k94", labelobs = 38:131) produces the output shown. (Note that observations 11–84 were omitted to conserve space.)

duces a patient's estimated survival probability. Repeat transplant patients are more than three times as likely to die within 30 days of surgery as are those receiving their first transplants [the estimated odds ratio is $3.6 = \exp(1.29)$, from Eq. (1)]. However, in the hospital-level model, one finds only an insignificant effect when fitting the percentage of repeat transplant patients (a "grouped" variable), as in Reg1.k94 of Section 5. These seemingly contradictory results actually are consistent with one another. The level one regression concludes that repeat transplant patients have lower survival probabilities, while the level two result is consistent with the programs being equally effective at treating repeat transplant patients. Bendel and Carlin (1990) discuss the Bayesian analysis of aggregate data and the related issue of "ecological correlation."

2.   Developing the exposure measure as in the level one model, Eq. (1), risks confounding some of the hospital differences with the exposure value. The consequent use of this measure in the level two analysis then could remove some of the hospital performance measurement that is meant to be part of the profile evaluation. This is discussed further by Gilks (1995) and Morris and Christiansen (1995) as a problem in ecological regression or Simpson's paradox (Simpson, 1951). We have recalculated the exposures $\{e_i\}$, controlling for hospital effects, and have found that this adjustment makes little difference for these data.

3.   Some generally available patient variables that may be predictive of mortality, such as time on the transplant waiting list, donor characteristics, and socioeconomic status, are not yet in our public-use data set. Missing variables can cause bias in the estimates of patient survival rates in hospitals. While the data used here adequately demonstrate hierarchical model checking techniques, in the absence of all available data, these analyses cannot provide a definitive study on heart transplant mortality. Despite this, the exercise just completed encourages us further to recommend using hierarchical models for medical profiling, with all relevant data.

APPENDIX A.1   PATIENT-LEVEL DATA AND ANALYSIS

Through Statlib, we provide hospital-level data for the reader's use (file "heart.dat"). However, one can obtain the public-use data set of the heart transplant patients (and of other organ transplant patients) for the October

1, 1987 through December 31, 1989 time period from UNOS. Their address is United Network for Organ Sharing, 1100 Boulders Parkway, Suite 500, P.O. Box 13770, Richmond, VA 23225.

Congenital heart disease afflicts 5% of the 3646 patients in this work, 2% of the patients have had a previous heart transplant, and 19% of the patients are female. The patients are described in one of six general health categories prior to transplantation: working full-time, 0.4%; working part-time, 2%; home-bound, 44%; hospitalized, 11%; in intensive care, 24%; on life support, 18%. For our purposes, we combine working full-time and working part-time into one category called "working." Blacks account for 8% of the patients, 87% of the patients are white, and the remaining 5% of the patients are either Native American, Oriental, or Hispanic ("other race"). Patient ages range from 0 (less than 6 months) to 70 years with a median of 50 years and a mean of 45 years.

The intercept term of Eq. (1) represents the estimated probability of mortality within 30 days after surgery, $\exp(-2.93)/(1 + \exp(-2.93)) = .05$, for a 45-year-old, male, white, "home-bound prior to transplant surgery" patient without congenital heart disease who received his first transplant. Patient ages are centered at 45 years to reduce the estimated partial correlation and to represent the full parabolic shape of the relationship between age and death. The estimated effects of "black," "other race," and health classifications "hospitalized" and "in intensive care" are not statistically different from zero (at the .05 level) but are included in the logistic model for predictive purposes. All variables except the age variable are binary indicators with 1 representing the named characteristic. A positive estimated coefficient of a characteristic means that characteristic tends to increase the probability of death. Note that patients are classified into one health status (prior to surgery) characteristic in line 3 of Eq. (1) or into the omitted category "home-bound," and into one race category, line 4 of Eq. (1), or into the omitted category "white."

APPENDIX A.2 HOSPITAL-LEVEL DATA

The heart transplant program data for the 131 programs are in the first four columns of Table A. The data are sorted by increasing number of patients, n_i. The programs' exposures to mortality, $\{e_i\}$, is the second sort. The random variable z_i is the number of deaths within 30 days after heart transplant surgery. These data span a 27-month period. The next six columns are derived for use in some of the informal model-checking techniques discussed in Section 4. The final column is discussed in Section

Table A Hospital-Level Data, Diagnostics, and Estimates

obs	e_i Data	z_i	n_i	$y_i \equiv z_i/e_i$	z_i^*	e_i^*	z_i^*/e_i^*	$\bar{e}_i \equiv e_i/n_i$	n_i^*	$\hat{\lambda}_i^{**}$
					*indicates the cumulative sum					
1	0.057	0	1	0.00	0	0.06	0.00	0.057	1	
2	0.064	0	1	0.00	0	0.12	0.00	0.064	2	
3	0.064	0	1	0.00	0	0.18	0.00	0.064	3	
4	0.066	1	1	15.15	1	0.25	3.98	0.066	4	
5	0.462	0	1	0.00	1	0.71	1.40	0.462	5	
6	0.086	0	2	0.00	1	0.80	1.25	0.043	7	
7	0.114	0	2	0.00	1	0.91	1.10	0.057	9	
8	0.117	0	2	0.00	1	1.03	0.97	0.059	11	
9	0.118	0	2	0.00	1	1.15	0.87	0.059	13	
10	0.119	0	2	0.00	1	1.27	0.79	0.059	15	
11	0.126	0	2	0.00	1	1.39	0.72	0.063	17	
12	0.231	1	2	4.33	2	1.62	1.23	0.116	19	
13	0.261	1	2	3.83	3	1.88	1.59	0.131	21	
14	0.211	1	3	4.74	4	2.10	1.91	0.070	24	
15	0.216	1	3	4.63	5	2.31	2.16	0.072	27	
16	0.218	0	3	0.00	5	2.53	1.98	0.073	30	
17	0.648	1	3	1.54	6	3.18	1.89	0.216	33	
18	0.143	0	4	0.00	6	3.32	1.81	0.036	37	
19	0.144	2	4	13.89	8	3.46	2.31	0.036	41	
20	0.210	1	4	4.76	9	3.68	2.45	0.053	45	
21	0.244	0	4	0.00	9	3.92	2.30	0.061	49	
22	0.530	3	6	5.66	12	4.45	2.70	0.088	55	
23	0.796	2	6	2.51	14	5.24	2.67	0.133	61	
24	1.622	4	6	2.47	18	6.87	2.62	0.270	67	
25	0.561	2	7	3.56	20	7.43	2.69	0.080	74	
26	0.762	0	7	0.00	20	8.19	2.44	0.109	81	
27	1.222	1	7	0.82	21	9.41	2.23	0.175	88	
28	0.513	4	8	7.80	25	9.93	2.52	0.064	96	
29	0.532	0	8	0.00	25	10.46	2.39	0.066	104	
30	0.936	1	8	1.07	26	11.39	2.28	0.117	112	
31	0.503	1	9	1.99	27	11.90	2.27	0.056	121	
32	0.564	0	9	0.00	27	12.46	2.17	0.063	130	
33	0.572	2	9	3.50	29	13.03	2.22	0.064	139	
34	0.583	0	9	0.00	29	13.62	2.13	0.065	148	
35	0.722	0	9	0.00	29	14.34	2.02	0.080	157	
36	0.771	2	9	2.59	31	15.11	2.05	0.086	166	
37	1.873	4	9	2.14	35	16.98	2.06	0.208	175	
38	0.532	0	10	0.00	35	17.51	2.00	0.053	185	1.135
39	0.584	0	10	0.00	35	18.10	1.93	0.058	195	1.123
40	0.672	2	10	2.98	37	18.77	1.97	0.067	205	1.457
41	0.722	1	10	1.38	38	19.49	1.95	0.072	215	1.270
42	0.904	1	11	1.11	39	20.40	1.91	0.082	226	1.218
43	1.236	0	11	0.00	39	21.63	1.80	0.112	237	1.002
44	0.950	0	12	0.00	39	22.58	1.73	0.079	249	1.032
45	1.405	1	12	0.71	40	23.99	1.67	0.117	261	1.124
46	0.776	3	13	3.87	43	24.76	1.74	0.060	274	1.550
47	1.013	0	13	0.00	43	25.78	1.67	0.078	287	1.012
48	0.739	0	14	0.00	43	26.51	1.62	0.053	301	1.046
49	1.770	1	14	0.56	44	28.28	1.56	0.126	315	1.054
50	0.821	0	15	0.00	44	29.10	1.51	0.055	330	1.023

Table A, continued

obs	e_i Data	z_i	n_i	$y_i \equiv z_i/e_i$	z_i^*	e_i^*	z_i^*/e_i^*	$\bar{e}_i \equiv e_i/n_i$	n_i^*	$\hat{\lambda}_i^{**}$
					*indicates the cumulative sum					
51	1.115	2	15	1.79	46	30.22	1.52	0.074	345	1.290
52	1.164	3	15	2.58	49	31.38	1.56	0.078	360	1.436
53	1.164	0	15	0.00	49	32.55	1.50	0.078	375	0.972
54	1.303	0	16	0.00	49	33.85	1.45	0.081	391	0.946
55	1.774	3	16	1.69	52	35.62	1.46	0.111	407	1.319
56	3.585	1	16	0.28	53	39.21	1.35	0.224	423	0.866
57	1.193	1	17	0.84	54	40.40	1.34	0.070	440	1.105
58	1.213	1	17	0.82	55	41.62	1.32	0.071	457	1.102
59	1.232	1	17	0.81	56	42.85	1.31	0.072	474	1.099
60	1.517	4	17	2.64	60	44.36	1.35	0.089	491	1.494
61	1.520	3	17	1.97	63	45.88	1.37	0.089	508	1.349
62	1.862	3	17	1.61	66	47.75	1.38	0.110	525	1.296
63	1.888	1	17	0.53	67	49.64	1.35	0.111	542	1.014
64	1.247	0	18	0.00	67	50.88	1.32	0.069	560	0.939
65	1.381	2	18	1.45	69	52.26	1.32	0.077	578	1.216
66	1.643	2	18	1.22	71	53.91	1.32	0.091	596	1.177
67	1.660	4	18	2.41	75	55.57	1.35	0.092	614	1.457
68	1.827	4	19	2.19	79	57.39	1.38	0.096	633	1.420
69	1.486	3	20	2.02	82	58.88	1.39	0.074	653	1.327
70	1.593	2	20	1.25	84	60.47	1.39	0.080	673	1.168
71	2.265	4	20	1.77	88	62.74	1.40	0.113	693	1.345
72	1.524	1	21	0.66	89	64.26	1.38	0.073	714	1.029
73	1.759	3	21	1.71	92	66.02	1.39	0.084	735	1.276
74	1.309	0	22	0.00	92	67.33	1.37	0.060	757	0.907
75	1.529	4	24	2.62	96	68.86	1.39	0.064	781	1.427
76	1.677	1	24	0.60	97	70.54	1.38	0.070	805	0.993
77	1.654	2	25	1.21	99	72.19	1.37	0.066	830	1.126
78	1.785	3	25	1.68	102	73.97	1.38	0.071	855	1.243
79	1.979	4	25	2.02	106	75.95	1.40	0.079	880	1.347
80	1.767	4	26	2.26	110	77.72	1.42	0.068	906	1.373
81	2.465	2	26	0.81	112	80.18	1.40	0.095	932	1.026
82	1.750	2	27	1.14	114	81.94	1.39	0.065	959	1.103
83	2.458	4	27	1.63	118	84.39	1.40	0.091	986	1.268
84	2.383	2	28	0.84	120	86.78	1.38	0.085	1014	1.024
85	2.717	3	28	1.10	123	89.49	1.37	0.097	1042	1.111
86	2.282	0	32	0.00	123	91.78	1.34	0.071	1074	0.772
87	2.115	0	33	0.00	123	93.89	1.31	0.064	1107	0.782
88	2.852	2	33	0.70	125	96.74	1.29	0.086	1140	0.957
89	2.856	5	33	1.75	130	99.60	1.30	0.087	1173	1.304
90	3.174	5	33	1.58	135	102.77	1.31	0.096	1206	1.265
91	2.369	1	35	0.42	136	105.14	1.29	0.068	1241	0.876
92	2.557	1	35	0.39	137	107.70	1.27	0.073	1276	0.860
93	3.859	3	35	0.78	140	111.56	1.25	0.110	1311	0.974
94	2.641	1	37	0.38	141	114.20	1.24	0.071	1348	0.847
95	2.741	1	38	0.36	142	116.94	1.21	0.072	1386	0.835
96	3.055	3	38	0.98	145	119.99	1.21	0.080	1424	1.034
97	3.513	1	38	0.28	146	123.51	1.18	0.092	1462	0.779
98	2.728	2	40	0.73	148	126.24	1.17	0.068	1502	0.945
99	3.354	6	40	1.79	154	129.59	1.19	0.084	1542	1.322
100	3.814	0	40	0.00	154	133.40	1.15	0.095	1582	0.651

Table A, continued

obs	e_i Data	z_i	n_i	$y_i = z_i/e_i$	z_i^*	e_i^*	z_i^*/e_i^*	$\bar{e}_i = e_i/n_i$	n_i^*	$\hat{\lambda}_i^{**}$
					*indicates the cumulative sum					
101	4.014	2	41	0.50	156	137.42	1.14	0.098	1623	0.841
102	2.612	2	42	0.77	158	140.03	1.13	0.062	1665	0.950
103	2.815	1	43	0.36	159	142.84	1.11	0.065	1708	0.816
104	4.294	2	43	0.47	161	147.14	1.09	0.100	1751	0.817
105	3.450	8	44	2.32	169	150.59	1.12	0.078	1795	1.506
106	3.628	6	45	1.65	175	154.22	1.14	0.081	1840	1.272
107	4.219	1	45	0.24	176	158.44	1.11	0.094	1885	0.719
108	3.932	6	47	1.53	182	162.37	1.12	0.084	1932	1.233
109	4.082	4	47	0.98	186	166.45	1.12	0.087	1979	1.020
110	4.203	1	48	0.24	187	170.65	1.10	0.088	2027	0.714
111	4.022	3	51	0.75	190	174.67	1.09	0.079	2078	0.916
112	4.636	5	51	1.08	195	179.31	1.09	0.091	2129	1.060
113	5.571	2	56	0.36	197	184.88	1.07	0.099	2185	0.719
114	6.436	4	57	0.62	201	191.32	1.05	0.113	2242	0.841
115	5.344	3	60	0.56	204	196.66	1.04	0.089	2302	0.812
116	4.445	4	61	0.90	208	201.11	1.03	0.073	2363	0.958
117	4.705	5	61	1.06	213	205.81	1.03	0.077	2424	1.031
118	5.039	2	61	0.40	215	210.85	1.02	0.083	2485	0.739
119	6.043	6	68	0.99	221	216.89	1.02	0.089	2553	1.007
120	5.121	8	69	1.56	229	222.01	1.03	0.074	2622	1.245
121	11.260	5	69	0.44	234	233.27	1.00	0.163	2691	0.696
122	5.789	0	73	0.00	234	239.06	0.98	0.079	2764	0.520
123	6.044	6	74	0.99	240	245.11	0.98	0.082	2838	0.996
124	5.569	8	75	1.44	248	250.68	0.99	0.074	2913	1.194
125	6.130	7	75	1.14	255	256.81	0.99	0.082	2988	1.069
126	6.249	3	79	0.48	258	263.06	0.98	0.079	3067	0.737
127	7.002	3	91	0.43	261	270.06	0.97	0.077	3158	0.690
128	7.851	9	99	1.15	270	277.91	0.97	0.079	3257	1.069
129	9.573	7	104	0.73	277	287.48	0.96	0.092	3361	0.842
130	12.050	18	133	1.49	295	299.53	0.98	0.091	3494	1.327
131	12.131	17	152	1.40	312	311.66	1.00	0.080	3646	1.247
wtd mean				1.00*				0.085**		
mean	2.38	2.38	27.8	1.38				0.089		
total	311.6	312	3646							

*weighted by the predicted number of deaths e_i
**weighted by the number of patients n_i

6 and represents the estimated posterior mean for programs $i = 38, \ldots, 131$ when the ith program treats n_i "average" patients, i.e., patients with probability of death equal to .05.

APPENDIX A.3   PRIMM S-PLUS PROGRAM AND HEART TRANSPLANT DATA

At the time of publication, the data and the S-PLUS program PRIMM are available through Statlib. Until that time, they are available via anonymous ftp to elixir@med.harvard.edu. We suggest obtaining the README document (Christiansen and Morris, 1994b) that accompanies the program functions. It serves as a program tutorial and describes the program functions and their capabilities. The data used in this work, columns 1–4 of Appendix A.2, are available similarly in the ASCII file "heart.dat".

Figure 7 contains manual-style instructions for using the PRIMM S-PLUS function. Partial PRIMM output for model Reg 1.k94, using the S-PLUS command

primm(e,z,log(e) − 2.379,title = "Model Reg 1.k94",labelobs = 38:131),

is shown in Figure 6. The output list contains a header ($header), information related to a likelihood ratio test ($lrt), a priori values for $z_0 = e_0 m_0$ ($e.mu0, in the heading e0*munot0 represents z_0), the estimate, the standard error, and the estimate divided by the standard error for the $r + 1$ elements of the hyperparameter ($hyperparameter), a hyperparameter correlation matrix ($correlation), and a matrix of observation-specific values ($output). The figure includes $output values for observations 38–47 (the 10 smallest programs in the Reg1.k94 model) and 122–131 (the 10 largest programs) only to conserve space (n_i = 10, 11, 12, 13 for the first 10 rows and n_i = 73 up to 152 for the final 10 rows).

The columns "y", "munot^", and "lam^" correspond to $y_i = z_i/e_i$ [see Eq. (4)], $\hat{\mu}_i$ [see Eq. (5)], and the estimated value for λ_i^* [see Eq. (13)]. The estimate of λ_i^* is always between the observed ratio y_i and the regression estimate $\hat{\mu}_i$. Its distance from y_i is $\hat{B}_i$ times the distance between y_i and $\hat{\mu}_i$; $\hat{\lambda}_i^* = y_i - \hat{B}_i(y_i - \hat{\mu}_i)$. Large shrinkage (values close to one) causes $\hat{\lambda}_i^*$ to be close to $\hat{\mu}_i$, whereas small $\hat{B}_i$ causes the $\hat{\lambda}_i^*$ estimate to be close to the observed value y_i. The values $\{\hat{\lambda}_i^*\}$ are calculated using the assumptions and the approximations detailed in Christiansen and Morris (1994a).

The last four columns of $output give information on the approximate gamma distributions for the transplant programs' performance that would be used in a profile analysis. For example, knowing the mean and the standard error of observation 85 are .4893 and .2205, respectively, the program performance distribution for program 85 is approximated as

$$\lambda_{85} \sim \frac{.2205^2}{.4893} \text{Gam}\left(\frac{.4893^2}{.2205^2}, 1\right) \tag{15}$$

Given that our model is well specified and that we have adequate data, this distributional information allows informative statements to be made regarding the probability that a transplant program performs within a certain interval. For example, the probability that program 85's true performance ratio lies between .1573 (from lam^0.025) and 1.007 (from lam^0.975

primm	Fit a hierarchical Poisson model	primm

primm(e, z, x, e0 = NA, munot0 = NA, zeta.h = NA, betaprev = NA,
srt = F, optional = F, title = "no title", labelx = NA, labelobs = NA, alpha = .05)

REQUIRED ARGUMENTS

e	a vector of exposures or sample sizes of length k = # observations,$e_i > 0$
z	a vector of Poisson counts of length k, z_i = integer

OPTIONAL ARGUMENTS

x	a regression matrix. The program adds a first column of 1's if this is missing from the x argument. For the exchangeable model, the x argument is omitted. Dim. of x: k x r where r < k and r = number of regression coefficients.
e0	the user's prior exposure value, if omitted, e0=min(e). CAUTION. If the exposures are equal the user should always specify this value.
munot0	the user's prior mean value, if omitted, munot0=mean(z/e)
zeta.h	a starting value for ζ defined as the convolution parameter of the prior Gamma distribution. "zeta-hat"
betaprev	a vector of starting values for β, the regression coefficients
srt	if T, the output data will be sorted by exposures, default is F
optional	if T, opt.out is is included in the value of the function
title	a title for the output in the form: title="Your title"
labelx	a label for the columns of x and the vector of regression coefficients, ie., labelx=c("name0", "name1") for a regression with one covariate
labelobs	labels for the observations
alpha	value for $(1-\alpha)\%$ 2-sided posterior intervals, default is .05

CREATED VALUES (in list format)

heading	contains optional title, date, program information
lrt	k = # observations, saturated log posterior value, maximum log posterior value, 2*posterior ratio, k-(r+1) = df where r = # regression coefficients
e.mu0	prior values, e0, munot0 and e0*munot0
hyperparameter	$\beta_0, \ldots \beta_{r-1}, \zeta$, and log($\zeta$) estimates, standard errors, and ratios
correlation	correlation matrix for the hyperaparmeter estimates
output	a k x 7 matrix with columns 1. obs #, 2. $y = z/e$, 3. $\hat{B}$ {estimates of the shrinkage factors}, 4. $\hat{\lambda}$ {estimates of the posterior means}, 5. $se(\hat{\lambda})$, 6. $\hat{\lambda}low$, 7. $\hat{\lambda}high$ {posterior interval values}
opt.out	optional output, a k x 9 matrix with columns 1. obs #, 2. $munot\wedge$ {estimates of the prior means}, 3. $se(munot\wedge)$, 4. $se(X'\beta)$, 5. $\hat{B}$, 6. $se(\hat{B})$, 7. a, 8. b {estimated Beta parameters for the approximated Beta distributions for the shrinkage values}, and 9. standarized marginal residuals.

Figure 7   This manual-style information on the use of PRIMM is part of the README LaTex file (Christiansen and Morris, 1994b) provided with the PRIMM function from Statlib.

is 95%. Other probabilities can be calculated easily using PRIMM by changing the S-PLUS function's default value alpha = .05. Alternatively, we can calculate the probability that a transplant program's performance ratio is above (or below) a policy-regulated value. This topic is addressed in Morris and Christiansen (1995).

APPENDIX A.4   STRATEGY FOR DETERMINING THE INTERVALS OF FIGURE 2

Because z_i is discrete, a rule is necessary for determining the end points of the intervals (dashed lines) drawn in Figure 2. We chose the following strategy, although other rules could have been used with similar results.

1. Define $E(Z_i) \equiv e_i \hat{\gamma}_i$.
2. Obtain the largest z_{i2} such that $P(Z_i \geq z_{i2}) \geq 1/6$.
3. Obtain the smallest z_{i1} such that $P(Z_i \leq z_{i1}) \geq 1/6$.
4. If $z_{i2} < e_i \hat{\gamma}_i$, increase z_{i2} to the smallest integer that is greater than $e_i \mu_i$.
5. If $e_i \hat{\gamma}_i < z_{i1}$, decrease z_{i1} to the largest integer that is less than $e_i \hat{\gamma}_i$.

This strategy forces the interval to fall above and below the regression estimate.

ACKNOWLEDGMENT

This work was supported by grant AHCPR 1RO1HS07118-01, Hierarchical Statistical Modeling in Health Policy Research. The authors appreciate the helpful comments on an earlier version of this paper from the editors, Glenn Chertow, Ree Dawson, Sarah Michalak, and Sue Rosenkranz.

REFERENCES

Bendel, R. B. and Carlin, B. P. (1990). Bayes methods in the ecological fallacy context: Estimation of individual correlation from aggregate data. *Communication in Statistics—Theory and Methods* 19(7): 2595–2623.

Christiansen, C. L. and Morris, C. N. (1994a). A Poisson Hierarchical Model., Report HCP-1994-2, Technical Series in Statistics, Department of Health Care Policy, Harvard Medical School, Boston.

Christiansen, C. L. and Morris, C. N. (1994b). README document, June 7,

1994, a seven-page LaTex tutorial for PRIMM, available through StatLib, statlib@lib.stat.cmu.edu.

Gilks, W. (1995). Discussion of "Hierarchical models for ranking and for identifying extremes, with applications," by C. N. Morris and C. L. Christiansen. In *Bayesian Statistics 5*, J. M. Bernardo, J. O. Berger, A. P. Dawid and A. F. M. Smith, eds. Oxford University Press, New York.

Hastie, T. and Tibshirani, R. (1986). Generalized additive models. *Stat. Sci.* 1: 297–318.

Health Resources and Services Administration. (1992). 1991 Report of Center-Specific Graft and Patient Survival Rates, 5 volume report. U.S. Department of Health and Human Services, Washington, DC.

Hosenpud, J. D., Breen, T. J., Edwards, E. B., Daily, O. P., and Hunsicker, L. G. (1994). "The effect of transplant center volume on cardiac transplant outcome." *JAMA* 271: 1844–1849.

Johnson, N. L. and Kotz, S. (1969). *Discrete Distributions.* Wiley, New York.

Kassirer, J. P. (1994). The use and abuse of practice profiles. *N Engl J Med* 330: 634–636.

Kolata, G. (1994). New frontier in research: Mining patient records. *New York Times*, August 9, 1994, A21.

McCullagh, P. and Nelder, J. A. (1989). *Generalized Linear Models.* Chapman & Hall, London.

Morris, C. N. and Christiansen, C. L. (1995) Hierarchical models for ranking and for identifying extremes, with applications. Discussion and rejoinder. In *Bayesian Statistics 5*, J. M. Bernardo, J. O. Berger, A. P. Dawid and A. F. M. Smith, eds. Oxford University Press, New York.

Simpson, E. H. (1951). The interpretation of interaction in contingency tables. *J. R. Stat. Soc. Ser. B* 13: 238–241.

StatSci. (1991). *S-PLUS Reference Manual*, Statistical Sciences, Inc., Seattle.

United Network for Organ Sharing. (1993). Monthly Data Summary. UNOS, Richmond, VA.

United Network for Organ Sharing Research Dept. (1994). Patients Waiting for Transplants. *Unos Update* 10(8): 38.

Weisberg, S. (1985). *Applied Linear Regression*, 2nd ed. Wiley, New York.

VII
OTHER TOPICS

19

Analyzing Rodent Tumorigenicity Experiments Using Expert Knowledge

Jane C. Lindsey   Harvard School of Public Health, Boston, Massachusetts

Louise M. Ryan   Harvard School of Public Health and Dana Farber Cancer Institute, Boston, Massachusetts

ABSTRACT

Rodent tumorigenicity experiments play an important role in assessing the carcinogenic risk of various compounds to which humans may be exposed. The analysis of data from such studies is complicated by several features, including the fact that tumors vary in their lethality. Although commonly used in practice, methods that make extreme lethality assumptions can be biased and inefficient. The only alternatives currently available require either cause-of-death information or sacrifice data to estimate the unknown lethality. Both of these approaches have shortcomings in applied settings. In this chapter, we propose a method of analysis that draws on the knowledge of pathologists who typically have extensive experience in this area and can reliably predict a feasible range for tumor lethalities. We use this information to construct a subjective prior distribution for the lethality parameter in a multiplicative three-state model. The model is compared to the standard analyses that make extreme assumptions about lethality and one that estimates lethality only from the current data.

1 INTRODUCTION

Tumorigenicity experiments are designed to assess and characterize the effect of dose on the tumor incidence rate of rodents exposed to potentially carcinogenic compounds (McKnight and Crowley, 1984). A typical experiment involves 600 animals, 50 in each of a control and two or three dose groups in both sexes of two species. The available data for each rodent are its death or sacrifice time and an indicator for tumor presence determined by autopsy. Animals surviving at the end of the 2-year experimental period are sacrificed.

Figure 1 provides a pictorial representation of the life history of an animal in dose group z in an experiment. The animal starts in a tumor-free state. It can either die without tumor, with hazard $\beta(t \mid z)$, or develop a tumor, with hazard $\lambda(t \mid z)$, and then die with hazard $\alpha(t \mid x, z)$. The lethality of the tumor is the difference between the hazards for death with and without tumor.

1.1 An Example

The examples used in this chapter are taken from a series of experiments conducted by the National Toxicology Program (NTP) on oxazepam, a commonly used antianxiety agent (National Toxicology Program, 1992). Experiments to assess the carcinogenicity of oxazepam were done on two strains of both male and female mice. The original design of the experiment on the Swiss-Webster mice was to take 60 animals in three dose groups (control, 2500 ppm, and 5000 ppm) and follow them for 2 years with an interim analysis at 66 weeks. By 40 weeks into the experiment, however, a large number of deaths had occurred in the exposed groups. This was due to a condition called systemic amyloidosis, which is common in the

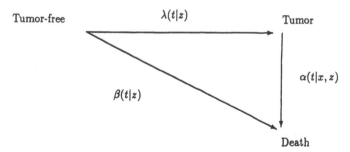

Figure 1   Three-state illness-death model.

breed of mice used and apparently enhanced by the experimental compound. By 57 weeks, 45 of the control animals were still alive, compared to only 10 of the animals in the high-dose group. In addition, the mean body weight of the exposed animals was significantly lower than that of the control animals. As a result, the experiment was ended at that time. The conclusions of the analysis in the NTP technical report were that the data showed clear evidence of increased liver carcinoma and increased incidence and severity of systemic amyloid deposition.

The subset of data used in this chapter will focus not on the tumor types that showed unequivocal evidence of carcinogenicity but on those with less clear-cut results. Data on the number of deaths and sacrifices with and without malignant lymphoma or histiocytic sarcoma in all organs are shown in Table 1 for the male Swiss-Webster mice in the control and high-dose groups only ($n = 120$). Data on alveolar/bronchiolar adenoma or carcinoma in the lung are shown in Table 2. Note the very low tumor incidence rate in the control group for the lung tumors.

1.2 Problems with the Analysis of Tumorigenicity Data

The simplest way to analyze data from a carcinogenicity experiment would be to ignore the time of death or sacrifice and to compare the proportion of animals developing tumor in the control group to that in the exposed group. Data from the oxazepam study (Table 2) illustrate why this can be misleading. While 14 of 60 (23%) of the control animals developed a tumor, only 7 of 60 (12%) developed a lung tumor in the exposed group. This result is marginally significant (p-value $= 0.074$ for Fisher's exact test), and it suggests a protective effect of the compound against development of tumors. If the additional fact is considered that only 16% of animals survived past the first 57 weeks in the exposed group compared to 75% in the control group, there is clear evidence that the toxic effects of the compound are killing animals in the exposed group before they have the opportunity to develop a tumor. A time-adjusted analysis is therefore required to control for the differential death rates. Time to tumor onset cannot be analyzed directly, however, as most tumors are occult (i.e., not detectable until the animal's death). A final complicating factor to consider is that tumors vary considerably in lethality, with some inducing death relatively quickly after onset and others having little effect.

Because of toxicity, unobservable tumor onset times, and varying tumor lethality, the process depicted in Figure 1 is nonidentifiable. For example, tumor onset for animals dying with tumor in month 10 could have occurred any time from month 1 to month 10. It is impossible to distinguish between lethal tumors that appear later in an animal's lifetime

Table 1 Oxazepam Data, Malignant Lymphoma and Histiocytic Sarcoma (All Organs)[a]

Day	Control dose group				High dose group			
	DNT	DWT	SNT	SWT	DNT	DWT	SNT	SWT
57	0	0	0	0	1	0	0	0
161	1	0	0	0	0	0	0	0
165	0	0	0	0	1	0	0	0
225	0	0	0	0	1	0	0	0
231	0	0	0	0	1	0	0	0
237	0	0	0	0	1	0	0	0
244	0	1	0	0	0	0	0	0
257	0	1	0	0	0	0	0	0
271	0	0	0	0	2	0	0	0
272	0	0	0	0	0	1	0	0
273	0	0	0	0	1	0	0	0
279	0	1	0	0	0	0	0	0
281	1	1	0	0	1	0	0	0
285	0	0	0	0	0	1	0	0
286	0	0	0	0	1	0	0	0
292	0	0	0	0	1	0	0	0
293	1	0	0	0	0	0	0	0
295	1	0	0	0	0	0	0	0
296	0	0	0	0	1	0	0	0
302	0	0	0	0	2	0	0	0
305	0	0	0	0	1	0	0	0
306	0	0	0	0	1	0	0	0
310	0	0	0	0	1	0	0	0
311	0	0	0	0	1	0	0	0
314	0	0	0	0	0	1	0	0
327	0	2	0	0	0	0	0	0
328	0	0	0	0	2	0	0	0
331	0	1	0	0	0	0	0	0
334	0	0	0	0	1	0	0	0
338	0	1	0	0	0	0	0	0
341	0	0	0	0	2	0	0	0
344	1	0	0	0	1	0	0	0
345	0	0	0	0	1	0	0	0
351	0	0	0	0	1	0	0	0
355	0	0	0	0	1	0	0	0
358	0	1	0	0	1	0	0	0
361	0	0	0	0	1	0	0	0
362	0	0	0	0	1	0	0	0
363	0	0	0	0	1	0	0	0
364	0	0	0	0	1	0	0	0
365	0	0	0	0	1	0	0	0
368	0	0	0	0	1	0	0	0
369	0	0	0	0	0	1	0	0
370	0	0	0	0	2	0	0	0
371	0	1	0	0	2	0	0	0
373	0	0	0	0	1	0	0	0
374	0	0	0	0	1	0	0	0
375	0	0	0	0	1	0	0	0
379	0	0	0	0	1	0	0	0
383	0	0	0	0	1	0	0	0
385	0	0	0	0	2	0	0	0
395	0	0	0	0	1	0	0	0
397	0	0	45	0	0	0	8	2
Total	5	10	45	0	46	4	8	2

[a] DWT and DNT, deaths with (without) tumor; SWT and SNT, sacrifices with (without) tumor ($n = 60$).

Table 2 Oxazepam Data, Alveolar/Bronchiolar Adenoma or Carcinoma (Lung)[a]

Day	Control dose group				High dose group			
	DNT	DWT	SNT	SWT	DNT	DWT	SNT	SWT
57	0	0	0	0	1	0	0	0
161	1	0	0	0	0	0	0	0
165	0	0	0	0	1	0	0	0
225	0	0	0	0	1	0	0	0
231	0	0	0	0	1	0	0	0
237	0	0	0	0	1	0	0	0
244	1	0	0	0	0	0	0	0
257	1	0	0	0	0	0	0	0
271	0	0	0	0	2	0	0	0
272	0	0	0	0	0	1	0	0
273	0	0	0	0	1	0	0	0
279	1	0	0	0	0	0	0	0
281	1	1	0	0	1	0	0	0
285	0	0	0	0	1	0	0	0
286	0	0	0	0	0	1	0	0
292	0	0	0	0	1	0	0	0
293	1	0	0	0	0	0	0	0
295	1	0	0	0	0	0	0	0
296	0	0	0	0	1	0	0	0
302	0	0	0	0	1	1	0	0
305	0	0	0	0	1	0	0	0
306	0	0	0	0	1	0	0	0
310	0	0	0	0	1	0	0	0
311	0	0	0	0	1	0	0	0
314	0	0	0	0	0	1	0	0
327	2	0	0	0	0	0	0	0
328	0	0	0	0	2	0	0	0
331	1	0	0	0	0	0	0	0
334	0	0	0	0	1	0	0	0
338	1	0	0	0	0	0	0	0
341	0	0	0	0	2	0	0	0
344	1	0	0	0	1	0	0	0
345	0	0	0	0	1	0	0	0
351	0	0	0	0	1	0	0	0
355	0	0	0	0	1	0	0	0
358	1	0	0	0	1	0	0	0
361	0	0	0	0	1	0	0	0
362	0	0	0	0	1	0	0	0
363	0	0	0	0	1	0	0	0
364	0	0	0	0	1	0	0	0
365	0	0	0	0	1	0	0	0
368	0	0	0	0	1	0	0	0
369	0	0	0	0	1	0	0	0
370	0	0	0	0	1	1	0	0
371	1	0	0	0	2	0	0	0
373	0	0	0	0	1	0	0	0
374	0	0	0	0	1	0	0	0
375	0	0	0	0	1	0	0	0
379	0	0	0	0	1	0	0	0
383	0	0	0	0	1	0	0	0
385	0	0	0	0	0	2	0	0
395	0	0	0	0	1	0	0	0
397	0	0	32	13	0	0	10	0
Total	14	1	32	13	43	7	10	0

[a] DWT and DNT, deaths with (without) tumor; SWT and SNT, sacrifices with (without) tumor ($n = 60$).

but kill the animal quickly, nonlethal tumors that initiate growth in young animals but have little impact on the death rate, and a toxic compound that is killing the animals earlier regardless of tumor presence. Valid tests for dose effects are not possible without either making simplifying assumptions or collecting additional information.

Additional data may be collected in three ways. Some experiments have sacrifices at interim points during the 2-year period that provide snapshot views of tumor prevalence among live animals. Others collect cause-of-death information where a pathologist assigns the additional indicator of whether or not the animal died from a tumor. Collecting either type of additional information allows nonparametric methods to be used to model differential mortality between animals with and without tumor (McKnight and Crowley, 1984; Dewanji and Kalbfleisch, 1986; Portier, 1986; Malani and Van Ryzin, 1988). Neither approach is totally satisfactory: cause of death is difficult to ascertain and subject to error; collecting interim sacrifice information increases the size of the experiments, and estimates can be highly variable, especially for rare tumor types.

Alternatively, parametric or semiparametric assumptions can be made about the cause-specific hazards for tumor onset and death. Fully parametric models have been proposed by Dinse (1988), Portier (1986), and Portier and Dinse (1987), among others. In these models, the hazards are assumed to follow a specific distribution, such as the Weibull. Assumptions about tumor lethality vary from the extremes at which tumors are assumed to have either no impact on the hazard for death or be instantly lethal to semiparametric restrictions such as those proposed by Dinse (1991) and Lindsey and Ryan (1993). Weaker restrictions can be placed on the underlying hazards when stronger assumptions are made about tumor lethality.

Another way to incorporate additional information is to use the pathologists' expert knowledge of tumor lethality. There is precedence in the literature for using historical information on control tumor incidence rates (Tarone, 1982; Hoel, 1983; Hoel and Yanagawa, 1986; Dempster et al., 1983). Only Dempster et al. (1983) use fully Bayesian methods. The other authors use "hybrid" Bayesian techniques that put prior distributions on control tumor incidence rates but perform inference using standard frequentist methods. Toxicologists are familiar with this hybrid approach, even though it is not necessarily conventional among statisticians. We will use a similar approach to incorporate expert opinion about tumor lethality. Before introducing the method, some standard analyses will be described in more detail.

1.3 Models Making Assumptions About Tumor Lethality

The two most established time-adjusted tests for carcinogenicity make extreme assumptions about tumor lethality. The prevalence or Hoel-Walburg test (Hoel and Walburg, 1972) assumes tumor presence has no impact on the death rate ($\alpha(t \mid x, z) = \beta(t \mid z)$), while a standard survival analysis using a log-rank test (Peto, 1974) assumes animals die the instant they acquire the tumor [$\alpha(t \mid x) = \infty$]. Full details on how to calculate these tests are shown in Appendix 1. Asymptotically equivalent results for the Hoel-Walburg test can also be obtained using logistic regression techniques (Dinse and Lagakos, 1983).

In reality however, most tumors are of intermediate lethality. When there are no toxic effects, both the log-rank and Hoel-Walburg tests are valid but may differ in their power to detect a carcinogenic effect (Lagakos and Ryan, 1985a). In the presence of life-shortening toxicity in the exposed group, the log-rank test tends to overestimate significance and the Hoel-Walburg test underestimates the true dose effect (Gart *et al.*, 1986). In practice, both tests are typically calculated (see any recent NTP technical report, for example). If both give consistent significant or nonsignificant results, the conclusion is that there is (or is not) evidence of a carcinogenic effect. The reasoning is that the true results would be bracketed by the two test statistics, although Lagakos and Louis (1988) show this is not necessarily true. When contradictory results are obtained, appropriate conclusions are unclear and often rely informally on experts' beliefs about the lethality of the tumor type in question.

To avoid extreme assumptions, Dinse (1991) and Lindsey and Ryan (1993) impose semiparametric constraints on the lethality of the tumor. Dinse (1991) recommends modeling the death rate with tumor as the death rate without tumor plus an additive constant [$\alpha(t) = \beta(t) + \Delta$]. Lindsey and Ryan (1993) use a multiplicative constraint [$\alpha(t) = \beta(t)\xi$] and model the underlying hazards using a piecewise constant assumption. This approach is described in detail in Section 2. Both approaches make a Markov assumption in which death hazards can change over time but the death hazard for animals with tumor does not depend on when the animal acquired the tumor. Both also control for the toxic effects of the compound on the lifetime of the animals. Although these methods provide a way to fit models of arbitrary lethality for experiments with as few as one sacrifice time, parameter estimates may still be unstable, especially when tumor incidence rates are low. This chapter describes a way to address this problem by incorporating expert opinion about tumor lethality.

1.4 Incorporating Historical Information on Tumor Lethality

As stated previously, a number of authors have incorporated historical information on control tumor incidence rates in tests for carcinogenicity. This can be particularly useful when control tumor rates are low and when marginal significance levels are obtained in a test for dose effects (Haseman *et al.*, 1984). Tarone (1982), Hoel (1983), and Hoel and Yanagawa (1986) propose methods to incorporate historical information about tumor incidence rates in control animals into a test for a dose-related trend in the proportion of animals acquiring tumor. These methods assume that the control response rates in each study follow a beta prior. The parameters of the beta prior are estimated using maximum-likelihood techniques, based on the historical information. Inference is done using either exact conditional tests (Hoel, 1983; Hoel and Yanagawa, 1986) or asymptotic results (Tarone, 1982) in a standard frequentist manner. Thus the methods are not fully Bayesian but what we term hybrid Bayesian. Dempster *et al.* (1983) take a similar approach, but assume the logits of the control tumor rates are normally distributed. In large samples, the results of the two methods are similar (Dempster *et al.*, 1983). In a somewhat different context, a fully Bayesian approach has been used by Meng and Dempster (1987) to deal with the issue of multiplicity when looking for dose effects over many tumor types. Ryan (1993) develops a method to use historical control information in the context of developmental toxicity.

Our approach is to combine prior information on the nuisance parameter tumor lethality with the time-adjusted semiparametric model proposed by Lindsey and Ryan (1993). We allow an expert in pathology to choose the parameters of the prior distribution. The approach is neither fully Bayesian nor empirical Bayes, since we do not integrate over the range of the prior parameters, nor are they estimated from previous data. Furthermore, inference about the parameter of interest (the dose effect on tumor onset) proceeds using classic frequentist methods. Our goal is to reduce bias and variability of the test for dose effects on tumor onset by exploiting the expert knowledge of pathologists, not to get posterior estimates of all parameters. The methods are described in detail in Sections 3 and 4.

2 ESTIMATING TUMOR LETHALITY FROM THE DATA

Consider first the control group ($Z = 0$). Let X denote time to the first event, let δ indicate whether or not the first event is tumor onset ($\delta = 1$)

or death ($\delta = 0$), and let T denote time to death. The baseline hazards in Figure 1 are defined as

$$\lambda(t) = \lim_{\epsilon \to 0^+} \Pr(t \leq X < t + \epsilon, \delta = 1 \mid X \geq t)/\epsilon$$

$$\beta(t) = \lim_{\epsilon \to 0^+} \Pr(t \leq X < t + \epsilon, \delta = 0 \mid X \geq t)/\epsilon$$

$$\alpha(t \mid x) = \lim_{\epsilon \to 0^+} \Pr(t \leq T < t + \epsilon \mid T \geq t, \delta = 1, X = x)/\epsilon$$

where $t \geq x > 0$. The hazard for death with tumor depends only on the age of the animal and not the age at which it acquired its tumor (a Markov assumption). As described in Lindsey and Ryan (1993), hazards for tumor onset and death are assumed to be piecewise constant with $\lambda(t)$ character-ized by U constant hazards, $\lambda_1, \ldots, \lambda_U$, and $\beta(t)$ by V constant hazards, $\beta_1, \ldots, \beta_V$. For notational convenience we break the time scale into J intervals $I_j = (\tau_{j-1}, \tau_j]$, so that each interval I_j contains a unique combina-tion of hazards, λ_j and β_j, where $\lambda_j \in \{\lambda_1, \ldots, \lambda_U\}$ and $\beta_j \in \{\beta_1, \ldots, \beta_V\}$. Choice of the number of intervals and cutpoints for the intervals is somewhat arbitrary, as is the case for the standard Hoel-Wal-burg analysis. Increasing the number of intervals relaxes the restrictive-ness of the piecewise constant hazards assumption. Choosing too many intervals can lead to numerical problems and makes the model less parsi-monious.

A number of different measures for lethality have been proposed. It will be convenient for this chapter to use the definition

$$\frac{\alpha(t) - \beta(t)}{\alpha(t)} \tag{1}$$

suggested by Archer and Ryan (1989). To reduce the number of param-eters, Lindsey and Ryan (1993) impose a multiplicative constraint on the death rate with tumor such that $\alpha(t) \equiv \beta(t)\xi$. The lethality function (1) then equals $(1 - 1/\xi)$ and is constrained to lie between 0 and 1. It can be interpreted as the proportion of the risk of death with tumor associated with tumor presence. An additional level of complexity could be added by allowing the lethality function to vary with dose and time (Dinse, 1988), although estimation then requires additional sacrifice information. The extreme case in which each interval has its own lethality parameter would yield a model analogous to those proposed by McKnight and Crowley (1984), Dewanji and Kalbfleisch (1986), and Malani and Van Ryzin (1988). For the purposes of this chapter, only dose effects on tumor onset and death will be used. The full vector of $(2J + 3)$ unknown parameters is η = $\{\lambda, \beta, \xi, \psi, \rho\}$.

Dose effects can be incorporated into the tumor onset rate and the death rates through proportional hazards assumptions. Let

$$\lambda(t, z) = \lambda(t)e^{\psi z}$$

$$\beta(t, z) = \beta(t)e^{\rho z}$$

where $\lambda(t)$ and $\beta(t)$ are the rates in the control group and Z denotes dose group. The parameter ψ represents the increased risk for tumor onset and ρ the increased risk for death due to toxicity of the compound in the exposed group over the control group.

Four types of events are possible at any observed event time: death with no tumor (DNT), death with tumor (DWT), sacrifice with no tumor (SNT), and sacrifice with tumor (SWT). The likelihood contributions of these events are shown in detail in Appendix 2. The log likelihood is the sum over the types of events at each event time. Lindsey and Ryan (1993) maximize the likelihood using an expectation–maximization (EM) algorithm (Dempster *et al.*, 1977). Alternatively, numerical routines such as those in the IMSL (1989) library can be used to find the maximum-likelihood estimates (MLEs) of the parameters.

3 INCORPORATING EXPERT KNOWLEDGE ON TUMOR LETHALITY

Experienced pathologists develop good intuition about the lethality of different tumor types. In studies requiring cause-of-death information, this knowledge may influence the decision of the pathologist. The probability that a pathologist will assign a particular tumor type as the cause of death can be thought of as imposing a subjective prior distribution on tumor lethality (Lagakos and Ryan, 1985b). For the purpose of eliciting prior information, we place a distribution on the quantity $0 \leq p = (1 - 1/\xi) \leq 1$, defined in Eq. (1) as the proportion of the risk of death with tumor associated with tumor presence. The quantity ξ, while mathematically convenient, is not easy for a pathologist to interpret. The beta distribution is chosen to characterize the distribution of p as it allows a flexible range of shapes.

A beta distribution with parameters (ν, γ) has the form:

$$\text{Beta}(p \mid \nu, \gamma) = \frac{p^{\nu - 1}(1 - p)^{\gamma - 1}}{B(\nu, \gamma)}, \qquad 0 < p < 1, \nu > 0, \gamma > 0$$

where

$$B(\nu, \gamma) = \frac{\Gamma(\nu)\Gamma(\gamma)}{\Gamma(\nu + \gamma)}, \qquad \Gamma(\nu) = \int_0^\infty x^\nu e^{-x}\, dx$$

To choose a prior distribution, a pathologist can be shown a template with a variety of beta distributions, as shown in Figure 2. The beta distributions show increasing lethality going down the rows and decreasing variance going across the columns. The pathologist chooses the prior distribution most closely corresponding to his or her intuition about the lethality of that tumor. If the pathologist's prior information does not favor any particular value of tumor lethality over any other, a uniform [beta(1, 1)] would be appropriate.

Given p, the likelihood contributions are the same as for the model not incorporating prior knowledge, taking into account the new parameterization with $\xi = 1/(1 - p)$. By applying the prior distribution on p to the likelihood, the marginal likelihood can be found by integrating over the range of p, i.e.,

$$L_{\lambda,\beta} = \int_p L\{\lambda,\beta \mid p\} f(p \mid \nu,\gamma)\, dp$$

Estimates for the parameters that maximize $L_{\lambda,\beta}$ can be obtained directly using established numerical routines for maximization and the required numerical integrations (Press *et al.*, 1986). The IMSL (1989) maximization routine DUMINF, which is a derivative-free method, and an adaptive integration routine based on an eight-panel Newton-Cotes rule are used for the examples in this chapter. Because the maximization routine is subject to scaling problems, the underlying tumor onset and death rates are reparameterized so that values of the estimates lie on the same scale, i.e., $\lambda^* = e^\lambda$. For the examples in this chapter, starting values are obtained from the piecewise exponential model with no prior information on lethality. As the number of intervals increases, finding good starting values becomes increasingly important. Maximization is computationally intensive as integrations have to be calculated at each event time for each iteration. Given the speed of modern computers, however, this is not a major disadvantage. Standard error estimates for the parameters are found using the numerical routine DFDHES (IMSL, 1989), which approximates the second-derivative matrix of the likelihood using forward differences (Press *et al.*, 1986).

Dose effects are incorporated in the same way as in Section 2 and are

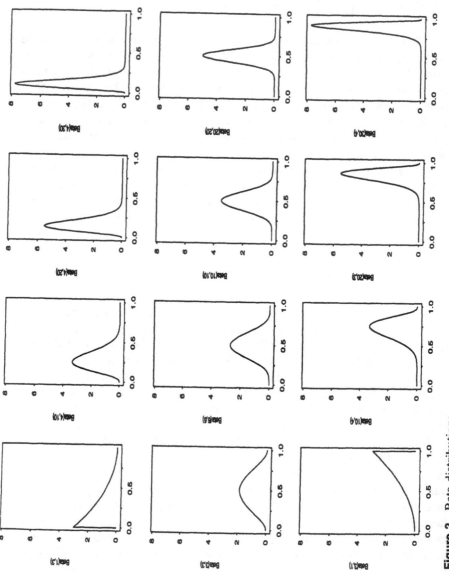

Figure 2 Beta distributions.

assessed using a standard frequentist approach by fitting models with and without the dose parameter and calculating significance using a likelihood ratio test. Wald tests are also possible, but because these are sometimes subject to numerical problems, the use of a likelihood ratio test is recommended. Additional covariates such as gender, weight, or cage location (Lagakos and Mosteller, 1981) can also be incorporated, if desired.

Finally, an estimate of the posterior mean of p can be found by conditioning on the data and substituting the maximum-likelihood estimates for the underlying hazards:

$$E(p \mid \hat{\eta}, t) = \int_0^1 p f(p \mid \hat{\eta}, t) \, dp$$

Although the posterior mode is more useful for comparing the results of the Bayesian analysis to the analysis estimating parameters from the data alone, in large samples the posterior distribution will approach normality and the mean and mode will be approximately the same.

4 EXAMPLE

The data set introduced at the beginning of the chapter will be used to explore differences between analyses that make extreme assumptions about tumor lethality and those that estimate it from the experimental data alone or in conjunction with prior knowledge. Direct comparisons between our approach and the established log-rank and Hoel-Walburg analyses are not straightforward because of the different parameterizations of the underlying hazards for tumor onset and death. These results will be given for completeness, but more attention will be paid to comparisons with piecewise constant hazards models making extreme lethality assumptions. A piecewise model with $\xi = 1$ ($p = 0$) is analogous to the Hoel-Walburg analysis and to a Bayesian model placing a point-mass prior at $p = 0$. When $\xi = \infty$ ($p = 1$), the model reduces to a simple piecewise exponential regression, with underlying tumor onset hazards estimated by the ratio of the number of events to the time at risk in each interval. This would be analogous to a point-mass prior at $p = 1$ and should yield results similar to those of the log-rank test. Maximizing these likelihoods is simple, as they factor into a component involving only the tumor onset hazards and a separate component involving the death hazards.

4.1 Extreme Lethality Assumptions

First, consider the standard methods that make extreme lethality assumptions. The Hoel-Walburg analyses are performed using a logistic regression with five intervals at 5, 7, 9, 11, and 13 months. The log-rank analyses are done using Cox regression (Cox, 1972). Significance probabilities (p-values) are calculated from likelihood ratio tests for dose effects on tumor onset. In the presence of life-shortening toxicity in the exposed group, the Hoel-Walburg test tends to underestimate differences between groups, and the log-rank test tends to overestimate significance (Gart et al., 1986). Results are shown in Table 3.

For the malignant lymphoma and histiocytic sarcoma data, the Hoel-Walburg analysis gives a significant result (p-value < 0.01), while the log-rank analysis leads to a nonsignificant result (p-value $= 0.83$). Both tests indicate a protective effect of the compound against the tumors of interest. For the lung tumors, although neither the Hoel-Walburg nor the log-rank test gives significant results, the parameter estimates lie in opposite directions. Because there is clear evidence of toxicity, the assumptions of the standard analyses are violated. Since Lagakos and Louis (1988) have shown that in the presence of toxicity, the Hoel-Walburg and log-rank tests do not necessarily bracket the true significance level, it could be incorrect to conclude a nonsignificant dose effect on tumor onset.

Results for the piecewise constant hazards model using extreme lethality assumptions are also shown in Table 3. Three intervals at 9, 11, and 15 months are used for the malignant lymphoma and histiocytic sarcoma data. In contrast to the Hoel-Walburg test, dose effects on tumor onset are not significant in the model imposing no tumor lethality. This discrepancy is discussed presently. Results are similar to those of the log-rank analysis in the model with extreme lethality. Only two intervals at 11 and 15 months are used for the lung data, as no animals died with

Table 3 Oxazepam Data Analyses Using Extreme Lethality Assumptions

	Lymphoma		Lung	
Test	ψ	p-value	ψ	p-value
Hoel-Walburg	-2.16	0.0013	-0.58	0.3554
Piecewise $p = 0$	-0.53	0.2987	-0.60	0.2148
Logrank	-0.11	0.8300	0.38	0.4419
Piecewise $p = 1$	-0.16	0.7634	-0.19	0.6757

tumor before 9 months and including intervals with no events can lead to numerical difficulties. The piecewise constant hazards models give similar results to the established analyses with no significant differences between dose groups. However, the sign of the parameter estimate with $p = 1$ is no longer positive.

These results can be understood by looking at the first four graphs in Figures 3 and 4, which show quantities analogous to cumulative hazards, $\hat{H}(t)$, for tumor onset. For the Hoel-Walburg analysis, this quantity is estimated by the proportion of animals dying with tumor among those dying in that interval and using the proportion sacrificed with tumor at the terminal sacrifice time (Finkelstein, 1986). For the log-rank analysis, the Nelson-Aalen estimator can be used:

$$\hat{H}_{\mathrm{L}}(t_i) = \sum_{i=1}^{t_i} \frac{b_i + n_i}{R_i}$$

where b_i and n_i are the numbers of animals dying and sacrificed with tumor and R_i is the number of animals at risk at event time i. For the piecewise model, an estimate of the cumulative tumor incidence function is given by

$$\hat{H}_{\mathrm{P}}(t) = \sum_{k=1}^{j-1} \hat{\lambda}_k(\tau_k - \tau_{k-1}) + \hat{\lambda}_j(t - \tau_{j-1})$$

Lymphoma is known to be a lethal tumor. The cumulative hazard estimates in both dose groups from the piecewise constant hazards model assuming $p = 1$ (Figure 3d) and the log-rank analysis (Figure 3b) are similar. There is a surprising difference between the cumulative hazard estimates from the Hoel-Walburg analysis (Figure 3a) and the piecewise model with $p = 0$ (Figure 3c), which is reflected in the results of the hypothesis tests. Imposing nonlethality forces the piecewise model to push back the start of tumor development into the first interval in both dose groups and, because the numbers of animals dying with tumor are similar, leads to a nonsignificant result. Another explanation is that the piecewise exponential model properly incorporates time at risk of acquiring a tumor. In contrast, the Hoel-Walburg test looks only at the proportion of animals dying with tumor out of the number of animals dying in that interval. With relatively few events early in the experiment, this appears to be giving unstable estimates of the hazard and an artificially extreme significance test.

Only one lung tumor occurs in the control group prior to the terminal sacrifice, an indication that the tumor is relatively nonlethal and/or occurs only in older animals. Tumors in the high-dose group occur only after 9

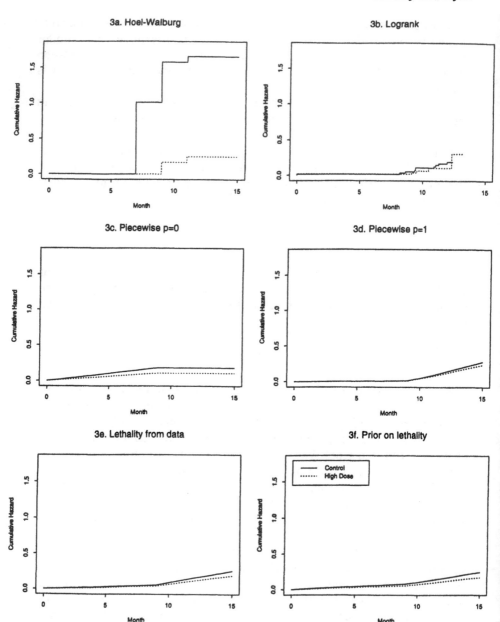

Figure 3 Cumulative hazard plots for tumor onset for oxazepam malignant lymphoma or histiocyctic sarcoma data based on analyses as shown.

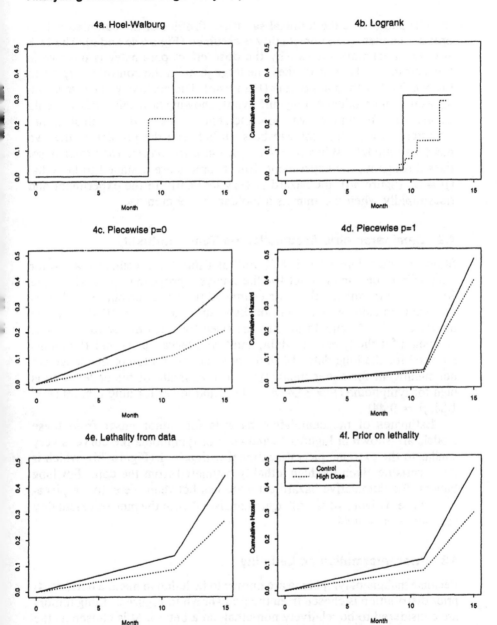

Figure 4 Cumulative hazard plots for tumor onset for oxazepam alveolar/bronchiolar adenoma or carcinoma data based on analyses as shown.

months and before the terminal sacrifice. Profiles of the cumulative hazards are similar when nonlethality is assumed (Figure 4a and c). The Nelson-Aalen estimate shows why the dose effect parameter is positive in the log-rank analysis as it takes one large jump in the control group when the one death with tumor occurs (Figure 4b). The positive value is misleading as it is the product of only one event, showing how this analysis could be unstable for tumors with low incidence rates in the control group. An interesting comparison can be seen between the profiles of the two piecewise models. When $p = 0$ (Figure 4c), the hazard for tumor onset rises throughout the experiment. But if tumors are assumed to be lethal ($p = 1$, Figure 4d), the hazard is very low early in the experiment and rises rapidly when the animals are older than 9 months.

4.2 Semiparametric Assumption on Tumor Lethality

Shown in the columns labeled "Data" of Table 4 are results for a test for dose effects on tumor onset for the analysis proposed by Lindsey and Ryan (1993) in which lethality is estimated from the current data. This is analogous to using a uniform prior distribution on tumor lethality. Three intervals at 9, 11, and 15 months are used for hazards for tumor onset and death for the lymphoma data, and two intervals at 11 and 15 months are used for the lung data. Tests for dose effects on tumor incidence are not significant for either tumor type. The estimate of tumor lethality is high for lymphoma ($\xi = 35.4$, $\hat{p} = 0.97$) and lower for lung tumors ($\xi = 1.80$, $\hat{p} = 0.44$).

Estimates of the cumulative hazards for tumor onset from these models are shown in Figures 3e and 4e. For lymphoma the curve is very similar to the extreme piecewise model with $p = 1$ (Figure 3d), which is not surprising given the high lethality estimated from the data. For lung tumors, the cumulative hazard estimate lies between the extreme piecewise curves (Figure 4c and d), again consistent with the moderate lethality estimate of $\hat{p} = 0.44$.

4.3 Prior Information on Lethality

Because malignant lymphoma are known to be lethal in mice, a beta(10, 4) prior distribution is chosen from the grid shown in Figure 2. Lung tumors are considered to be relatively nonlethal, so a beta(6, 6) is chosen as the prior. In practice, these priors would be chosen by an experienced pathologist, based presumably on the number of animals the pathologist had seen with a particular tumor type and how lethal the pathologist had judged the tumor to be in his or her previous experience. Since the variance of

Table 4 Oxazepam Data Analyses
Estimating Lethality from the Data and
Using a Prior Distribution on Lethality[a]

Parameter	Lymphoma Data	beta(10,4)	Lung Data	beta(6,6)
$\hat{\lambda}_1$	-5.07	-4.70		
$\hat{\lambda}_2$	-3.36	-3.71	-4.37	-4.51
$\hat{\lambda}_3$	-3.44	-3.54	-2.57	-2.44
$\hat{\psi}$	-0.30	-0.34	-0.49	-0.45
se($\hat{\psi}$)	0.53	0.52	0.49	0.48
p-value	0.57	0.51	0.31	0.34
$\hat{\beta}_1$	-6.69	-6.49		
$\hat{\beta}_2$	-3.85	-3.63	-4.78	-4.82
$\hat{\beta}_3$	-3.06	-2.87	-2.55	-2.61
$\hat{\xi}$	35.42	14.29	1.80	2.17
se($\hat{\xi}$)	31.15	0.45	0.95	0.25
Post. p	0.97	0.93	0.44	0.54
$\hat{\rho}$	2.15	1.95	1.64	1.67
se($\hat{\rho}$)	0.41	0.35	0.31	0.30

[a] p-value (two-sided) from likelihood ratio test. Tumor onset and death hazards are shown on the logged scale.

a beta distribution is a function of the inverse of the quantity $(\nu + \gamma)$, increasing this sum can be thought of as increasing the number of animals seen with the tumor, which in turn corresponds to imposing a more precise and influential prior distribution on tumor lethality. The mean of a beta distribution is $\nu/(\nu + \gamma)$, so as ν increases (holding γ fixed), it is analogous to increasing the proportion of the risk of death with tumor associated with tumor presence.

Complete results are shown in the columns labeled "beta" of Table 4. The same intervals for the hazards for tumor onset and death are used as in the analysis estimating lethality from the data alone. For both lymphoma and lung, estimates from the analysis incorporating prior information are similar to those from the model estimating lethality from the current data alone. Cumulative hazard estimates for tumor incidence from the model incorporating prior information are shown in Figures 3f and 4f. For both tumor types, they are nearly indistinguishable from the model estimating lethality from the current data alone (Figures 3e and 4e).

5 SENSITIVITY ANALYSIS

To investigate the sensitivity of the test for dose effects on tumor onset to the choice of prior, additional beta prior distributions are applied to

Table 5 Estimates of Dose Effects, Posterior Estimate of Tumor Lethality, and p-Value from Likelihood Ratio Test on Dose Effect on Tumor Onset Rate for Different Beta Prior Distributions on Lethality of Lymphoma and Lung Tumors

α	β	Prior p	$\hat{\psi}$	p-value	$\hat{\rho}$	Post. p
			Lymphoma			
		$p=0$	-0.5283	0.2987		
4	30	0.12	-0.5286	0.2986	1.6071	0.16
1	1	0.50	-0.3035	0.5581	2.1125	0.97
		Data	-0.2985	0.5663	2.1497	0.97
6	6	0.50	-0.4160	0.4205	1.7312	0.81
10	4	0.71	-0.3431	0.5083	1.9486	0.93
30	4	0.88	-0.3222	0.5340	2.0327	0.95
80	4	0.95	-0.2907	0.5787	2.1648	0.97
300	4	0.99	-0.2731	0.5981	2.2418	0.98
1000	4	1.00	-0.2648	0.6096	2.2765	0.98
		$p=1$	-0.1558	0.7634		
			Lung			
		$p=0$	-0.5967	0.2148		
4	1000	0.00	-0.5946	0.2154	1.5742	0.01
4	300	0.01	-0.5923	0.2167	1.5755	0.02
4	80	0.05	-0.5668	0.2223	1.5795	0.05
4	30	0.12	-0.5668	0.2363	1.5891	0.13
		Data	-0.4845	0.3102	1.6384	0.44
1	1	0.50	-0.4358	0.3540	1.6885	0.55
6	6	0.50	-0.4486	0.3398	1.6716	0.54
30	4	0.88	-0.3651	0.4310	1.8477	0.82
80	4	0.95	-0.3375	0.4651	2.0252	0.90
300	4	0.99	-0.3033	0.5113	2.4490	0.96
1000	4	1.00	-0.2820	0.5406	2.8333	0.98
		$p=1$	-0.1933	0.6757		

the lymphoma and lung data. Estimates of dose effects on tumor onset and death and significance tests for the former are shown in Table 5. Also shown are results from the piecewise constant hazards models with extreme assumptions on tumor lethality and the results from the model estimating lethality from the data.

One would expect the analysis incorporating prior information to be similar to the piecewise models with extreme assumptions when the priors place almost all mass at $p = 0$ or $p = 1$. Similar results are also expected for the model estimating lethality from the data and the one with a uniform [beta(1, 1)] prior. This is the case for both tumor types. For the lymphoma data, a uniform prior on the lethality parameter gives results very close to the analysis based on current data alone. The posterior estimate of $p = 0.97$ is the same as that estimated by the analysis not incorporating prior information. The nonlethal prior [beta(4, 30)] also gives results similar to the piecewise model with p forced to equal 0. Going from a prior with $p = 0.5$ [beta(6, 6)] to the extreme case of $p = 0$, the estimate of ψ becomes more negative and the p-value becomes smaller. In contrast, going from a beta(6, 6) toward the extreme of $p = 1$, the estimates get closer to 0 and the p-value increases. Similar patterns are seen for the lung data.

6 DISCUSSION

Testing for dose effects in rodent tumorigenicity experiments is complicated by several intrinsic features of the data. The most commonly used tests make extreme assumptions about tumor lethality. Results can be misleading in the presence of life-shortening toxic effects of the compound on the exposed animals. Methods that do not impose extreme assumptions may rely on the collection of interim sacrifice or cause-of-death information, which is not always available. Imposing incorrect parametric assumptions on the underlying tumor onset and death rates can result in misleading conclusions. We propose a time-adjusted method of analyzing the data that draws on historical and expert knowledge of tumor lethality. Our model assumes that hazards for tumor onset and death follow a piecewise exponential distribution, and covariate effects are incorporated using proportional hazards assumptions. The model adjusts for toxic effects of the compound without imposing strict parametric restrictions on the underlying hazards.

In the example used in this chapter, the results from the model using prior information on tumor lethality were very similar to those of the method estimating lethality from the current data alone. Preliminary sensitivity analyses show that the prior can have a moderate impact on the test for dose effects. Using a prior distribution on tumor lethality may be particularly helpful in situations in which the standard analyses give contradictory results or an analysis estimating lethality from the data gives a marginal significance level.

There are a number of ways in which this approach could be extended. In the rodent tumorigenicity context, a beta prior is sensible. Many processes can be represented using three-state models, and other prior distributions may be more appropriate than the beta in different contexts. For modeling exponential data, there is some precedent for the use of gamma priors (Turnbull *et al.*, 1974). A gamma prior on ξ is less appealing in the rodent tumorigenicity context as it allows tumors to be protective.

In reality, tumor lethality is a complex phenomenon that is likely to vary over the lifetime of an animal (Dinse, 1988; Lagakos and Louis, 1988) and may also depend on the exposure level and on characteristics such as size and extent of the tumor (Ryan and Orav, 1988). Simulation techniques could be used to assess the adequacy of the simpler constant lethality models used in the examples in this chapter on the test for dose effects. If found to be inadequate, the model could be extended by allowing pathologists to propose different priors, $f(\xi_{jz})$, for each time interval j and dose group z. Unless pathologists were very confident of the accuracy of their prior distributions, sacrifice information should ideally be available in each of the time intervals I_j.

A further extension would be to use a fully Bayesian analysis and to base inference for all the parameters on posterior distributions. Turnbull *et al.* (1974) discuss an analogous model in the context of survival in a clinical trial. They place a prior distribution on the hazard for death and base their inference on the posterior distributions. Theirs is only a three-dimensional problem, so maximum-likelihood estimates are relatively easy to find. In the analysis presented in this chapter, the parameters of the prior were chosen by the pathologist. This was done to make use of the pathologist's expert knowledge about tumor lethality. Although the estimates of tumor lethality or toxicity of the compound are not of direct interest, incorporating what experts know about these parameters allows us to avoid making extreme lethality assumptions or imposing a point estimate for these parameters.

In addition to lethality, historical information is generally available on underlying tumor incidence and death rates, on how these rates change over time, and on the extent of toxicity for certain compounds. It would be useful to consider ways to incorporate this information into the piecewise exponential multiplicative model for all parameters.

APPENDIX 1. HOEL-WALBURG AND LOG-RANK SURVIVAL TESTS

Hoel-Walburg Test

The time scale is broken into J intervals. The data can be summarized in a series of 2×2 tables (see Table 6) with b_{zj} indicating a death with tumor and a_{zj} a death without tumor in the jth interval in dose group Z ($Z = 0$ for the control group and $Z = 1$ for the exposed group). Additional tables are formed for each sacrifice time, $s = 1, \ldots, S$. Sacrifices with and without tumor at time s are denoted by n_{zs} and m_{zs}, respectively. The test statistic has an "observed minus expected" form:

$$z = \sqrt{\frac{O - E}{V}} \sim N(0, 1)$$

where:

$$O_{j/s} = \begin{cases} b_{1j} & \text{for deaths} \\ n_{1s} & \text{for sacrifices} \end{cases}$$

$$E_{j/s} = \begin{cases} \dfrac{b_j(b_{1j} + a_{1j})}{b_j + a_j} & \text{for deaths} \\ \dfrac{n_s(n_{1s} + m_{1s})}{n_s + m_s} & \text{for sacrifices} \end{cases}$$

Table 6   Tables for the Hoel-Walburg Test

Deaths, $j = 1, \ldots, J$

	Control	Exposed	Total
Deaths with tumor	b_{0j}	b_{1j}	b_j
Deaths without tumor	a_{0j}	a_{1j}	a_j
Deaths	$b_{0j} + a_{0j}$	$b_{1j} + a_{1j}$	$b_j + a_j$

Sacrifice times, $s = 1, \ldots, S$

	Control	Exposed	Total
Sacrifices with tumor	n_{0s}	n_{1s}	n_s
Sacrifices without tumor	m_{0s}	m_{1s}	m_s
Sacrifices	$n_{0s} + m_{0s}$	$n_{1s} + m_{1s}$	$n_s + m_s$

Table 7 Tables for the Log-Rank Test

	Control	Exposed	Total
Deaths and sacrifices	d_{0i}	d_{1i}	d_i
No event	$Y_{01} - d_{0i}$	$Y_{1i} - d_{1i}$	$Y_i - d_i$
At risk	Y_{0i}	Y_{1i}	Y_i

$$V_{j/s} = \begin{cases} \dfrac{b_j a_j (b_{0j} + a_{0j})(b_{1j} + a_{1j})}{(b_j + a_j)^2(b_j + a_j - 1)} & \text{for deaths} \\[2ex] \dfrac{n_s m_s (n_{0s} + m_{0s})(n_{1s} + m_{1s})}{(n_s + m_s)^2(n_s + m_s - 1)} & \text{for sacrifices} \end{cases}$$

and

$$O = \sum_{j,s} O_{j/s}, \qquad E = \sum_{j,s} E_{j/s}, \qquad V = \sum_{j,s} V_{j/s}$$

Log-Rank Test

Data for the log-rank analysis can also be summarized in a series of I tables (see Table 7), where i indexes each event time. Deaths or sacrifices with tumor in dose group Z are denoted by d_{zi} and numbers at risk of death or sacrifice at each event time by Y_{zi}. The test statistic has the same "observed minus expected" form, where the observed, expected, and variance components are given by

$$O_i = d_{1i}$$

$$E_i = \frac{d_i Y_{1i}}{Y_i}$$

$$V_i = \frac{d_i(Y_i - d_i)Y_{0i} Y_{1i}}{Y_i^2(Y_i - 1)}$$

APPENDIX 2. OBSERVED DATA LIKELIHOOD FOR CONTROL GROUP

The likelihood contributions for death or sacrifice with and without tumor for the ith animal can be written as follows:

$$\text{SNT:} \qquad p(t_i) = \exp\left\{-\int_0^{t_i} [\lambda(u) + \beta(u)]\, du\right\}$$

DNT: $\beta(t_i)p(t_i)$

SWT: $\displaystyle\int_0^{t_i} q(x, t_i)\, dx$

$$= \int_0^{t_i} \lambda(x) \exp\left\{ - \int_0^x [\lambda(u) + \beta(u)]\, du \right\} \exp\left\{ - \int_x^{t_i} \alpha(u)\, du \right\} dx$$

DWT: $\displaystyle\int_0^{t_i} \alpha(t_i) q(x, t_i)\, dx$

Suppose that the death time t_i falls in I_j defined in the piecewise exponential model. Then these contributions reduce to

SNT: $\displaystyle\exp\left[- \sum_{l=1}^{j-1} (\lambda_l + \beta_l)(\tau_l - \tau_{l-1}) \right] \times$

$$\exp[- (\lambda_j + \beta_j)(t_i - \tau_{j-1})]$$

SWT: $\displaystyle\sum_{k=1}^{j-1} \int_{\tau_{k-1}}^{\tau_k} q(x, t_i)\, dx + \int_{\tau_{j-1}}^{t_i} q(x, t_i)\, dx$

$$= \sum_{k=1}^{j-1} \left\{ g_k(t \mid \boldsymbol{\lambda}, \boldsymbol{\beta}, \xi) \frac{1}{\Delta_k} \underbrace{\left[e^{-\Delta_k \tau_{k-1}} - e^{-\Delta_k \tau_k} \right]}_{} \right\}$$

$$+ \left\{ g_j(t \mid \boldsymbol{\lambda}, \boldsymbol{\beta}, \xi) \frac{1}{\Delta_j} \underbrace{\left[e^{-\Delta_j \tau_{j-1}} - e^{-\Delta_j \tau_i} \right]}_{} \right\}$$

where

$$\Delta_k = \lambda_k + \beta_k(1 - \xi)$$

$$g_k(t \mid \boldsymbol{\lambda}, \boldsymbol{\beta}, \xi) = \lambda_k \exp\left[- \sum_{l=1}^{k-1} (\lambda_l + \beta_l)(\tau_l - \tau_{l-1}) \right] \exp[- \beta_j \xi(t_i - \tau_{j-1})]$$

$$\times \exp\left[- \sum_{l=k+1}^{j-1} \beta_l \xi(\tau_l - \tau_{l-1}) \right] \exp[(\lambda_k + \beta_k)\tau_{k-1} \exp[- \beta_k \xi \tau_k]]$$

$$g_j(t \mid \boldsymbol{\lambda}, \boldsymbol{\beta}, \xi) = \lambda_j \exp\left[- \sum_{l=1}^{j-1} (\lambda_l + \beta_l)(\tau_l - \tau_{l-1}) \right] \exp[(\lambda_j + \beta_j)(\tau_{j-1})] \exp[- \beta_j \xi t_i]$$

REFERENCES

Archer, L. E. and Ryan, L. M. (1989). On the role of cause-of-death data in the analysis of rodent tumorigenicity experiments. *Applied Statistics* 38: 81–93.

Cox, D. R. (1972). Regression models and life-tables (with discussion). *Journal of the Royal Statistical Society Series B* 34: 187–220.

Dempster, A. P., Laird, N. M., and Rubin, D. B. (1977). Maximum likelihood from incomplete data via the EM algorithm. *Journal of the Royal Statistical Society Series B* 39: 1–38.

Dempster, A. P., Selwyn, M. R., and Weeks, B. J. (1983). Combining historical and randomized controls for assessing trends in proportions. *Journal of the American Statistical Association* 78: 221–227.

Dewanji, A. and Kalbfleisch, J. D. (1986). Non-parametric methods for survival/sacrifice experiments. *Biometrics* 42: 325–341.

Dinse, G. E. (1988). Simple parametric analysis of animal tumorigenicity data. *Journal of the American Statistical Association* 83: 638–649.

Dinse, G. E. (1991). Constant risk differences in the analysis of animal tumorigenicity data. *Biometrics* 47: 681–700.

Dinse, G. E. and Lagakos, S. W. (1983). Regression analysis of tumor prevalence data. *Appl. Stat.* 32: 236–248.

Finkelstein, D. M. (1986). A proportional hazards model for interval-censored failure time data. *Biometrics* 42: 845–854.

Gart, J. J., Krewski, D., Lee, P. N., Tarone, R. E., and Wahrendorf, J. (1986). *Statistical Methods in Cancer Research*, Vol. 3: *The Design and Analysis of Long-Term Animal Experiments*. International Agency for Research on Cancer, Lyon.

Haseman, J. K., Huff, J., Boorman, G. A. (1984). Use of historical control data in carcinogenicity studies in rodents. *Toxicologic Pathology* 12: 126–135.

Hoel, D. G. (1983). Conditional two-sample tests with historical controls. In *Contributions to Statistics: Essays in Honor of Norman L. Johnson*, ed. Sen, P. K. North Holland, Amsterdam.

Hoel, D. G. and Walburg, H. E. (1972). Statistical analysis of survival experiments. *Journal of the National Cancer Institute* 49: 361–372.

Hoel, D. G. and Yanagawa, T. (1986). Incorporating historical controls in testing for a trend in proportions. *Journal of the American Statistical Association* 81: 1095–1099.

IMSL Math Library. (1989). *FORTRAN Subroutines for Mathematical Applications*, Version 1.1. IMSL, Houston, Texas.

Lagakos, S. W., and Louis, T. A. (1988). Use of tumor lethality to interpret tumorigenicity experiments lacking cause-of-death data. *Applied Statistics* 37: 169–179.

Lagakos, S. W., and Mosteller, F. (1981). A case study of statistics in the regulatory process: The FD&C Red No. 40 experiments. *Journal of the National Cancer Institute* 66: 197–212.

Lagakos, S. W., and Ryan, L. M. (1985a). On the representativeness assumption in prevalence tests of carcinogenicity. *Applied Statistics* 34: 54–62.

Lagakos, S. W., and Ryan, L. M. (1985b). Statistical analysis of disease onset and lifetime data from tumorigenicity experiments. *Environmental Health Perspectives* 63: 211–216.

Lindsey, J. C., and Ryan, L. M. (1993). A three state multiplicative model for rodent tumorigenicity experiments. *Applied Statistics* 42: 283–300.

Malani, H. M., and Van Ryzin, J. (1988). Comparison of two treatments in animal carcinogenicity experiments. *Journal of the American Statistical Association* 83: 1171–1177.

McKnight, B., and Crowley, J. (1984). Tests for differences in tumor incidence based on animal carcinogenesis experiments. *Journal of the American Statistical Association* 79: 639–648.

Meng, C. Y. K. and Dempster, A. P. (1987). A Bayesian approach to the multiplicity problem for significance testing with binomial data. *Biometrics* 43: 301–311.

National Toxicology Program. (1992). Toxicology and carcinogenesis studies of oxazepam in Swiss-Webster and B6C3F$_1$ mice (feed studies). NIH Publication No. 92-3359. U.S. Dept. of Health and Human Services, National Institutes of Health, Springfield, Virginia.

Peto, R. (1974). Guidelines on the analysis of tumor rates and death rates in experimental animals. *Brit. J. Cancer* 29: 101–105.

Portier, C. J. (1986). Estimating the tumor onset distribution in animal carcinogenesis experiments. *Biometrika* 73: 371–378.

Portier, C. J. and Dinse, G. E. (1987). Semiparametric analysis of tumor incidence rates in survival/sacrifice experiments. *Biometrics* 43: 107–114.

Press, W. H., Flannery, B. P., Teukolsky, S. A., and Vetterling, W. T. (1986). *Numerical Recipes*. Cambridge University Press, Cambridge, UK.

Ryan, L. M. (1993). Using historical controls in the analysis of developmental toxicity data. *Biometrics* 49: 1126–1135.

Ryan, L. M., and Orav, E. J. (1988). On the use of covariates for rodent bioassay and screening experiments. *Biometrika* 75: 631–637.

Tarone, R. E. (1982). The use of historical control information in testing for a trend in proportions. *Biometrics* 38: 215–220.

Turnbull, B. W., Brown, B. W., and Hu, M. (1974). Survivorship analysis of heart transplant data. *Journal of the American Statistical Association* 69: 74–80.

20

Assessing Drug Interactions: Tamoxifen and Cyclophosphamide

Keith Abrams University of Leicester, Leicester, England

Deborah Ashby University of Liverpool, Liverpool, England

Joan Houghton and Di Riley King's College School of Medicine and Dentistry, London, England

ABSTRACT

This chapter considers the problem of making inferences about the efficacy of tamoxifen and cyclophosphamide, and their possible interaction, in the treatment of patients with early stage breast cancer. Synthesis of beliefs based on previous trial evidence and on the results of a specific factorial designed trial is used to assess the current weight of evidence for the use of tamoxifen and cyclophosphamide. This is achieved by using a Bayesian analysis of a fully parametric proportional hazards regression model in which the baseline hazard is assumed to be constant over time. This example illustrates how the Bayesian philosophy is able to formalize the assessment of the current clinical evidence.

1 INTRODUCTION

Over the past few years increased interest has been shown in the application of Bayesian methodology in clinical trials. In particular, Spiegelhalter and Freedman (1988) and Hughes (1993) have argued persuasively for the use of a Bayesian approach to making inferences in clinical trials. There have been a number of practical applications of Bayesian methodology

(Racine *et al.*, 1986; Berry *et al.*, 1992; Carlin *et al.*, 1993; Spiegelhalter *et al.*, 1993 and Chapter 2). These authors have stressed the intuitive manner in which the Bayesian paradigm can assess the weight of evidence for a particular treatment. In the example discussed here we focus on the interaction of two treatments.

Tamoxifen by the mid-1980s had been shown to be beneficial for the treatment of early stage breast cancer in terms of both overall survival and disease-free survival (Early Breast Cancer Trialists' Collaborative Group, 1990, 1992a,b). In particular, the Nolvadex Adjuvant Trial Organisation (NATO) had been one of the earliest to report moderate benefits in terms of overall and disease-free survival with tamoxifen treatment (Nolvadex Adjuvant Trial Organisation, 1983). Prior to this evidence regarding the efficacy of tamoxifen, evidence had emerged regarding the efficacy of various cytotoxic agents (Early Breast Cancer Trialists' Collaborative Group, 1990). In particular, cyclophosphamide-based treatment regimens were being shown to be effective for early stage breast cancer, just as cyclophosphamide-based treatment regimens had previously been shown to be effective for metastatic breast cancer (Henderson, 1987). The SACS-1 trial in southern Sweden (Nissen-Meyer *et al.*, 1978) was one of the earliest trials to demonstrate beneficial effects of cyclophosphamide used alone. The logical next question was, "What happens when tamoxifen is given in combination with cytotoxic agents, and in particular cyclophosphamide?"

In response to this question, the Cancer Research Campaign (CRC) in the United Kingdom initiated a 2×2 factorial designed trial (CRC2) in which women with early stage operable breast cancer were randomized to either tamoxifen, cyclophosphamide, both tamoxifen and cyclophosphamide, or neither tamoxifen nor cyclophosphamide. In order to estimate precisely even small health gains for either treatment, especially given that few adverse effects had been reported, large numbers of events were necessary. The CRC2 trial was therefore designed to be as pragmatic as possible and to dovetail with clinical practice. This pragmatic nature, it was hoped, would appeal to clinicians in a country where only approximately 7% of all cancer patients are entered into clinical trials (Chouillet *et al.*, 1993).

The analysis of data from the CRC2 trial would require the use of survival models, given that the two major end points of the trial were overall survival and disease-free survival; i.e., death or recurrence is treated as an event. While randomization should ensure that the treatment groups are balanced with respect to covariates, we would also wish to model the treatment interactions and therefore some form of regression model would be required. Traditionally the semiparametric proportional hazards regression model (Cox, 1972) has been used extensively in the analysis of survival data, especially in a medical context. It assumes that covariates act multiplicatively on a time constant hazard ratio, although this appears to be verified only rarely in practice. Algebraically, the model is of the form

$$\lambda(t \mid z_i) = \lambda_0(t) \, e^{\beta^T z_i} \tag{1}$$

where t is time since randomization, $\lambda_0(t)$ is a baseline hazard function, z_i is a vector of covariates for the ith patient, and β is a vector of unknown regression parameters that are estimated via a partial likelihood (Cox, 1975). However, the semiparametric nature of the model is not the only option, and an alternative proportional hazards model is to assume that $\lambda_0(t)$ is known conditional on a vector of parameters, θ (Collett, 1994). In the simplest case $\lambda_0(t \mid \theta)$ is just a constant, say θ, and this corresponds to assuming that the survival times also follow an exponential distribution in addition to the covariates acting multiplicatively on the baseline hazard. While this imposes assumptions on the model, there is the possibility that other perhaps more flexible baseline hazard functions could be assumed if more appropriate (Kalbfleisch and Prentice, 1980; Cox and Oakes, 1984; Collett, 1994).

To place this trial in the context of the already available information, especially for the NATO trial and the SACS-1 trial, we used a Bayesian approach. Following such an approach, evidence from other trials available before the inception of the CRC2 trial is formally quantified in terms of some, or possibly all, the parameters in the regression model. Beliefs based on this prior data–based evidence are then formally updated in the light of the CRC2 trial data using Bayes' theorem. Specifically, the evidence from other trials is used to form a joint prior distribution for the model parameters, which is then updated by multiplying it by the likelihood function of the current data to form a joint posterior distribution for the model parameters. Inferences about particular model parameters or functions of parameters can then be derived from the joint posterior distribution.

The rest of this chapter is organized as follows. Section 2 considers the evidence for the use of both tamoxifen and cyclophosphamide in more detail and in particular quantifies the evidence available prior to the start of the CRC-sponsored trial. Section 3 considers how this evidence can be updated using the results of the CRC2 trial using Bayes' theorem within the setting of a fully parametric proportional hazards regression model. Section 4 presents the results of the analysis of the CRC2 trial, and finally Section 5 discusses some of the issues raised by this synthesis of evidence regarding tamoxifen and cyclophosphamide, from both a clinical perspective and a statistical one.

2 TAMOXIFEN AND CYCLOPHOSPHAMIDE IN EARLY STAGE BREAST CANCER

2.1 Tamoxifen

Tamoxifen is a form of hormone treatment for patients with breast cancer. Tamoxifen works by opposing the effects of estrogen, the hormone that is thought to be a stimulant of some breast cancers (Henry, 1991). As the effect of tamoxifen is specific there are fewer immediate reactions with tamoxifen than with standard cytotoxic agents. The most common short-term side effects with tamoxifen are hot flushes, irregular vaginal bleeding, and swelling of certain joints (Love *et al.*, 1991). Concern has been expressed about the long-term effects of tamoxifen use, especially at high doses, with regard to eyesight problems (Bentley *et al.*, 1992) and the development of uterine cancer (Fornander *et al.*, 1989).

By the early-1980s a number of trials had suggested a benefit for tamoxifen in the management of early stage breast cancer (Early Breast Cancer Trialists' Collaborative Group, 1990). In particular, the NATO trial (Nolvadex Adjuvant Trial Organisation, 1983) had reported an approximate 25% reduction (95% CI: 41% reduction to 5% reduction) in the risk of all-cause mortality for those women receiving tamoxifen for 2 years after surgery and had reported an approximate 21% reduction (95% CI: 39% reduction to 1% increase) in the risk of recurrence or death.

In order to evaluate formally the weight of evidence for the efficacy of tamoxifen and cyclophosphamide and their possible interaction in light of the CRC2 trial, the information available prior to the inception of the CRC2 trial in 1980 needs to be quantified. This will be most usefully done on a log-hazard ratio scale, which corresponds to the β terms in a model such as Eq. (2).

For tamoxifen the data-based information prior to 1980 is the results of the NATO trial (Nolvadex Adjuvant Trial Organisation, 1983). For overall survival, 166 deaths out of 564 patients were observed on the tamoxifen arm, and 210 deaths out of 567 patients were observed on the control arm of the trial. Thus, the log hazard ratio may be estimated by $\log_e[(O_t/E_t)/(O_c/E_c)]$ where O_t and O_c are the observed numbers of deaths in the tamoxifen and cyclophosphamide groups, respectively, and E_t and E_c are the expected numbers of deaths. An estimate of the variance of this estimator is $[r^2(r-1)]/[r_t r_c f(r-f)]$ where r is the total number of patients in the trial, f is the total number of deaths in the trial, and r_t and r_c are the numbers of patients in the tamoxifen and cyclophosphamide-based arms, respectively (Simon, 1986). For overall survival this corresponds to an estimate for the log-hazard ratio of tamoxifen of -0.29 with a standard deviation of 0.12.

Considering disease-free survival, 208 events were observed among 564 patients on the tamoxifen arm, and 274 events were observed on the control arm in the NATO trial. Using the same method as for overall survival, an estimate of the log-hazard ratio is -0.24 with a standard deviation of 0.13.

2.2 Cyclophosphamide

Cyclophosphamide is a cytotoxic alkylating agent that acts within a malignant cell's nucleus and destroys the cellular DNA material, thus preventing the cell from multiplying (Henry, 1991). The main short-term side effects of cyclophosphamide are nausea/vomiting, alopecia, and irregular menstruation (Henry, 1991). Long-term use of the drug can be associated with a reduction in the production of red blood cells; combined with irregular bleeding, this can lead to opportunistic infections (Henry, 1991).

By the early 1980s evidence was emerging regarding the beneficial effect of using cyclophosphamide to treat women with early stage breast cancer. In particular, the SACS-1 trial (Nissen-Meyer et al., 1978) suggested that the use of cyclophosphamide could lead to a 20% reduction (95% CI: 38% reduction to 4% increase) in the risk of all-cause mortality, with a 24% reduction (95% CI: 41% reduction to 2% reduction) in risk of recurrence or death.

For cyclophosphamide the available prior information is the results from the SACS-1 trial (Nissen-Meyer et al., 1978). In this trial in terms of overall survival, 146 deaths out of 507 patients were observed on the cyclophosphamide arm and 196 deaths out of 519 patients were observed on the control arm. Using the same methodology as for the NATO trial above, we can estimate the log-hazard ratio as -0.22 with standard deviation 0.13. Similarly, for disease-free survival 175 events out of 507 patients were observed on the cyclophosphamide arm, and 234 events were observed on the control arm; these correspond to a log-hazard ratio of -0.27 with standard deviation 0.13.

2.3 Combined Use of Tamoxifen and Cyclophosphamide

The rationale for the combination of tamoxifen and chemotherapy was that many breast cancers are not thought to be composed of a homogeneous group of malignant cells with identical properties and that whereas some malignant cells may show hormonal dependence others may not. Therefore, the hope was that by giving both a hormonal and a cytotoxic agent the chances of achieving a cellular response would be greatly increased. Some evidence had already been reported regarding the synergism between hormonal and cytotoxic therapy for advanced disease (Heuson, 1976; Carbone and Tormey, 1977).

Unfortunately, no phase III trials comparing explicitly tamoxifen and cyclophosphamide had been initiated by 1980. One trial that had reported results of comparing tamoxifen with a cyclophosphamide-based regimen was the Case Western A trial (Hubay *et al.*, 1980). This trial had reported results comparing cyclophosphamide, methotrexate, and 5-fluorouracil (CMF) and CMF plus 1 year of tamoxifen. Although an approximate assessment of the combined effect of tamoxifen and a cyclophosphamide-based regimen could be derived using the results of the Case Western A trial, it would be conditional on the effect of tamoxifen. In addition, such a derived data-based prior distribution for the interaction effect would be conditional on the effect of a cyclophosphamide-based treatment regimen, with the effect attributable to cyclophosphamide itself unknown. Therefore, the analysis in this chapter assumes a noninformative uniform prior distribution for the interaction effect, on a log hazard scale, of tamoxifen and cyclophosphamide, although we return to the issue of prior distributions for interaction effects in the discussion section.

2.4 CRC2 Trial of Tamoxifen and Cyclophosphamide

The entry criteria of the CRC2 trial were kept suitably wide so that a relatively large number of women may be recruited, in order to increase the precision of the estimated treatment effects and to reflect the pragmatic nature of the trial (Cancer Research Campaign Clinical Trials Centre, 1980). Primary treatment was either total mastectomy with axillary sampling followed by radiotherapy for node-positive patients or total mastectomy with axillary clearance.

Figure 1 shows the trial schema, in which 2230 women under the age of 70 with early stage operable breast cancer were randomized to receive either a control treatment, tamoxifen alone, cyclophosphamide alone, or both drugs in combination. Women randomized to a tamoxifen group received 10 mg of tamoxifen twice daily for 2 years or until first confirmation of distant recurrence, and women randomized to a cyclophosphamide group received 30 mg per kg body weight intravenously over a 6-day period.

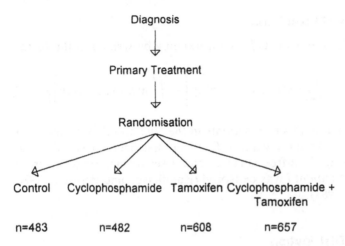

Figure 1 Cancer Research Campaign 2 × 2 factorial trial of tamoxifen and cyclophosphamide for the treatment of early breast cancer—trial schema.

3 STATISTICAL MODELING

3.1 A Parametric Proportional Hazards Regression Model

As the outcome in the trial is either overall survival or disease-free survival, a model that can allow for differential follow-up and censoring is required. Often either a semiparametric or fully parametric proportional hazards regression model is used. We will consider a fully parametric model of the form

$$\lambda(t \mid x_{1i}, x_{2i}) = e^{\alpha + \beta x_{1i} + \gamma x_{2i} + \tau x_{1i} x_{2i}} \tag{2}$$

where α, β, γ, and τ are unknown regression coefficients, x_{1i} is an indicator variable of whether the ith patient received tamoxifen, and x_{2i} is an indicator variable of whether the ith patient received cyclophosphamide. Thus, e^{β} is the hazard ratio for tamoxifen and e^{γ} is the hazard ratio for cyclophosphamide, and τ is an interaction term. In the case $\tau = 0$ the two treatments are termed "additive" on a log-hazard scale. When $\tau > 0$ the treatments are termed "subadditive"; i.e., their combined effect is less than the sum of their independent effects. When $\tau < 0$ they are termed "synergistic"; i.e., their combined effect is greater than the sum of their independent effects. Note that the hazards are assumed to act multiplicatively.

3.2 Formation of Likelihood

The likelihood for a model such as Eq. (2) may be written in the form

$$L(\alpha, \beta, \gamma, \tau) = \prod_{i=1}^{n} \left\{ \lambda(t_i \mid x_{1i}, x_{2i})^{d_i} \exp\left[- \int_0^{t_i} \lambda(u \mid x_{1i}, x_{2i}) du \right] \right\} \quad (3)$$

where n is the total number of patients in the trial and d_i is an indicator variable of death, taking the value 1 if patient i has been observed to die, or in the case of disease-free survival had a recurrence or died, and 0 otherwise, i.e., if patient i was censored (Kalbfleisch and Prentice, 1980; Cox and Oakes, 1984).

3.3 Posterior Distribution

To make inferences about the parameters of the model in the light of both the data in the CRC2 trial and the prior distributions of β, γ, and τ we need to obtain the posterior distribution:

$$P(\alpha, \beta, \gamma, \tau \mid \text{data}) \propto L(\alpha, \beta, \gamma, \tau)P(\alpha, \beta, \gamma, \tau) \quad (4)$$

If α, β, and γ are assumed to be independent of one another, and τ is assumed independent of α, then the joint posterior distribution, up to the constant of proportionality, may be written in the form

$$P(\alpha, \beta, \gamma, \tau \mid \text{data}) \propto L(\alpha, \beta, \gamma, \tau)P(\alpha)P(\beta)P(\gamma)P(\tau \mid \beta, \gamma) \quad (5)$$

Therefore we are able to specify prior distributions for the unknown model parameters individually for α, β, and γ and conditionally for τ. We wish to make posterior inferences about the model parameters separately, especially β, γ, and τ. To do this we require the *marginal* posterior distributions for the parameters of interest. For example, to obtain the marginal posterior distribution for β, the log-hazard ratio for tamoxifen, we integrate out α, γ, and τ, thus

$$P(\beta \mid \text{data}) \propto \int_{-\infty}^{+\infty} \int_{-\infty}^{+\infty} \int_{-\infty}^{+\infty} P(\alpha, \beta, \gamma, \tau, \mid \text{data}) \, d\alpha \, d\gamma \, d\tau \quad (6)$$

Having obtained the marginal posterior distributions for parameters of interest, we can also obtain posterior summary statistics such as the posterior mean and standard deviation.

3.4 Parameter Estimation

In simple cases the evaluation of the integrals in Eq. (6) required to obtain the marginal posterior distributions may be performed analytically. More often, however, they are not analytically tractable and either some form of approximation technique or a numerical integration method will be required.

Models such as Eq. (2), in which the prior distributions are either uniform or normal, lend themselves to *Laplace approximations*. These approximations assume that the joint posterior distribution is unimodal and that it is *peaked* near its mode. Essentially, the closer the joint distribution is to multivariate normality, the better the approximation.

In practice, the use of Laplace approximations means that the multiple integrals that have to be evaluated in order to obtain the marginal posterior distributions, Eq. (6), may be replaced by optimization over the parameters that need to be integrated over. This is very often a computationally easier task; a number of efficient algorithms for optimization are available. For further details on Laplace approximations see Tierney and Kadane (1986) and Tierney, *et al.* (1989). Practical implementations of these methods are available in XLISP-STAT (Tierney, 1990) or in S-PLUS (Statistical Sciences Inc., 1991) using sbayes, a suite of S-PLUS functions. For a general discussion of strategies for integration see Thisted (1988, Chapter 5) or Bernardo and Smith (1993, Chapter 5).

4 RESULTS

Using the fully parametric proportional hazards regression model of Section 3 together with the data-based prior distributions of Section 2, we are able to assess the current weight of evidence generated by the CRC2 trial for the use of tamoxifen and cyclophosphamide, both independently and in combination, for patients with early stage breast cancer. Figure 2a and b present the Kaplan-Meier survival curves from the study.

4.1 Mortality

Consider first the effects of the two drugs on all-cause mortality. Table 1 and Figure 3 show the prior and posterior distributions for the model parameters of interest, that is, β, the log-hazard ratio for tamoxifen; γ, the log hazard ratio for cyclophosphamide; and τ, a measure of the interaction of the two drugs on a log-hazard ratio scale.

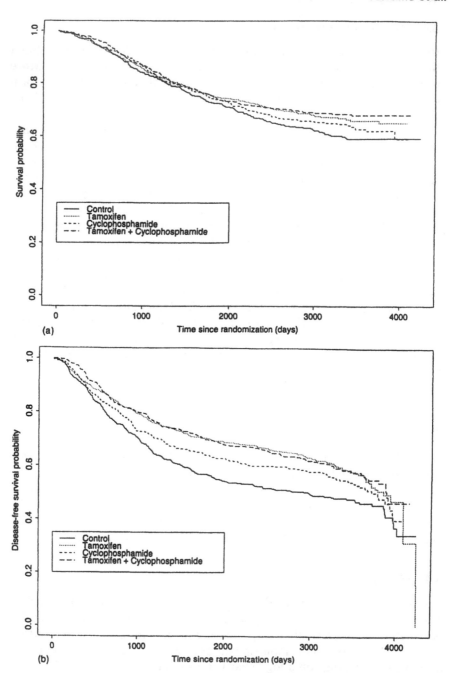

Figure 2 Kaplan-Meier estimated survival curves with (a) death from any cause and (b) recurrence or death from any cause as outcome measure.

Table 1 Prior to Posterior Analysis for Overall Survival Data, Using Log Hazard Ratio Scale[a]

Parameter	Distribution	Mean	SD	95% CI	Prob. < 0
Beta (β) (tamoxifen)	Reference posterior	−0.160	0.216	(−0.592,0.272)	0.771
	Data-based prior	−0.293	0.120	(−0.739,0.153)	0.906
	Data-based posterior	−0.256	0.214	(−0.684,0.172)	0.884
Gamma (γ) (cyclophosphamide)	Reference posterior	−0.071	0.217	(−0.505,0.363)	0.628
	Data-based prior	−0.222	0.132	(−0.666,0.222)	0.841
	Data-based posterior	−0.133	0.197	(−0.527,0.261)	0.750
Tau (τ) (interaction)	Reference posterior	0.077	0.242	(−0.407,0.561)	0.375
	Data-based posterior	0.129	0.217	(−0.305,0.563)	0.276

[a] Negative coefficients imply increased overall survival.

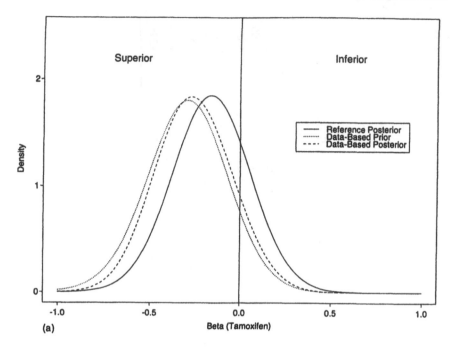

(a)

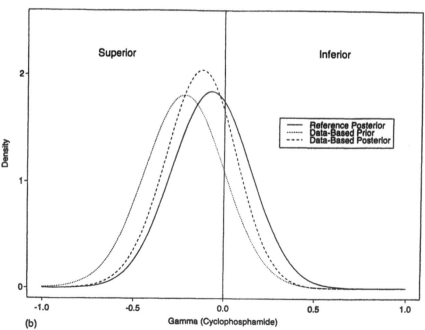

(b)

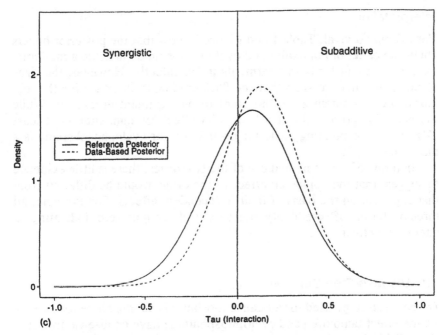

Figure 3 Prior and posterior distributions for (a) β (log-hazard ratio for tamoxifen), (b) γ (log-hazard ratio for cyclophosphamide), and (c) τ (interaction effect), with death from any cause as outcome measure.

The reference posterior distribution refers to the case in which we assume a uniform prior distribution over the whole real line for all the parameters in the model. This corresponds to the case in which we wish to use a noninformative prior distribution, i.e., assuming that all values in the parameter space are equally likely and allowing the data to dominate through the likelihood.

The data-based posterior distribution is the posterior distribution for the model parameters in the case in which we assume a uniform prior distribution over the whole real line for α and τ, the baseline parameter and the interaction parameter, respectively, but assume normal prior distributions for β and γ with the means and variances obtained in Section 3 from the NATO and SACS-1 trials, respectively.

Interpretation

For overall survival, Table 1 and Figure 3 show that the posterior beliefs about the effect of tamoxifen and cyclophosphamide indicate a moderate treatment benefit for both treatments independently. However, the sizes of these benefits are such that the 95% credibility intervals for the log-hazard ratios contain zero, the point of no treatment difference. While the posterior probability of a beneficial effect for tamoxifen is at least 77%, the corresponding posterior probability for cyclophosphamide is a more modest 63%.

In terms of the combined use of the two drugs, there is little evidence to suggest that the combined effect of the drugs would be different from that expected on the basis of their independent effects. For example, all three posterior 95% credibility intervals easily contain zero, indicating no interaction effect.

4.2 Disease-Free Survival

The same strategy used for all-cause mortality is adopted here for assessing the effect tamoxifen and cyclophosphamide have on disease-free survival. Table 2 and Figure 4 show the prior and posterior distributions of the main model parameters.

Interpretation

As with the overall survival results, there is evidence to suggest that tamoxifen yields a larger reduction in the risk of death or recurrence than cyclophosphamide. Defining $(e^\alpha - e^{\alpha+\beta})/e^\alpha = 1 - e^\beta$ as the risk reduction provided by the treatment, here is a $1 - e^{-0.335} = 28\%$ reduction from tamoxifen compared to a $1 - e^{-0.224} = 20.1\%$ reduction for cyclophosphamide. In the case of tamoxifen the 95% credibility interval barely contains the point of no treatment difference, with a posterior probability of a beneficial treatment effect of over 96%. For cyclophosphamide the posterior 95% credibility intervals all easily contain the point of no treatment difference.

In terms of the combined effects of tamoxifen and cyclophosphamide although there is some evidence to suggest a reduction in the combined effect of the two drugs, given their independent effects, the strength of this evidence is relatively weak, with the posterior 95% credibility intervals all easily containing zero, i.e., no interaction effect.

Table 2 Prior to Posterior Analysis for Disease-Free Survival Data, Using Log Hazard Ratio Scale[a]

Parameter	Distribution	Mean	SD	95% CI	Prob. < 0
Beta (β) (tamoxifen)	Reference posterior	−0.356	0.200	(−0.748, 0.036)	0.962
	Data-based prior	−0.267	0.120	(−0.502, −0.032)	0.987
	Data-based posterior	−0.335	0.187	(−0.701, 0.032)	0.963
Gamma (γ) (cyclophosphamide)	Reference posterior	−0.220	0.217	(−0.645, 0.205)	0.845
	Data-based prior	−0.266	0.127	(−0.515, −0.017)	0.982
	Data-based posterior	−0.224	0.216	(−0.647, 0.199)	0.850
Tau (τ) (interaction)	Reference posterior	0.208	0.226	(−0.235, 0.651)	0.179
	Data-based posterior	0.184	0.217	(−0.241, 0.609)	0.198

[a] Negative coefficients imply increased disease-free survival.

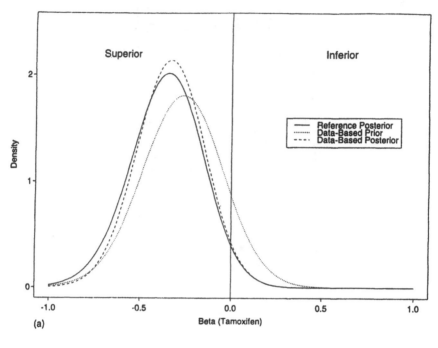

(a)

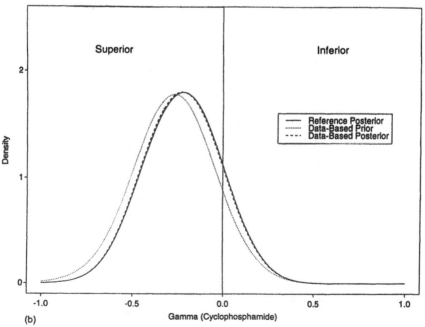

(b)

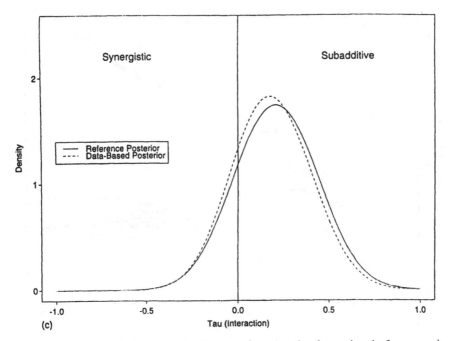

Figure 4 Prior and posterior distributions for (a) β (log-hazard ratio for tamoxi-fen), (b) γ (log-hazard ratio for cyclophosphamide), and (c) τ (interaction effect), with recurrence or death from any cause as outcome measure.

5 MODEL CHECKING

The results in this chapter, and in turn the inferences drawn, have of course assumed that models such as Eq. (2) can adequately describe the data. Chaloner (1991) has considered the use of residuals within a Bayesian framework to assess model fit. Various forms of Bayes factors have also been advocated as a method for model comparison (Berger and Pericchi, 1994; Kass and Raftery, submitted).

The two key assumptions in this model are (1) that the effects of covariates act multiplicatively on a baseline hazard and (2) that using an exponential distribution to model the survival times is adequate. A simple graphical check of these assumptions is to plot the log-log survivor function against log time for the four treatment groups. If an exponential distribution is appropriate, then the lines should be linear with an approximate

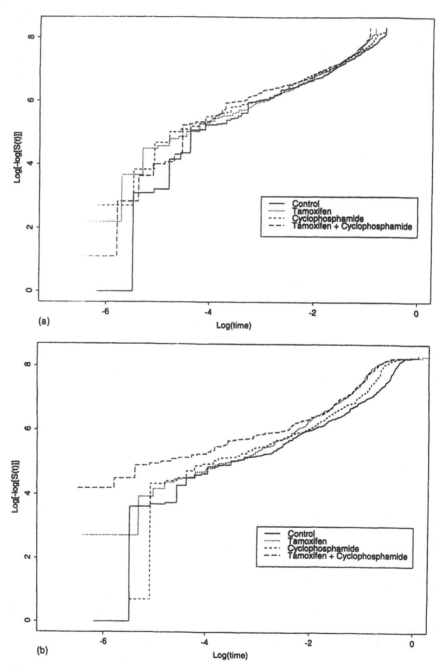

Figure 5 $\log_e[-\log_e[S(t)]]$ against log time with (a) death from any cause and (b) recurrence or death from any cause as outcome measure.

gradient of one. Similarly, if the assumption of proportional hazards is appropriate, the lines should also be parallel. Figure 5 shows these plots for both overall and disease-free survival in the CRC2 trial. It can be seen that the assumption of exponential survival would appear to be a reasonable one for both outcome measures, and the assumption of proportional hazards may warrant further investigation, perhaps within the framework of an additive hazards model (Gore *et al.*, 1984).

6 DISCUSSION

In this chapter we have seen how by using the Bayesian paradigm we have been able to obtain an assessment of the current weight of evidence for the use of tamoxifen and cyclophosphamide for the treatment of early breast cancer. In particular we have been able to place the results of the CRC2, the only trial so far to consider this particular combination of drugs, in their clinical context by formally incorporating the data-based beliefs using the results of other previously conducted relevant trials.

In summary, there appears to be considerable evidence to suggest that the use of tamoxifen in the treatment of early stage breast cancer leads to a beneficial increase in disease-free survival, and to a lesser extent there is evidence to support a more modest increase in overall survival. Although the same trend is seen with cyclophosphamide, the magnitude of the actual benefits is smaller. In terms of the combined use of the drugs there appears very little evidence to suggest that the benefit achieved by giving both tamoxifen and cyclophosphamide is less than that achieved by giving either of the drugs independently. There appears slight evidence, however, that the combination of the two drugs does produce a benefit that is smaller than might be anticipated from each drug's independent action. However, placing this in its clinical context, there is only slight evidence for this effect, and the magnitude of this effect is small compared to the further risk reduction achieved by the use of combined therapy.

Whereas further evidence has emerged during the 1980s regarding the possible beneficial effects of tamoxifen, little further evidence has emerged about the use of cyclophosphamide as a single chemotherapeutic agent (Early Breast Cancer Trialists' Collaborative Group, 1990). Although we have not attempted to do so, the Bayesian paradigm allows the synthesis of this further information to be formally included in the assessment of tamoxifen and cyclophosphamide.

The inferences drawn from this modeling process, would further benefit from having elicited clinical demands available for the magnitude of an effect for either tamoxifen and cyclophosphamide (Freedman, 1987). However, in this particular situation it is not unreasonable to use a zero treatment effect as being clinically significant, as neither drug has severe side effects associated with it, and therefore even small survival gains would incur little health cost.

The issue of a prior distribution for the interaction effect, τ, is an important one. It is unlikely that beliefs about a possible interaction effect could be assumed independent of the main effects. However, in practice, either the elicitation of or the derivation, using previous trial data, of a prior distribution for the interaction effect conditional upon beliefs about the main effects is difficult. Although the strategy adopted in this chapter of assuming a noninformative prior distribution for the interaction effect conditional on the main effects would appear reasonable, a prior distribution derived from the results of the NATO trial and the Case Western A trial should be treated with extreme caution and for this reason was not used.

Although many, both classical and Bayesian, would argue that a properly conducted randomized trial removes the need to consider covariates other than treatment in any model (Pocock, 1983), we may still wish to explore the interaction between known prognostic covariates such as menstrual status, nodal status, and estrogen status and treatments (Berry, 1994). In theory, models such as Eq. (2) can accommodate further covariates, although in practice a number of difficulties arise. First is the question of the specification of prior distributions for the model parameters. As models become more complicated, it becomes harder to extract suitable prior information from other study results, and similarly it becomes difficult to elicit prior beliefs from clinical experts. The second difficulty arises in terms of parameter estimation. In practice, the Laplace approximations of Section 3 work well for well-behaved models with relatively small numbers of parameters. More complex models with relatively large number of parameters require more sophisticated methods of parameter estimation such as Gibbs sampling (Gilks et al., 1993).

A number of other Bayesian survival models have been proposed. Among the earliest were those of Cornfield and Detre (1977) and Kalbfleisch (1978) who both considered a multiplicative model in which the baseline hazard function was composed of a finite number of disjoint time intervals, the hazard being assumed constant within each one. By assuming that prior information regarding the baseline hazard rate followed a gamma process, posterior inference regarding the treatment effect was relatively straightforward. Gamerman (1991) has also considered a model in which the baseline hazard was a piecewise constant function, but he also assumed that there was an autoregressive structure on the parameters between successive intervals. This enables a sufficiently flexible baseline hazard to be assumed, while at the same time allowing prior information about the regression parameters and parameter estimation via linear Bayes methods (West and Harrison, 1989). An alternative approach has been considered by Sweeting (1987) and Achcar et al. (1985) in which the survival times are transformed yielding a "location-scale" regression model. Clayton (1991) considered a "frailty model" in which a random term, the frailty, was introduced into model (1) to account for familial associated risk. By using a gamma distribution for this risk Clayton showed that parameter estimation via Gibbs sampling is relatively straightforward. A number of authors have extended parametric proportional hazards models such as Eq. (2) to cases in which the baseline hazard varies over time, e.g., by assuming that the survival times follow a Weibull distribution (Dellaportas and Wright, 1991; Greenhouse, 1992; Abrams et al., 1995). Extensions of the fully parametric survival model (2) to the case in which there may also be a large number of classification units as well as patients, for example, centers in multicenter trials, have also been developed (Gray, 1994; Stangl, 1995). Finally, the increased awareness of Markov chain Monte Carlo methods, and Gibbs sampling in particular, has enabled Bayesian approaches to a semiparametric model, in the spirit of the Cox proportional hazards regression model (1), to be implemented (Dellaportas and Smith, 1993).

In conclusion, regarding an assessment of the possible interaction between tamoxifen and cyclophosphamide, further evidence is required before we can obtain one with adequate precision. However, our assessment of the current evidence suggests that the benefit, in terms of both overall and disease-free survival, achieved when both drugs are used in combination is less than expected, given their independent effects, and that these conclusions are strengthened when formally viewed in the light of other evidence.

REFERENCES

Abrams, K. R., Ashby, D., and Errington, R. D. (1995). Bayesian parametric survival models—an application to a cancer clinical trial. Technical Report 95-04, University of Leicester, Leicester, England.

Achcar, J. A., Brookmeyer, R., and Hunter, W. G. (1985). An application of Bayesian analysis to medical follow-up data. Statistics in Medicine 4: 509–520.

Bentley, C. R., Davies, G., and Aclimandos, W. A., (1992). Tamoxifen retinopathy: A rare but serious complication. British Medical Journal 304: 495–496.

Berger, J. O. and Pericchi, L. R. (1994). The intrinsic Bayes factor for linear models. In Bayesian Statistics 5, ed. Bernardo, J., Berger, J., Dawid, A., and Smith, A. Oxford University Press, Oxford.

Bernardo, J. M. and Smith, A. F. M. (1993). *Bayesian Theory*. Wiley, Chichester, England.

Berry, D. A. (1994). Scientific inference and predictions, multiplicities and convincing stories: a case study in breast cancer research. In *Bayesian Statistics 5*, ed. Bernardo, J., Berger, J., Dawid, A., and Smith, A. Oxford University Press, Oxford.

Berry, D. A., Wolff, M. C., and Sack, D., (1992). Public health decision making: A sequential vaccine trial. In *Bayesian Statistics 4*, ed. Bernardo, J., Berger, J., Dawid, A., and Smith, A. Oxford University Press, Oxford.

Cancer Research Campaign Clinical Trials Centre. (1980). *Protocol: Cancer Research Campaign Adjuvant Trial*. Cancer Research Campaign, King's College School of Medicine and Dentistry.

Carbone, P. P. and Tormey, D. C. (1977). Combination chemotherapy for advanced disease. In *Breast Cancer 1—Advances in Research and Treatment Current Approaches to Therapy*, ed. McGuire, W. L. Churchill Livingstone, London.

Carlin, B. P., Chaloner, K. M., Louis, T. A., and Rhame, F. S. (1993). Elicitation, Monitoring, and Analysis for an AIDS Trial. Technical Report, University of Minnesota, Minneapolis, Minnesota.

Chaloner, K. M. (1991). Bayesian residual analysis in the presence of censoring. *Biometrika* 78: 637–644.

Chouillet, A. M., Bell, C. M. J., and Hiscox, J. G. (1993). Management of breast cancer in southeast England. *British Medical Journal* 308: 168–171.

Clayton, D. G. (1991). A Monte Carlo method for Bayesian inference in frailty models. *Biometrics* 47: 467–485.

Collett, D. (1994). *Modelling Survival Data in Medical Research* Chapman & Hall, London.

Cornfield, J. and Detre, K. (1977). Bayesian life table analysis. *Journal of the Royal Statistical Society B)* 39: 86–94.

Cox, D. R. (1972). Regression models and life-tables. *Journal of the Royal Statistical Society (B)* 39: 187–220.

Cox, D. R. (1975). Partial likelihood. *Biometrika* 62: 269–276.

Cox, D. R. and Oakes, D. (1984). *Analysis of Survival Data*. Chapman & Hall, London.

Dellaportas, P. and Smith, A. F. M. (1993). Bayesian inference for generalized linear and proportional hazards models via Gibbs sampling. *Applied Statistics* 42: 443–459.

Dellaportas, P. and Wright, D. (1991). Numerical prediction for the two parameter Weibull distribution. *Statistician* 40: 365–372.

Early Breast Cancer Trialists' Collaborative Group. (1990). *Treatment of Early Breast Cancer—Worldwide Evidence 1985–1990*, Vol. 1. Oxford University Press, Oxford.

Early Breast Cancer Trialists' Collaborative Group. (1992a). Systemic treatment of early breast cancer by hormonal, cytotoxic, or immune therapy: 133 randomised trials involving 31,000 recurrences and 24,000 deaths among 75,000 women. *Lancet* 339: 1–15.

Early Breast Cancer Trialists' Collaborative Group. (1992b). Systemic treatment of early breast cancer by hormonal, cytotoxic, or immune therapy: 133 randomised trials involving 31,000 recurrences and 24,000 deaths among 75,000 women. *Lancet* 339: 71–85.

Fornander, T., Cedermark, B., Mattsson, A., *et al.* (1989). Adjuvant tamoxifen in early breast cancer: occurrence of new primary tumours. *Lancet* 1: 117–120.

Freedman, B. (1987). Equipoise and the ethics of clinical research. *New England Journal of Medicine* 317: 141–145.

Gamerman, D. (1991). Dynamic Bayesian models for survival data. *Applied Statistics* 40: 63–79.

Gilks, W. R., Clayton, D. G., Spiegelhalter, D. J., Best, N. G., McNeil, A. J., Sharples, L. D., and Kirby, A. J. (1993). Modelling complexity: Applications of Gibbs sampling in medicine. *Journal of the Royal Statistical Society (B)* 55: 39–52.

Gore, S. M., Pocock, S. J., and Kerr, G. R. (1984). Regression models and non-proportional hazards in the analysis of breast-cancer survival. *Applied Statistics* 33: 176–195.

Gray, R. J. (1994). A Bayesian analysis of institutional effects in a multicenter cancer clinical trial. *Biometrics* 50: 244–253.

Greenhouse, J. (1992). On some applications of Bayesian methods in cancer clinical trials. *Statistics in Medicine* 11: 37–53.

Henderson, I. C. (1987). Chemotherapy for advanced disease. In *Breast Diseases*, eds. Harris, J. R., Hellman, S., Henderson, I. C., and Kinne, D. W. pp. 428–479. J. B. Lippincott, Philadelphia.

Henry, J. E. (1991). *The British Medical Association Guide to Medicines and Drugs*. The British Medical Association, London.

Heuson, J. C. (1976). Editorial. *Cancer Treatment Reports* 60: 1463.

Hubay, C. A., Pearson, O. H., Marshall, J. S., Rhodes, R. S., Debanne, S. M., Rosenblatt, J., Mansour, E. G., Hermann, R. E., Jones, J. C., Flynn, W. J., Eckert, C., and McGuire, W. L. (1980). Adjuvant chemotherapy, antiestrogen therapy and immunotherapy for stage II breast cancer: 45 months follow-up of a prospective randomized clinical trial. *Cancer* 46: 2805–2808.

Hughes, M. D. (1993). Reporting Bayesian analyses of clinical trials. *Statistics in Medicine* 12: 1651–1663.

Kalbfleisch, J. D. (1978). Non-parametric Bayesian analysis of survival time data. *Journal of the Royal Statistical Society (B)* 40: 214–221.

Kalbfleisch, J. D. and Prentice, R. L. (1980). *The Statistical Analysis of Failure Time Data*. Wiley, New York.

Kass, R. E. and Raftery, A. E. (1995). Bayes factors. *Journal of the American Statistical Association* 90: 773–795.

Love, R. R., Cameron, L., Connell, B., *et al.* (1991). Symptoms associated with tamoxifen treatment in postmenopausal women. *Archives of Internal Medicine* 151: 1842–1847.

Nissen-Meyer, R., Kjellgren, K., Malmio, K., Mansson, B., and Norin, T. (1978). Surgical adjuvant chemotherapy—results with one short course with cyclophosphamide after mastectomy for breast cancer. *Cancer* 41: 2088–2098.

Nolvadex Adjuvant Trial Organisation. (1983). Controlled trial of tamoxifen as adjuvant agent in the management of early breast cancer. *Lancet* 1: 257–261.

Pocock, S. J. (1983). *Clinical Trials—A Practical Approach*. Wiley, Chichester, UK.

Racine, A., Grieve, G. P., Fluehler, H., and Smith, A. F. M. (1986). Bayesian methods in practice: Experiences in the pharmaceutical industry (with discussion). *Applied Statistics* 35: 93–150.

Simon, R. (1986). Confidence intervals for reporting the results of clinical trials. *Annals of Internal Medicine* 105: 429–435.

Spiegelhalter, D. J. and Freedman, L. S. (1988). Bayesian approaches to clinical trials. In *Bayesian Statistics 3*, ed. Bernardo, J., DeGroot, M., Lindley, D., and Smith, A., editors, pp. 453–477. Oxford University Press, Oxford.

Spiegelhalter, D. J., Freedman, L. S., and Parmar, M. K. B. (1993). Applying Bayesian thinking in drug development and clinical trials. *Statistics in Medicine* 12: 1501–1511.

Spiegelhalter, D. J., Freedman, L. S., and Parmar, M. K. B. (1994). Bayesian analysis of randomised trials (with discussion). *Journal of the Royal Statistical Society (A)* 157: 357–416.

Stangl, D. K. (1995). Modelling and decision making using Bayesian hierarchical models. *Statistics in Medicine* 14(20): 2173–2190.

Statistical Sciences Inc. (1991). *S-PLUS Users Manual*. Version 3.0. Statistical Sciences Inc., Seattle.

Sweeting, T. J. (1987). Approximate Bayesian analysis of censored survival data. *Biometrika* 74: 809–816.

Thisted, R. A. (1988). *Elements of Statistical Computing—Numerical Computation*. Chapman & Hall, New York.

Tierney, L. (1990). *Lisp-Stat: An Object-Orientated Environment for Statistical Computing and Dynamic Graphics*. Wiley, New York.

Tierney, L. and Kadane, J. B. (1986). Accurate approximations for posterior moments and marginal densities. *Journal of the American Statistical Association* 81: 82–86.

Tierney, L., Kass, R. E., and Kadane, J. B. (1989). Fully exponential Laplace approximations of expectations and variances of non-positive functions. *Journal of the American Statistical Association* 84: 710–716.

West, M. and Harrison, P. J. (1989). *Bayesian Forecasting and Dynamic Models*. Springer Series in Statistics. Springer-Verlag, New York.

21

Bayesian Subset Analysis of a Clinical Trial for the Treatment of HIV Infections

Richard Simon and Dennis O. Dixon   National Institutes of Health, Bethesda, Maryland
Boris Freidlin   The Emmes Corporation, Potomac, Maryland

1   INTRODUCTION

Subset analysis has been a controversial area in the conduct of modern clinical trials. Physicians are often interested in examining whether treatment effects differ among subsets of patients, because they are sensitive to the heterogeneity of patient populations. Many new treatments are toxic or expensive, and hence it is desirable to determine which patients benefit from a new treatment. Because there are often prognostic factors that may influence patient compliance, tolerance to treatment, and pharmacokinetic or pharmacodynamic effects, subset analysis is often viewed as scientifically and medically important.

Statisticians are sensitive to the multiple-comparison aspects of subset analysis. They are also concerned that pressure to identify "positive results" causes overinterpretation of statistical significance. In spite of these statistical concerns, subset analyses play a prominent role in the reports of many major clinical trials. Unfortunately, in many cases only rudimentary statistical analyses are presented. Often subsets of interest are not specified in advance and frequentist methods available for addressing the multiple comparison issues (e.g., Simon, 1988) are not used. The scientific difficulties involved in either ignoring or performing subset analyses have been discussed by Berry (1990).

Dixon and Simon (1991) developed a Bayesian method for subset analysis in the context of generalized linear models. This approach uses noninformative priors for the specification of prior beliefs and hence is relatively "objective." Our purposes here are to describe an extension of the method for use in clinical trials with more than two treatment groups and to apply the extension to the analysis of a clinical trial for the treatment of patients with advanced human immunodeficiency virus (HIV) disease. We demonstrate how the posterior probability of "qualitative interactions" (Gail and Simon, 1985) can be computed. We also indicate how to compute the posterior probability that a combination regimen is better than each of its single-agent components.

2 THE CLINICAL TRIAL

The AIDS Clinical Trials Group (ACTG) is a national clinical trials organization sponsored by the AIDS Division of the National Institute of Allergy and Infectious Diseases. Study ACTG 155 was started in December 1990 to compare three antiretroviral therapy regimens for persons with advanced HIV disease. The regimens were zidovudine (ZDV, also known as azathioprine, or AZT 200 mg three times per day, zalcitabine (ddC), 0.75 mg three times per day, and combined ZDV and ddC. The primary objective was to compare the treatment groups in terms of times to occurrence of an AIDS-defining event or death.

Randomizations were stratified on the basis of HIV disease status (symptomatic or asymptomatic), length of previous treatment with ZDV (up to 1 year or more than 1 year), and type of prophylaxis for *Pneumocystis carinii* pneumonia (PCP) (local only versus systemic only versus neither or both local and systemic). One thousand one volunteers were enrolled from 51 sites by the time accrual stopped in August 1991. Follow-up ended on January 15, 1993. The randomization was weighted 2:2:3 in favor of the combination therapy group.

A detailed presentation of study design and results has been published (Fischl *et al.*, 1995). Groups were well balanced with respect to pretreatment level of CD4-positive T cells as well as all stratification factors. Investigators intended from the start to examine results in subsets of patients defined by these four characteristics. Figure 1 shows the distribution of time without progression or death for the three treatment groups. The log-rank test of the homogeneity of these three curves gives a nonsignificant result ($p = 0.28$).

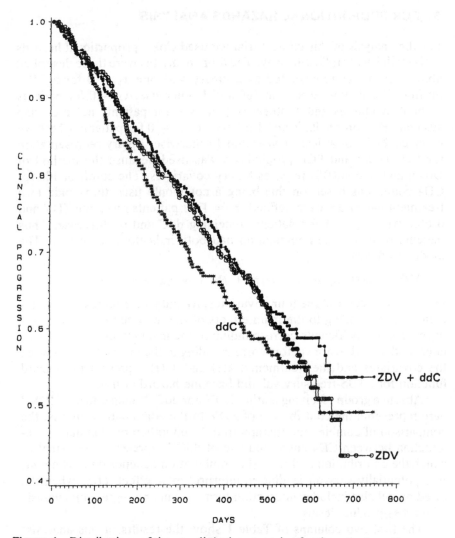

Figure 1 Distributions of time to clinical progression for three treatment groups. Open circles are used for ZDV, asterisks for ddC, and filled circles for the combination of ZDV and ddC.

3   COX PROPORTIONAL HAZARDS ANALYSIS

For the analysis of this clinical trial we used Cox's proportional hazards (PH) model in the following way. The four covariates were those described above. Each was represented as a binary variable: $x_1 = 0$ for a CD4-positive T cell count less than 100 and 1 otherwise; $x_2 = 0$ for patients without symptoms and 1 otherwise; $x_3 = 0$ for patients not receiving systemic PCP prophylaxis and 1 otherwise; $x_4 = 0$ for patients who have received ZDV for at least 1 year and 1 otherwise. Binary representation for CD4 count and PCP prophylaxis was used because the method of Dixon and Simon (DS) requires binary covariates. The cutoff of 100 for CD4 count was based on this being a commonly used threshold. Two treatment indicators were defined; $\tau_1 = 1$ for patients receiving ZDV and 0 otherwise; $\tau_2 = 1$ for patients receiving ddC and 0 otherwise. Thus patients receiving the combination had both indicators equal to 1. The model used was

$$\lambda(t) = \lambda_0(t) \exp(\alpha_1 \tau_1 + \alpha_2 \tau_2 + \beta x + \gamma_1 \tau_1 x + \gamma_2 \tau_2 x)$$

where x is a vector of the four covariates, α_1 and α_2 are regression coefficients corresponding to the main effects of the treatments, β is a vector of regression coefficients corresponding to the main effects of the covariates, and γ_1 and γ_2 are vectors corresponding to the interactions between the covariates and the treatments. $\lambda(t)$ and $\lambda_0(t)$ represent the hazard function for AIDS-free survival and baseline hazard at time t.

Absent a group receiving neither ZDV nor ddC, "main effect of ZDV" here represents the contribution of ZDV to the combination, that is, the comparison of combination therapy to ddC monotherapy. Similarly, "interaction between CD4 count and use of ddC" represents the extent to which the contribution of ddC to the combination depends on CD4 count. It is potentially very misleading to interpret main-effect terms when the fitted model also includes interaction terms, as in any regression analysis with cross-product terms.

The first two columns of Table 1 show the results of this analysis. The values of regression coefficients are shown with statistical significance levels in parentheses. The only regression coefficients that are statistically significant at the .05 two-sided level are the main effect of CD4 cell count (x_1) and the interaction between CD4 count and use of ddC. We performed a likelihood ratio test of the global hypothesis that all treatment-by-covariate interactions are zero. The value of the test statistic is 10.937, which has a chi-squared distribution with 8 degrees of freedom under the null hypothesis. This yields a significance level greater than 0.25 and hence the hypothesis of homogeneity would not be rejected.

Table 1 Proportional Hazards Regression Models

Variable	Full model	Main-effects model	Reduced model
τ_1	$-.243\ (.47)^a$	$-.256\ (.032)$	$-.261\ (.029)$
τ_2	$-.358\ (.29)$	$-.154\ (.20)$	$.047\ (.75)$
x_1	$-1.09\ (.0009)$	$-1.43\ (.0001)$	$-1.04\ (.0001)$
x_2	$.312\ (.46)$	$.436\ (.002)$	$.445\ (.002)$
x_3	$-.078\ (.80)$	$.003\ (.98)$	$-.009\ (.93)$
x_4	$-.606\ (.08)$	$-.123\ (.28)$	$-.124\ (.28)$
$\tau_1 * x_1$	$.041\ (.88)$	—	—
$\tau_1 * x_2$	$-.061\ (.86)$	—	—
$\tau_1 * x_3$	$-.077\ (.75)$	—	—
$\tau_1 * x_4$	$.198\ (.46)$	—	—
$\tau_2 * x_1$	$-.543\ (.036)$	—	$-.577\ (.012)$
$\tau_2 * x_2$	$.241\ (.48)$	—	—
$\tau_2 * x_3$	$.166\ (.50)$	—	—
$\tau_2 * x_4$	$.471\ (.10)$	—	—

[a] Regression coefficient (statistical significance level).

The terms in the reduced model without interaction terms are shown in the third column of Table 1. For this model, the main effect of CD4 count (x_1) is highly significant, the main effect of HIV symptoms (x_2) is highly significant, and the main effect of ZDV is also significant. For the model without interactions, α_1 represents the effect of the combination relative to ddC alone.

The last column of Table 1 shows the terms when the CD4-by-ddC interaction is retained in the model. Here we obtain significant main effects of CD4 count, HIV symptoms, and the combination versus ddC contrast, as well as a significant interaction between CD4 count and the combination versus ZDV contrast. The results of both reduced models must be interpreted with caution, however, because they are models selected based on the data. Hence, the regression coefficients and significance levels may be distorted. Regression coefficients and significance levels for these models are the types of results that are often shown for PH model analysis.

Because of multiple-comparison issues, dependence of the regression coefficients on variable selection, and the difficulty of interpreting regression coefficients for models containing interactions, the conclusion to be reached from these analyses is somewhat ambiguous. If we accept the results of the main-effects model, then the conclusion seems to be that the combination is superior to ddC but not superior to ZDV and there is no statistically significant evidence of treatment effect specificity. The full model, on the other hand, appears to indicate that the only treatment difference is one between the combination and ZDV and that difference depends on the CD4 cell count. The benefit of the combination over ZDV is greater for patients with CD4 cell counts above 100. If we accept the results of the reduced model, however, we would conclude that the combination is superior to ddC and that there is also evidence that the relative benefit of the combination over ZDV depends on the CD4 cell count.

4 THE DIXON-SIMON HIERARCHICAL MODEL

Dixon and Simon considered models of the type shown above for the case of two treatments. In that case there is a single treatment indicator and a single vector of interaction effects. Our approach here will be a direct extension of the DS method. It will be described in the context of a three-treatment-group trial, but application to more than three treatments is immediate. Let

$$\theta = (\alpha_1, \alpha_2, \beta, \gamma_1, \gamma_2)$$

denote the vector of unknown parameters. From the usual PH model analysis we assume that we have an estimate

$$\hat{\theta} \mid \theta \sim N(\theta, C)$$

where C is the covariance matrix of the estimates. We assume that we have prior distributions

$$(\alpha_1, \alpha_2) \sim \text{NID}(0, \sigma_\alpha^2)$$
$$\beta \sim N(0, I\sigma_\beta^2)$$

and

$$(\gamma_1, \gamma_2) \mid \xi^2 \sim N(0, I\xi^2)$$

where NID denotes normal and independently distributed. The limiting case $\xi^2 = 0$ corresponds to the absence of treatment-by-covariate interactions, representing absence of treatment effect variation from subset to subset. Instead of specifying the value of ξ^2, we use a second level prior of

$$p(\xi^2) = \frac{1}{\max(\xi^2, \epsilon)}$$

where ϵ is a small positive quantity. Modification of the standard indifference prior is necessary to achieve a proper posterior density for θ. We have used a value of $\epsilon = .005$, but results are relatively insensitive to choices that differ from this by three orders of magnitude. It is actually unnecessary to specify prior distributions for σ_α^2 and σ_β^2 because we are interested only in locally uniform priors for the corresponding parameters, so we allow the respective variances to grow without limit.

It may not be clear why the means of the prior distributions for the parameters are all zero or whether they would also be zero if the coding of the covariates were changed from $(0, 1)$ to $(-1, 1)$. Remember that each of these parameters represents a comparison of one kind or another; the model is deliberately neutral on the question of whether each α is positive or negative, and similarly for the β's and γ's. Changing the coding would appear to require a change in the location of the priors, but all of these locations can be absorbed in the baseline hazard function.

The prior distribution assumed for $\gamma = (\gamma_1, \gamma_2)$ is an example of the exchangeability exploited by Lindley and Smith (1972) and others. It is a form of symmetry that seems natural in this setting. That is, none of the interaction parameters is considered, a priori, more likely to be positive (or negative) than any other interaction parameter. If there is a single interaction of particular interest a priori, such an interaction should be modeled separately. The assumption that all of the covariates are binary is needed to make the regression coefficients commensurate.

Let $\eta = d\theta$ denote a linear combination of the parameters. Then analogous to the results of DS for the two-treatment case, the posterior distribution of η can be approximated by

$$p(\eta \mid \hat{\theta}) \propto \int p(\hat{\gamma} \mid \xi^2)p(\eta \mid \hat{\theta}, \xi^2)p(\xi^2) \, d\xi^2 \tag{1}$$

where the first factor of the integrand is the marginal distribution of the vector of interaction estimates given the variance component, the second factor is the posterior distribution of the linear combination given the variance component, and the final factor is the prior for the variance component. This approximation is based on the asymptotic sufficiency and normality of the maximum likelihood estimator $\hat{\theta}$. Since $\hat{\gamma} \mid \gamma$ is normal and $\gamma \mid \xi^2$ is normal, it follows that

$$p(\hat{\gamma} \mid \xi^2) \sim N(0, C_{22} + I\xi^2) \tag{2}$$

where C_{22} denotes the submatrix of C corresponding to the interaction terms.

It can be shown from the results of Lindley and Smith (1972) that

$$p(\eta \mid \hat{\theta}, \xi^2) \sim N(dBb, dBd) \tag{3}$$

where

$$B^{-1} = C^{-1} + \begin{bmatrix} 0 & 0 \\ 0 & I/\xi^2 \end{bmatrix}$$

and $b = C^{-1}\hat{\theta}$.

Consequently, the posterior distribution of any linear combination of the parameters can be computed as a single-dimensional integral. Since the posterior density must integrate to one, the constant of integration can be determined by computing the single-dimensional integral on a grid of points and then numerically integrating the resulting function. Dixon and Simon (1991) demonstrated how the efficiency of computing the posterior densities can be substantially improved by using eigenvalue-eigenvector decompositions. The same approach was used for the calculations presented here but the derivations will not be repeated.

Three types of linear combinations were of interest. In one type the vector η contains a single 1 and the other components are zero. This is used for computing the posterior distribution of a particular parameter. The second type of linear combination is for evaluating the posterior distribution of a treatment contrast within an elementary subset of patients. An *elementary subset* is a subset defined by the simultaneous specification of all four covariates. For example, to evaluate the posterior distribution of the log-hazard ratio of failure for the combination versus ddC for patients with CD4 count greater than 100, with no HIV symptoms, receiving systemic PCP prophylaxis who have received ZDV for at least 1 year the linear combination is (1,0, 0,0,0,0, 1,0,1,0, 0,0,0,0). This is because for such a patient the covariates are $x = (1, 0, 1, 0)$ and the treatment indicators are (1, 1) for the combination and (0, 1) for ddC alone. There are 2^4 elementary subsets.

We are also interested in evaluating treatment contrasts for subsets determined by each covariate separately, for example, patients with CD4 count greater than 100. Such quantities are not uniquely determined without specifying the values of the other covariates or the distribution of those values. For example, let w_i denote the proportion of cases of those with $x_1 = 1$ that also have $x_i = 1$. Then the average treatment effect of the combination versus ddC alone for cases with CD4 > 100 is taken as $d\theta$ with $d = (1,0, 0,0,0,0, 1,w_2,w_3,w_4, 0,0,0,0)$. This type of linear combination is also used to evaluate treatment contrasts for the patient sample as a whole. The average treatment effect of the combination versus ddC alone overall is taken as $d\theta$ with $d = (1,0, 0,0,0,0, w_1,w_2,w_3,w_4, 0,0,0,0)$ where (w_1,w_2,w_3,w_4) are the average values of the covariates for the sample overall.

5 QUALITATIVE INTERACTIONS

Two treatments exhibit a qualitative interaction over a class of subsets if one treatment is preferable for some of the subsets and the other treatment is preferable for other of the subsets. Peto (1982) has argued that only qualitative interactions are important because the usual quantitative interactions are scale dependent and there is no reason to expect that treatment effects should be exactly the same for different subsets. Gail and Simon (1985) and Piantadosi and Gail (1993) have developed significance tests of the hypothesis that there is no qualitative interaction for disjoint subsets. Russek-Cohen and Simon (1993) have developed such tests for multiway classifications. Here we shall show how to calculate the probability that a qualitative interaction does or does not exist for a specified treatment contrast and class of subsets. We shall derive this for the case of two subsets, but the results generalize directly to any number of subsets.

Let η_1 and η_2 denote linear combinations that represent the same treatment contrast for two different subsets. The subsets may be of any type, either disjoint elementary subsets or composite subsets each determined by the level of a single covariate (e.g., individuals with CD4 > 100). A qualitative interaction for the treatment contrast over these two subsets is said to exist if the linear combinations are not of the same sign. The probability that both linear combinations are positive is

$$\int \Pr[\eta_1, \eta_2 \geq 0, \xi^2 \mid \hat{\theta}] \, d\xi^2 = \int \Pr[\eta_1, \eta_2 \geq 0 \mid \xi^2, \hat{\theta}] p(\xi^2 \mid \hat{\theta}) \, d\xi^2$$

$$= \frac{1}{p(\hat{\theta})} \int \Pr[\eta_1, \eta_2 \geq 0 \mid \xi^2, \hat{\theta}] p(\hat{\theta} \mid \xi^2) p(\xi^2) \, d\xi^2$$

On taking the limit as the variance components of the main effects become infinite, this approaches

$$\frac{1}{p(\hat{\gamma})} \int \Pr[\eta_1, \eta_2 \geq 0 \mid \xi^2, \hat{\theta}] p(\hat{\gamma} \mid \xi^2) p(\xi^2) \, d\xi^2 \qquad (4)$$

similar to Eq. (1). The second factor in the integrand is (2) and the third factor is the modified Jeffreys prior for the variance component. For computing the first factor we use the result that η_1 and η_2 are jointly normal with means and variances given by Eq. (3) and covariance $d_1 B d_2$ where $\eta_1 = d_1 \theta$ and $\eta_2 = d_2 \theta$. Expression (4) gives the probability that both linear combinations are positive. Similar expressions are obtained for the probability that they are both negative and for the probability that one is positive and the other is negative. Since these four expressions must sum to one, the constant of integration before the integral can be determined by performing four single-dimensional numerical integrations.

The approach described above can also be used to calculate the probability that the combination is better than both single agents either overall or for a particular subset of patients. To do this we define the two linear combinations to represent the contrast of the combination versus ZDV and the combination versus ddC for the same group of patients. For example, to determine whether the combination is better than both single agents on the average for the overall population, we use the linear combinations (1,0, 0,0,0,0, .56,.83,.41,.27, 0,0,0,0) and (0,1, 0,0,0,0, 0,0,0,0, .56,.83,.41,.27).

6 RESULTS

Figure 2 shows the posterior density of the treatment contrasts averaged over the patient sample overall. The solid line is the posterior density for the contrast of the combination versus ddC alone The point estimate of the mode of the posterior distribution is -0.245. This is negative, indicating that combination therapy achieves a lower hazard rate than ddC alone. The 95% highest posterior density interval (-0.485, 0.005) includes zero, but just barely. The posterior probability that the linear combination is positive is 0.026. Consequently, the posterior probability that the average hazard for the combination is less than the average hazard for ddC alone is 0.974. In this case the posterior density is almost normal and the highest posterior density interval is almost an equal-tail confidence interval, but not quite.

The long-dashed line in Figure 2 shows the posterior density for the comparison of the combination versus ZDV averaged over all the patients. The mode of the posterior distribution of the average log hazard for the combination minus that for ZDV alone is -0.203. The probability that the average log hazard for the combination is less than that for ZDV alone is 0.947.

The short-dashed line in Figure 2 shows the posterior density for the linear combination that represents the average log hazard for ddC alone minus that for ZDV alone. In the tables that will be presented below, the contrasts of the two single agents always represent ddC alone minus ZDV alone. As can be seen from this curve, there is no evidence that one single agent is better than the other on average overall based on these data. Using the methods described in Section 5, we computed the posterior probability that the linear combinations representing contrasts of the combination regimen versus ZDV and the combination regimen versus ddC

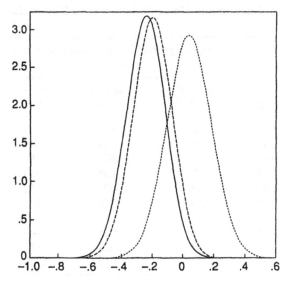

Figure 2 Posterior density functions for treatment contrasts. Each treatment contrast represents a difference in log hazard averaged over the entire sample of patients. Solid line: combination versus ddC. Mode = −0.245, 95% highest posterior density interval (−0.485, 0.005), posterior probability contrast is positive = 0.026. Long-dashed line: combination versus ZDV. Mode = −0.203, 95% highest posterior density interval (−0.448, 0.047), posterior probability contrast is positive = 0.053. Short-dashed line: ddC alone versus ZDV alone. Mode = 0.043, 95% highest posterior density interval (−0.227, 0.303), posterior probability contrast is positive = 0.611.

overall are both negative. This was 0.866 and represents the probability that the combination is better than both single agents overall for the patient sample.

The results of Figure 2 are summarized in the first row entry in Table 2. This table also provides the posterior probability that a linear combination is positive for linear combinations representing treatment contrasts for subsets of patients determined by the level of a single covariate. With four binary covariates there are eight such subsets and there are three pairwise contrasts of the three treatments. The middle column shows results for the combination versus ddC for these subsets. It can be seen that there is relatively strong evidence that the combination is more effective than ddC alone for each of these subsets. The first column of numbers is for the comparison of the combination versus ZDV. Here there are

Table 2 Treatment Effects for Subsets Determined by One Covariate
$\epsilon = 0.005$

Subset	n patients	Posterior probability of treatment effect > 0		
		Combination vs. ZDV	Combination vs. ddC	ddC vs. ZDV
All patients	991	.053	.026	.611
CD4 < 100	310	.213	.014	.850
CD4 > 100	681	.035	.074	.394
No symptoms	172	.073	.095	.483
Symptoms	819	.061	.025	.652
No systemic PCP prophylaxis	570	.045	.050	.497
Systemic PCP prophylaxis	421	.132	.031	.734
Prior ZDV > 1 year	726	.037	.029	.541
Prior ZDV < 1 year	265	.233	.072	.733

some subsets for which the evidence is much weaker for superiority of the combination—for example, patients with CD4 counts less than 100, patients receiving systemic PCP prophylaxis, and patients who have not received prior ZDV or who have received it for less than 1 year. It should be noted, however, that there is no evidence that ZDV is better than the combination for these subsets.

Figure 3 shows the posterior density for the contrast of the combination versus ZDV for patients with CD4 count <100 separately from that for patients with CD4 count >100. The mode in one case is half that in the other case, but the 95% highest posterior density intervals overlap substantially. We computed the probability that both linear combinations are of the same sign using the methods described in Section 5. The probability that both linear combinations are positive is 0.035. The probability that both are negative is 0.588. The probability that they differ in sign, i.e., that a qualitative interaction exists, is 0.377. Hence, we cannot exclude the possibility of a qualitative interaction involving CD4 count for this treatment contrast. However, the probability that the linear combination is greater than 0.10 for patients with CD4 count less than 100 is 0.07. As a difference of 0.10 on the log-hazard scale represents a quite small effect, it is unlikely that ZDV alone is better than the combination to a medically significant degree. On the other hand, the combination may not be superior

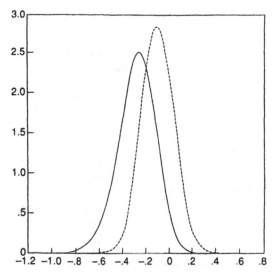

Figure 3 Posterior density functions for combination versus ZDV based on initial CD4 count. Solid line: combination versus ZDV for patients with CD4 count <100. Mode = −0.252, 95% highest posterior density interval (−0.607, 0.038), posterior probability contrast is positive = 0.035. Dashed line: combination versus ZDV for patients with CD4 count >100. Mode = −0.118, 95% highest posterior density interval (−0.383, 0.167), posterior probability contrast is positive = 0.213.

to ZDV by a medically significant degree for this subset. For comparison, the probability that the linear combinations contrasting the combination regimen to ddC alone are both negative is 0.822. The probability of a qualitative interaction is 0.159.

Table 2 was computed using $\epsilon = 0.005$ for the modified Jeffreys prior for the variance component. Table 3 presents the same results using $\epsilon = 5.0$. Qualitatively, the conclusions are unchanged. In cases in which the posterior distribution is inconclusive as to whether the linear combination is positive, the actual posterior probability can change substantially, however.

Table 4 shows summary results for pairwise treatment contrasts in each of the 16 elementary subsets. The covariate values are coded "y" for yes and "n" for no in an attempt to simplify reading this complex table. The column indicating results for comparisons of the single agents is relatively simple because there are few indications of the superiority of either treatment. For the four elementary subsets that include patients

Table 3 Treatment Effects for Subsets Determined by One Covariate
$\epsilon = 5.0$

		Posterior probability of treatment effect > 0		
Subset	n patients	Combination vs. ZDV	Combination vs. ddC	ddC vs. ZDV
All patients	991	.040	.034	.531
CD4 < 100	310	.361	.016	.920
CD4 > 100	681	.018	.120	.220
No symptoms	172	.072	.169	.346
Symptoms	819	.057	.032	.609
No systemic PCP prophylaxis	570	.032	.079	.357
Systemic PCP prophylaxis	421	.177	.044	.745
Prior ZDV > 1 year	726	.019	.035	.407
Prior ZDV < 1 year	265	.393	.143	.749

Table 4 Treatment Effects for Elementary Subsets

CD4 > 100	sx (symp-toms)	Systemic PCP prophylaxis	ZDV < 1 year	Posterior probability of treatment effect > 0		
				Comb. vs. ZDV	Comb. vs. ddC	ddC vs. ZDV
y	y	y	y	.202	.113	.641
y	y	y	n	.060	.068	.494
y	y	n	y	.123	.145	.496
y	y	n	n	.030	.093	.343
y	n	y	y	.169	.169	.525
y	n	y	n	.063	.132	.408
y	n	n	y	.106	.203	.412
y	n	n	n	.036	.163	.294
n	y	y	y	.463	.050	.887
n	y	y	n	.260	.020	.853
n	y	n	y	.368	.071	.818
n	y	n	n	.137	.030	.746
n	n	y	y	.381	.103	.783
n	n	y	n	.200	.068	.707
n	n	n	y	.286	.132	.692
n	n	n	n	.117	.087	.573

with CD4 counts <100 and HIV symptoms, there is some evidence that ZDV alone is more effective than ddC alone. The column of these linear combinations is calculated for the log hazard of ddC minus the log hazard of ZDV. A positive value for this linear combination means that the hazard for ddC is greater than that for ZDV. The posterior probabilities of positivity for these four subsets, however, do not constitute conclusive evidence.

Table 5 supplements the information given in Table 4 for the contrasts of the combination versus each of the single agents. Table 5 shows the number of patients in each elementary subset as well as the posterior mode and 95% highest posterior density interval for each contrast. For the contrasts between the combination and ddC alone the results are relatively uniform favoring the combination. The posterior modes are very uniform across subsets. The results appear less conclusive in terms of posterior tail areas for patients with CD4 counts >100 who are asymptomatic, but these subsets contain relatively few patients. Of course, it is a consequence of the use of a model that one is able to obtain estimates of effects for subsets that contain very few, or perhaps even no, patients. The more sparse the data, the more one depends on the model itself.

Table 5 Posterior Mode and 95% Highest Posterior Density (HPD) Intervals for Elementary Subsets

CD4 >100	sx	Systemic PCP prophylaxis	ZDV < 1 year	n patients	Comb vs. ZDV Mode	HPD interval	Comb vs. ddC Mode	HPD interval
y	y	y	y	43	−.16	(−.54, .25)	−.24	(−.61, .17)
y	y	y	n	134	−.23	(−.63, .08)	−.25	(−.59, .10)
y	y	n	y	77	−.20	(−.59, .17)	−.22	(−.57, .20)
y	y	n	n	208	−.27	(−.71, .03)	−.22	(−.54, .12)
y	n	y	y	8	−.19	(−.69, .27)	−.23	(−.66, .29)
y	n	y	n	33	−.24	(−.81, .11)	−.23	(−.64, .22)
y	n	n	y	8	−.22	(−.75, .18)	−.22	(−.62, .33)
y	n	n	n	39	−.27	(−.89, .06)	−.22	(−.59, .24)
n	y	y	y	55	−.08	(−.40, .49)	−.27	(−.62, .06)
n	y	y	n	119	−.12	(−.41, .23)	−.29	(−.6, −.01)
n	y	n	y	52	−.10	(−.42, .38)	−.25	(−.58, .09)
n	y	n	n	131	−.16	(−.45, .14)	−.26	(−.55, .02)
n	n	y	y	7	−.11	(−.48, .47)	−.26	(−.68, .18)
n	n	y	n	22	−.16	(−.54, .24)	−.27	(−.66, .11)
n	n	n	y	15	−.14	(−.52, .36)	−.24	(−.63, .22)
n	n	n	n	40	−.20	(−.60, .16)	−.24	(−.61, .13)

For the contrasts between the combination and ZDV alone, the results appear less uniform. Evidence for the superiority of the combination is strongest for patients with CD4 counts >100 who have received ZDV for more than 1 year. Although conclusions favoring a particular treatment cannot be stated for the other subsets, the posterior probability that the log hazard for the combination is greater than for ZDV is less than 0.5 for all of them. Consequently, the posterior probability that the overall average log hazard for the combination is greater than that for ZDV alone is quite small, specifically 0.053.

7 COMPUTATIONAL ASPECTS

Calculations of the posterior distribution of linear combinations of parameters were performed using an SAS macro written to implement the Dixon-Simon method. Integration was performed using the trapezoid rule. Integration with regard to ξ^2 in expression (1) was done over the range from 0 to 20 using a 4000-point grid. Numerical integration of the posterior density for calculating highest posterior density intervals and the posterior probability that a linear combination η is positive was performed in the following manner. The estimate ξ^2_{ml} that maximized the marginal likelihood was obtained. Then the posterior mean and variance of η given ξ^2 = ξ^2_{ml} were calculated as described by Dixon and Simon (1991, p. 876). The range of numerical integration was taken as the estimated mean plus or minus six times the square root of the estimated variance. The calculation and graphical display of the posterior distribution of a linear combination required approximately 3 minutes on a personal computer with a 66-mHz Intel 486 microprocessor running PC-SAS under DOS. The SAS macro for this analysis is available from the authors and on StatLib. It has been tested on PC compatibles running DOS and on an IBM RS6000 workstation running AIX.

Posterior probabilities associated with qualitative interactions were calculated using a program written in S-PLUS which calls the FORTRAN subroutine for computing the multivariate normal integral described by Schervish (1984). Integration with respect to ξ^2 was performed using the trapezoid rule with 4400 grid points.

8 DISCUSSION

One of our objectives has been to extend the method of analysis introduced by Dixon and Simon (1991, 1992) for use with clinical trials having more than two treatment groups. Both of the clinical trials previously used to illustrate this method were multiarm trials, although the analysis was limited to only two of the arms. The ability to analyze all arms in one unified model is important for efficiently estimating the main effects of covariates and for providing a consistent interpretation of treatment contrasts. With three treatment groups, there are two independent contrasts and hence two indicator variables were used for treatment. In general, $K - 1$ indicator variables should be used with K treatment groups. The coding of these indicator variables is not critical. Although the coding determines the interpretation of individual regression coefficients, such interpretations are problematic in any case. Average treatment effects either overall or for subsets can be obtained by appropriate specification of linear combinations for any coding. Because we place a locally uniform prior on the regression coefficients associated with the treatment indicators, a similar locally uniform prior distribution results for all linear combinations of these indicators. Hence, the parameterization is only a matter of convenience. For special designs where locally uniform priors may not be desired, the parameterization is more important and other approaches may be needed. This is the case, for example, with factorial designs if a priori one expects small interactions or for dose-response designs if monotone relationships are expected.

We considered several approaches to modeling the treatment-by-covariate interactions. One approach involved assuming that the interaction effects associated with a specific treatment indicator were exchangeable with each other but not with other interaction effects. In the three-treatment case investigated here, we would assume that the components of the vector γ_1 were $NID(0, \xi_1^2)$ conditional on ξ_1^2, the components of the vector γ_2 were $NID(0, \xi_2^2)$ conditional on ξ_2^2, and the two variance components were independently and identically distributed from the modified Jeffreys prior. Alternatively, one could assume exchangeability for interactions of the same covariate with the different treatment indicators but not with different covariates. For the clinical trials we have considered, however, these assumptions did not seem more appropriate than the assumption that all of the treatment-by-covariate interactions are exchangeable. We were able to obtain results for the posterior distribution of linear combinations for the restricted types of exchangeability described above, but multiple integration with regard to the several variance components was required rather than the single-dimensional integration of expression (1). Although eigenvalue-eigenvector decompositions were still possible for reducing the amount of computation involved, the resulting methods were much more computationally demanding than that based on Eq. (1).

Our analysis suggests that for averages over all patients, the combination appears more effective than either single agent. The posterior probability that the average log hazard for the combination is lower than that for ddC is 0.974; for the combination versus ZDV the figure is 0.947. The posterior probability that the average log hazard for the combination is lower than that for both single agents is 0.866. For the comparison of the combination to ddC the effect appeared consistent across the subsets of patients. For the comparison of the combination to ZDV the effect was conclusive only for patients with CD4 counts greater than 100 who had received ZDV for more than 1 year. The results were less conclusive in the other subsets, but in none was there an indication that ZDV was more effective than the combination. Because the Bayesian model indicated some evidence of superiority of the combination over ZDV consistently across subsets, the average over all patients indicated fairly strong support for superiority of the combination over ZDV.

Has this analysis produced greater insight or different conclusions than the Cox proportional hazards analyses described in Section 3? This insight is primarily for the reader to decide but the original authors "found no overall benefits of zalcitabine used alone or with zidovudine." The usual presentation of the full model provides little information and invites erroneous conclusions. One is tempted to conclude that there are no main effects of either ZDV or ddC because neither regression coefficient approaches statistical significance. One is cautioned from this interpretation by the nominally significant interaction between the ddC effect and CD4 cell count, but there are no interactions that approach significance involving the ZDV effect.

Usually frequentist analyses do not stop with full models retaining numerous nonsignificant variables. Model reduction and variable selection procedures are quite varied and ad hoc, however (e.g., Miller 1990). One approach often used in clinical trials in which treatment-by-subset interactions are not expected is to test the global null hypothesis that all interactions are zero. This provides protection against the possibility that at least one interaction will appear significant by chance. As noted in Section 3, this approach results in the main-effect model. In this model there is a significant main effect of ZDV that was not apparent in the full model. There is also a significant main effect of symptoms that was not apparent in the full model. Alternatively, one may ignore the multiple-comparison issue of there being eight interaction terms and eliminate all interactions except for the one showing nominal significance in the full model. This model has significant main effects of ZDV, CD4, and symptoms as well as the retained ddC-by-CD4 interaction. One might have obtained a similar model from one of the many types of variable selection regression procedures. But results may have depended on whether forward addition or backward elimination was used, the nominal significance level cutoffs used for determining whether variables are retained, rules for whether main effects are permitted to be eliminated if interactions are retained, and whether variables are tested in groups for elimination. Consequently, one can have little confidence in the appropriateness of a model reduced from the full model using variable selection procedures or in the statistical properties of the regression coefficients and covariance matrix of the selected model.

For any selected model, the regression coefficients, standard errors, and statistical significance values usually presented provide limited information. In our Bayesian analysis we have emphasized the presentation of posterior distributions of treatment contrasts for subsets of patients or for averages across subsets. In fact, we have not even presented the posterior distributions or modes or "significance" of individual regression coefficients in our model. Although we have rarely seen such presentations, frequentist analyses could present point estimates and confidence intervals for such linear combinations. Table 6 shows such results for the elementary subsets using the full model. In computing such confidence intervals, one must decide how to deal with the multiple-comparison problem that there are numerous subsets and averages of interest. Table 6

Table 6 Maximum-Likelihood Estimates, 95% Confidence Intervals, and Bonferroni Adjusted 95% Confidence Intervals for Comparing the Combination to ZDV in Elementary Subsets

CD4 >100	sx	Systemic PCP prophylaxis	ZDV < 1 year	MLE	Unadjusted confidence interval	Adjusted confidence interval
y	y	y	y	−.02	(−.70, .65)	(−1.16, 1.11)
y	y	y	n	−.49	(−1.01, .02)	(−1.36, .38)
y	y	n	y	−.19	(−.82, .44)	(−1.24, .87)
y	y	n	n	−.66	(−1.13, .19)	(−1.45, .13)
y	n	y	y	−.26	(−1.19, .67)	(−1.83, 1.30)
y	n	y	n	−.73	(−1.52, .05)	(−2.06, .59)
y	n	n	y	−.43	(−1.30, .44)	(−1.90, 1.04)
y	n	n	n	−.90	(−1.63, −.17)	(−2.13, .33)
n	y	y	y	.52	(−.06, 1.10)	(−.46, 1.50)
n	y	y	n	.05	(−.38, .48)	(−.68, .78)
n	y	n	y	.35	(−.21, .92)	(−.59, 1.30)
n	y	n	n	−.12	(−.54, .30)	(−.82, .59)
n	n	y	y	.28	(−.56, 1.12)	(−1.13, 1.69)
n	n	y	n	−.19	(−.90, .51)	(−1.37, .99)
n	n	n	y	.11	(−.68, .91)	(−1.23, 1.45)
n	n	n	n	−.36	(−1.02, .31)	(−1.48, .76)

shows just elementary subsets for contrasts between the combination and single agent ZDV, but there are also contrasts comparing the single agents and the combination to ddC as well as combinations representing composite subsets and overall average effects. There is, of course, an extensive literature on multiple-comparison procedures for linear contrasts in analysis of variance problems. In Table 6 we show two types of confidence intervals. One column gives intervals unadjusted in any way for multiplicity. The other column gives confidence intervals incorporating a Bonferroni adjustment for the 48 combinations resulting from three treatment contrasts for 16 elementary subsets. Clearly, there are at least this many contrasts of interest.

The differences between the entries in Table 6 and the columns of Table 5 corresponding to the combination versus ZDV are striking. The maximum-likelihood estimates (MLEs) of the treatment effects are much more variable among subsets than the posterior modes. The values of the posterior mode are "shrunk" toward the overall posterior mode. This generally, but not always, means being shrunk toward zero. Some of the MLE values are quite extreme. Another difference is that the 95% highest posterior density intervals are considerably narrower than the unadjusted confidence intervals. The Bonferroni adjusted confidence intervals are so broad as to be useless.

There are obvious limitations to this method of analysis. Currently, we still require that the covariates be binary. This was a limitation for the modeling of CD4 count, and future research will investigate relaxing this requirement. The method is also based on assumptions of exchangeability and normality. The appropriateness of the exchangeability assumption must be examined in each case. For this clinical trial we believed that this assumption was appropriate. The normality assumption is made for convenience. With more covariates, the method outlined in Dixon and Simon (1991) could be used to test this assumption. In spite of these limitations, however, we believe that the method has been useful for evaluating the results of this clinical trial and hope that others will be encouraged to utilize it for other trials.

Gray (1994) used a Bayesian proportional hazards model for studying treatment-by-institution interactions in clinical trials. He also used essentially flat priors for the main effects but did not approximate the likelihood function using the asymptotic sufficiency of the maximum-likelihood estimator as we have done. Although Gray did not require the covariates to be binary, he did not incorporate treatment-by-covariate interactions. His treatment-by-institution effects were assumed to be normal with mean zero, but he used a Wishart distribution for their covariance matrix instead of the modified Jeffreys prior used here. His approach also differed in that he placed a prior on the baseline hazard function and required the Gibbs sampler to approximate the posterior distributions.

Evaluating treatment effects for heterogeneous populations of patients is a complex endeavor that requires a variety of good tools. Issues of specificity of effects are of increased importance for several reasons, however. First is the development of molecular and genetic characterizations of the differences in disease characteristics among patients. There is increased expectation that these covariates will be important treatment selection factors and increased emphasis on evaluating such interactions in clinical trials. Second is the increased emphasis on evaluating whether there are gender or minority group differences in treatment effects. Recently, the U.S. Congress has mandated that all clinical trials supported by the National Institutes of Health address this issue. Finally is the movement toward larger clinical trials with broader patient eligibility criteria. Because of the increased heterogeneity of such samples, there may be a greater interest in subset analyses. We are encouraged that the model examined here may be found useful in other clinical trials.

ACKNOWLEDGMENT

The authors thank the AIDS Clinical Trials Group ACTG 155 investigators and study team who conducted the clinical trial that produced the data set used in our analyses. The AIDS Clinical Trials Group is sponsored by the National Institute of Allergy and Infectious Diseases.

REFERENCES

Berry, D. A. (1990). Subgroup analyses. *Biometrics* 46: 1227–1230.

Dixon, D. O. and Simon, R. (1991). Bayesian subset analysis. *Biometrics* 47: 871–882.

Dixon, D. O. and Simon, R. (1992). Bayesian subset analysis in a colorectal cancer clinical trial. *Statistics in Medicine* 11: 13–22.

Fischl, M. A., Stanley, K., Collier, A., et al. (1995). Combination and monotherapy with zidovudine and zalcitabine in patients with advanced HIV disease. *Annals of Internal Medicine* 122: 24–32.

Gail, M. and Simon, R. (1985). Testing for qualitative interactions between treatment effects and patient subsets. *Biometrics* 41: 361–372.

Gray, R. J. (1994). A Bayesian analysis of institutional effects in a multicenter cancer clinical trial. *Biometrics* 50: 244–253.

Lindley, D. V. and Smith, A. F. M. (1972). Bayes estimates for the linear model (with discussion). *Journal of the Royal Statistical Society Series B* 34: 1–41.

Miller, A. J. (1990). *Subset Selection in Regression.* Chapman & Hall, London.

Peto, R. (1982). Statistical aspects of cancer trials. In *Treatment of Cancer*, ed. Halnan, K. E., pp. 867–871. Chapman & Hall, London.

Piantadosi, S. and Gail, M. H. (1993). A comparison of the power of two tests for qualitative interactions. *Statistics in Medicine* 12: 1239–1248.

Russek-Cohen, E. and Simon, R. (1993). Qualitative interactions in multifactor studies. *Biometrics* 49: 467–477.

Schervish, M. J. (1984). Algorithm AS 195. Multivariate normal probabilities with error bound. *Applied Statistics* 33: 81–94.

Simon, R. (1988). Statistical tools for subset analysis in clinical trials. In *Recent Results in Cancer Research*, Vol. 111, ed. Baum, M., Kay, R., and Scheurlen, H. Springer-Verlag, New York.

22

Bayesian Modeling of Binary Repeated Measures Data with Application to Crossover Trials

James Albert Bowling Green State University, Bowling Green, Ohio
Siddhartha Chib Washington University, St. Louis, Missouri

1 INTRODUCTION

This chapter is concerned with the Bayesian statistical analysis of binary data. Such data arise in medicine, economics, and other scientific fields. A considerable literature devoted to the proper analysis of such data has emerged. The logit model, for example, is available in most statistical software packages. It is straightforward to fit by the method of maximum likelihood, and it leads to log odds that are linear in the covariates. This latter property makes it easy to interpret the effect of a given covariate on the experimental response. Albert and Chib (1993a) have developed a new approach to the analysis of binary data models (with probit and t-links) by introducing latent random variables following a scale mixture of normal distributions. This latter approach has opened up new avenues for Bayesian inference in binary, binomial, multinomial, and longitudinal discrete data settings.

The purpose of this chapter is to explain the Albert and Chib approach in the setting of longitudinal binary data and a crossover trial design in which binary measurements y_{it} for patient i at time t are related to covariates x_{it}. A crossover design is used to compare treatments or conditions for chronic diseases in medical and pharmaceutical research.

The most commonly used crossover design has two periods. We will consider a more complicated three-period crossover in the next section.

An advantage of crossover designs is that the treatments are used on the same patients and this minimizes variability. A disadvantage is that the patients are in the treatment phase of the trial for a longer time and may tire. Another disadvantage is that treatments may have a residual effect and so confound treatment comparisons. See Grieve (1989) or Senn (1993) for descriptions of crossover trials and their advantages and disadvantages.

1.1 Crossover Trial Data

Consider the trial described by Kenward and Jones (1987). The goal was to compare the effects of a placebo (labeled treatment A) with an analgesic at low (treatment B) and high doses (treatment C) for the relief of pain in primary dysmenorrhea. Each patient receives all of the treatments in various orders. For example, one patient may receive treatment A during the first period, treatment C during the second period, and treatment B during the third. There are six possible treatment orderings, and the 86 patients are divided randomly into the six groups that correspond to the different orderings. Each patient is given one treatment for a specific time interval and, at the end of the interval, the patient rates the treatment as giving either no relief (0) or some relief (1). The observed data are presented in Table 1, where columns correspond to the six groups, the rows to the eight possible response combinations for the three periods, and the entries of the table to the number of patients with that particular response for a group. The numbers of patients for each response combination and

Table 1 Number of Patients for Each Treatment Sequence and Response Pattern

	ABC	ACB	BAC	BCA	CAB	CBA	Total	Number relief
000	0	2	0	0	3	1	6	0
001	2	0	1	1	0	0	4	4
010	2	0	1	1	0	0	4	4
011	9	9	3	0	2	1	24	48
100	0	1	1	1	0	5	8	8
101	0	0	8	0	7	3	18	36
110	1	0	0	8	1	4	14	28
111	1	4	1	1	1	0	8	24
Total	15	16	15	12	14	14	86	152

each treatment order are placed as row and column totals, respectively, in the table. In addition, for each response combination, the total number of times the patients received relief is recorded. Note from the table that there is a total of $3 \times 86 = 258$ measurements with 152 experiencing relief with an overall proportion of $152/258 = 59\%$ of relief.

The researchers are interested in determining whether the analgesic provides more relief than the placebo and whether the level of the dose of the analgesic makes a difference in the degree of relief. To see briefly from the data if there appears to be a treatment effect, Table 2 gives marginal proportions of relief for each treatment and treatment sequence. Looking at the "Total" column, it appears that there may be an important dose effect but a relatively small difference between the levels of the dose.

In this experimental situation, the responses may be influenced by a *period effect*; this is said to arise when patients experience more relief during one of the treatment periods. Table 3 gives marginal proportions of relief for each period for each of the treatment sequences. Note that there appears to be a small increase in the proportions from period 1 through period 3. Also, the responses may be influenced by carryover (or residual) effects when a patient experiences relief during the second period not because of that period's treatment but because there is a carryover effect from the treatment in period 1. Finally, the responses can be expected to be influenced by a *heterogeneity effect* that arises from the differing tolerances of the individual patients in the experiment. This example will be used to illustrate Bayesian inference in a binary longitudinal random effects model.

This chapter is organized as follows. Section 2 provides the general model and the simulation fitting algorithm that is used in the estimation. A detailed description of the general method, in the context of the example above, is given in Section 3. The distributions that have to be simulated are derived and methods for simulating these distributions are included. In addition, Section 3 considers various implementation issues that arise

Table 2 Observed Marginal Proportions of Relief for Each Treatment and Treatment Sequence

Treatment	ABC	ACB	BAC	BCA	CAB	CBA	Total
A	2/15	5/16	5/15	2/12	4/14	4/14	22/86
B	13/15	13/16	10/15	10/12	10/14	5/14	61/86
C	12/15	13/16	13/15	10/12	9/14	12/14	69/86

Table 3 Observed Marginal Proportions of Relief for Each Period and Treatment Sequence

Period	ABC	ACB	BAC	BCA	CAB	CBA	Total
1	2/15	5/16	10/15	10/12	9/14	12/14	48/86
2	13/15	13/16	5/15	10/12	4/14	5/14	50/86
3	12/15	13/16	13/15	2/12	10/14	4/14	54/86

in iterative simulation-based methods such as the one used in this chapter. The ideas are used to obtain a sample of draws from the joint posterior distribution of all parameters of interest, and this sample is used to estimate treatment effects and the probabilities of relief of different patients across treatments. Finally, Section 3 is concluded by considering methods for evaluating the goodness of a particular random-effects model. These methods include the computation of residual posterior distributions that identify unusual patients and the computation of Bayesian measures of evidence to detect the significance of treatment, period, or carryover effects. In Section 4, the main advantages of the proposed approach are discussed in the context of alternative non-Bayesian methods for fitting binary random effect models.

1.2 Notation

In the sequel, we will use $\mathcal{N}(\mu, \sigma^2)$ to denote the univariate normal distribution with mean μ and variance σ^2, and $\mathcal{N}_p(\mu, D)$ to denote the p-variate normal distribution with mean vector μ and covariance matrix D. The $\mathcal{N}(\mu, \sigma^2)$ distribution truncated to the interval (a, b) will be denoted as $\mathcal{TN}_{(a,b)}(\mu, \sigma^2)$. Finally, $\mathcal{IG}(\nu, \delta)$ will be used to denote the inverse gamma distribution with shape parameter ν and scale parameter δ.

2 BINARY LONGITUDINAL RANDOM-EFFECTS MODEL

2.1 Introduction

In this section, a simulation-based procedure is described to estimate a probit longitudinal binary random-effects model. This model, which Searle *et al.* (1992) refer to as the probit-normal model, is similar to those discussed by Gilmour *et al.* (1985) and Ochi and Prentice (1984). Similar models are also considered by Stiratelli *et al.* (1984), Zeger and Liang (1986), and Zeger and Karim (1991) for the logit link.

Suppose for the ith patient, binary measurements $Y_i = (y_{i1}, \ldots, y_{iT})'$, $y_{it} \in \{0, 1\}$ are taken at T specific time points. The probability of obtaining a positive response on the ith patient at occasion t ($1 \le i \le n$, $1 \le t \le T$) is modeled by the function

$$\Pr(y_{it} = 1 \mid b_i) = \Phi(x_{it}'\beta + b_i), \tag{1}$$

where Φ is the standard normal cumulative distribution function (cdf), β is a p-vector of regression (fixed-effects) parameters, b_i is a patient-specific random effect, and x_{it} is a p-vector of known regression covariates. The n random effects $\{b_i\}$ are assumed to be a random sample from a normal distribution with mean 0 and unknown variance σ^2. [Albert and Chib (1993b) consider a model in which there are multiple random effects for each i.]

In the common frequentist approach to fitting this model, it is necessary to construct the likelihood function for (β, σ^2) by integrating out the random effects from the joint density of the data Y_i and b_i and then obtaining parameter estimates for (β, σ^2) by maximizing the likelihood. However, the computation of the likelihood function of (β, σ^2) is difficult in this case. The contribution of the ith observation to the likelihood function is given by

$$\int \left\{ \prod_{t=1}^{T} [\Phi(x_{it}'\beta + b_i)]^{y_{it}}[1 - \Phi(x_{it}'\beta + b_i)]^{1-y_{it}} \right\}$$
$$|\sigma^{-2}|^{1/2} \exp\left(-\frac{b_i}{2\sigma^2}\right) db_i$$

and so each likelihood value requires the evaluation of an intractable integral.

2.2 A Data Augmentation Strategy

Consider the estimation of the parameters (β, σ^2) from a Bayesian viewpoint. Suppose that β and σ^2 have independent prior distributions with β distributed $\mathcal{N}_p(\beta_0, B_0^{-1})$ and σ^2 distributed $\mathcal{IG}(\nu, \delta)$. The posterior density of the parameters, which is proportional to the product of the prior and the likelihood function, is difficult to simplify due to the inconvenient form of the likelihood function given above. For this reason, a different approach to this estimation problem is necessary.

The approach we present is based on the work of Albert and Chib (1993a), who introduce latent random variables to simplify the implementation of the Gibbs sampling algorithm. Then the simulation algorithm proceeds by iterating between the simulation of the parameters given the latent data and the simulation of the latent data given the parameters. The theory of Markov chain Monte Carlo (Tanner and Wong, 1987; Gelfand and Smith, 1990; Tierney, 1994) can be used to show that, after an initial transient phase, the draws of (β, σ^2) obtained by this procedure are distributed according to the posterior distribution. This simulated sample provides the basis for computing marginal posterior densities, posterior point estimates, and other inferential summaries.

To fix ideas, let z_{it} denote independent latent variables such that

$$z_{it} \mid b_i \text{ is distributed } \mathcal{N}(x'_{it}\beta + b_i, 1), \qquad 1 \leq t \leq T, 1 \leq i \leq n \qquad (2)$$

and let the observed response y_{it} indicate whether the latent variable z_{it} is positive or negative:

$$y_{it} = \begin{cases} 1 & \text{if } z_{it} > 0 \\ 0 & \text{if } z_{it} \leq 0 \end{cases} \qquad (3)$$

Then, it can be seen that the y_{it} satisfy the model (1). With the introduction of the latent data, the objective is to simulate from the joint posterior distribution of $(\{z_{it}\}, \beta, \{b_i\}, \sigma^2)$. This is straightforward to implement by means of the Gibbs sampling algorithm. This algorithm obtains a sample value from the joint distribution by simulating a sequence of values from the *full conditional distributions*, the distributions of a parameter (or a block of parameters) given the data and the remaining parameters. In this setting, consider the following collection of full conditional distributions:

(D1) $\{z_{it}\}$ given Y, β, $\{b_i\}$
(D2) β given $\{z_{it}\}$, $\{b_i\}$
(D3) $\{b_i\}$ given $\{z_{it}\}$, β, σ^2
(D4) σ^2 given $\{z_{it}\}$, $\{b_i\}$

With the introduction of the latent data, all of the conditional distributions have simple forms. For fixed values of the latent data $\{z_{it}\}$, the posterior distributions (D2), (D3), and (D4) follow from standard results for normal error models. In addition, the conditional posterior distribution of the latent data (D1) follows independent truncated normal distributions.

The simulation algorithm starts with initial guesses at the parameters and then produces simulated draws from the distributions (D1), (D2), (D3), and (D4) in that order, where the conditioning variates in each case are set to the most recently simulated values. One repeats this simulation process for a large number of cycles. After a brief "burn-in" period in which the variates are sufficiently mixed, the collection of simulated values of the parameters is taken as a sample from the joint posterior distribution. This algorithm is described in detail in Section 3 for the crossover trial application.

3 ANALYZING THE CROSSOVER TRIALS DATA

In the crossover trial described in Section 1.1, it is convenient to think of z_{it} as representing the patient's relief from the ailment measured on a continuous scale. Then the mean of z_{it} in Eq. (2) is given by

$$\text{mean}(z_{it} \mid b_i) = x'_{it}\beta + b_i, \qquad 1 \le i \le 86, \, 1 \le t \le 3 \qquad (4)$$

where $x'_{it}\beta$ captures the effects due to the particular treatment, period, and crossover, and b_i is the patient-specific tolerance. (If patient-specific information such as age and sex was available, it could be incorporated through x_{it}.) As there are three treatments, three periods, and three potential crossover effects, let the regression parameters corresponding to these effects be (T_A, T_B, T_C), (P_1, P_2, P_3), and (C_A, C_B, C_C), respectively. To obtain a unique estimate for the vector of regression parameters, one typically imposes some restriction on the values of the effects. Here the first level of each effect will be set equal to zero; that is, $T_A = P_1 = C_A = 0$. So the model for the mean of z_{it} in Eq. (4) can be rewritten as

$$\text{mean}(z_{it} \mid b_i) = \beta_0 + \overbrace{x_{it1}T_B + x_{it2}T_C}^{\text{Treatment effects}} + \overbrace{x_{it3}P_2 + x_{it4}P_3}^{\text{Period effects}} + \overbrace{x_{it5}C_B + x_{it6}C_C}^{\text{Crossover effects}} + b_i$$

$$(5)$$

where the x_{itk}, $k = 1, \dots, 6$, are covariates that indicate the presence of a particular treatment, period, or carryover effect.

This model for the ith patient can be written in matrix form as

$$\text{mean}(Z_i \mid b_i) = X_i\beta + i_T b_i \qquad (6)$$

where $Z_i = (Z_{i1}, Z_{i2}, Z_{i3})'$, $\beta = (\beta_0, T_B, T_C, P_2, P_3, C_B, C_C)'$, X_i is a three-by-seven design matrix in which each row corresponds to the covariate indicators for a particular period, and i_T is $(1, 1, 1)'$. For illustration on the construction of this design matrix, consider an individual in the group ABC. In that case,

$$x'_{i1}\beta = \beta_0$$
$$x'_{i2}\beta = \beta_0 + T_B + P_2$$
$$x'_{i3}\beta = \beta_0 + T_C + P_3 + C_B$$

which leads to the form

$$
X_i\beta \equiv \begin{bmatrix} 1 & 0 & 0 & 0 & 0 & 0 & 0 \\ 1 & 1 & 0 & 1 & 0 & 0 & 0 \\ 1 & 0 & 1 & 0 & 1 & 1 & 0 \end{bmatrix} \begin{bmatrix} \beta_0 \\ T_B \\ T_C \\ P_2 \\ P_3 \\ C_B \\ C_C \end{bmatrix}
$$

To complete the Bayesian model, prior distributions need to be assigned to the regression vector β and the random-effects variance σ^2. A constant prior will be assigned to β and σ^2 will be given an $\mathcal{IG}(\nu, \delta)$ distribution with density proportional to $\sigma^{-2(\nu+1)} \exp\{-\delta/\sigma^2\}$.

In summary, the model has four components:

(M1) $\{z_{it}\}$ are independent where z_{it} is distributed $\mathcal{N}(x_{it}'\beta + b_i, 1)$.

(M2) $\{y_{it}\}$ are the observed indicators of the $\{z_{it}\}$, where $y_{it} = 1$ if $z_{it} > 0$, and $y_{it} = 0$ if $z_{it} < 0$.

(M3) β is distributed uniform, $b_1, \ldots, b_n$ is a random sample from $\mathcal{N}(0, \sigma^2)$.

(M4) σ^2 is distributed $\mathcal{IG}(\nu, \delta)$.

After the data $\{y_{it}\}$ have been observed, the unknown quantities are the latent data $\{z_{it}\}$, the regression parameter β, the random effects $\{b_i\}$, and the random-effects variance σ^2, and all information about these quantities is contained in the joint posterior distribution of $(\{z_{it}\}, \beta, \{b_i\}, \sigma^2)$. For example, one can learn about the size of the treatment effects by means of the marginal posterior distribution on the components of β that correspond to the treatments.

A simulated sample from the joint posterior distribution is obtained by simulating through the set of full conditional distributions. For this problem, all the distributions required for the Gibbs sampling algorithm have convenient functional forms.

First, consider the distribution of the latent data $\{z_{it}\}$ in (D1). Due to (M2) and the fact that the parameters $(\beta, \{b_i\}, \sigma^2)$ are held fixed, one can see that the $\{z_{it}\}$ have independent truncated normal distributions, $\mathcal{TN}_{(0,\infty)}(x_{it}'\beta + b_i, 1)$ if $y_{it} = 1$ or $\mathcal{TN}_{(-\infty,0)}(x_{it}'\beta + b_i, 1)$ if $y_{it} = 0$.

Second, the distribution (D2) of the regression vector β conditional on the latent data and random effects is $\mathcal{N}(\hat{\beta}, B_1^{-1})$, where $\hat{\beta} = B_1^{-1} \sum_{i=1}^n X_i'(Z_i - i_T b_i)$ and $B_1 = \sum_{i=1}^n X_i'X_i$.

Third, the distribution (D3) of the random effects $\{b_i\}$, conditional on the latent data, regression vector, and random-effects variance, has an independent structure with b_i distributed $\mathcal{N}(\hat{b}_i, V_1^{-1})$, where $\hat{b}_i = V_1^{-1} \sum_{t=1}^{3}(z_{it} - x'_{it}\beta)$ and $V_1 = 3 + \sigma^{-2}$.

Fourth, the posterior distribution of the random effects variance σ^2 in (D4) is $\mathcal{IG}(\nu + n/2, \delta + \sum_{i=1}^{n} b_i^2/2)$.

3.1 Fitting the Model by Gibbs Sampling

To describe one cycle of the Gibbs sampling algorithm, let $(\{z_{it}^{(k)}\}, \beta^{(k)}, \{b_i^{(k)}\}, \sigma^{2(k)})$ denote values of the latent data and parameters at iteration k of the algorithm. The steps below produce the $(k + 1)$st value of these unobservables. In the description of the cycle, methods for simulating the random variables are outlined.

(C1) Simulate $z_{it}^{(k+1)}$ from $\mathcal{TN}_{(0,\infty)}(x'_{it}\beta^{(k)} + b_i^{(k)}, 1)$ if $y_{it} = 1$ or from $\mathcal{TN}_{(-\infty,0)}(x'_{it}\beta^{(k)} + b_i^{(k)}, 1)$ if $y_{it} = 0$. Random variates can be obtained from these distributions using the inversion simulation method. Suppose, in general, that one wishes to simulate a variate X with distribution function F. Then the inversion method simulates X as $F^{-1}(U)$, where F^{-1} is the inverse distribution function and U is a uniform random variate on the unit interval. In this case, if $z \sim \mathcal{TN}_{(a,b)}(\mu, 1)$, then the inverse distribution function of z is given by

$$F^{-1}(p) = \mu + \Phi^{-1}[\Phi(a - \mu) + p(\Phi(b - \mu) - \Phi(a - \mu))]$$

where Φ^{-1} is the inverse cdf of the standard normal distribution. Thus a simulated value from this truncated distribution is given by $F^{-1}(U)$.

(C2) Simulate $\beta^{(k+1)}$ from a $\mathcal{N}(\hat{\beta}, B_1^{-1})$, where

$$B_1 = \sum_{i=1}^{n} X'_i X_i, \qquad \hat{\beta} = B_1^{-1} \sum_{i=1}^{n} X'_i(Z_i^{(k+1)} - b_i^{(k)})$$

This multivariate normal random variate can be generated using simulated variates from independent standard normal distributions. Let W denote a q-vector of random $\mathcal{N}(0, 1)$ variates and let $B_1^{-1} = A'A$ denote the Cholesky triangular decomposition of the variance-covariance matrix. Then the simulated regression vector is given by

$$\beta^{(k+1)} = \hat{\beta} + A'W$$

(C3) Simulate the random effects $\{b_i^{(k+1)}\}$ independently, where

$b_i^{(k+1)}$ is $\mathcal{N}(\hat{b}_i, V_1^{-1})$, where $V_1 = 3 + \sigma^{-2(k)}$ and $\hat{b}_i = V_1^{-1} \sum_{t=1}^{3}(z_{it}^{(k+1)} - x_{it}'\beta^{(k+1)})$. If W is a simulated standard normal variate, then the simulated ith random effect is given by

$$b_i^{(k+1)} = \hat{b}_i + V_1^{-1/2}W$$

(C4) Simulate the variance of the random effects, σ^2, from an inverse gamma distribution with parameters $\nu + n/2$ and $\delta + \sum_{i=1}^{n}(b_i^{(k+1)})^2/2$. If W is a random variate from a gamma distribution with shape parameter $\nu + n/2$, then the simulated value of the variance is

$$\sigma^{2(k+1)} = \frac{\delta + \frac{1}{2}\sum_{i=1}^{n}(b_i^{(k+1)})^2}{W}$$

Starting with an initial value $(\beta^{(0)}, \{b_i^{(0)}\}, \sigma^{2(0)})$ of the parameters, the above simulation cycle is continued for m iterations, producing the matrix of simulated values

$$\begin{bmatrix} \{z_{it}^{(1)}\} & \beta^{(1)} & \{b_i^{(1)}\} & \sigma^{2(1)} \\ \{z_{it}^{(2)}\} & \beta^{(2)} & \{b_i^{(2)}\} & \sigma^{2(2)} \\ \vdots & \vdots & \vdots & \vdots \\ \{z_{it}^{(m)}\} & \beta^{(m)} & \{b_i^{(m)}\} & \sigma^{2(m)} \end{bmatrix}$$

As the number of cycles m approaches infinity, the distribution of $(\{z_{it}^{(m)}\}, \beta^{(m)}, \{b_i^{(m)}\}, \sigma^{2(m)})$ will converge to the joint posterior of $(\{z_{it}\}, \beta, \{b_i\}, \sigma^2)$.

3.2 Implementation Issues

The fact that a variate from the joint posterior distribution can be obtained by simulating variates repeatedly from the conditional posterior distributions does not help in the practical implementation of this algorithm. In practice, one uses the Markov Chain Monte Carlo (MCMC) algorithm by deciding on the iterate m_0 in which $(\{z_{it}^{(m_0)}\}, \beta^{(m_0)}, \{b_i^{(m_0)}\}, \sigma^{2(m_0)})$ is approximately distributed according to the joint posterior distribution and deciding on the number $m - m_0$ of additional simulated values of the chain to collect to obtain accurate estimates of all posterior quantities of interest. In this section, some guidelines regarding these issues are presented. These appear to work well for this particular application of the Gibbs sampler.

First, the choice of starting values for the regression vector, random effects, and variance parameter does not appear crucial in this setting.

The convergence of the iterates to the posterior distribution appears to be quick for a wide range of initial parameter estimates. One can initially set β to the maximum-likelihood value found by fitting the binary data to the probit covariate model ignoring the random effects. The random-effects variance σ^2 can be set to its prior mean, and the random effects $\{b_i\}$ can be assigned simulated values from a $\mathcal{N}(0, \sigma^2)$ distribution, where σ^2 is set to its initial value.

Next, from experience in fitting this model, the algorithm appears to converge quickly to the joint posterior. Because the number of cycles of the burn-in period m_0 will be small compared with the total number of iterates collected, little accuracy is lost by taking $m_0 = 0$. Then one can assume that the m simulated values of the parameters represent a sample from the posterior distribution and focus on the problem of collecting enough simulated values to obtain a good approximation to the posterior.

To help understand how many iterates to take, suppose that one is interested in the posterior mean of one of the regression parameters, say P_2. From the sample of simulated values $\{P_2^{(k)}, k = 1, \ldots, m\}$, one can estimate the posterior mean by the sample mean $(1/m)\sum_{k=1}^{m} P_2^{(k)}$. If the simulated values were independent, the accuracy of this estimate could be measured by $s/\sqrt{m}$, where s is the standard deviation of the simulated values. However, this measure overstates the accuracy of the estimate, as the simulated values are correlated.

An alternative method of gauging the accuracy of the estimate of the posterior mean is provided by the batch means method (Bratley *et al.*, 1987). This method groups the simulated values (in order) in batches of 2, 4, 8, etc. and computes the sample means of the individual batches. If $M_1, \ldots, M_q$ represent the batch means in sequence, then define the lag-k correlation to be the correlation in the pairs $\{(M_j, M_{j+k}), j = 1, \ldots, q - k\}$. When the lag-one correlation between the batch means is sufficiently small, say under 0.1, there is approximate independence between the means and the accuracy of the overall mean estimate is given by the standard deviation of the batch means divided by the square root of the number of batches.

This method can be used to measure the accuracy of the sample mean estimate of any posterior mean of interest. It works when the correlation between the successive simulated values is relatively small. When there is a strong correlation between the iterates, little information is provided about the parameters, even for a large simulation sample size, and alternative simulation methods need to be used. In the next section, graphs will be used to illustrate the correlation structure of the MCMC run for particular parameters. These graphs are helpful in understanding when the simulation procedure has converged to the joint posterior.

3.3 Summarizing the Posterior Distribution

For the crossover trial data set, the MCMC algorithm described in Section 3.2 was run for $m = 2000$ iterations. The MATLAB function that was used to perform this simulation is listed in the Appendix. To investigate the convergence of this algorithm, Figure 1 shows the sequence of simulated values for two parameters, the regression coefficient T_B, which can be interpreted as the difference of the response (measured on the inverse probit scale) between treatments B and A, and the logarithm of the random-effects variance $\log(\sigma^2)$. Two features of these plots of simulated values are relevant to the issue of convergence. First, if the Gibbs sampler takes a long time to converge, there will be a general drift in the location of the simulated values across iterates. Here the general location of the simulates for both parameters appears constant across iterates, indicating that the number of iterations needed for convergence is relatively small. A second concern is the degree of correlation between successive iterates. The simulated values for T_B look roughly independent because of the

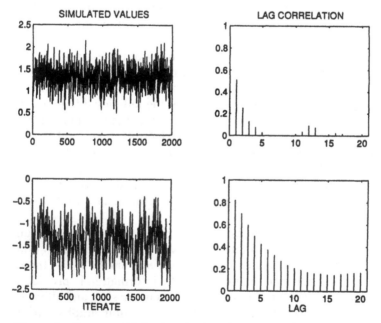

Figure 1 Sequence of simulated values from Gibbs sampling run and autocorrelation plots of simulated values for regression parameter T_B (top graphs) and logarithm of random-effects variance $\log(\sigma^2)$ (bottom graphs).

"white noise" pattern. In contrast, the simulated values for $\log(\sigma^2)$ show more instability, indicating that there is a significant amount of correlation between successive iterates.

To measure more precisely the degree of autocorrelation reflected in the graphs, Figure 1 also shows the lag correlation of the simulated values for lag values from 1 to 20. In the Gibbs sampler, one expects correlation between simulated values for small lag values. However, to obtain good estimates of posterior quantities of interest, it is desirable that the lag values fall to zero as the size of lag increases. Note that the lag correlations for T_B decrease quickly as a function of the lag; in contrast, the lag-1 correlation of $\log(\sigma^2)$ is approximately .8 and there is still a significant correlation for a lag size of 20. The interpretation of these plots is that a greater number of $\log(\sigma^2)$ simulated values must be collected to estimate, say, a posterior mean.

To summarize the model fitting, Figure 2 gives histograms for the 2000 simulated values of the marginal posterior distribution for each of the regression parameters and the random-effects variance. Table 4 summarizes these histograms by giving posterior means and standard deviations. In this table, the numerical standard error row is a measure of accuracy of the posterior means computed by the batch means method described in Section 3.2. So the reported posterior means appear to be accurate to within .01 for all of the regression parameters. The estimated posterior mean of $\log(\sigma^2)$ is less accurate due to the higher autocorrelation structure observed in Figure 1.

What is learned from Figure 2 and Table 4? First, there appear to be important treatment effects, because all of the simulated values of T_B and T_C are positive. There is some evidence that the degree of relief is higher for periods 2 and 3 compared with period 1, as the posterior means of P_2 and P_3 are approximately 1 standard deviation above zero. In addition, because the posterior mean of C_C is larger in absolute value than C_B, there is some evidence that the carryover effect for treatment C is larger than the corresponding carryover effect for treatment B.

The posterior distribution of the random-effects variance σ^2 is concentrated away from 0, indicating that patients experience different tolerances to the treatments. These varying tolerances can be expressed using $\Pr(y_{it} = 1) = \Phi(x'_{it}\beta + b_i)$, the probability that the ith individual experiences relief during the tth period. These probabilities are functions of the regression parameters and random effects, and the posterior means of these probabilities can be estimated as the sample means of the functions of the simulated values of β and $\{b_i\}$.

Although there are 86 patients, the estimated probabilities of relief will be the same for patients with the same treatment sequence and response

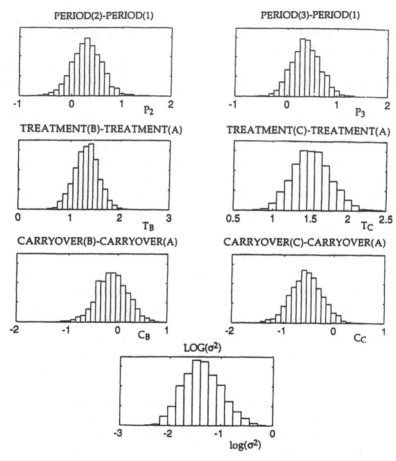

Figure 2 Histograms of simulated values of the posterior distribution of six regression coefficients and logarithm of random-effects variance $\log(\sigma^2)$.

Table 4 Posterior Moments for Regression Parameters and Random-Effects Variance

	P_2	P_3	T_B	T_C	C_B	C_C	$\log \sigma^2$
Mean	.307	.383	1.321	1.502	−.118	−.553	−1.38
Numerical standard error	.012	.012	.009	.012	.014	.012	.04
Standard deviation	.293	.297	.244	.252	.315	.318	.371

pattern. Thus there are 32 distinctive probabilities to estimate. Table 5 gives the posterior means of these probabilities for all of the cells of the table. In each cell, the probability estimates are given for the placebo, low dose, and high dose, respectively. The numbers in the table confirm the early observation that there is a large dose effect but a small difference between the two levels of dose. Although there appears to be a significant treatment effect, the values in the table reflect period and carryover effects and patient effects. To learn about the sizes of the random effects, the posterior means of the b_i are equal to $-.6$, $-.3$, $.1$, $.4$ for patients with total numbers of relief equal to 0, 1, 2, and 3, respectively. On a probability scale, the patients experiencing no relief had an estimated probability of relief approximately 10% smaller than those with one relief. The patients who experienced relief for all three periods had an estimated probability approximately 10% larger than those with two reliefs.

To look more carefully at the variation in these relief probabilities across patients, Figure 3 plots the change in the estimated probabilities of relief from placebo to low dose against the change in estimated probabilities of relief from low to high dose for all individuals. As in Table 5, each point represents all of the patients in a particular cell. This graph again illustrates the disparity in the degree of effect of the analgesic; some people experienced only a .3 increase in relief (as measured on the probability scale), while others experienced a .6 increase in relief. One interesting aspect of the graph is that there is a downward tilt in the scatter of points. This indicates that patients who experienced a substantive degree of relief due to the analgesic at low dose experienced a modest change in relief due to the higher dose of the drug. Patients who showed a modest drug

Table 5 Posterior Means of Probabilities of Relief[a]

	ABC	ACB	BAC	BCA	CAB	CBA
000	—	(12,45,67)	—	—	(8,62,57)	(19,44,60)
001	(18,70,75)	—	(22,60,78)	(16,64,75)	—	—
010	(18,70,75)	—	(22,61,78)	(16,62,74)	—	—
011	(27,80,84)	(28,68,84)	(32,71,86)	—	(21,82,78)	(35,65,78)
100	—	(19,58,78)	(21,59,77)	(16,62,73)	—	(26,54,69)
101	—	—	(32,71,86)	—	(22,82,78)	(36,65,78)
110	(26,79,83)	—	—	(24,73,82)	(21,82,78)	(36,65,78)
111	(38,87,90)	(39,78,91)	(43,80,91)	(35,82,89)	(33,89,87)	—

[a] Each cell contains 100 times the probability of relief for treatments A, B, and C, respectively.

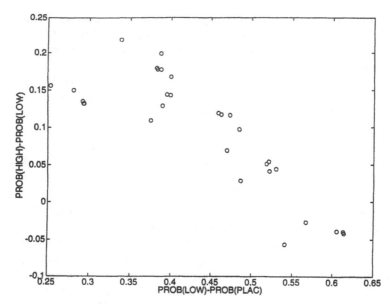

Figure 3 Difference in estimated probabilities of relief between placebo and treatment at low dose plotted against difference in estimated probabilities of relief at low and high doses.

over placebo effect showed a relatively high dose effect. For most patients the difference between the placebo and the high dose was approximately 58% with a larger uncertainty about the placebo–low dose and low dose–high dose effects.

3.4 Model Evaluation

The foregoing analysis focused on the fitting and interpretation of the probit random-effects model. In this section, some methods are introduced for criticizing the model and for selecting alternative models. One standard method of evaluating a particular model is to compute a set of residuals. The residuals can help in detecting patients who show unusual patterns of response that are not explained very well by the fitted model. Also, there are concerns about the best choice of variables. The posterior analysis described above suggests that the only important variables were the treatment effects. In the following, a Bayesian measure of evidence is developed using a hierarchical model that indicates the significance of the period, treatment, or carryover effects.

One standard definition of a residual analogous to that of normal regression models is $r_{it} = y_{it} - \Phi(x'_{it}\beta - b_i)$, where β is estimated using the fitted model and b_i is set equal to its expected value of 0. This residual is difficult to interpret from a frequentist perspective. The sampling distribution of r_{it} is only a two-point distribution, corresponding to the two possible values of the binary response y_{it}, so it is not possible to assess whether a given residual value is large.

The Bayesian views residuals differently. The observed responses $\{y_{it}\}$ are fixed and the random quantity in r_{it} is the regression parameter vector β. Following Chaloner and Brant (1988), one can perform a residual analysis by contrasting the prior and posterior distributions of the set of residuals $\{r_{it}\}$, which are real-valued functions of β. If the marginal posterior distribution for a particular residual is very different from its corresponding prior distribution, there is reason to believe that the corresponding observation is outlying.

In our Gibbs sampling fitting procedure, latent data $\{z_{it}\}$ were introduced, and alternative definitions of residuals can be defined in terms of the entire set of unobserved quantities β, $\{b_i\}$, and $\{z_{it}\}$. In particular, following Albert and Chib (1995), consider the latent-data residual

$$r^L_{it} = \frac{z_{it} - x'_{it}\beta}{\sqrt{1 + \sigma^2}}$$

These residuals are random quantities as functions of the $\{z_{it}\}$, the regression vector β, and the prior hyperparameter σ^2. This particular residual definition is attractive because the $\{r^L_{it}\}$ have independent standard normal prior distributions. One can assess the size of a particular residual by comparing its posterior distribution with the standard normal prior distribution. In particular, one can decide whether a particular residual is unusually large by computing the posterior probability that the residual exceeds a percentile of the standard normal distribution. For the crossover trial example, the posterior outlying probabilities $\Pr(|r^L_{it}| > 2)$ are given in Table 6 for a single patient in each cell of the table. Since the prior distribution of r_{it} is standard normal, the prior probability that a residual is large is .05 and we compare the posterior probabilities in Table 6 with this number. Note that most of the posterior probabilities are smaller than .05. The "large" outlying probabilities in Table 6 correspond to individuals who have unusual responses after accounting for covariate and individual-specific effects. A few individual responses appear to stand out. For example, an individual in the upper left corner of the table had a response pattern (for treatments A, B, C) of 0, 0, 0, had corresponding estimated probabilities of response of .12, .45, .67 (see Table 5), and, as a result,

Table 6 Posterior Outlying Probabilities[a]

	ABC	ACB	BAC	BCA	CAB	CBA
000	—	(6,10,16)	—	—	(7,17,13)	(7,10,12)
001	(3,10,2)	—	(3,7,1)	(6,7,10)	—	—
010	(4,2,13)	—	(4,8,15)	(3,7,1)	—	—
011	(2,3,3)	(3,3,2)	(6,4,2)	—	(10,2,5)	(5,3,7)
100	—	(5,6,13)	(4,2,15)	(4,1,13)	—	(3,5,1)
101	—	—	(2,3,2)	—	(2,2,3)	(5,4,3)
110	(7,2,11)	—	—	(2,4,2)	(9,8,3)	(3,4,2)
111	(11,5,5)	(11,5,4)	(9,4,4)	(13,6,6)	(15,5,5)	—

[a] Each cell contains 100 times the probability that the residual r_{ti}^* exceeds 2 in absolute value for treatments A, B, and C, respectively.

had outlying probabilities of .06, .10, and .16, respectively. A particular pattern of large residuals may suggest a more complicated model that includes interaction terms between the covariates or more random-effects parameters.

To measure the significance of a particular set of effects, a hierarchical prior is used. In our earlier analysis, vague prior distributions were chosen for all of the covariates. These priors reflected little information about the size of any of the covariates. In this setting, a proper prior distribution will be used for a particular covariate, reflecting the prior belief that the particular covariate may be insignificant and should be set equal to zero. This prior distribution will introduce a new variance parameter that determines the importance of the covariate. By performing a Bayesian test on this variance parameter, one can measure the importance of this covariate.

Suppose, for example, that one was interested in whether patients experienced the same degree of relief for the three periods. One can model the belief that the period regression parameters P_2 and P_3 are possibly equal to zero using the hierarchical prior

1. P_2, P_3 independent $\mathcal{N}(0, \gamma)$
2. γ distributed $\mathcal{IG}(\nu_0, \delta_0)$

The key parameter with respect to model evaluation is γ, which indicates the strength of the belief that the period effects are equal to zero. If $\gamma = 0$, one has a regression model with the period effects deleted. One "tests" whether the period effects are important by comparing the posterior probabilities of the hypotheses $H: \gamma \leq \epsilon$, $K: \gamma > \epsilon$. These posterior probabilities will depend on the choice of prior distribution. A measure of evidence that does not depend on the prior probabilities of the hypotheses is the Bayes factor, the ratio of the posterior odds of the hypothesis K, $\Pr(K \mid \text{data})/\Pr(H \mid \text{data})$, to the prior odds $\Pr(K)/\Pr(H)$. If, for example, the Bayes factor is equal to 1000, then the data suggest that the covariate is important and should be retained in the model.

The parameter ϵ is determined from the user by assessing the size of a "practically significant" effect. For example, suppose that the period effect is not felt to be significant unless a patient experiences a 10% increase in the probability of relief. For probability values between .2 and .8, a change of .1 on the probability scale is approximately equal to a change of .256 on the covariate scale. By matching the value of .256 to the standard deviation $\sqrt{\gamma}$ of P_2 and P_3, one obtains the value $\epsilon = (.256)^2 = .0655$. Then the hypothesis H that the values of P_2 and P_3 are "insignificant" will correspond to the values $\gamma \leq .0655$.

The specification of the hierarchical prior is completed by assigning values to the hyperparameters ν_0 and δ_0. In this example, $\nu_0 = 1$ and $\delta_0 = 1$, reflecting vague prior information about the locations of these parameters. The fitting of this new model is analogous to the procedure described in Section 3.1. The uniform prior on β is replaced by the hierarchical prior and the Gibbs sampling cycle contains an additional step corresponding to draws from the new hyperparameter γ.

The hierarchical prior distribution was fit three times, reflecting possible nonsignificance of the period, treatment, and carryover effects. The posterior probability $\Pr(K \mid \text{data})$ of the hypothesis K is estimated using the proportion of simulated values that were greater than ϵ. From this value, one can compute the posterior odds $\Pr(K \mid \text{data})/\Pr(H \mid \text{data})$. The prior odds $\Pr(K)/\Pr(H)$ are computed directly from the prior distribution that is given. In the example, the Bayes factors for the period, carryover, and treatment effects are computed to be .27, 1.0, and 130,000, respectively. These values reflect the strengths of the conclusions of the posterior analysis of Section 3.3. There is strong evidence of differences between the three treatments and there is little evidence of period or carryover effects.

4 CONCLUDING REMARKS

Most of the literature on longitudinal data analysis uses maximum-likelihood methods to fit models. Laird (1991) provides a survey of two general approaches for modeling binary longitudinal data: multivariate methods, which model the response Y_i directly, and the use of random effects such as the probit-normal model considered here. However, implementing likelihood-based models presents a number of difficulties. First, as mentioned in Section 2, there can be substantive computational problems in obtaining maximum-likelihood estimates. Second, the accuracy of inferences based on maximum likelihood depends on how accurately one can approximate the likelihood function by a normal distribution. Finally, the emphasis in the literature is primarily on maximum-likelihood estimates of the regression parameter, and there is relatively little discussion on estimating probabilities or model criticism based on residuals or goodness-of-fit measures.

In contrast, the Bayesian approach presented here provides a very easy-to-implement method of fitting a probit random-effects model. With the introduction of the latent data, it is easy to write a short program (in a matrix language such as MATLAB or S-PLUS) that produces a simulated sample from the joint posterior distribution of all unobservables. This simulated sample can be used to approximate accurately the exact posterior distribution of any function of the parameters of interest. For example, the posterior distribution of the entire set of Bayesian residuals computed in Section 3.4 is obtained as a simple by-product of the output from the simulation. In addition, the Bayesian approach is very flexible because one can assign a multivariate normal prior or a hierarchical prior to the regression vector β to reflect presample beliefs about the importance of the variables. In Section 3.4, a hierarchical prior was used to test the plausibility of various effects.

Although this chapter has focused on binary response, the fitting procedure described can be generalized to the case in which the response is ordinal with more than two categories. (Actually, the patients in our example described the amount of relief as none, moderate, or complete, which is an ordinal response with three categories. Here, the latter two responses were combined to obtain a binary response.) Albert and Chib (1993a) describe the generalization of the sampling to accommodate ordinal and multinomial responses. As in the binary case, the bin boundaries (here unknown) are used to partition the continuous latent data, with each interval corresponding to one categorical response. Gibbs sampling is then used to estimate the regression parameters together with the unknown bin boundaries.

APPENDIX: DESCRIPTION OF MATLAB FUNCTION GIBBS-CROSSOVER

Below is a listing of a program to implement the Gibbs sampling algorithm using the mathematics software MATLAB.* The data, following Table 1,

---

* For information about MATLAB, contact The MathWorks, Inc., 24 Prime Park Way, Natick, MA 01760-1500. MATLAB is a registered trademark of The MathWorks, Inc.

```
function [Mb,Md1,Md2,Mp1,Mp2,Ms2,Mr1,Mr2,Mr3]=gibbs_crossover(x,y,n,nu,delta,m)

%-------------------------------------------------------------------------
% MATLAB FUNCTION TO FIT CROSSOVER MODEL VIA GIBBS SAMPLING
% INPUT:
%        x         COVARIATE MATRIX
%        y         MATRIX OF POSSIBLE BINARY RESPONSES
%        n         NUMBER OF OBSERVATIONS FOR EACH SET OF BINARY RESPONSES
%        nu        INVERSE GAMMA SHAPE HYPERPARAMETER FOR SIGMA^2
%        delta     INVERSE GAMMA SCALE HYPERPARAMETER FOR SIGMA^2
%        m         NUMBER OF ITERATIONS OF GIBBS SAMPLER
% OUTPUT:
%        Mb        MATRIX OF SIMULATED VALUES OF BETA
%        Md1       POSTERIOR MEANS OF RANDOM EFFECTS
%        Md2       POSTERIOR STANDARD DEVIATIONS OF RANDOM EFFECTS
%        Mp1       POSTERIOR MEANS OF PROBABILITIES OF RESPONSE
%        Mp2       POSTERIOR STANDARD DEVIATIONS OF PROBS OF RESPONSE
%        Ms2       VECTOR OF SIMULATED VALUES OF SIGMA^2
%        Mr1       POSTERIOR MEANS OF RESIDUALS R(I)
%        Mr2       POSTERIOR STANDARD DEVIATIONS OF RESIDUALS R(I)
%        Mr3       POSTERIOR PROBS THAT |R(I)|>2
%-------------------------------------------------------------------------

N=sum(n);                              % N = NUMBER OF SUBJECTS
[nr,np]=size(y);                       % np = NUMBER OF PERIODS
[nr,npq]=size(x); q=npq/np;            % q = NUMBER OF REGRESSION PARAMETERS

xx=[]; for i=1:length(n),xx=[xx;ones(n(i),1)*x(i,:)]; end     % xx AND yy CONTAIN
yy=[]; for i=1:length(n),yy=[yy;ones(n(i),1)*y(i,:)]; end     % UNGROUPED VALUES OF x AND y

k1=1:q; k2=(q+1):(2*q); k3=(2*q+1):(3*q);
X=[];for i=1:N,X=[X;xx(i,k1);xx(i,k2);xx(i,k3)];end
Y=yy'; Y=Y(:);

beta=ones(q,1); d=zeros(N,1); s2=1;                          % STARTING VALUES
                                                            % FOR GIBBS SAMPLER

Mb=[]; Md1=zeros(size(d)); Md2=Md1;
Mp1=zeros(N,np); Mp2=Mp1; Ms2=[];
Mr1=zeros(N,np); Mr2=Mr1; Mr3=Mr1;

for i=1:m                                                   % GIBBS SAMPLING ITERATIONS
    D=reshape((d*ones(1,np))',N*np,1);
    lp=X*beta+D;                                            %
    bb=erf(-lp);                                            % SIMULATE
    tt=(bb.*(1-Y)+(1-bb).*Y).*rand(N*np,1)+bb.*Y;           % LATENT DATA Z
    z=ipsi(tt)+lp;                                          %

    mn=X\(z-D); v=inv(X'*X); a=chol(v);                     % SIMULATE
    beta=a'*randn(q,1)+mn;                                  % BETA

    sm=sum(reshape(z-X*beta,np,N))';                        % SIMULATE
    d=randn(N,1)/sqrt(np+1/s2)+sm/(np+1/s2);                % RANDOM EFFECTS

    s2=(delta+sum(d.^2)/2)/rgam(1,nu+N/2);                  % SIMULATE SIGMA^2

    p=reshape(1-bb,np,N)';
    r=reshape((z-lp)/sqrt(1+s2),np,N)';

    Mb=[Mb;beta']; Ms2=[Ms2;s2];                            % STORE
    Md1=Md1+d; Md2=Md2+d.^2; Mp1=Mp1+p; Mp2=Mp2+p.^2;       % SIMULATED
    Mr1=Mr1+r; Mr2=Mr2+r.^2; Mr3=Mr3+(abs(r)>2);            % VALUES
```

```
      if i/10==fix(i/10),i,end
end

Md1=Md1/m; Md2=sqrt(Md2/m-Md1.^2);              % COMPUTE POSTERIOR
Mp1=Mp1/m; Mp2=sqrt(Mp2/m-Mp1.^2);              % MEANS AND STANDARD
Mr1=Mr1/m; Mr2=sqrt(Mr2/m-Mr1.^2); Mr3=Mr3/m;  % DEVIATIONS
```

are contained in three matrices y, n, and x. The matrix y is a 48-by-3 matrix, where each row of the matrix corresponds to a sequence of three binary responses. Data are recorded by rows in Table 1, so the first six rows of y contain the elements 0, 0, 0, the next six rows 0, 0, 1, etc. The matrix n, of 48 rows and 1 column, contains the numbers of patients in the 48 cells of Table 1. The first 12 elements of n for this example are 0, 2, 0, 0, 3, 1, 2, 0, 1, 1, 0, 0. The matrix x for this example is a 48-by-21 design matrix, where each row consists of $(x'_{i1}\ x'_{i2}\ x'_{i3})$, where x'_{i1} includes the covariate indicators for the first period and x'_{i2} and x'_{i3} for the second and third periods, respectively. In Section 3, the covariate matrix X_i was defined as the 3-by-7 matrix with rows x'_{i1}, x'_{i2}, and x'_{i3}. The remaining inputs are the hyperparameters for the random-effects variance σ^2 and the number of iterations for the Gibbs sampler m. The outputs, as explained in the program, are the simulated sequences of the regression vector β and the variance σ^2 and posterior means and standard deviations of the random effects b_i, the probabilities of response $\Pr(y_{it} = 1)$, and the residuals r^t_{it}. In addition, the posterior probabilities that the r^t_{it} exceed 2 in absolute value are recorded.

In addition to basic matrix functions that are contained in the current version of MATLAB, this program requires the use of several utility functions. The function erf( ) is the standard normal cdf, ipsi( ) is the inverse normal cdf, randn(m, n) simulates a matrix of normal variates with m rows and n columns, and rgam(n, a) simulates n values of a gamma variate with shape parameter a.

REFERENCES

Albert, J. and Chib, S. (1993a). Bayesian analysis of binary and polychotomous response data. *Journal of the American Statistical Association* 88: 657–667.
Albert, J. and Chib, S. (1993b). A Practical Bayes Approach for Longitudinal Probit Regression Models with Random Effects. Technical Report, Department of Mathematics and Statistics, Bowling Green State University, Bowling Green, OH.
Albert, J. and Chib, S. (1995). Bayesian residual analysis for binary response regression models. *Biometrika,* in press.

Bratley, P., Fox, B., and Schrage, L. (1987). *A Guide to Simulation*. Springer-Verlag, New York.

Chaloner, K. and Brant, R. (1988). A Bayesian approach to outlier detection and residual analysis. *Biometrika* 75: 651–659.

Gelfand, A. E. and Smith, A. F. M. (1990). Sampling-based approaches to calculating marginal densities. *Journal of the American Statistical Association* 85: 398–409.

Gilmour, A. R., Anderson, R. D., and Rae, A. L. (1985). The analysis of binomial data by a generalized linear model. *Biometrika* 72: 593–599.

Grieve, A. (1989). Bayesian analysis of crossover trials. In D. A. Berry, *Statistical Methodology in the Pharmaceutical Sciences*, ed. Berry, D. A. Marcel Dekker, New York.

Kenward, M. G. and Jones, B. (1987). Modelling binary data from a theoretical crossover trial. *Statistics in Medicine* 6: 555–564.

Kenward, M. G. and Jones, B. (1991). The analysis of categorical data from crossover trials using a latent variable model. *Statistics in Medicine* 10: 1607–1619.

Laird, N. M. (1991). Topics in likelihood-based methods for longitudinal data analysis. *Statistica Sinica* 1: 33–50.

Ochi, Y., and Prentice, R. L. (1984). Likelihood inference in a correlated probit regression model. *Biometrika* 71: 531–543.

Searle, S. R., Casella, G., and McCulloch, C. E. (1992). *Variance Components*. Wiley, New York.

Senn, S. (1993). *Cross-over Trials in Clinical Research*. Wiley, New York.

Stiratelli, R., Laird, N. M., and Ware, J. H. (1984). Random-effects models for several observations with binary response. *Biometrics* 40: 961–971.

Tanner, M. A. and Wong, W. H. (1987). The calculation of posterior distributions by data augmentation. *Journal of the American Statistical Association* 82: 528–549.

Tierney, L. (1994). Markov chain for exploring posterior distributions. *Annals of Statistics*, 22: 1701–1762.

Zeger, S. L. and Karim, M. R. (1991). Generalized linear models with random effects: A Gibbs sampling approach. *Journal of the American Statistical Association* 86: 79–86.

Zeger, S. L. and Liang, K. (1986). Longitudinal data analysis using generalized linear models. *Biometrika* 73: 13–22.

23

A Comparative Study of Perinatal Mortality Using a Two-Component Mixture Model

Petros Dellaportas   Department of Statistics, Athens University of Economics, Athens, Greece
David A. Stephens and Adrian F. M. Smith   Department of Mathematics, Imperial College, London, England
Irwin Guttman   Statistics Department, University of Toronto, Toronto, Ontario, Canada

ABSTRACT

A simple two-component mixture density f takes the form $f(y \mid \theta) = \pi f_1(y \mid \theta_1) + (1 - \pi)f_2(y \mid \theta_2)$ for densities f_1, f_2 and $0 \leq \pi \leq 1$. Given a sample of size n from such a density, the likelihood is effectively the sum of 2^n component terms, and the implementation of Bayesian methods runs up against considerable computational difficulties. Recently, however, it has been shown that the Gibbs sampler can be used to remove this problem via the introduction of latent variables, thus allowing fully Bayesian inference to be performed where previously this was not feasible. In this chapter we study a version of the two-component mixture model with particular reference to the following problem in medical statistics. Low birth weight is widely considered to be an important factor in perinatal mortality, but it is also recognized that it is not sufficient to take action on the basis of the birth weight figure alone. It is also necessary

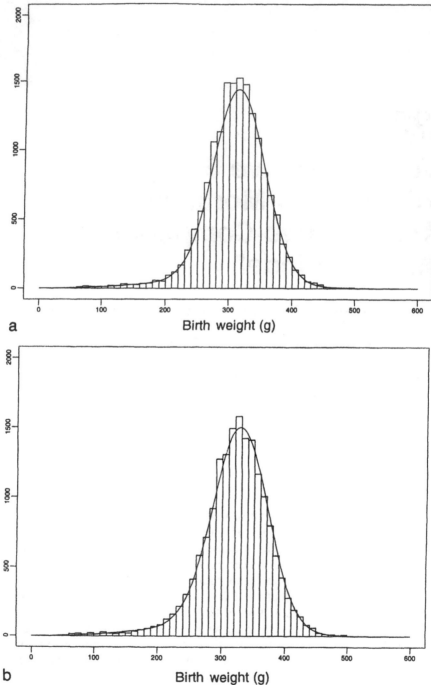

a

Birth weight (g)

b

Birth weight (g)

Figure 1 (a) Data and predictive for Asian females. (b) Data and predictive for Asian males.

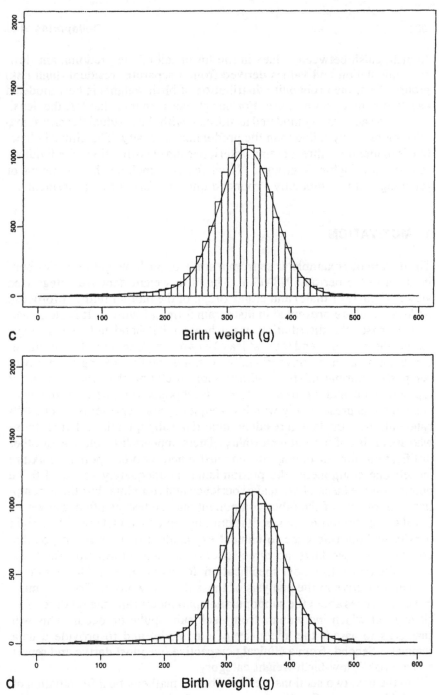

c   Birth weight (g)

d   Birth weight (g)

Figure 1 *(continued)*. (c) Data and predictive for Moroccan females. (d) Data and predictive for Moroccan males.

to distinguish between values in the lower tail of the predominant (low-risk) population and values derived from a separate, residual (high-risk) group. Thus, the probability distribution of birth weights is best modeled as a two-component mixture. For our purposes in this chapter, the densities can be adequately modeled as normal, with the residual density somewhat more heavy-tailed than the predominant density. The clinical objective is to identify a thresholding criterion to aid classification of individuals into the low/high-risk groups. Novel features include the treatment of rounding and the implicitly "rounded integer" form of measurements.

1 MOTIVATION

The data in our example arise from a study of birth weight of babies born in Israel in the period 1978–1980. The birth weight data are categorized via sex (female or male) and ethnic origin of the mother (either Moroccan or Asian) and are presented in histogram form in Figure 1. It is clear that, in each case, the distribution has a heavier left-hand tail than a single normal distribution, and thus (at least) a two-component mixture is plausible. Such data have conventionally been modeled as arising from a two-component normal mixture, with the interpretation that one component is predominant and represents a low-risk subpopulation, whereas the other component is present only with low frequency and represents (via its left-hand tail) individuals in a residual, high-risk subpopulation, that is, those who are at risk of perinatal mortality. There appears little clinical or other justification for assuming an unconstrained two-component mixture model; one component (the predominant) is adequately assumed to be normal, on the basis of clinical experience and q-q plots, but the assumption of normality of the other component may be less justified. However, it will suffice for our purposes. It is our objective to model the birth weight density without reference to clinical explanation or relation to possible covariates (as we have no direct access to such information for these data). We return to model specification details in our final discussion.

Our objective in the analysis of these data is twofold. First, we must make inferences about the subpopulation parameters and the relative "frequency" at which individuals from each subpopulation occur. This will enable us to assess the adequacy of our model and to provide a data summary. Second, from a clinical perspective, we must derive and report a "high-risk", low-birth-weight category.

In the next two sections we describe the mathematical formulation of the problem, and in Section 4 we present the basic idea of analysis using the Gibbs sampler algorithm.

2 MATHEMATICAL BACKGROUND

Suppose that the probability density of a (possibly vector) random variable Y, $f(Y)$, can be expressed (via the theorem of total probability) in the form

$$f(Y) = \sum_{i=1}^{K} \pi_i f_i(Y) \tag{1}$$

for some component densities $f_i(\cdot)$ and probabilities π_i, $i = 1, \ldots, K$, $\pi_1 + \pi_2 + \cdots + \pi_K = 1$. Then Y is said to have a K-component finite mixture density. More explicitly, if the component densities are functions of (possibly vector) parameters θ_i, then we can more fully express Eq. (1) as

$$f(Y \mid \theta_1, \ldots, \theta_K) = \sum_{i=1}^{K} \pi_i f_i(Y \mid \theta_i) \tag{2}$$

Often, the component densities can be assumed to have the same parametric form, enabling Eq. (2) to be simplified still further.

Densities having forms such as those in Eqs. (1) and (2) have been used to model data in many practical problems, for a comprehensive survey, see Titterington *et al.* (1985). Inference problems associated with such models include estimation of the parameters θ_i, identification of the number of components, and investigation of modality. However, conventional likelihood-based methods are inhibited by the nature of the sampling density: for a sample $y_1, \ldots, y_n$ from Eq. (2), the complete likelihood is a product of n terms, each one being the sum of K distinct elements, and even for moderate values of n there are considerable computational difficulties. Diebolt and Robert (1994) and West (1992) have developed Markov chain Monte Carlo algorithms that overcome the computational problems.

3 TWO-COMPONENT MIXTURE PROBLEMS

When $K = 2$ in Eq. (2), the sampling density of variable Y reduces to

$$f(Y \mid \theta_1, \theta_2) = \pi f_1(Y \mid \theta_1) + (1 - \pi) f_2(Y \mid \theta_2) \tag{3}$$

for densities f_1, f_2, and $0 \le \pi \le 1$. For a sample $y_1, \ldots, y_n$ from Eq. (3), the likelihood is readily seen to be

$$l(y, \theta_1, \theta_2) = \prod_{i=1}^{2} \{\pi f_1(y_i \mid \theta_1) + (1 - \pi) f_2(y_i \mid \theta_2)\} \tag{4}$$

which is a sum of 2^n terms. Clearly, conventional likelihood-based/Bayesian inference techniques are difficult to implement, and Titterington *et al.* (1985) survey a number of analytic approximations and nonparametric and graphical procedures. The version of the two-component mixture problem we shall adopt assumes the densities f_1 and f_2 to be univariate normal with means and standard deviations μ_i and σ_i for $i = 1, 2$, respectively.

4 SAMPLING-BASED INFERENCE

The key to the Gibbs sampling approach to inference for mixture models is the introduction of a set of parameters that enable the structure of the likelihood in Eq. (4) to be simplified considerably, albeit at the cost of introducing more unknowns. Consider the binary variables ϕ_i, $i = 1, \ldots, n$ where

$$f(Y_i \mid \phi, \theta_1, \theta_2) = \begin{cases} f_1(Y_i \mid \theta_1), & \phi_i = 1 \\ f_2(Y_i \mid \theta_2), & \phi_i = 2 \end{cases} \tag{5}$$

Then, clearly, conditional on the vector of ϕ_i's, the likelihood factorizes as

$$l(y, \phi, \theta_1, \theta_2) = \prod_{\phi_i = 1} \pi f_1(y_i \mid \theta_1) \prod_{\phi_i = 2} (1 - \pi) f_2(y_i \mid \theta_2) \tag{6}$$

so that, conditional on ϕ, inference about the other parameters is more straightforwardly implemented.

It can now readily be seen that the Gibbs sampler (see, for example, Smith and Roberts, 1993) is straightforwardly implemented in this problem. We need to write down the full conditional posterior distributions for the parameters $(\phi_1, \ldots, \phi_n, \theta_1, \theta_2, \text{and } \pi)$ and sample iteratively with suitable updating of the conditioning parameters. From suitably long runs of the chain, marginal/joint posterior quantities of interest may be estimated; for a general discussion, see, for example, Smith and Roberts (1993); for details of the mixture model case, see Diebolt and Robert (1994). In the two-component normal model described above, the complete likelihood is of the form

$$l(y, \phi, \mu_1, \sigma_1, \mu_2, \sigma_2, \pi) \propto \prod_{\phi_i = 1} \pi \frac{1}{\sigma_1} \exp\left\{ \frac{1}{2\sigma_1^2} (y_i - \mu_1)^2 \right\}$$

$$\cdot \prod_{\phi_i = 2} (1 - \pi) \frac{1}{\sigma_2} \exp\left\{ \frac{1}{2\sigma_2^2} (y_i - \mu_2)^2 \right\} \tag{7}$$

Assuming independent prior distributions conjugate to this likelihood, it is easily seen that the full conditional posterior distributions for μ_1 and μ_2 are normal, for σ_1^2 and σ_2^2 are inverse gamma, and for π is beta. The full conditional posterior distribution for ϕ_i, $i = 1$ to n, is given directly by Eq. (6), so that the conditional posterior probability that $\phi_1 = 1$ is proportional to π multiplied by the value of the $N(\mu_1, \sigma_1^2)$ density evaluated at y_1, and so on. We discuss the particular choices of hyperparameters for these conjugate priors, in relation to our chosen example described in Section 1.

Returning now to the particular example we are interested in, note that the question posed in Section 1, namely the derivation of a high-risk low-birth-weight category, can be done in either an estimative or predictive sense. In an estimative context, we might report a cutoff value, γ say, that minimizes the proportion of misclassifications of individuals from one subpopulation to the other. In effect, this corresponds to assuming that most of the high-risk distribution is concentrated on $(-\infty, \gamma)$. It is easily seen that, conditional on the population parameters, γ defined in this sense satisfies $\pi f_1(\gamma \mid \mu_1, \sigma_1) = (1 - \pi)f_2(\gamma \mid \mu_2, \sigma_2)$. Thus, using the Gibbs sampler, we may produce a sample from the posterior distribution of γ simply as a function of the sampled population parameters. From this sample, we may produce a suitable estimate of γ, say the posterior sample mean. In a predictive context, we might report the posterior predictive probability (as a function of birth weight) that an as yet unrecorded individual is actually from the residual high-risk subpopulation. This predictive probability function, $q(x)$ say, conditional on the population parameters, is given by

$$q(x) = \frac{\pi f_1(x \mid \mu_1, \sigma_1)}{\pi f_1(x \mid \mu_1, \sigma_1) + (1 - \pi)f_2(x \mid \mu_2, \sigma_2)} \tag{8}$$

Again, via the Gibbs sampler, it is possible to produce an estimate of the (marginal) predictive probability function by averaging the function in Eq. (8) over successive iterations/replicates.

In addition to the posterior quantities mentioned above, we may be interested in the posterior predictive density for as yet unobserved data, Z say. The formal Bayesian predictive density for Z, denoted here by $p(Z \mid y)$ is given by

$$p(Z \mid y) = \int_\theta f(Z \mid \theta)p(\theta \mid y) \, d\theta \tag{9}$$

where f is the sampling density function conditional on parameters θ, and $p(\theta \mid y)$ is the posterior density of θ given the realization y. The calculation in Eq. (9) is not always analytically tractable. However, in the Gibbs sampler context, it is possible to produce a sample from the predictive density by regarding Z as a new "parameter" and sampling from its conditional posterior density in addition to the usual sampling cycle. An estimate of the predictive density is also available through the Gibbs sampler procedure. As Eq. (9) suggests, given a set of samples from $p(\theta \mid y)$ (over successive iterations/replicates), an estimate of $p(Z \mid y)$ is calculated in the usual Monte Carlo sense, pointwise, over a range of values of Z, by averaging the sampling density function. Furthermore, as pointed out by Escobar and West (1995), the predictive density is the formal Bayesian equivalent of a kernel density estimate (given the Bayesian model specification), and therefore it may be used (albeit in an informal way) as a model validation diagnostic; see also Gelfand et al. (1992a). For example, we might compute the predictive density and then overlay it on the data histogram or possibly use other exploratory techniques.

Returning to the birth weight example, we now consider a prior specification. For the subpopulation parameters, we choose proper and relatively uninformative priors: for μ_1 and μ_2 we place normal densities with means 2.0 and 3.0, respectively, and standard deviations equal to 2; for σ_1 and σ_2 we place inverse gamma densities with mean 1 and standard deviation 100. Furthermore, we impose the constraints that the normal prior densities above are constrained at zero and that $\mu_1 < \mu_2$ and $\sigma_1 > \sigma_2$. For π, we choose a conjugate beta prior. On the basis of quite strong expert prior opinion that π lies between 0 and 0.2 with high probability and has its prior modal value at around 0.1, we choose a beta density with parameters 11.0 and 91.0. This complete specification is sufficient to ensure identifiability of all the parameters.

6. ANALYSIS OF BIRTH WEIGHT DATA

In order to make inferences for these data, we inspect the marginal posterior densities for the parameters of interest, computed as ergodic averages of the conditional densities during single (but subsequently repeated and validated) Gibbs sampler runs. For example, to estimate (pointwise) the marginal posterior density for θ_1, we average, over a grid of values for θ_1, the form of $p(\theta_1 \mid \phi_1, \ldots, \phi_n, \theta_2, \sigma_1, \sigma_2, \text{data})$ with respect to successive drawings of the conditioning variables. Convergence was diagnosed to have occurred when the marginal density estimates were not perceptibly different when examined several hundred iterations apart. In each case, the marginal plots were actually reasonably stable after about 2500 iterations and extremely stable after about 10,000 iterations.

First, we note (see Figure 1) that, for each of the subpopulations, the predictive density provides a close overlay of the observed data histogram, indicating, at least informally, that the two-component normal mixture model is adequate as a representation (if not an explanation) of the data. Second, on inspection of the marginal plots for π, the proportion of high-risk babies, it is evident that this proportion is not significantly different across the subpopulations; see Figure 2. More strikingly, there does appear to be a significant difference (150 g) in the low-risk population means (μ_2) between males and females in each ethnic group and, perhaps more surprising, between ethnic groups for both males and females, with the male babies and the babies of Moroccan origin being the heavier in the two cases. It would be of subsequent interest to investigate possible clinical/social reasons for such a difference. The posterior marginals for the means of high-risk groups in each subpopulation provide less insight. It was also discovered that the standard deviations in the low-risk/high-risk populations were approximately in the ratio 1:2 for each of the subpopulations (posterior modal values having the range 400–500 g/800–1000 g,

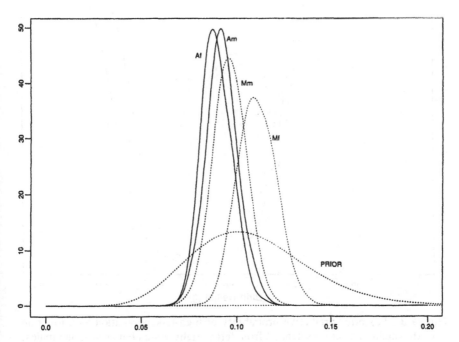

Figure 2 Marginal posterior densities for p.

respectively). We have not here reported details of inference for ϕ_i, $i =$ 1, . . . , n, although summaries are readily available from the simulation output. This information provides the basis for classifying individuals into the two groups if this were of interest.

One important by-product of our analysis is that it allows predictive assessment of risk of future newborns as a function of birth weight, x, via the q function in Eq. (8), a function of the population parameters that can be averaged ergodically in the usual way; see Figure 3. The predictive probability of the high-risk population as a function of birth weight decreases from 1, reaches a minimum, and eventually increases again as birth weight increases over the studied range. This is an inherent feature of two-component normal discrimination problems in which the standard deviation in one population (here the high-risk population) is presumed larger than the standard deviation in the other. As an example, note that in the case of Asian males, a birth weight less than 2.20 kg would indicate that the individual should be allocated to the high-risk group.

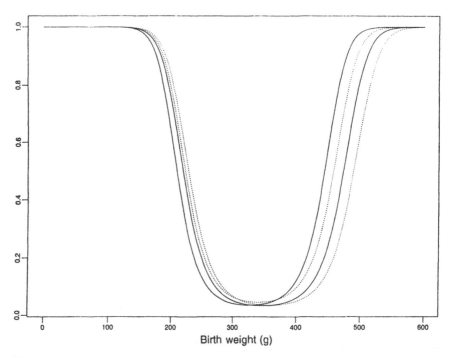

Figure 3 Posterior predictive probability of high-risk population as a function of birth weight. The curves denote, from left to right, Asian females, Asian males, Moroccan females, and Moroccan males.

7   INTEGER ROUNDING OF DATA

The two-component mixture model appears to have provided an adequate fit to the data as presented in earlier sections. In particular, the predictive densities overlay closely the data histograms. However, in Figure 1 the data are binned in 200-g intervals. If we replot the histograms with, say, 10-g bins, as in Figure 4, another effect is evident. There are pronounced "spikes" at regular intervals across the birth weight range. These spikes can be seen to occur on multiples of 50 g and 100 g, with the effect greater in the latter case. The explanation for this is probably either unconscious rounding or, perhaps more plausibly, the accuracy of weighing equipment. In either case, we can regard the recorded data as being "integer rounded," although at this stage it is not clear whether this rounding has any significant influence on the inferences to be made.

Once we recognize that the rounding has occurred and that the data that have been recorded are not the "actual" data, the inference problem becomes more complicated. Conventional Bayesian analysis would re-

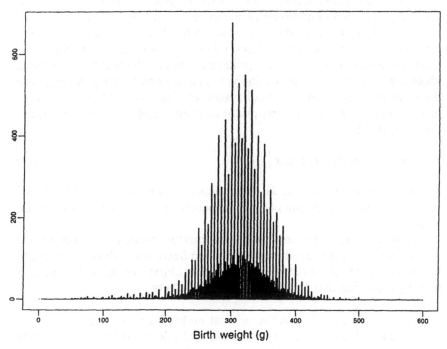

Figure 4   Birth weights of Asian females.

quire that an integral of the form

$$p(\theta \mid y) = \int p(\theta \mid x, y) \, p(x \mid y) \, dx \qquad (10)$$

be carried out, where dummy variable x can be interpreted as the actual, unobserved data, so that $p(\theta \mid x, y) \equiv p(\theta \mid x)$ is merely the putative posterior density for θ conditional on the actual data, and $p(x \mid y)$ is a probability density relating the actual and observed versions of the data. This latter density can either be specified directly, or itself computed via Bayes theorem, if the conditional density $p(y \mid x)$ is more readily or naturally available. Often, the computations required to evaluate the necessary quantities in the conventional Bayesian approach are analytically intractable and numerically complex.

The Gibbs sampler is of considerable use in problems of this type (see, for example, Gelfand *et al.*, 1992b). The sampling approach is similar to the conventional one, in that a dummy variable representing the actual data is introduced. However, this variable is then regarded as a parameter of interest and included in the Gibbs sampler iterative cycle, thus simplifying not only the formulation but also the computations involved in the inference. The Gibbs sampler then allows marginal inference about θ and x given y. The full conditional posterior densities here are $p(\theta \mid x, y) \equiv p(\theta \mid x)$ and $p(x \mid y, \theta)$, which, in line with what was discussed in the previous paragraph, are either specified directly or evaluated using Bayes theorem. Which of these is appropriate depends on the assumptions made about the data, that is, whether the y_i are conditionally independent given the x_i, or vice versa, and the mechanism relating them. If there is "structure" in the x's it is most appropriate to evaluate $p(x \mid y, \theta)$ using Bayes theorem, that is,

$$p(x \mid y, \theta) \propto l(y \mid x, \theta) p(x \mid \theta) \qquad (11)$$

It remains to specify $l(y \mid x, \theta)$, the function that models the "rounding" mechanism, as the function $p(x \mid \theta)$ is merely the sampling density of the actual data.

In the context of the two-component mixture modeling, therefore, we need to augment the procedure described in Section 4 with steps designed to take into account the rounding mechanism. First, we model the mechanism by assuming that $l(y \mid x, \theta) \equiv l(y \mid x)$, where

$$l(y \mid x) = \begin{cases} p_1, & y \text{ is the nearest multiple of 10 g to } x \\ p_2, & y \text{ is the nearest multiple of 50 g to } x \\ p_3, & y \text{ is the nearest multiple of 100 g to } x \end{cases} \qquad (12)$$

This represents a rounding mechanism that treats values of y symmetrically around x, uniformly on three discontinuous intervals. Under such a model, for example, if $x = 144$ g, $y = 140$ g with probability p_1, $y = 150$ g with probability p_2, and $y = 100$ g with probability p_3. The form in Eq. (12), being symmetric in x and y, is readily viewed as a function of x as required when computing the conditional density in Eq. (11). Recalling that, as parameter θ also incorporates the current class membership ϕ for the datum concerned, $p(x \mid \theta)$ is simply one of the component densities in Eq. (3), and so it is easy to see that $p(x \mid y, \theta)$ given by

$$p(x \mid y, \theta) \propto p_1 p(x \mid \theta), \qquad |x - y| \le 5 \text{ g} \quad \text{if } y \text{ is a multiple of 10 g}$$

$$p(x \mid y, \theta) \propto \begin{cases} p_1 p(x \mid \theta), & |x - y| \le 5\,\text{g} \\ p_2 p(x \mid \theta), & 5\,\text{g} < |x - y| \le 25\,\text{g} \end{cases} \quad \text{if } y \text{ is a multiple of 50 g}$$

$$\tag{13}$$

$$p(x \mid y, \theta) \propto \begin{cases} p_1 p(x \mid \theta), & |x - y| \le 5 \text{ g} \\ p_2 p(x \mid \theta), & 5 \text{ g} < |x - y| \le 25 \text{ g} \\ p_3 p(x \mid \theta), & 25 \text{ g} < |x - y| \le 50 \text{ g} \end{cases} \quad \text{if } y \text{ is a multiple of 100 g}$$

from which we can readily generate x's using rejection sampling. The values of p_1, p_2, and p_3 must be specified at the outset, and we return to this point in the discussion of Section 6.

The joint conditional density $p(\theta \mid x)$ has precisely the same form as that described in Section 4 and is computed in the same way. Thus, all we have to do to amend our original Gibbs sampler implementation is to include a preliminary step in the iterative cycle in which a variate, x_i, say, is sampled via Eq. (13), corresponding to the actual value of the recorded datum y_i, thereafter proceeding as in Section 4.

7.1 Reanalysis of Birth Weight Data

Having taking the rounding mechanism into account, we reanalyzed the birth weight data, with $p_1 = 0.55$, $p_2 = 0.325$, and $p_3 = 0.125$, the latter reflecting our subjective assessment of a plausible mechanism. The only effect that this appeared to have was to inflate the marginal variances of each of the parameters, without radically changing the inferences stated

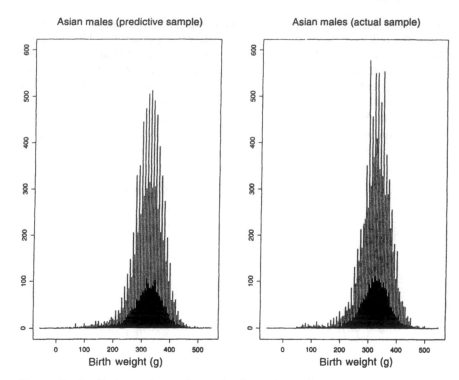

Figure 5   Predictive and actual samples for Asian males.

in Section 6. To demonstrate the adequacy of the model specification in terms of the rounding mechanism, we generated in Figure 5a a typical (predictive) data set that would have been observed under the assumed rounding model. It is clear, when compared with Figure 5(b), that our model now captures many of the features of the data.

8   DISCUSSION

We have used a two-component normal mixture model to analyze birth weight data and possibly gain important clinical insight. However, as mentioned in Section 6, there is no clear reason for assuming the residual component as modeled via density $f_1(Y \mid \theta_1)$ to be normally distributed.

Although this model appears to explain the data adequately, as indicated in the predictive density/histogram plots, it is perhaps a little intellectually unsatisfactory. We are not, of course, tied to the normality assumption—any suitable density may be proposed—and the ease of the Gibbs sampler–based analysis indicates that inferences may just as readily be made for any other form of residual density, although there is no real justification for any alternative choice. Ultimately, we might wish to introduce a nonparametric density to reflect the fact that the residual population arises from a number of clinically distinguishable groups.

In relation to the specification of the rounding mechanism probabilities, they were chosen in our analysis by means of trial and error, by inspection of typical predictive data sets. Sampling-based methods lend themselves readily to such model criticism/validation exercises. A further step could involve including p_1, p_2, and p_3 as unknowns and performing full Bayesian inference for these parameters also.

Finally, it would be interesting to relate our inferences to other aspects of the data that might be forthcoming, such as the actual mortality rate in our designated high-risk group, or covariate information. At the moment we have no access to any such further information, so we must be satisfied with empirically modeling the observed data as described above.

ACKNOWLEDGMENTS

The first two authors were supported by the SERC Complex Stochastic Systems Initiative during part of the period of this work. We are grateful to the late Paul Y. Wax for providing the data and initial motivation for this work.

REFERENCES

Diebolt, J. and Robert, C. (1994). Estimation of finite mixture distributions through Bayesian sampling. *J. R. Statist. Soc. B* 56: 363–375.

Escobar, M. D. and West, M. (1995). Bayesian density estimation and inference using mixtures. *J. Am. Statist. Assoc.* 90: 577–588.

Gelfand, A., Dey, D. K., and Chang, H. (1992a). Model determination using predictive distributions with distributions with implementation via sampling-based methods. In *Bayesian Statistics 4*, ed. Bernardo, J. M., Berger, J. O., Dawid, A. P., and Smith, A. F. M. Oxford University Press, New York.

Gelfand, A., Smith, A. F. M., and Lee, T.-M. (1992b). Bayesian analysis of constrained parameter and truncated data problems using Gibbs sampler. *J. Am. Statist. Assoc.* 87: 523–532.

Smith, A. F. M. and Roberts, G. O. (1993). Bayesian computation via the Gibbs sampler and related Markov chain Monte Carlo methods. *J. R. Statist. Soc. B* 55: 3–23.

Titterington, D. M., Smith, A. F. M., and Makov, U. E. (1985). *Statistical Analysis of Finite Mixture Distributions*. Wiley, New York.

West, M. (1992). Modeling with mixtures. In *Bayesian Statistics 4*, eds. Bernardo, J. M., Berger, J. O., Dawid, A. P., and Smith A. F. M., pp. 503–524. Oxford University Press, New York.

24

Change-Point Analysis of a Randomized Trial on the Effects of Calcium Supplementation on Blood Pressure

Lawrence Joseph   Montreal General Hospital and McGill University, Montreal, Quebec, Canada
David B. Wolfson   McGill University, Montreal, Quebec, Canada
Roxane du Berger   Montreal General Hospital, Montreal, Quebec, Canada
Roseann M. Lyle   Purdue University, West Lafayette, Indiana

ABSTRACT

Evidence from observational and laboratory studies suggests a possible link between dietary calcium and blood pressure. To investigate this issue, a randomized placebo-controlled clinical trial was carried out in a group of 75 normotensive males. The design of the trial included a 4-week baseline period during which weekly blood pressure readings were taken on all subjects and a 12-week treatment period during which biweekly blood pressure measurements were recorded for 37 calcium supplementation patients and 38 control subjects. The reduction in blood pressure of some subjects, in response to calcium supplementation, may be clinically important, whereas in others there may be little or no response. For subjects who do respond, it is possible that there are delays from the time calcium supplementation is commenced to the time effects are manifest, because the metabolism may need to adjust to the higher levels of calcium intake.

Any analysis of the data must account for these features. In this chapter, we present a change-point model to estimate the proportion of subjects who respond to calcium supplementation, as well as the time delay and the magnitude of these effects (if any).

1 INTRODUCTION

Both epidemiological and laboratory investigations have suggested a possible link between calcium intake and blood pressure. Several independent analyses of the data collected for the Health and Nutrition Examination Surveys (NHANES) I and II from the National Center for Health Statistics in the United States found an inverse relationship between dietary calcium and blood pressure. For example, Grouchow et al. (1985) showed that blood pressure decreased with increased dietary calcium intake. Similar results were reported by Harlan et al. (1984) and McCarron et al. (1984). Other studies support these findings only among black males but not among other sex and ethnic groups, Sempos et al. (1986).

These studies were cross-sectional in design and thus are subject to several possible limitations. First, hypertension is thought to develop slowly over time, so that a cross-sectional study may be unsuitable for finding causal relationships. Second, approximate calcium intake in the NHANES I and II surveys was based on only a single 24-hour recall questionnaire and may not be an adequate measure of diet history or even diet at that time. These studies also did not account for nondietary sources of calcium, such as calcium in the water supply or calcium supplements through multivitamins. Furthermore, high variation in calcium intake within individuals compared with variation in intake between individuals may attenuate effect measures (see Sempos et al., 1986).

Although laboratory studies demonstrated that increased calcium intake may lower the blood pressures of rats (Schleiffer et al., 1984), the question of whether this effect carries over to humans remained unresolved.

To address these issues, Lyle et al. (1987) designed a randomized double-blind placebo-controlled clinical trial to examine the effects of calcium supplementation on the blood pressures of 75 normotensive white and black males, aged 19 to 52 years. The men were followed over a period of 16 weeks. Weekly blood pressure measurements were taken during the first 4 weeks. At the end of this baseline period, patients were randomized into either a calcium supplementation group (37 men) or a placebo group (38 men). The randomization was carried out separately within each racial group, resulting in 10 and 11 blacks in the calcium and placebo groups, respectively, along with 27 whites in each of these groups. In both groups, 500-mg tablets were distributed to be taken three times daily with meals.

Biweekly blood pressure measurements were then taken for the next 12 weeks. There were therefore four baseline measurements per subject and six after the tablets were distributed. At each visit, diastolic and systolic blood pressures were recorded from both the supine and seated positions. Other variables collected included age, height, race, and an approximate dietary calcium intake based on a food questionnaire. Data for systolic and diastolic blood pressures from both the supine and seated positions are given in Tables 1 through 8.

Lyle *et al.* (1987) used a repeated measures analysis of covariance to test for differences in mean blood pressure between the calcium supplementation and placebo groups. Tests were carried out for both diastolic and systolic blood pressures, in each of the supine and seated positions. The analysis of covariance controlled for possible group differences in mean baseline blood pressure (average of the four weekly measurements), age, race, skin folds, calcium intake, and compliance. Each analysis assessed differences between the mean blood pressure vectors formed from the six posttreatment and placebo measurements.

The results showed statistically significant decreases in average diastolic and systolic blood pressures from both the seated and supine positions. The overall average decreases were reported to be approximately 2 to 3 mm Hg. The p-values ranged from 0.008 to 0.02. There were no statistically significant differences found between the two races. Adjusted values were given for mean blood pressure for each group at baseline and at each of the 6 biweekly measurements. Table 9 summarizes the estimated baseline means and the estimated postsupplementation means based on a repeated measures analysis of covariance, adjusted as described above.

Given the small p-values for treatment differences and nonsignificant tests for differences in racial groups, one might expect to be able to observe consistent decreases in blood pressures from baseline means across treatment groups. However, for only 6 out of 8 comparisons within the calcium group (four types of blood pressure measurements in the two racial groups) is the mean blood pressure postbaseline lower than baseline, while 5 out of 8 have lower pressures postbaseline in the placebo group. In fact, diastolic pressure in the black subjects seemed to *increase* in the treatment group. With a mean decrease in blood pressure over the 8 comparisons of only 2.5 mm Hg, it is tempting to dismiss the significant results as clinically uninteresting. Nevertheless, if there is heterogeneity

Table 1 Individual Patient Data for Supine
Systolic Blood Pressure in the Placebo Group[a]

Patient #	Weeks									
	1	2	3	4	6	8	10	12	14	16
Placebo group, black subjects										
1	114	124	129	124	116	125	127	131	119	124
2	109	110	103	113	108	109	108	104	98	97
3	111	115	110	112	98	112	106	112	111	113
4	98	109	97	105	105	93	107	100	107	105
5	99	107	101	86	93	98	94	92	94	95
6	114	112	119	112	117	115	122	114	117	119
7	120	128	111	116	114	112	123	111	106	114
8	119	114	111	111	114	117	118	110	113	112
9	116	110	107	109	110	112	107	118	105	121
10	119	118	113	118	125	126	115	122	114	118
11	136	135	129	120	124	117	124	120	124	133
Placebo group, white subjects										
12	102	104	109	108	106	101	97	104	105	108
13	105	106	103	103	117	113	101	103	105	102
14	120	126	118	129	117	118	115	122	114	123
15	120	122	120	123	123	113	121	117	115	119
16	112	108	114	109	115	114	113	119	110	105
17	117	117	112	103	118	114	112	111	115	120
18	102	111	106	107	110	107	115	109	114	114
19	120	126	122	113	116	125	117	118	115	113
20	110	100	99	98	98	97	97	97	95	100
21	105	107	111	100	111	108	99	108	101	100
22	108	113	107	119	122	104	108	108	116	110
23	110	113	97	116	113	106	112	120	125	116
24	123	116	116	110	105	113	114	107	116	100
25	105	106	106	103	109	112	108	110	101	107
26	113	108	108	106	105	111	101	110	113	105
27	120	136	134	133	120	124	125	121	136	129
28	139	134	136	139	130	134	135	130	122	135
29	129	128	128	121	127	130	130	118	125	124
30	112	111	112	103	107	111	108	107	106	105
31	104	106	112	112	99	99	110	117	101	99
32	110	105	105	109	99	99	96	110	99	106
33	139	135	105	135	127	133	122	122	114	127
34	115	105	112	106	104	102	108	98	107	94
35	131	126	120	124	119	118	123	117	119	118
36	120	114	121	114	119	114	117	118	120	119
37	122	125	119	108	117	114	119	123	119	119
38	119	118	124	119	121	128	107	113	112	118

[a] Weeks 1 through 4 are baseline measurements. Weeks 6
through 16 are posttreatment measurements. The units are mm
Hg.

Table 2 Individual Patient Data for Supine
Systolic Blood Pressure in the Calcium Group[a]

Patient #	Weeks									
	1	2	3	4	6	8	10	12	14	16
	Calcium group, black subjects									
39	104	110	109	103	97	99	108	96	105	100
40	104	112	110	112	108	111	104	105	110	113
41	132	114	125	119	117	119	114	110	106	105
42	130	120	137	130	124	116	117	136	116	112
43	107	107	118	117	111	116	107	114	111	115
44	107	109	117	110	113	117	105	109	123	116
45	103	110	109	106	99	94	110	100	106	106
46	107	104	112	124	110	109	98	109	100	102
47	136	135	134	140	132	130	129	128	126	125
48	101	105	99	102	111	100	102	104	112	104
	Calcium group, white subjects									
49	135	122	128	115	114	111	118	117	102	108
50	106	106	103	102	121	112	105	113	98	105
51	108	129	131	121	101	131	122	118	119	115
52	100	99	103	117	104	107	100	94	96	93
53	125	116	119	114	113	118	103	105	109	107
54	102	114	115	112	107	112	107	104	117	113
55	122	119	123	122	111	115	115	117	102	105
56	112	134	114	113	110	106	114	108	122	111
57	115	119	116	109	114	118	118	106	112	111
58	105	111	111	104	106	105	98	110	105	116
59	125	123	137	129	121	122	123	123	127	125
60	119	121	108	123	119	110	128	109	100	103
61	101	105	105	105	111	94	104	104	104	108
62	111	128	125	121	126	124	119	132	116	110
63	126	125	115	115	115	115	116	113	118	117
64	127	120	132	130	129	116	117	116	129	115
65	116	112	111	122	115	120	114	110	115	117
66	128	130	125	131	125	114	119	117	120	115
67	107	107	117	112	114	105	114	111	123	114
68	112	120	109	116	116	100	116	111	96	105
69	110	120	118	121	106	113	109	101	115	108
70	101	101	108	104	97	105	104	108	100	102
71	123	116	127	114	116	112	109	119	114	113
72	109	121	118	116	123	111	116	111	105	116
73	115	109	112	115	109	111	102	115	116	117
74	120	117	115	116	110	105	109	122	111	109
75	107	109	102	98	105	99	97	106	105	102

[a] Weeks 1 through 4 are baseline measurements. Weeks 6
through 16 are posttreatment measurements. The units are mm
Hg.

Table 3 Individual Patient Data for Seated
Systolic Blood Pressure in the Placebo Group[a]

Patient #	Weeks									
	1	2	3	4	6	8	10	12	14	16
Placebo group, black subjects										
1	125	125	123	127	112	126	126	117	124	125
2	115	111	115	123	115	112	103	115	102	103
3	122	121	117	125	110	119	117	119	106	115
4	99	106	109	115	117	107	105	105	106	112
5	92	102	103	103	93	97	103	92	97	95
6	118	121	122	117	118	115	115	125	124	121
7	122	123	119	114	117	110	123	113	110	105
8	126	124	122	125	122	129	127	120	118	125
9	115	115	110	109	107	111	104	105	112	116
10	118	123	117	115	121	123	119	125	115	114
11	135	142	127	128	127	125	129	126	131	133
Placebo group, white subjects										
12	109	104	115	112	115	107	108	104	110	110
13	109	112	104	100	117	104	102	109	110	99
14	117	117	116	119	116	113	112	120	120	119
15	122	119	118	116	134	108	118	105	121	112
16	106	117	118	112	114	110	113	112	116	111
17	127	118	122	122	116	114	116	116	114	121
18	117	113	121	109	114	110	118	116	109	110
19	136	118	123	122	117	128	126	117	116	113
20	107	102	100	98	100	99	101	102	100	101
21	109	110	121	112	105	106	104	108	105	102
22	115	119	118	113	112	106	110	114	111	113
23	116	110	111	110	126	112	112	134	113	122
24	110	111	109	107	105	113	106	106	113	103
25	107	113	110	103	105	105	104	109	100	109
26	124	117	103	112	113	111	108	110	115	117
27	126	125	125	124	126	130	129	125	123	136
28	126	128	140	131	133	128	128	120	129	124
29	121	130	132	135	136	137	137	130	134	129
30	116	108	112	116	110	107	109	114	102	113
31	105	109	115	122	109	107	109	111	96	99
32	105	100	106	96	108	98	105	111	102	101
33	122	138	111	137	121	131	124	121	112	122
34	117	113	111	114	115	108	107	113	105	105
35	128	115	120	119	116	114	118	124	115	107
36	129	122	114	113	116	114	117	115	109	110
37	120	129	118	122	130	117	126	117	128	119
38	117	112	112	113	112	109	112	105	128	112

[a] Weeks 1 through 4 are baseline measurements. Weeks 6
through 16 are posttreatment measurements. The units are mm
Hg.

Table 4 Individual Patient Data for Seated Systolic Blood Pressure in the Calcium Group[a]

Patient #	Weeks									
	1	2	3	4	6	8	10	12	14	16
Calcium group, black subjects										
39	110	111	111	109	109	102	114	110	102	115
40	107	117	108	111	116	112	105	112	110	113
41	130	120	120	142	124	124	126	108	117	116
42	132	137	135	130	112	128	111	120	124	125
43	115	116	117	109	114	115	107	117	116	107
44	125	126	128	119	126	118	124	123	122	118
45	105	111	108	111	104	105	108	109	98	101
46	119	117	118	118	110	112	118	112	115	109
47	139	135	138	137	133	135	127	136	131	131
48	114	103	108	117	114	113	98	105	112	99
Calcium group, white subjects										
49	124	138	123	120	117	118	124	116	112	112
50	107	103	112	105	108	98	109	107	105	100
51	116	130	138	131	117	132	129	130	125	127
52	98	102	101	112	104	105	101	97	93	93
53	118	114	126	116	117	118	108	102	109	111
54	124	104	115	114	112	107	108	113	117	112
55	136	122	123	120	115	113	117	116	109	112
56	124	133	117	124	117	124	122	113	133	115
57	113	121	115	117	111	109	118	107	113	121
58	112	111	111	112	116	119	102	107	108	113
59	125	114	127	127	121	118	121	121	123	120
60	121	131	112	114	121	118	132	114	116	112
61	104	108	104	104	99	104	99	101	99	102
62	117	117	121	116	114	126	108	119	117	116
63	120	122	108	110	116	114	108	112	111	110
64	120	118	140	139	138	123	109	122	127	110
65	126	121	114	121	112	118	114	117	104	108
66	125	126	122	125	116	117	118	121	118	117
67	127	117	113	111	108	108	114	116	114	113
68	118	120	111	121	120	112	120	115	108	116
69	120	114	122	117	115	112	114	111	116	108
70	107	104	111	103	100	102	99	98	109	104
71	126	112	123	116	118	118	118	124	122	115
72	114	117	118	108	110	117	117	115	106	107
73	108	112	117	114	120	110	109	120	116	108
74	118	117	128	117	110	108	117	118	114	112
75	108	110	114	104	105	101	108	103	114	108

[a] Weeks 1 through 4 are baseline measurements. Weeks 6 through 16 are posttreatment measurements. The units are mm Hg.

Table 5 Individual Patient Data for Supine Diastolic Blood Pressure in the Placebo Group[a]

Patient #	Weeks									
	1	2	3	4	6	8	10	12	14	16
Placebo group, black subjects										
1	71	74	85	69	71	89	85	80	77	85
2	66	71	67	66	72	74	66	69	60	65
3	68	71	67	69	68	80	77	72	90	78
4	56	64	66	66	67	62	65	68	65	64
5	58	82	69	61	72	69	64	73	64	64
6	72	80	80	73	77	72	70	78	78	77
7	66	77	70	66	74	71	73	71	69	72
8	85	78	75	76	72	83	81	82	80	82
9	66	78	85	77	70	78	63	58	76	74
10	74	80	81	68	79	81	72	83	80	76
11	46	80	54	70	71	65	70	65	72	80
Placebo group, white subjects										
12	75	65	77	65	78	76	64	70	70	81
13	68	74	74	64	80	78	65	69	70	68
14	84	89	78	79	80	85	84	87	81	81
15	64	71	76	79	71	73	82	73	75	69
16	76	78	79	74	77	72	73	79	75	75
17	64	69	79	75	75	69	74	78	74	78
18	65	67	73	66	82	71	68	73	73	78
19	68	74	75	72	63	76	67	68	61	66
20	72	69	63	70	62	64	66	71	56	64
21	59	67	76	70	70	69	68	69	73	68
22	72	74	77	80	84	72	81	71	71	66
23	72	76	67	61	83	77	62	67	73	74
24	83	82	85	77	78	81	80	75	78	75
25	77	81	80	77	72	79	80	77	75	76
26	68	63	69	69	69	71	75	72	75	63
27	82	77	95	88	82	80	85	88	91	89
28	92	97	91	91	94	79	80	87	86	99
29	91	89	86	90	86	88	88	80	84	87
30	77	72	77	69	68	74	72	73	74	69
31	69	69	77	79	70	68	72	74	65	73
32	75	68	72	68	69	71	67	72	69	70
33	89	87	73	87	89	93	90	84	76	89
34	76	72	76	76	80	65	75	74	73	76
35	52	69	64	66	69	60	68	73	64	69
36	73	65	68	76	75	73	86	75	83	76
37	76	87	73	72	77	80	82	81	72	79
38	72	73	69	67	83	78	71	83	76	80

[a] Weeks 1 through 4 are baseline measurements. Weeks 6 through 16 are posttreatment measurements. The units are mm Hg.

Table 6 Individual Patient Data for
Supine Diastolic Blood Pressure in the
Calcium Group[a]

Patient #	Weeks									
	1	2	3	4	6	8	10	12	14	16
Calcium group, black subjects										
39	67	73	72	74	62	71	60	68	71	72
40	78	83	84	85	83	86	76	75	85	84
41	88	72	81	81	88	87	84	77	70	78
42	74	79	81	91	79	82	80	90	80	77
43	70	59	69	70	65	63	69	64	66	78
44	70	60	66	66	78	73	74	76	71	71
45	71	60	49	73	73	54	69	61	51	70
46	74	71	75	75	73	65	73	65	65	74
47	56	60	69	48	56	59	53	72	73	78
48	62	58	60	56	57	60	58	67	63	71
Calcium group, white subjects										
49	60	67	72	75	69	69	75	70	63	71
50	70	74	69	72	82	73	71	74	76	79
51	86	95	83	80	81	90	84	82	83	76
52	67	66	71	86	70	72	71	65	67	67
53	85	84	89	84	83	94	75	71	79	76
54	59	69	75	59	58	65	65	68	69	69
55	88	80	74	76	69	71	65	73	69	69
56	76	81	71	82	74	73	75	73	76	75
57	76	89	80	86	81	82	85	75	87	77
58	71	71	58	64	69	72	64	69	67	75
59	83	86	67	69	67	68	80	77	60	67
60	74	70	69	78	74	68	78	73	66	72
61	74	78	75	77	75	74	71	72	67	75
62	82	92	87	82	81	84	81	84	86	82
63	74	68	73	70	77	69	66	62	68	65
64	79	77	89	92	74	72	75	76	78	79
65	71	70	73	76	83	75	78	80	76	77
66	72	72	78	82	72	74	74	73	84	79
67	68	65	75	70	73	61	78	77	75	78
68	74	78	71	78	72	66	72	75	68	71
69	62	82	73	77	75	72	78	68	80	71
70	66	65	67	59	63	61	64	70	69	70
71	83	85	89	91	96	88	80	82	87	79
72	75	76	80	77	75	75	78	67	65	74
73	66	72	69	68	65	58	67	73	73	72
74	57	67	51	63	45	73	59	49	68	66
75	62	62	59	60	56	55	51	62	64	66

[a] Weeks 1 through 4 are baseline measurements.
Weeks 6 through 16 are posttreatment measurements.
The units are mm Hg.

Table 7 Individual Patient Data for
Seated Diastolic Blood Pressure in the
Placebo Group[a]

Patient #	Weeks									
	1	2	3	4	6	8	10	12	14	16
Placebo group, black subjects										
1	69	78	82	79	68	88	81	83	80	89
2	69	69	73	66	70	73	63	73	68	73
3	80	70	75	73	75	88	83	81	83	81
4	59	69	68	74	69	51	55	61	71	66
5	65	82	80	76	72	70	76	71	70	70
6	60	84	72	55	64	69	67	77	79	79
7	69	73	64	67	77	68	72	78	66	71
8	84	78	77	85	77	79	83	83	82	77
9	57	79	79	81	78	72	67	59	69	63
10	70	77	88	68	76	82	77	85	73	73
11	36	50	52	70	54	62	74	78	75	80
Placebo group, white subjects										
12	76	68	75	70	74	72	71	67	70	68
13	71	80	76	75	79	76	68	73	78	70
14	82	80	72	85	82	80	76	86	78	77
15	63	68	77	78	60	68	78	77	81	72
16	76	75	80	76	74	68	71	71	76	81
17	64	64	71	80	70	64	81	75	73	75
18	64	61	79	64	68	68	73	74	76	64
19	74	70	74	85	88	70	65	61	72	59
20	70	75	67	68	69	69	67	66	61	72
21	64	76	74	74	67	68	68	70	69	70
22	75	82	75	77	73	72	77	80	74	72
23	81	66	63	60	75	71	65	76	74	75
24	75	73	76	76	73	77	71	76	73	76
25	78	84	78	68	71	74	79	74	68	72
26	66	68	70	69	68	70	76	70	73	69
27	62	71	74	78	62	74	82	86	86	84
28	79	87	91	85	89	79	83	86	82	87
29	82	83	87	90	86	86	86	88	79	87
30	78	71	74	73	70	73	77	73	73	75
31	62	65	63	65	68	66	63	68	68	66
32	72	62	74	72	68	69	73	78	74	68
33	83	93	69	92	86	89	88	86	86	89
34	79	77	85	80	80	73	72	77	77	75
35	41	50	57	46	56	42	47	60	51	59
36	80	62	70	72	74	72	77	77	73	70
37	77	83	73	70	71	83	79	86	71	85
38	67	70	71	49	75	72	71	67	72	75

[a] Weeks 1 through 4 are baseline measurements.
Weeks 6 through 16 are posttreatment measurements.
The units are mm Hg.

Table 8 Individual Patient Data for Seated Diastolic Blood Pressure in the Calcium Group[a]

Patient #	Weeks									
	1	2	3	4	6	8	10	12	14	16
Calcium group, black subjects										
39	72	77	79	72	63	69	59	88	72	71
40	76	88	81	84	84	81	81	87	90	88
41	91	81	80	90	85	81	92	74	71	79
42	62	77	88	79	67	80	72	73	81	70
43	72	65	73	61	62	64	63	63	66	70
44	72	64	73	62	64	58	70	70	65	75
45	75	51	61	51	72	62	66	48	54	73
46	66	79	71	75	73	74	79	70	81	69
47	63	64	69	53	58	60	63	69	77	86
48	74	64	67	62	58	61	65	60	70	70
Calcium group, white subjects										
49	54	70	63	64	52	68.	66	61	58	64
50	73	66	73	71	76	70	75	73	69	69
51	78	93	81	80	81	95	91	86	72	78
52	64	66	73	77	69	71	68	64	66	68
53	84	82	88	78	82	84	83	80	79	81
54	63	55	68	54	61	61	69	71	66	68
55	85	82	65	72	73	68	64	76	73	58
56	73	79	78	82	70	81	76	76	80	80
57	81	87	87	89	76	83	88	70	86	86
58	74	70	66	67	75	71	72	72	70	66
59	67	80	64	72	62	75	75	80	74	76
60	67	58	72	73	67	78	69	71	73	71
61	75	80	74	75	73	74	70	68	71	75
62	78	79	83	80	79	82	79	75	84	84
63	62	59	57	60	68	53	48	41	73	59
64	79	83	84	84	66	88	75	62	61	73
65	74	80	74	68	80	76	74	82	72	70
66	67	62	67	76	61	69	73	65	69	77
67	82	64	63	70	69	65	70	74	72	62
68	78	80	74	77	73	72	73	74	75	73
69	72	63	63	73	76	68	75	69	68	61
70	62	72	68	67	66	65	70	64	53	64
71	83	80	80	86	88	89	81	85	87	80
72	68	80	80	73	68	78	73	75	67	66
73	75	75	81	78	72	65	74	78	75	72
74	54	54	47	48	63	62	49	58	67	62
75	63	67	66	62	46	64	52	53	73	65

[a] Weeks 1 through 4 are baseline measurements. Weeks 6 through 16 are posttreatment measurements. The units are mm Hg.

Table 9 Baseline Means and Adjusted Means from Repeated Measures Analysis of Covariance

	Supine Systolic				Seated Systolic			
	Placebo		Calcium		Placebo		Calcium	
	White	Black	White	Black	White	Black	White	Black
Baseline	115.9	117.8	117.2	119.6	114.9	113.5	115.8	114.7
Post-Baseline	114.8	114.5	113.2	112.4	113.4	113.5	111.1	110.4

	Supine Diastolic				Seated Diastolic			
	Placebo		Calcium		Placebo		Calcium	
	White	Black	White	Black	White	Black	White	Black
Baseline	72.9	71.2	72.2	71.6	74.7	71.0	74.3	70.2
Post-Baseline	72.5	75.3	71.2	72.1	73.9	75.6	72.1	73.0

Source: From Lyle *et al.* (1987).

in response, it is possible that nonresponders may have diluted the overall means, and that the effects in some subjects may be higher, and of clinical interest.

Classical repeated measures analysis assumes that the effects occur immediately after commencement of therapy. It is, however, conceivable that the effects in this trial may not be immediate, as the concentration of calcium increases, and the metabolism slowly adjusts to these new levels. This is especially relevant since the mechanism by which calcium may lower blood pressure is not known (Chockalingam *et al.*, 1990). By not taking into account this delay in response, the treatment effect may be diluted, leading to a loss of power to detect any real differences. Only after estimating the proportion of positive responders in the population as well as the delay in response to treatment can one properly interpret the mean effects. Power is in any case a concern. For example, no statistically significant differences were found for blacks versus whites. However, Table 9 indicates that in all four cases (seated and supine and in both the placebo and calcium groups) for diastolic blood pressures, values for blacks increased while those for whites decreased. Finally, there are modeling assumptions that may be restrictive or difficult to verify, such as how covariates should enter the model.

Consequently, several questions of considerable importance to a clinician attempting to interpret the impact of the results on clinical practice remain unanswered.

1. What proportion of the population would respond to calcium supplementation? What proportion of these would have decreases in the clinically interesting range of 5 mm Hg or greater?

2. What is the distribution of the change in blood pressure in those who respond?
3. What is the distribution of the time to response (if any) from the start of treatment?
4. Is there any evidence that the effect of calcium supplementation is different in blacks than in whites?

As will be shown below, the change-point analysis proposed here allows comparisons of the proportion with decreased blood pressure in each group (which may be small in the placebo group), as well as estimation of the time to reaction and magnitude of the decrease in those who respond. Information concerning clinically interesting differences can also be obtained from the analysis by tracking the within-subject estimated mean differences, so that individuals changing less than a given amount can be counted as nonresponders. In addition, the model allows each patient to serve as his own control, so that the modeling of covariates affecting mean blood pressures are no longer a concern. However, it is still important to account for covariates that may interact with the blood pressure–lowering effects of calcium. These can be investigated by subgroup analyses.

Change-point models assume abrupt changes in parameter values, whereas it may be more realistic to assume a gradual change in blood pressure. In this case one may prefer a trilinear regression model with two change-points, indicating the beginning and end of the transition period. A single change-point model is simple to implement, whereas it is not feasible to fit a trilinear regression model with only eight blood pressure readings per patient. Our approach is conservative in the sense that estimated pre- and posttreatment mean differences may be attenuated if the true change is gradual. This phenomenon will occur whenever measurements are taken during the period of gradual change but falsely included in either the pre- or posttreatment mean estimate. The change point will be estimated to occur at the time point that best divides the data into two sets with apparently identically distributed variables. Whether such a change will be found will depend on specific features of the data, such as the magnitude and the rate of change, and how many observations there are for each subject. Growth curves could also be used, but in order to answer the specific questions listed above, one would need to estimate where the derivative of the curve changes. This is a complex problem, and reliable estimates may require much more data than does our simpler model. However, Müller and Rosner (1994) have fit a hierarchical Bayesian model to white blood cell count data with a gradual change. This analysis is feasible because typical white blood cell counts do not show much fluctuation over short periods of time.

A naive change-point model might look for changes in the data from each patient separately by first testing for a change-point and then estimating its location and the change in mean blood pressure if the test is positive. A histogram of the estimated change points and mean differences can then be used to describe the patterns and magnitudes of changes (Joseph and Wolfson, 1992). Here, a hierarchical change-point model is employed to address the above questions, shifting the focus from the individual patient to a population of patients. Rather than classifying the subjects into responders and nonresponders based on a test, the data for each subject contribute to the estimation of population change-point parameters. In estimating both individual and population parameters, the model is able to "borrow strength" from the ensemble of patients.

The results of the analyses of the blood pressure data are deferred to Section 4, after a summary of change-point models and the multipath modification are provided in Section 2. A Bayesian approach to estimating the parameters of the multipath change-point model is described in Section 3. Finally, Section 5 contains further discussion.

2 CHANGE-POINT MODELS

A change-point is said to have occurred at $T = \tau$, $1 \leq \tau \leq N - 1$ in a sequence of random variables $X_1, X_2, \ldots, X_N$, if $X_1, X_2, \ldots, X_\tau$, are identically distributed with common distribution F_1, which is different from the common distribution F_2 of $X_{\tau+1}, X_{\tau+2}, \ldots, X_N$. If $T = N$, then no change has occurred. Usually, the change-point, τ, and some parameters of the distributions F_k, $k = 1, 2$, are unknown and are to be estimated from the data.

Hinkley's early work on maximum-likelihood estimation has been followed by a burgeoning literature in the field. Pettitt (1979) offered a nonparametric solution to test for change, and Picard (1985) addressed change-point problems in time series. More recently, Yao (1987, 1990) has also discussed maximum-likelihood change-point inference, and Carlin et al. (1992) have used the Gibbs sampler to find marginal posterior distributions in a hierarchical change-point model.

Often in a medical experiment, data are collected on several occasions on each of a number of subjects. If a drug is introduced during the trial that may affect the distribution of the variable under study, then one may be interested in estimating the proportion of the population that would undergo a change or the distribution to the time of change after the administration of the drug.

Inferring time to effect is especially interesting if it depends on the patient, as may be the case in a trial of a treatment designed to lower blood pressure, as described in Section 1. Classical methods, such as the repeated measures analysis of variance also outlined in Section 1, are, as has been pointed out, not well suited to data that undergo a possible change at some unknown time; they usually require that the time of change be specified in advance and assumed fixed for all subjects. Often, however, little is known about the reaction time, and individual times may vary, leading to erroneous inference from a considerable loss of power. As we shall show, multipath change-point methods will permit answers to both traditional hypothesis-testing questions, which can be loosely phrased as "overall, has this group of patients exhibited a change in mean level?" and the more subtle questions alluded to above.

2.1 Description of the Multipath Change-Point Model

In a multipath change-point model, we allow more than one sequence of data, which considerably broadens the applicability of change-point models. In this case, we assume that the data are in the form of the $M \times N$ array

$$
X = \begin{pmatrix}
X_{11} & X_{12} & \cdots & X_{1\tau_1} & X_{1\tau_1+1} & \cdots & X_{1N} \\
X_{21} & X_{22} & \cdots & X_{2\tau_2} & X_{2\tau_2+1} & \cdots & X_{2N} \\
\vdots & \vdots & & \vdots & \vdots & & \vdots \\
X_{M1} & X_{M2} & \cdots & X_{M\tau_M} & X_{M\tau_M+1} & \cdots & X_{MN}
\end{pmatrix}
\tag{1}
$$

A change-point will be said to have occurred at $T_i = \tau_i$ in sequence or row i, $i = 1, 2, \ldots, M$ and $1 \le \tau_i \le N - 1$, if $X_{i1}, X_{i2}, \ldots, X_{i\tau_i}$ are identically distributed with common distribution F_{i_1}, which is different from the common distribution F_{i_2} of $X_{i\tau_i+1}, X_{i\tau_i+2}, \ldots, X_{iN}$. If $T_i = N$, then no change has occurred in row i. Usually, the points of change, τ_i, and some parameters of the distributions F_{i_k}, $i = 1, \ldots, M$, $k = 1, 2$, are unknown and are to be estimated from the data matrix (1).

It is assumed here that the times of change, T_i, in each row are independent and identically distributed in a given population, following a distribution $g(t) = \Pr\{T_i = t\} = \pi_t$, $i = 1, \ldots, M$, $t = 1, \ldots, N$. If $\pi_N > 0$, there are some members of the population who do not change. The primary goal is to estimate $g(\cdot)$, a secondary goal being to estimate other unknowns in the model, arising from the parameters of the F_{i_k}, $i = 1, \ldots, M$, $k = 1, 2$. It is important to emphasize that $g(\cdot)$ provides the pattern of probabilities for the location of the change-point in the population, and its introduction does not necessarily mean that each subject in the population has exactly the same change-point.

The multipath extension also introduces many interesting theoretical questions. In the single-path setup, interest usually focuses on the probabilities of where the change has occurred for the particular sequence. While it is possible to focus on the individual sequences in the multipath case, inference can also be made about the process governing the change in a particular population, here represented by the distribution $g(\cdot)$. Powerful statistical techniques that have not been effectively employed in the single-path context are the bootstrap (Efron, 1979, 1982) and empirical Bayes methods (Robbins, 1964), but see Hinkley and Schechtman (1987) for a conditional bootstrap approach. It has been shown in Joseph and Wolfson (1992) that both of these techniques, as well as a bootstrap–empirical Bayes combination (Laird and Louis, 1987), may be utilized in the multipath context. Whereas it is well known that the single-path maximum-likelihood estimator of the change-point is not consistent (see Hinkley, 1970), the nonparametric estimator of $g(\cdot)$ has been shown in Joseph and Wolfson (1993) and Joseph et al. (1994) to be consistent under certain conditions in the multipath case. However, consistency does not hold unless either $\pi_N = 0$ or $F_{i1} = F_1$ and $F_{i2} = F_2$ for all $i = 1, 2, \ldots, M$. This is a serious limitation of the frequentist methods, because in most medical trials one might expect patients to have different baseline and/or final distributions, and it is often the case that not all subjects respond to treatment.

From a practical point of view, the implementation of a multipath change-point model offers substantial challenges. Assuming only one unknown parameter from each F_{ik} and $g(\cdot)$, in a Bayesian approach one must calculate posterior distributions of at least $2 \times M + 1$ interrelated parameters. Similarly, directly finding the maximum-likelihood estimators even in problems of moderate dimension requires a considerable and often infeasible number of computations. Fortunately, the structure permits formulation as a "missing data" problem, so that data augmentation (Tanner and Wong, 1987) or the closely related Gibbs sampler (Geman and Geman, 1984; Gelfand and Smith, 1990; Gelfand et al., 1990) and EM (expectation-maximization) algorithm (Dempster et al., 1977) approaches are possible. Joseph and Wolfson (1993) and Joseph et al. (1994a) calculated maximum-likelihood estimates using the EM algorithm. The present discussion represents an extension of the Gibbs sampler methods of Carlin et al. (1992), where interest centered on estimating the location of a single change-point. Here, the focus is on estimating the distribution of the change-point in a given population. Accordingly, the output is a multivariate posterior distribution for the parameters of $g(\cdot)$, rather than posterior probabilities for the location of a single change-point.

Throughout this chapter, $f(\cdot)$ is used to denote a probability density or probability function. The random variables to which these distributions refer will be clear from their arguments and the context in which they appear.

3 CHANGE-POINT ANALYSIS OF THE TRIAL

The trial of Lyle *et al.* (1987) provides eight data sets of the form of Eq. (1), one for each of diastolic and systolic blood pressure measurements, from both the seated and supine positions, each of these with $M = 38$ placebo and $M = 37$ calcium supplementation patients (see Tables 1–8). Throughout, $N = 10$ is the number of observations on each subject. The main analysis consists of comparing the change-point posterior distributions in the two treatment arms for each of the four types of blood pressure measurements. In addition, to examine the plausibility of differences in reactions to calcium supplementation in whites compared with blacks, submatrices of the above will be created and analyzed separately. Here the responses of the $M = 27$ whites in each of the placebo and calcium-treated groups will be contrasted with results from the $M = 10$ or $M = 11$ blacks in the treated and placebo groups, respectively.

3.1 The Likelihood and Model Parameters

The likelihood for the data (1) given θ_1 and θ_2 of the model described in Section 2 is given by

$$f(x \mid \theta_1, \theta_2, \pi) = \prod_{i=1}^{M} \sum_{t=1}^{t} \left\{ \prod_{j=1}^{t} f_1(x_{ij} \mid \theta_1) \right\} \left\{ \prod_{j=t+1}^{N} f_2(x_{ij} \mid \theta_2) \right\} \pi_t \quad (2)$$

where θ_1 and θ_2, possibly vector valued, are the parameters of the before and after change-point densities, f_1 and f_2, respectively, and $\pi_t = \Pr\{T_i = t\}$, $t = 1, 2, \ldots, N$ are the parameters of $g(\cdot)$. Inference using this likelihood is difficult because it takes the form of a mixture. However, conditional on the "latent data" τ_i, $i = 1, 2, \ldots, M$ (see Tanner and Wong, 1987), the change-points in each data sequence, the likelihood simplifies to

$$f(x \mid \theta_1, \theta_2, \tau_1, \ldots, \tau_M)$$

$$= \prod_{i=1}^{M} \left\{ \prod_{j=1}^{\tau_i} f_1(x_{ij} \mid \theta_1) \right\} \left\{ \prod_{i=\tau_i+1}^{N} f_2(x_{ij} \mid \theta_2) \right\} \quad (3)$$

a much easier problem. The Gibbs sampler algorithm described below takes advantage of this simplification by alternating between imputing a set of change points τ_i, $i = 1, \ldots, M$, based on the data and estimating other model parameters using the simplified likelihood (3).

As is implicit in the repeated measures analysis, we assume that the within-patient blood pressure readings are normally distributed, that is, $X_{ij} \sim N(\theta_{ik}, [\rho_{ik}]^{-1})$, $i = 1, \ldots, M$, $j = 1, \ldots, N$, where $k = 1, 2$ represents the before and after change indices. The parameters in the model then are:

1. $\boldsymbol{\mu}_1 = (\mu_{11}, \ldots, \mu_{M1})$ and $\boldsymbol{\mu}_2 = (\mu_{12}, \ldots, \mu_{M2})$, vectors of the mean blood pressure for each patient, before and after the change-point, respectively.

2. $\boldsymbol{\rho}_1 = (\rho_{11}, \ldots, \rho_{M1})$ and $\boldsymbol{\rho}_2 = (\rho_{12}, \ldots, \rho_{M2})$, vectors of the precision (the reciprocal of the variance) of the blood pressure distribution before and after the change-point, respectively, for each patient.

3. $\boldsymbol{\pi} = (\pi_1, \ldots, \pi_N)$, the multinomial probably vector, where π_i is the probability that a change occurs at time $j, j = 1, \ldots, N$.

4. $\boldsymbol{\tau} = (\tau_1, \ldots, \tau_M)$, the unobserved latent data representing the change-point locations for each patient.

In this model, different means and precisions are allowed for each subject. All of the above parameters are in general unknown, although occasionally the structure of the problem may suggest values for certain parameters. For example, here one may wish to set $\pi_1 = \pi_2 = \pi_3 = 0$, since one may be convinced that no change in blood pressure will occur before commencing treatment in week 4.

The analysis focusses on estimating and comparing π_t and π_c, the posterior change-point distributions in the treatment and control groups. Of particular interest is estimating

$$\mathscr{P} = \Pr\{\pi_{N_t} < \pi_{N_c}\} \qquad (4)$$

The parameter $\mathscr{P}$ represents the probability that subjects on calcium supplementation are more likely to change than control subjects. $\mathscr{P} = 0.5$ has the interpretation that there is no difference in the proportion of placebo and treatment patients that exhibit changes, and $\mathscr{P} < 0.5$ (> 0.5) indicates that a smaller (greater) proportion of treatment subjects exhibit changes. Also of interest is the distribution of $(\mu_{i1} - \mu_{i2})$ in each group, higher values indicating larger decreases. Of course, μ_{i2} is defined only for patients who undergo changes. The parameters $\boldsymbol{\rho}_1$ and $\boldsymbol{\rho}_2$ are of interest in that low values of ρ_{i1} and/or ρ_{i2} (i.e., high within-patient variability) relative to $(\mu_{i1} - \mu_{i2})$ would tend to make the detection of a change point more difficult.

The following subsections provide the details necessary for deriving marginal posterior densities of the above quantities.

3.2 Choice of Prior Distributions

Because in this analysis conjugate prior distributions seem plausible, they will be used throughout, although nonconjugate priors can also be accommodated (Müller, 1991). Following DeGroot (1970), the conjugate family in the case of a likelihood proportional to a multinomial distribution is Dirichlet, and a conjugate family for normal distributions with unknown mean and precision is the normal/gamma. Accordingly, the priors can be written as

$$f(\pi_1, \ldots, \pi_N) = \frac{\Gamma(p_0)}{\prod\limits_{j=1}^{N} \Gamma(p_j)} \prod_{j=1}^{N} \pi_j^{p_j - 1} \tag{5}$$

$$f(\mu_{ik} \mid \rho_{ik}) = \frac{1}{\sqrt{2\pi}} \sqrt{w_{ik}\rho_{ik}} \exp\left(- \frac{w_{ik}\rho_{ik}(\mu_{ik} - m_{ik})^2}{2}\right) \tag{6}$$

$$f(\rho_{ik}) = \frac{1}{\Gamma(a_{ik})b_{ik}^{a_{ik}}} \rho_{ik}^{a_{ik} - 1} \exp\left(- \frac{\rho_{ik}}{b_{ik}}\right), \quad i = 1, \ldots, M, k = 1, 2 \tag{7}$$

where $p_0 = \sum_{j=1}^{N} p_j$, $p_j > 0, j = 1, \ldots, N$. The quantities p_j/p_0 are the means of the marginal prior distributions of the π_j. The m_{ik}'s are prior mean blood pressure values, and the a_{ik}'s and b_{ik}'s are the shape and scale of the gamma prior distribution for the precision of the corresponding data x_{ij}. Finally, for given w_{ik} and ρ_{ik}, $w_{ik}\rho_{ik}$ is the precision of the prior distribution of the mean μ_{ik}. Throughout, $k = 1$ (2) indicates a before (after) change-point parameter. All fixed parameters have been suppressed on the left-hand side of Eqs. (5)–(7).

The values of the p_j's, a_{ik}'s, b_{ik}'s, m_{ik}'s, and w_{ik}'s are chosen according to the available prior information. Because treatment commenced at week 4, $p_1 = p_2 = p_3 = 0$ for all analyses. Equal values were selected for p_j, $j = 4, \ldots, 10$, that is, $p_4 = p_5 = p_6 = \cdots = p_{10} = 0.1$. As $p_0 = \sum_{j=1}^{10} p_j = 0.7$, the prior information about π was equivalent to "observing data from less than one subject" (Lee, 1989). We used a noninformative prior despite the results suggested by the previously cited studies. The uncertainty of the utility of calcium supplementation that led to the proposed trial was taken as evidence that clinicians were divided on the issue. Even less was known about the timing of effects. Another alternative would be to run several analyses with different prior parameters representing a range of prior opinions.

Ideally, the physician treating each patient in the experiment would choose the prior parameters for that specific patient. A common problem in epidemiologic studies is that the patients are often not well known by the experimenters, so that prior parameters cannot be ascertained. Further, data analysis may often be done remotely in both time and location from the center where the experiment is carried out, hindering communication between the physicians in the study and the data analysts. Another difficulty with prior parameter selection is the large number of prior parameters that sometimes must be evaluated. For all of these reasons, vague, although not entirely flat, prior parameters were selected in this study. In particular, prior means were selected by using sample means across all patients and at all times both at baseline and after baseline, and standard deviations were chosen to cover the entire range of observations and converted to precisions. The prior values were also selected to be conservative, in the sense that they cover a wider interval than the data would suggest. In this way, the entire range of possible values for the parameters was included, with a relatively flat prior over this range. An alternative to this naive empirical Bayes approach was the solicitation of expert opinion for the mean and within-subject variation for systolic and diastolic blood pressures of normotensive men. This gave values that were within 2 mm Hg in all cases. A sensitivity analysis indicated that the differences had a negligible effect on all posterior distributions.

Completely noninformative priors are not realistic in this setting, because the range and variation of blood pressures occur within well-known bounds. Keeping within the feasible range also avoids numerical problems. This is because the Gibbs sampler draws random variables from the updated conditional distributions of the unknown parameters (see Section 3.3), and one wishes to avoid unreasonable choices that can retard convergence of the algorithm. For example, in the case that $\tau_i = N$ is selected, the prior parameters for μ_{i2} will not be updated by the data. Thus the Gibbs sampler will draw the next value for μ_{i2} directly from its prior distribution. If this is a noninformative distribution, it is possible to draw a sample value very far from all of the observed data. The next iteration then would also be very likely to choose $\tau_i = N$, since the unlikely value for μ_{i2} would lead to a very small posterior probability for any other value of τ_i. In this way, long sojourns in this state will slow down convergence.

Table 10 summarizes all of the prior parameter values used in this study. Gamma distributions used for the prior precision of the data had means of 0.04 and standard deviations of 0.01. These correspond to a range

Table 10 Parameter Values for the Prior Distributions
Used in the Analysis

Parameters for $\underset{\sim}{\pi}$									
p_1	p_2	p_3	p_4	p_5	p_6	p_7	p_8	p_9	p_{10}
0.0	0.0	0.0	0.1	0.1	0.1	0.1	0.1	0.1	0.1

Parameters for $\underset{\sim}{\mu_1}$ and $\underset{\sim}{\mu_2}$							
Systolic Blood Pressure				Diastolic Blood Pressure			
m_{i1}	ω_{i1}	m_{i2}	ω_{i2}	m_{i1}	ω_{i1}	m_{i2}	ω_{i2}
115	0.25	115	0.25	73	0.25	73	0.25

Parameters for $\underset{\sim}{\rho_1}$ and $\underset{\sim}{\rho_2}$							
Systolic Blood Pressure				Diastolic Blood Pressure			
a_{i1}	b_{i1}	a_{i2}	b_{i2}	a_{i1}	b_{i1}	a_{i2}	b_{i2}
16	0.0025	16	0.0025	16	0.0025	16	0.0025

for the standard deviations for the within-patient data of approximately 4 to 7 mm Hg. Setting $\omega_{ik} = 0.25$ corresponds to a very small prior precision on the mean.

3.3 Practical Implementation

Implementing the Gibbs sampler requires specifying the full conditional distribution of the parameters. That is, the conditional distribution of each parameter given the values of all the other parameters must be evaluated. As is often the case, the full conditional distribution of each parameter does not depend on all the other parameters, which leads to some further simplifications.

Let $\bar{x}_{1,\tau_i}$ and $\bar{x}_{\tau_i+1,N}$ be the before and after change-point sample means for each subject, i.e.,

$$\bar{x}_{1,\tau_i} = \frac{\sum_{j=1}^{\tau_i} x_{ij}}{\tau_i}$$

and

$$\bar{x}_{\tau_i+1,N} = \frac{\sum_{j=\tau_i+1}^{N} x_{ij}}{N-\tau_i}, \qquad i = 1, 2, \ldots, M$$

Then the conditional distributions are

$$\Pr\{\tau_i = t \mid \mu_1, \mu_2, \rho_1, \rho_2, \pi, x\}$$

$$= \frac{\left\{\prod_{j=1}^{t} N(x_{ij}; \mu_{i1}, \rho_{i1})\right\}\left\{\prod_{j=t+1}^{N} N(x_{ij}; \mu_{i2}, \rho_{i2})\right\}\pi_t}{\sum_{k=1}^{N}\left\{\prod_{j=1}^{k} N(x_{ij}; \mu_{i1}, \rho_{i1})\right\}\left\{\prod_{j=k+1}^{N} N(x_{ij}; \mu_{i2}, \rho_{i2})\right\}\pi_k} \tag{8}$$

$$f(\pi \mid \tau) \sim \text{Dirichlet}(p') \tag{9}$$

$$f(\rho_{i1} \mid x, \tau_i, m_{i1})$$

$$\tag{10}$$

$$\sim \text{Gamma}\left(a_{i1} + \frac{\tau_i}{2}, \left[b_{i1} + \frac{1}{2}\sum_{j=1}^{\tau_i} (x_{ij} - \bar{x}_{1,\tau_i})^2 + \frac{w_{i1}\tau_i(\bar{x}_{1,\tau_i} - m_{i1})^2}{2(w_{i1} + \tau_i)}\right]^{-1}\right)$$

$$f(\rho_{i2} \mid x, \tau_i, m_{i2})$$

$$\tag{11}$$

$$\sim \text{Gamma}\left(a_{i2} + \frac{(N - \tau_i)}{2}, \left[b_{i2} + \frac{1}{2}\sum_{j=1}^{(N-\tau_i)} (x_{ij} - \bar{x}_{\tau_i+1,N})^2\right.\right.$$

$$\left.\left. + \frac{w_{i2}(N - \tau_i)(\bar{x}_{\tau_i+1,N} - m_{i2})^2}{2(w_{i2} + (N - \tau_i))}\right]^{-1}\right)$$

$$f(\mu_{i1} \mid x, \tau_i, \rho_{i1}) \sim N\left(\frac{w_{i1}m_{i1} + \tau_i\bar{x}_{\tau_i}}{w_{i1} + \tau_i}, [(w_{i1} + \tau_i)\rho_{i1}]^{-1}\right) \tag{12}$$

and

$$f(\mu_{i2} \mid x, \tau_i, \rho_{i2})$$

$$\tag{13}$$

$$\sim N\left(\frac{w_{i2}m_{i2} + (N - \tau_i)\bar{x}_{\tau_i+1,N}}{w_{i2} + (N - \tau_i)}, [(w_{i2} + (N - \tau_i))\rho_{i2}]^{-1}\right)$$

where $N(a; b, c)$ represents a normal density, with mean b and precision c, evaluated at a, and p_k, the kth element of p', is given by $p_k + \sum_{i=1}^{M} I_{\{\tau_i = k\}}$, where $I_{\{y\}}$ is the indicator function for the set $\{y\}$. In other words, 1 is added to p_k each time a $\tau_i = k$.

The Gibbs sampler operates as follows: Arbitrary starting values are chosen for each parameter. (Although arbitrary, certain "bad" choices may influence convergence rates. See note in Section 3.2.). A sample of size m is then drawn from each full conditional distribution, in turn. The sampled parameters from the previous iterations are used in the conditional distributions for subsequent iterations. A cycle of the algorithm is completed when samples from all conditional distributions are available. The entire cycle is repeated a large number of times. The random samples thus generated for each parameter can be regarded as a random sample from the correct posterior marginal distribution (Gelfand and Smith, 1990).

In our model, each τ_i, $i = 1, \ldots, M$, is drawn from Eq. (8), using the starting values for all other variables, and π is sampled from Eq. (9) using these newly sampled values of τ. Then ρ_{i1} and ρ_{i2}, $i = 1, \ldots, M$, are sampled from Eqs. (10) and (11) using the previously sampled values of τ and the starting values of all other variables. The first cycle is completed by generating μ_{i1} and μ_{i2}, $i = 1, \ldots, M$, from Eqs. (12) and (13) using the previously sampled values for τ, ρ_{i1}, and ρ_{i2}, and the next cycle begins by sampling τ given the variables sampled in the first cycle.

One must decide on a value for m, the number of random variables to draw at each iteration, which need not be fixed from iteration to iteration. Tanner and Wong (1987) suggest gradually increasing m as the number of iterations increases. Values of the order of $m = 1000$ were used in their examples. Geyer (1992a) has countered that small values of m are more efficient, with values as small as $m = 1$ often proving optimal. Convergence issues of Gibbs sampling have been avoided here by carefully monitoring the sequences produced and repeating each analysis from several starting points. Convergence was assumed only if all runs provided very similar posterior distributions. Geyer (1992a, 1992b), Casella and George (1992), and Gelman and Rubin (1992) provide informative discussions concerning this and other issues surrounding the convergence of the Gibbs sampler.

3.4 Inference Using the Output from the Gibbs Sampler

The "steady-state" output from the Gibbs sampler is a sequence of random variables drawn from the joint posterior distributions of μ_1, μ_2, ρ_1, ρ_2, τ, and π given x. These variables can then be used in several ways:

1. The marginal posterior densities of any component of μ_1, μ_2, ρ_1, or ρ_2 can be approximated.
2. The marginal posterior densities of functions of μ_1, μ_2, ρ_1, or ρ_2 can be approximated. For example, it is of interest to generate summary figures for $\mu_{i1} - \mu_{i2}$ and to examine the differences in these plots between treatment and control. Mean or median values of these distributions are also of interest.

3. The output generated from the full conditional distribution for π is a multivariate sample from a Dirichelet distribution in N dimensions. Although this by itself is hard to visualize, several interesting summaries are available.

   (a) Bar graphs of the means of the marginal Dirichlet posterior distributions for the change-point probabilities may be constructed.

   (b) Posterior marginal densities for selected change-point probabilities may be plotted. They display the variability about the above Dirichlet means and are calculated as a mixture of beta densities over the random samples generated by the Gibbs algorithm.

   (c) The probabilities $\mathcal{P}$ can be estimated using a Mann-Whitney-type statistic

$$\frac{\displaystyle\sum_{\text{all}(\pi_{N_t},\pi_{N_c})} I_{\{\pi_{N_{t_i}}<\pi_{N_{c_j}}\}}}{k^2} \tag{14}$$

   where k is the total number of Gibbs iterates generated for π_{N_t} and π_{N_c}.

4. Each subject in each iteration may have $\tau_i < N$ or $\tau_i = N$. Another interesting statistic is then

$$\frac{\{\text{number of times } \tau_i < N\}}{\text{number of iterations}} \tag{15}$$

This approximates the subject-specific probability of change.

Examples of all of the above forms of output are given in the next section. Other inferences from the Gibbs iterates are also possible.

4 RESULTS

Throughout, 10,000 Gibbs cycles were run, 3000 to ensure convergence of the algorithm and the last 7000 for inference. Figure 1 displays the means of the marginal Dirichlet posterior distributions for the change point for both the placebo ($M = 38$) and calcium supplementation ($M = 37$) groups for the four types of blood pressure measurements. The heights of the bars labeled 1 through 14 in Figure 1 therefore represent point estimates of the probability that a change in blood pressure occurs at that week. For example, there is an estimated 30% chance that subjects receiving calcium supplementation will experience a change in supine sys-

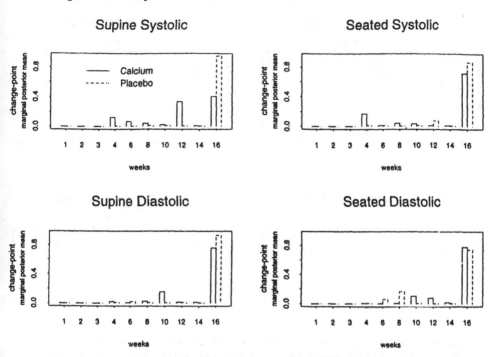

Figure 1 Means of the marginal posterior distributions for the components of $\pi = (\pi_1, \pi_2, \pi_3, \pi_4, \pi_6, \pi_8, \pi_{10}, \pi_{12}, \pi_{14}, \pi_{16})$ for both calcium and placebo groups, for supine and seated, systolic and diastolic blood pressure. Weeks 1 through 14 represent change-point locations. Week 16 represents no change.

tolic blood pressure at the 12th week, i.e., 8 weeks after starting the supplement. The heights of the bars labeled 16 are estimates of the probabilities of no change, π_N. For example, approximately 40% of the calcium group and almost all in the placebo group are estimated not to experience changes in supine systolic blood pressure. Figure 2 exhibits the marginal posterior densities of π_N for each group, together with the probabilities $\mathcal{P}$ given by Eq. (4). For example, for supine systolic blood pressure, virtually all of the probability mass for π_N for the placebo group is concentrated on the interval (0.8, 1.0), indicating a high probability that at least 80% of control subjects do not change. This is not surprising, because it would be expected that few subjects receiving placebo would change. On the other hand, the same figure shows that most of the probability mass in the calcium group is concentrated on the interval (0.1, 0.8), indicating

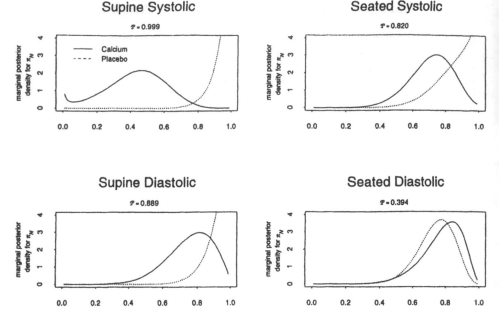

Figure 2 Marginal posterior densities for the probability of no change (π_N) for both calcium and placebo groups, for supine and seated, systolic and diastolic blood pressure.

high likelihood that calcium supplementation affects at least 20% of subjects.

Box plots of the individual subject mean differences over all subjects for all groups and blood pressure types appear in Figure 3. These means are calculated as follows. Since μ_{i2} is meaningful only when there is a change, for each subject i at each iteration of the Gibbs sampler, it is noted whether or not $\tau_i < N$. When $\tau_i < N$ the posterior mean difference is found by subtracting the means of the normal densities defined by Eqs. (12) and (13). Next, an average value for each subject i is found by dividing the sum of these posterior mean differences by the number of times $\tau_i < N$. The ensemble of these values over all M subjects is used in constructing the box plots.

Different subjects have different probabilities of changing. Therefore, to estimate the overall mean treatment effect for the patients who experienced a change, a weighted average of the values calculated for each subject as described above was formed using weights defined by

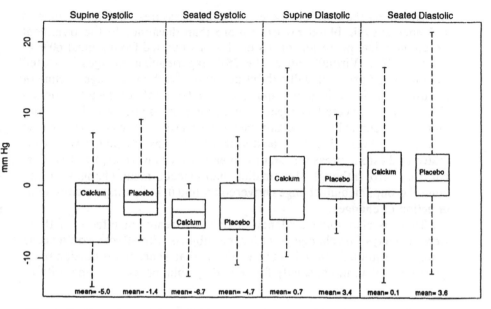

Figure 3 Box plots of the individual posterior mean blood pressure differences (before – after, or $\mu_{i1} - \mu_{i2}$) for the calcium and placebo groups, for supine and seated, systolic and diastolic blood pressure. The five horizontal lines from top to bottom represent the 100%, 75%, 50% (median), 25%, and 0% points of the estimated differences in each group.

Eq. (15). These weighted averages appear at the bottom of each box plot in Figure 3.

Figures 1, 2, and 3 reveal that in all four types of blood pressure measurements, subjects receiving calcium supplements perform better than those receiving a placebo. For both seated and supine systolic blood pressures and for supine diastolic blood pressures, the treatment group changes more often and has larger decreases (systolic) or smaller increases (diastolic) in mean pressures. The only exception was for seated diastolic pressure. Here the control group will experience change points slightly more often than the treatment group, but the changes tend to be increases in blood pressure.

In all cases, mean group differences over all subjects, including those without changes, were 2 to 4 mm Hg, agreeing with Lyle *et al.* (1987). In addition, however, we are now able to address the first three questions posed in Section 1. The proportion of calcium subjects exhibiting changes varied among the different types of blood pressures. Calcium seems to influence systolic blood pressure more than diastolic. In the treatment group, marginal posterior means of $(1 - \pi_N)$ varied from almost 60% to a low of 22%. Virtually all of the 28% experiencing changes in seated systolic blood pressure will exhibit decreases, with an average decline of almost 7 mm Hg. At least one-quarter of the 60% (15% of the total population) of those expected to experience a change in supine systolic pressure will have a clinically interesting mean decrease of at least 5 mm Hg as indicated by Figure 3. Lower posterior probabilities of changes were estimated in diastolic pressure, and the mean changes were negligibly different from zero, although still smaller than placebo mean changes. In the control group, median changes hovered around 0, indicating no consistent direction of changes.

There is also some evidence of delayed treatment effects. If there were no delays, the change point for the subjects with effects would occur in week 4. Although there is some evidence that a certain proportion react quickly to calcium, especially for seated systolic pressure, a majority of

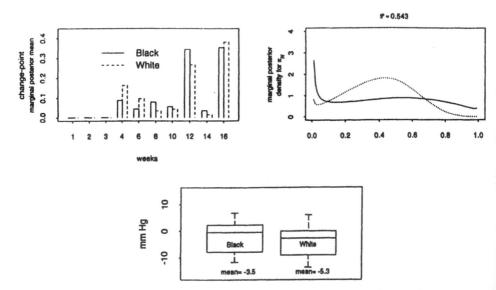

Figure 4 Clockwise from the upper left corner: means of the marginal posterior distributions for the components of π, marginal posterior densities for the probability of no change, and box plots of the posterior mean blood pressure differences (after − before), for the black (solid lines) and white (dashed lines) subgroups of the calcium group for supine systolic blood pressure.

changes occur later, typically 6 to 8 weeks after the start of calcium supplementation (weeks 10 to 12 of the experiment). See Figure 1.

For hypothesis testing, a prior that assigns half the probability to π_N and the rest uniformly distributed seems appropriate. The posterior probabilities for a change in supine systolic blood pressure were then 0.51 versus 0.03 in the calcium and placebo groups, respectively, indicating a high probability of a change in the calcium group compared with the probability of change in the placebo group.

Finally, subgroup analyses were run to search for evidence of response differences between black and white subjects. For systolic pressure, results for both ethnic groups were nearly identical to those already reported for all subjects in Figures 1, 2, and 3, with $\mathscr{P} = 0.54$ and $\mathscr{P} = 0.68$ for supine and seated measurements, respectively. Here and in the paragraph below, the parameter $\mathscr{P}$ represents the probability that black subjects are more likely to change than white subjects. This is analogous to the comparisons between calcium and placebo subjects given previ-

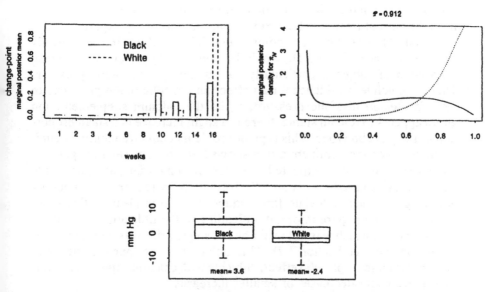

Figure 5 Clockwise from the upper left corner: means of the marginal posterior distributions for the components of π, marginal posterior densities for the probability of no change, and box plots of the posterior mean blood pressure differences (after − before), for the black (solid lines) and white (dashed lines) subgroups of the calcium group for seated diastolic blood pressure.

ously by Eq. (4). Figure 4 presents the results for supine systolic blood pressure.

Large differences were found for diastolic pressure, with $\mathcal{P} = 0.91$ and 0.92 for seated and supine pressures, respectively, with blacks having larger probabilities of change, 0.69 or 0.66, and lower probabilities in whites, 0.18 and 0.16, for supine and seated, respectively. Again, all changes were manifest after some delay, usually 10 or 12 weeks from the start of the experiment. In both cases, mean blood pressure of whites decreases by 2 to 3 mm Hg, while the mean blood pressure of blacks rises (inexplicably) by approximately 4 mm Hg. Figure 5 presents the results for seated diastolic pressure.

5 DISCUSSION

Multipath change-point methods are useful for interpreting repeated measures data. They offer estimates of the proportion of subjects in a given population who would respond to the treatment, as well as the magnitude of and time delay to the change. From the data on calcium supplementation presented here, delays for some subjects of approximately 2 months from the start of treatment (12 weeks from the start of the experiment) are expected. After taking into account these delays, effects in the clinically interesting range of 5 mm Hg would be consistently observed in a small proportion of subjects, particularly for supine systolic blood pressure. There is much less evidence for an effect on diastolic blood pressure. In fact, for seated diastolic pressure, while the calcium supplementation group appears to remain stable, there is a slight increase in a small proportion of the placebo group. This type of problem occurs frequently in clinical trials, when the treatment group shows little response and the placebo group responds in an unexpected direction. In some trials, an appropriate choice of prior based on historical information on responses in patients in both groups may alleviate this conundrum. To implement this here, however, would require information leading to the utilization of subject-specific priors, which was not available. A single prior was therefore used for all subjects (see Section 3.2). Although there does appear to be some evidence for a small placebo effect, subjects with calcium supplementation experience larger decreases or smaller increases.

A classical repeated measures analysis would not provide this type of information. A clinician may benefit from the extra level of detail provided by a multipath change-point analysis. First, knowing that a certain proportion of the population responds in the clinical range may encourage the use of calcium supplementation as a first-line therapy in cases of mild hypertension. Knowing that the effects may be manifest only after 2 months will help determine the length of time before switching to an alternative therapy. Of course, the data analyzed here are for normotensive men only and represent only one of several trials on calcium supplementation and blood pressure. A review of the relevant literature is given in Chockalingam et al. (1990).

The analysis provides moderate evidence of a difference between the races for diastolic pressure and no evidence for systolic. The small sample sizes within subgroups—especially blacks—are reflected in the larger variance in the posterior distributions. Further observation would decrease the variance and would probably change the mean but in an unknown direction.   In the above, it has been assumed that $X_{ij}, j = 1, 2, \ldots, N$ were independent for each fixed $i = 1, 2, \ldots, M$. It may be that observations within a sequence are correlated and should be modeled accordingly (Joseph et al., 1994a). If, however, the measurements are sufficiently separated in time or represent averages or rates over long periods of time, the correlation may be negligible, allowing the change-point distribution to be investigated using a much simpler estimation procedure. If a Bayesian time series approach is required, one can adapt the methods of Marriot et al. (1992) to the change-point situation, where again Gibbs sampling may be employed to estimate the parameters of the model. This idea is being investigated (Joseph et al., 1994b).

Programs may obtained by mailing the one-line email message: "send mpcpn from general", to "statlib@lib.stat.cmu.edu".

ACKNOWLEDGMENTS

The authors wish to thank Shanshan Wang for her computer expertise and Marie-Pierre Aoun for her skills in preparing this chapter. This work was supported in part by the Natural Science and Engineering Council of Canada and the Fonds pour la Formation de chercheurs et l'aide à la recherche, Gouvernement du Québec.

REFERENCES

Carlin, B. P., Gelfand, A. E., and Smith, A. F. M. (1992). Hierarchical Bayesian analysis of changepoint problems. *Applied Statistics* 41: 389–405.
Casella, G. and George, E. I. (1992). Explaining the Gibbs sampler. *American Statistician* 46: 167–174.
Chockalingam, A., Abbott, D., Bass, M., Battista, R., et al. (1990). Recommendations of the Canadian consensus conference on non-pharmacological ap-

proaches to the management of high blood pressure. *Canadian Medical Association Journal* 142: 1397–1409.

DeGroot, M. H. (1970). *Optimal Statistical Decisions*, Chapter 9. McGraw-Hill, New York.

Dempster, A. P., Laird, N. M. and Rubin, D. B. (1977). Maximum likelihood from incomplete data via the EM algorithm. *Journal of the Royal Statistical Society Series B* 39: 1–38.

Efron, B. (1979). Bootstrap methods: Another look at the jackknife. *Annals of Statistics* 7: 1–26.

Efron, B. (1982). The jackknife, the bootstrap, and other resampling plans. *Society for Industrial and Applied Mathematics, Conference Board of the Mathematical Sciences, National Sciences Foundation Monograph* 38.

Gelfand, A. E. and Smith, A. F. M. (1990). Sampling-based approaches to calculating marginal densities. *Journal of the American Statistical Association* 85: 398–409.

Gelfand, A. E., Hills, S. E., Racine-Poon, A., and Smith, A. F. M. (1990). Illustration of Bayesian inference in normal data using Gibbs sampling. *Journal of the American Statistical Association* 85: 972–985.

Gelman, A. and Rubin, D. B. (1992). Inference from iterative simulation using multiple sequences. *Statistical Science* 7: 457–472.

Geman, S. and Geman, D. (1984). Stochastic relaxation, Gibbs distributions and the Bayesian restoration of images. *IEEE Transactions on Pattern Analysis and Machine Intelligence* 6: 721–741.

Geyer, C. J. (1992a). Markov chain Monte Carlo maximum likelihood. *Computer Science and Statistics: Proceedings of the 23rd Symposium on the Interface*, Fairfax Station, VA., pp. 156–163.

Geyer, C. J. (1992b). Practical Markov chain Monte Carlo. *Statistical Science* 7: 473–511.

Grouchow, H. W., Sobocinski, K. A., and Barboriak, J. J. (1985). Alcohol, nutrient intake, and hypertension in US adults. *Journal of the American Medical Association* 253: 1567–1570.

Harlan, W. R., Hull, A. L., Schmouder, R. L., Landis, J. R., *et al.* (1984). Blood pressure and nutrition in adults. *American Journal of Epidemiology* 120: 17–28.

Hinkley, D. V. (1970). Inference about the change-point in a sequence of random variables. *Biometrika* 57: 1–16.

Hinkley, D. V. and Schechtman, E. (1987). Conditional bootstrap methods in the mean shift model. *Biometrika* 74: 85–93.

Joseph, L. and Wolfson, D. B. (1992). Estimation in multi-path change-point problems. *Communications in Statistics, Theory and Methods* 21: 897–913.

Joseph, L. and Wolfson, D. B. (1993). Maximum likelihood estimation in the multi-path change-point problem. *Annals of the Institute of Statistical Mathematics* 45: 511–530.

Joseph, L., Vandal, A., and Wolfson, D. B. (1994a). Repeated measures for correlated data with a change-point: Maximum likelihood estimation. Submitted.

Joseph, L., Wolfson, D. B., and Vandal, A. (1994b). Bayesian approaches to correlated data with a change-point. In preparation.

Laird, N. M. and Louis, T. A. (1987). Empirical Bayes confidence intervals based on bootstrap samples. *Journal of the American Statistical Association* 82: 739–757.

Lee, P. M. (1989). *Bayesian Statistics: An Introduction*, p. 81. Halsted Press, New York.

Lyle, R. M., Melby, C. L., Hyner, G. C., Edmonson, J. W., Miller, J. Z., and Weinberger, M. H. (1987). Blood pressure and metabolic effects of calcium supplementation in normotensive white and black men. *Journal of the American Medical Association* 257: 1772–1776.

Marriot, J., Ravishanker, N., Gelfand, A., and Pai, J. (1992). Bayesian Analysis of ARMA Processes: Complete Sampling Based Inference Under Full Likelihoods. Research Report, Department of Statistics, University of Connecticut, Storrs.

McCarron, D. A., Morris, C. D., and Cole, C. (1984). Dietary calcium in human hypertension. *Science* 217: 267–269.

Müller, P. (1991). A Generic Approach to Posterior Integration and Gibbs Sampling. Technical Report 91-09, Department of Statistics, Purdue University, West Lafayette, IN.

Müller, P. and Rosner, G. L. (1994). A semiparametric Bayesian model with hierarchical mixture priors. Preprint.

Pettitt, A. N. (1979). A nonparametric approach to the change-point problem. *Applied Statistics* 28: 126–135.

Picard, D. (1985). Testing and estimating change-points in time series. *Advances in Applied Probability* 17: 841–867.

Robbins, H. (1964). The empirical Bayes approach to statistical decision problems. *Annals of Mathematical Statistics* 35: 1–20.

Schleiffer, R., Pernot, F., Berthelot, A. and Gairard, A. (1984). Low calcium diet enhances development of hypertension in the spontaneously hypertensive rat. *Clinical and Experimental Hypertension, Part A, Theory and Practice* 6: 783–793.

Sempos, C., Cooper, R., Kovar, M. G., Johnson, C., et al. (1986). Dietary calcium and blood pressure in National Health and Nutrition Examination Surveys I and II. *Hypertension* 8: 1067–1074.

Smith, A. F. M. (1975). A Bayesian approach to inference about a change-point in a sequence of random variables. *Biometrika* 62: 407–416.

Tanner, M. A. (1991). *Tools for Statistical Inference*. Springer-Verlag, New York.

Tanner, M. A. and Wong W. H. (1987). The calculation of posterior densities by data augmentation (with discussion). *Journal of the American Statistical Association* 82: 528–550.

Yao, Y.-C. (1987). Approximating the distribution of the maximum likelihood estimate of the change-point in a sequence of independent random variables. *Annals of Statistics* 15: 1321–1328.

Yao, Y.-C. (1990). On the asymptotic behaviour of a class of nonparametric tests for a change-point problem. *Statistics and Probability Letters* 9: 173–177.

25

Bayesian Predictive Inference for a Binary Random Variable: Survey Estimation of the Quality of Care that Radiation Therapy Patients Receive

Michael Racz   State University of New York at Albany, Albany, New York

J. Sedransk   Case Western Reserve University, Cleveland, Ohio

ABSTRACT

Given binary data from a two-stage cluster sample, we outline and apply a method of Bayesian predictive inference for a finite population proportion, P. Our probabilistic specification should be useful for many surveys of this type and yields simple analytical expressions for the moments of the prior and posterior distributions. To facilitate routine analysis of survey data, we investigate approximations for the posterior distribution of P. We contrast our inferences with those using empirical Bayes techniques and design-based methods. We apply our methodology to a survey designed to investigate the quality of care that cancer patients receive. By appropriate specification of the prior distribution we show that there is agreement between the observed data and proposed model.

1   INTRODUCTION

The Patterns of Care Studies (PCS) investigate the quality of treatment received by cancer patients whose primary treatment modality is radiation

651

therapy. These studies, initiated in 1974, evaluate the level of care being delivered to cancer patients. Development of the study evolved from beliefs that (1) all patients experiencing cancer deserve the same high level of care, (2) more information is needed about radiation services to assess gaps and accurately project the need for facilities, and (3) the structure and process of proper care need to be delineated so that financial resources are channeled in a manner appropriate to providing this care. Kramer and Herring (1976) provide an excellent summary of the early history, objectives, and methods of the PCS.

Three examples of the use of the data from a PCS survey are in Cooper *et al.* (1979), Coia *et al.* (1994), and Maclean *et al.* (1981). Cooper *et al.* report estimated proportions of patients with cervix cancer who are given specific tests and relate these estimates to the clinical importance of the tests. Coia *et al.* used the 1989 PCS to determine the national practice standards of radiation oncologists in evaluating and treating adenocarcinoma of the rectum and sigmoid. They explored the frequency and scheduling of radiation, chemotherapy, and surgical procedures. Maclean *et al.* show how compliance with practice patterns differs among subpopulations of radiation therapy facilities.

Each PCS sample design uses stratified random sampling of U.S. radiation therapy facilities possessing megavoltage equipment with simple random sampling of patients within sampled facilities. For the 1983 survey, considered in this chapter, strata 1 and 2 consist of research facilities whose members presumably have access to the latest advances in research (full members of the radiation therapy oncology research group, comprehensive cancer centers, and other facilities having ongoing resident training programs with more than two residents). The other strata consist of nonresearch facilities and are differentiated by facility size (number of new patients seen per year). It is believed that smaller facilities are more likely to provide a lower quality of patient care.

At each sampled facility, and for each disease site under study, a simple random sample of patients was selected from the population of new patients who were given definitive treatment. In most cases five patients were sampled from each facility for each disease site. Complete enumeration was used for facilities where less than five patients were treated. (The prostate cancer portion of the 1983 data set contains 46 facilities and 240 patients.) The sources of data were the patient charts, port films, and additional information supplied by the attending radiation therapist.

The variables under study here relate to the *processes* involved in radiation therapy practice and include indicators of the quality of both the pretreatment evaluation and the planning and actual delivery of therapy. Here, therapy refers not only to the type of irradiation (e.g., external beam irradiation, intracavitary irradiation) but also to the total dose and fractionation schedule administered. The quality of the pretreatment evaluation is important because the selection of the appropriate therapy is based on this evaluation. Clearly, the goal is to provide therapy that is optimally planned and delivered. Table 1 is a list of the most important variables, all binary, identified as "pretreatment" or "therapy."

To satisfy the principal objective of providing national estimates of quality of care, estimates of a set of finite population proportions are required. In addition, estimates for subpopulations are increasingly important. Currently, for the PCS, routine inference is made using design-based methodology as presented, for example, in Cochran (1977, Chapter 11). This framework has insufficient structure to permit satisfactory inferences for subpopulations (e.g., age × race groups, or type of radiation therapy facility) when sample sizes are small and for the other objectives. In this chapter, a Bayesian alternative is proposed; the methodology is sketched in Section 2. Inferences for a very large ensemble of binary variables are

Table 1 Important Pretreatment and Therapy Variables, Prostate Cancer

Pretreatment[a]	Therapy[a]
KPS	Rx plan recorded
History/physical	CT for Rx planning
IVP	Simulation
Pelvic CT	Port films
Bone scan	Ports/beginning of Rx
Ultrasound	Ports/when fields change
Chemistry screen	Ports/regular schedule
Alkaline phosphatase	Photo/diagram of field
Serum PSA	Primary dose/PCS
Differentiation recorded	Radiographs of implant
Gleason recorded	Isodose/implant
Nodal status recorded	Isodose/external beam
Seminal vesicle involvement recorded	Adequate margins around target
Involved surgical margins recorded	Cystourethrogram/sim
Capsule penetration recorded	Three-point Rx setup for positioning
Extracapsular extension recorded	Cradle for positioning
Stage recorded	Number of fields per day recorded

[a] Abbreviations: KPS, Karnofsky performance status; Rx, treatment; CT, computed tomography; IVP, intravenous pyelogram; PSA, prostate specific antigen; PCS, Patterns of Care Study.

required, and this precludes intensive modeling for any variable. Also, the inferential methodology should be as straightforward as possible. Thus, we wish to develop somewhat automatic methods for drawing inferences, including the use of approximations for the posterior distributions. To do this we consider several variables typical of the ensemble of binary variables to be analyzed. Section 3 describes how we chose the prior specifications and shows the concordance of the observed data with the proposed probabilistic specifications. A second part describes the calculation of the posterior distribution of the finite population proportion, P, and exhibits an excellent approximation. There is also a comparison of hierarchical Bayes, empirical Bayes, and design-based estimates. Concluding remarks are in Section 4.

2 METHODOLOGY

Using an approach similar to the one (for normally distributed random variables) proposed by Malec and Sedransk (1985), Nandram and Sedransk (1993) developed a technique that is appropriate for the PCS binary variables and will provide inferences more sound than those of the methodology currently used. Nandram and Sedransk (1993) assume that n clusters (i.e., facilities) are sampled from the N clusters in the population. Denote by M_k and m_k the known number of units (i.e., patients) and sample size in cluster $k(0 \leq m_k \leq M_k)$. Letting Y_{ki} denote the Bernoulli random variable corresponding to the ith unit in cluster k, it is assumed that, given θ_k, $\{Y_{ki}: i = 1, \ldots, M_k\}$ are independent with

$$\Pr(Y_{ki} = 1 \mid \theta_k) = \theta_k \tag{1}$$

Moreover, it is assumed that the sample design is not informative, in the following sense: the probability of selecting sample s, $p(s)$, does not depend on the Y_{ki}, and if $p(s)$ depends on design variables they are assumed to be known for all units in the population. For weaker conditions and further discussion see Sugden and Smith (1984). Finally, it is assumed that any lack of knowledge about the labels associated with the units can be ignored (see Scott and Smith, 1973).

Letting $M_k' = \sum_{i=1}^{M_k} Y_{ki}$, $\Delta_k = M_k'/M_k$, and $\rho_k = M_k/\sum_{k=1}^{N} M_k$, we wish to make inference about

$$P = \frac{\sum_{k=1}^{N} M_k'}{\sum_{k=1}^{N} M_k} = \sum_{k=1}^{N} \rho_k \Delta_k \tag{2}$$

We may also wish to make inference about weighted sums of the Δ_k corresponding to subpopulations of interest.

For the prior distribution, given β and τ, we take $\theta_1, \ldots, \theta_N$ to be distributed independently with beta density function

$$p(\theta \mid \beta, \tau) = B(\beta, \tau - \beta)^{-1}\theta^{\beta-1}(1 - \theta)^{\tau-\beta-1} \tag{3}$$

where $B(a, b) = \Gamma(a)\Gamma(b)\Gamma(a + b)^{-1}$. For $\beta \in \{a_r: 0 < a_1 < a_2 < \cdots < a_R < \tau\}$

$$\Pr(\beta = a_r) = \omega_r \tag{4}$$

where $\sum_{r=1}^{R} \omega_r = 1$. It is assumed throughout that τ and the ω_r are fixed quantities; methods for assigning values are discussed in Section 3.

The PCS has a stratified two-stage sample design, and we include two stages in the hierarchical specification. Data analyses (Calvin and Sedransk, 1991) have shown that the stratification effect for the 1983 survey is very small, and we do not include this component in our specification.

The key results needed to obtain the posterior distribution and moments of P [see Eq. (2)] are outlined in the Appendix; additional results and discussion are in Nandram and Sedransk (1993).

3 ANALYSIS

3.1 Selection of Prior Distribution and Model Fit

The first step in the analysis is to select values for τ, $\underline{a} = (a_1, \ldots, a_R)'$, and $\underline{\omega} = (\omega_1, \ldots, \omega_R)'$ defined in Eqs. (3) and (4). Given $\tau, \underline{a}, \underline{\omega}, \underline{m}'$, and $\underline{m}$, the second step is to assess the fit of the model given by Eqs. (1), (3), and (4). When we do not have a specific alternative in mind we assess the model fit using the procedure suggested by Box (1980, Section 2), i.e., evaluate

$$\alpha = \Pr\{p(\underline{m}' \mid \underline{m}, \tau) < p(\underline{m}'_{\text{obs}} \mid \underline{m}, \tau) \mid (\underline{m}', \underline{m}, \tau)\} \tag{5}$$

where $\underline{m}'_{\text{obs}}$ is the observed value of $\underline{m}'$, and $p\,(\underline{m}' \mid \underline{m}, \tau)$, the prior predictive distribution of $\underline{m}'$ from Eqs. (1), (3), and (4), is

$$p(\underline{m}' \mid \underline{m}, \tau) = \sum_{r=1}^{R} \omega_r \prod_{k=1}^{n} \left\{ \binom{m_k + \tau - 1}{m_k}^{-1} \right.$$

$$\left. \cdot \binom{m'_k + a_r - 1}{m'_k} \binom{m_k - m'_k + \tau - a_r - 1}{m_k - m'_k} \right\} \tag{6}$$

Table 2 Values of Population and Sample Sizes, Sample Proportions, and Posterior Means for the Sampled Facilities[a]

		IVP		History and physical	
M_k	m_k	$\hat{\Delta}_k$	$E(\Delta_k \mid \underline{m}, \underline{m}', \tau = 4)$	$\hat{\Delta}_k$	$E(\Delta_k \mid \underline{m}, \underline{m}', \tau = 4)$
22	5	0.00	0.17	1.00	0.99
9	5	1.00	0.90	1.00	0.99
32	5	0.60	0.57	1.00	0.99
48	5	0.60	0.56	1.00	0.99
12	5	0.40	0.43	1.00	0.99
12	5	0.00	0.13	1.00	0.99
23	5	0.00	0.18	1.00	0.99
14	5	0.40	0.43	0.80	0.85
13	5	0.20	0.28	1.00	0.99
15	5	0.60	0.57	1.00	0.99
3	2	0.50	0.50	1.00	0.99
16	5	0.60	0.57	1.00	0.99
6	5	0.60	0.59	1.00	1.00
21	5	1.00	0.83	1.00	0.99
18	5	0.00	0.16	1.00	0.99
45	9	0.22	0.29	1.00	0.99
61	5	0.00	0.21	1.00	0.99
7	5	0.40	0.41	1.00	1.00
4	4	0.50	0.50	0.75	0.75
13	5	0.40	0.43	1.00	0.99
51	5	0.80	0.68	0.80	0.87
39	10	0.90	0.82	1.00	0.99
31	5	0.60	0.57	1.00	0.99
32	5	0.60	0.57	1.00	0.99
26	5	0.20	0.31	1.00	0.99
19	5	1.00	0.84	1.00	0.99
3	3	0.67	0.67	1.00	1.00
34	5	1.00	0.81	1.00	0.99
25	5	1.00	0.82	0.80	0.86
4	4	0.75	0.75	1.00	1.00
14	5	0.40	0.43	1.00	0.99
3	3	0.67	0.67	1.00	1.00
16	5	0.80	0.71	1.00	0.99
18	5	0.40	0.43	1.00	0.99
20	5	0.20	0.30	1.00	0.99
36	5	0.80	0.69	1.00	0.99
40	10	0.60	0.58	1.00	0.99
16	5	0.40	0.43	1.00	0.99
19	5	0.20	0.30	1.00	0.99
37	10	0.90	0.82	1.00	0.99

Table 2   *(continued)*

M_k	m_k	IVP		History and physical	
		$\hat{\Delta}_k$	$E(\Delta_k \mid \underline{m}, \underline{m}', \tau = 4)$	$\hat{\Delta}_k$	$E(\Delta_k \mid \underline{m}, \underline{m}', \tau = 4)$
13	5	0.20	0.28	1.00	0.99
16	5	0.20	0.29	1.00	0.99
30	5	0.40	0.44	1.00	0.99
11	5	0.60	0.58	1.00	0.99
7	5	0.40	0.41	1.00	1.00
5	5	0.80	0.80	1.00	1.00

[a] Note: $\hat{\Delta}_k = m_k'/m_k$, the observed proportion of individuals having an IVP or a history and physical examination, and $E(\Delta_k \mid \underline{m}, \underline{m}', \tau = 4)$ uses (for IVP) the prior distribution of β/τ with SD = 0.05.

In Eq. (6),

$$\binom{h}{g} \equiv \Gamma(h + 1)\{\Gamma(g + 1)\Gamma(h - g + 1)\}^{-1}$$

for any nonnegative real numbers, g, h, $g \leq h$. Although frequentist in nature, Box's procedure is akin to a calculation (Box and Tiao, 1973, Section 2.8.3) used to determine whether a specific value of a parameter θ is included in the highest posterior density region for θ. In Eq. (5), $p(\underline{m}'_{obs} \mid \underline{m}, \tau)$ is a constant and α is estimated by repeated sampling of $\underline{m}'$ using Eqs. (1), (3), and (4).

To illustrate our approach we consider two representative binary variables indicating whether or not (1) an intravenous pyelogram (IVP) and (2) a history and physical examination were done for a prostate cancer patient. For both variables we present in Table 2 the values of m_k and $\hat{\Delta}_k = m_k'/m_k$ for the sample of $n = 46$ facilities. Without any prior data for the IVP variable, our initial choices for τ, $\underline{a}$, and $\underline{\omega}$ correspond to a "noninformative" prior distribution. For a sample from facility k, the likelihood of θ_k is proportional to $\theta_k^{m_k'}(1 - \theta_k)^{m_k - m_k'}$ and the prior density for θ_k is proportional to $\theta_k^{\beta - 1}(1 - \theta_k)^{\tau - \beta - 1}$ [see Eqs. (1) and (3)]. We regard this prior distribution as the posterior distribution resulting from an initial uniform prior distribution for θ_k together with a pilot sample of size $\tau - 2$ (with $\beta - 1$ "successes") and take $\tau = 2$. As our initial choice for β/τ we used a uniform distribution from 0.10 to 0.90; that is, $\Pr(\beta = a_r \mid \tau) = R^{-1}$ for $r = 1, \ldots, R$ with $R = 41, \tau = 2, a_1 = 0.20,$

and $a_R = 1.80$. Throughout, we take $a_{i+1} - a_i = c$ to be small. If c is taken to be too large the posterior distribution of the finite population proportion, P, can be multimodal, which would usually be inappropriate. It also seems sensible to choose R sufficiently large that the distribution of β is a good approximation to the continuous uniform distribution over the range $[a_1, a_R]$. Performing Box's test, $\alpha = 0.01$, indicating that this specification is unsatisfactory. Although a more concentrated distribution for $\Pr(\beta = a_r \mid \tau = 2)$ yields somewhat better results (the Bayes factor for the more concentrated distribution vs. the uniform distribution is 2.95), the new distribution is still unsatisfactory. It is easily shown that the prior variance of θ_k, $\text{var}(\theta_k \mid \tau)$, is

$$\text{var}(\theta_k \mid \tau)$$
$$= E(\beta/\tau) \{1 - E(\beta/\tau)\} (\tau + 1)^{-1} + \tau\{\text{var}(\beta/\tau)\}(\tau + 1)^{-1} \quad (7)$$

Clearly, $\tau = 2$ is too small. In our case, $E(\theta_k \mid \tau) = \beta/\tau \doteq 0.5$. Since $\tau = 2$, it follows that $\{\text{var}(\theta_k \mid \tau)\}^{1/2} \geq 0.29$, which is too large for a variable centered at 0.5 and with range $[0, 1]$.

From Table 2 it is seen that 37 of the 46 sampled facilities have $m_k = 5$, and $m_k \leq 10$ for all 46 facilities. This suggests that $\tau = 4$ (corresponding to a prior sample of size 2) is a reasonable choice. With $\Pr(\beta = a_r \mid \tau = 4) = R^{-1}$ with $R = 41$, $a_1 = 0.40$, and $a_R = 3.60$, we found $\alpha = 0.073$ for Box's test, suggesting that we consider alternatives. (The Bayes factor corresponding to this uniform prior vs. the *least* appropriate of the alternatives considered below is 0.33.) One can decrease $\text{var}(\theta_k \mid \tau = 4)$ by decreasing $\text{var}(\beta/\tau)$ in Eq. (7). We assigned a normal distribution to β/τ with mean 0.50 and standard deviation (SD) 0.09, and derived $\Pr(\beta = a_r \mid \tau = 4)$ using equally spaced values, $a_1, \ldots, a_R$, with $a_{i+1} - a_i = 0.08$. We repeated this using SDs of 0.05 and 0.02. With SD $= 0.09$, only the values of $\beta/\tau \in [0.28, 0.72]$ have a prior probability exceeding 0.01, and for SD $= 0.05$ and SD $= 0.02$ the corresponding ranges are $[0.38, 0.62]$ and $[0.44, 0.56]$. Letting $B(i, j)$ denote the Bayes factor for the case in which the prior distribution has SD $= i$ versus the prior with SD $= j$, $B(.05, .09) = 1.63$ and $B(.02, .05) = 1.28$. We prefer the specification having SD $= 0.05$ because the range for SD $= 0.02$, $[0.44, 0.56]$, appears to be too small, and $B(.05, .09)$ indicates a small preference for SD $= 0.05$.

It is expected that a history and physical examination will be done for almost all patients. This variable is included in our study because it is typical of many PCS variables ($\hat{\Delta}_k = 1$ for many sampled facilities; see Table 2), and finding a good approximation for the posterior distribution of P may not be straightforward. We consider three uniform distributions for β, each with $a_{i+1} - a_i = 0.04$, $a_R = 3.96$, and $\tau = 4$: (1) $a_1 = 3.4$, (2) $a_1 = 2.8$, and (3) $a_1 = 2.0$. Letting $B(k, m)$ denote the Bayes factor for the case in which the prior distribution for β uses $a_1 = k$ versus $a_1 = m$, $B(3.4, 2.8) = 2.00$, $B(2.8, 2.0) = 1.66$, and $B(3.4, 2.0) = 3.33$. We chose $a_1 = 3.4$, which is the most plausible value. In the current case, the posterior distribution of β is the same for each of the (uniform) prior distributions for β that we considered.

3.2 Inference for P

To evaluate P in Eq. (2), the values of the M_k must be specified. For the sampled facilities the mean of the M_k is about 20 (Table 2). To simplify our analysis we took $M_k = 20$ for the nonsampled facilities.

For each variable the posterior mean, $E_{HB} = E(P \mid \underline{m}, \underline{m}', \tau = 4)$, and posterior variance, $V_{HB} = \text{var}(P \mid \underline{m}, \underline{m}', \tau = 4)$, are given in Table 3. An alternative is to use empirical Bayes (EB) rather than the hierarchical Bayes (HB) specification in Eqs. (3) and (4); that is, for EB, take $\Pr(\beta = a_r) = 1$ for some r. Taking $\beta/\tau = E_{HB}$ with certainty, the values of the posterior mean, E_{EB}, and posterior variance, V_{EB}, are given in Table 3. Comparing these results with those from the HB analysis, the posterior means are essentially the same, but the posterior variances using HB are about 25 times larger than the posterior variances using EB. For the IVP variable, this difference is of practical importance; for example, $2\sqrt{V_{HB}} = .07$ whereas $2\sqrt{V_{EB}} = .01$. Such disparities between hierarchical and empirical Bayes are consistent with those found in Calvin and Sedransk (1991) and in recent analyses of data from the National Health Interview Survey (Malec and Sedransk 1994).

The design-based estimator of P, $\hat{P} = \hat{Y}_R/M_0$, where $M_0 = \sum M_k$, and associated variance estimator, $v(\hat{P}) = v(\hat{Y}_R)/M_0^2$, are given by Cochran (1977, Section 11.8); the estimates are given in Table 3. As expected, $|\hat{P} - E_{HB}|$ is small. The values of $v(\hat{P})/V_{HB}$ are large (4.7 for IVP, 8.9 for

Table 3 First Two Moments of the Finite Population Proportion, P

Variable	Point estimate			Measure of variability		
	Hierarchical Bayes	Empirical Bayes	Design based	Hierarchical Bayes	Empirical Bayes	Design based
IVP	.5078	.5078	.5170	.001083	.000052	.005139
History and physical	.9705	.9703	.9800	.000173	.000006	.001547

history and physical). This may be indicative of the "gaining of strength" for the Bayesian method, or that the two variances are measuring two different things. Unfortunately, analytical comparisons of the two variances are difficult because of the complexity of the specification in Eqs. (1), (3), and (4). For example, one may obtain a simple expression for $E_{\underline{m}'}\{v(\hat{P})\}$ but not for $E_{\underline{m}'}\{\text{var}(P \mid \underline{m}, \underline{m}', \tau)\}$.

Using hierarchical Bayes, smoothed posterior densities of P for the IVP and history and physical variables, produced using the SPLINE function with PLOT in S-PLUS, are shown in Figures 1 and 2. Each figure also has an approximation for the distribution of P, i.e., a beta distribution with expected value and variance equal to E_{HB} and V_{HB}, respectively. Clearly, the approximations are excellent. If, as in these cases, a fitted beta distribution can be used, simulations are unnecessary because there are analytical formulas for E_{HB} and V_{HB} [see Eqs. (A.7) and (A.8)]. This would greatly simplify the analysis of the multitude of variables in the PCS.

3.3 Inference for Subpopulations

Inference is often desired for a subset of the facilities (e.g., facilities with a resident training program). Examples of this can be found in Simon *et*

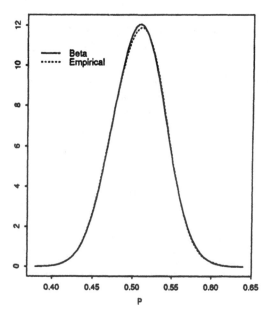

Figure 1   Posterior distributions of P for IVP: empirical and beta approximation.

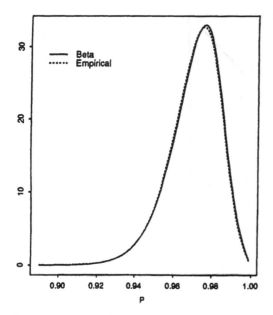

Figure 2 Posterior distributions of P for history and physical: empirical and beta approximation.

al. (Chapter 21), Stangl (Chapter 16), and Christiansen and Morris (Chapter 18). To investigate this, we consider the special case in which inference is desired for the individual Δ_k (i.e., subpopulations of size 1).

Table 2 has the values of the standard estimate of Δ_k, $\hat{\Delta}_k = m'_k/m_k$, together with $E(\Delta_k \mid \underline{m}, \underline{m}', \tau = 4)$, and Figure 3 is a plot for IVP of $\hat{\Delta}_k$ against $E(\Delta_k \mid \underline{m}, \underline{m}', \tau = 4)$. As $|\hat{\Delta}_k - E_{HB}|$ increases, the deviations of the points $[\hat{\Delta}_k, E(\Delta_k \mid \underline{m}, \underline{m}', \tau = 4)]$ from the 45° line tend to increase, as one would hope. In particular, for this variable $\hat{\Delta}_k = 1$ may be an unrealistic estimate.

For these two variables, inference about the individual Δ_k is essentially the same for hierarchical and empirical Bayes. The HB expression for $\text{var}(\Delta_k \mid \underline{m}, \underline{m}', \tau)$ is given by Eq. (A.6), while the EB expression has only the right term in Eq. (A.6) (i.e., $\hat{\eta}_2 = 0$). For IVP, $\hat{\eta}_2 = 0.0011$, $(1 - \lambda_k)^2 \doteq 1/9$, and the right term has a mean value of about 0.015. Thus, the left term, $(1 - \lambda_k)^2 \hat{\eta}_2$, is much smaller than the right term, and replacing $(1 - \lambda_k)^2 \hat{\eta}_2$ with 0 (as for EB) has only a small effect.

As noted in Section 3.2, $V_{HB}/V_{EB} \doteq 21$ for this example. The reason is that for HB the left term in Eq. (A.8) is 0.001031 and the right term is 0.000052. The left term in Eq. (A.8) is the dominant one, and replacing

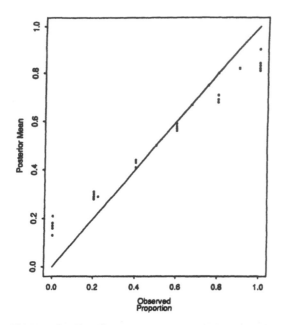

Figure 3 Comparison of observed proportions, $\hat{\Delta}_k$, with posterior means, $E(\Delta_k \mid \underline{m}, \underline{m}', \tau)$, for the 46 sampled facilities: IVP.

it by 0 (as is done for EB) grossly understates the variance. These results, replicated for the history and physical variable, can be explained by examining $\text{var}(P \mid \underline{m}, \underline{m}', \tau)$ in Eq. (A.8). For this example, $\rho_k \doteq 1/N$. Also, $\lambda_k = 0$ and $\hat{v}_k^2 = \hat{\eta}_3$ for almost all of the facilities (because they are not sampled). Then

$$\text{var}(P \mid \underline{m}, \underline{m}', \tau) \doteq \hat{\eta}_2 + \frac{\tau^{-1} + \overline{M}^{-1}}{N} \hat{\eta}_3 \qquad (8)$$

where $\overline{M} = \sum_{k=1}^{N} M_k/N$.

When N is large the right term in Eq. (8) will typically be small, the left term will dominate, and EB will understate the HB posterior variance. If $N = 1$, $P \equiv \Delta_k$ and the right term may dominate the left term (as in the examples considered here). Then the posterior variance of Δ_k will be similar for EB and HB.

Typically, inference is desired for the entire population of facilities or for a subset of the facilities. In either case, it is likely that the left term of $\text{var}(P \mid \underline{m}, \underline{m}', \tau)$ in Eq. (A.8) will be much larger than the right term.

4 COMMENTS

The probabilistic specification in Eqs. (1), (3), and (4) provides the basis
for Bayesian predictive inference for a finite population proportion when
there are binary data from a two-stage cluster sample, a common sample
design. Formulas (3) and (4) provide a flexible prior distribution and permit
a fully Bayes analysis. As shown in this chapter, posterior variances of
finite population parameters may be grossly understated when empirical
Bayes is used.

   We have applied this methodology to a survey whose objective is to
investigate the quality of care that cancer patients receive. By appropriate
specification of τ and the distribution of β, there is concordance between
the observed data and the proposed model.

   Because sample surveys typically include too many variables to permit individualized study, a routine analysis is usually desired. Here, this
implies that approximate posterior distributions are needed. We have
shown that good approximations are available. This should make the proposed methodology attractive for use in settings where it is infeasible to
implement sampling-based methods.

APPENDIX

Let m_k' denote the number of sampled units in cluster k having the desired
characteristic, $\underline{m}' = (m_1', \ldots, m_n')'$ and $\underline{m} = (m_1, \ldots, m_n)'$, the vector
of sample sizes. Also, s denotes the set of clusters in the sample. Using
Eqs. (1), (3), and (4), the posterior distribution of $\underline{\theta} = (\theta_1, \ldots, \theta_N)'$
given $\underline{m}$, $\underline{m}'$, and τ is

$$p(\underline{\theta} \mid \underline{m}, \underline{m}', \tau) = \sum_{r=1}^{R} \omega_r^* \prod_{k=1}^{N} B\{a_r + I_k m_k', \tau - a_r + I_k(m_k - m_k')\}^{-1}$$
$$\cdot \theta_k^{a_r + I_k m_k - 1}(1 - \theta_k)^{\tau - a_r + I_k(m_k - m_k') - 1} \quad \text{(A.1)}$$

where

$$\omega_r^* = \text{Pr}(\beta = a_r \mid \underline{m}, \underline{m}', \tau) = \omega_r \prod_{k \in s} \binom{\tau - 2}{a_r - 1}$$
$$\cdot \binom{m_k + \tau - 2}{m_k' + a_r - 1}^{-1} \left\{ \sum_{r=1}^{R} \omega_r \prod_{k \in s} \binom{\tau - 2}{a_r - 1}\right.$$
$$\left.\cdot \binom{m_k + \tau - 2}{m_k' + a_r - 1}^{-1}\right\}^{-1} \quad \text{(A.2)}$$

$r = 1, 2, \ldots, R$, and $I_k = 1$ if cluster k is in the sample and $I_k = 0$ otherwise.

Given $\underline{m}$, $\underline{m}'$, and θ, $M_k' - I_k m_k'$ $(k = 1, \ldots, N)$ are independent binomial random variables [see Eq. (1)]. Using Eq. (A.1),

$$p(\{M_k' - I_k m_k': k = 1, \ldots, N\} \mid \underline{m}, \underline{m}', \tau)$$

$$= \sum_{r=1}^{R} \omega_r^* \prod_{k=1}^{N} \binom{M_k + \tau - 1}{M_k - I_k m_k}^{-1} \binom{M_k' + a_r - 1}{M_k' - I_k m_k'}$$

$$\cdot \binom{M_k - M_k' + \tau - a_r - 1}{M_k - M_k' - I_k(m_k - m_k')} \quad \text{(A.3)}$$

Using Eq. (A.2),

$$E(\beta/\tau \mid \underline{m}, \underline{m}', \tau) = \hat{\eta}_1 = \sum_{r=1}^{R} \omega_r^* a_r \tau^{-1}$$

$$\text{var}(\beta/\tau \mid \underline{m}, \underline{m}', \tau) = \hat{\eta}_2 = \sum_{r=1}^{R} \omega_r^* (a_r \tau^{-1} - \hat{\eta}_1)^2$$

and

$$E\left\{\frac{\beta(\tau - \beta)}{\tau(\tau + 1)} \mid \underline{m}, \underline{m}', \tau\right\} = \hat{\eta}_3 = (\tau + 1)^{-1} \sum_{r=1}^{R} \omega_r^* a_r (1 - a_r \tau^{-1}) \quad \text{(A.4)}$$

Defining

$$\lambda_k = \begin{cases} \{1 + (\tau/M_k)\}\{1 + (\tau/m_k)\}^{-1}, & k \in s \\ 0, & k \notin s \end{cases}$$

it can be shown that

$$E(\Delta_k \mid \underline{m}, \underline{m}', \tau) = \lambda_k \frac{m_k'}{m_k} + (1 - \lambda_k)\hat{\eta}_1 \quad \text{(A.5)}$$

and

$$\text{cov}(\Delta_k, \Delta_l \mid \underline{m}, \underline{m}', \tau)$$

$$= \begin{cases} (1 - \lambda_k)^2 \hat{\eta}_2 + (1 - \lambda_k)(\tau^{-1} + M_k^{-1})\hat{v}_k^2, & k = l \\ (1 - \lambda_k)(1 - \lambda_l)\hat{\eta}_2, & k \neq l \end{cases} \quad \text{(A.6)}$$

where

$$
\hat{v}_k^2 = \begin{cases} \displaystyle\sum_{r=1}^{R} \omega_r^*(a_r + m_k') \dfrac{\{\tau + m_k - (a_r + m_k')\}}{(\tau + m_k)(\tau + m_k + 1)}, & k \in s \\[2ex] \hat{\eta}_3, & k \notin s \end{cases}
$$

In Eq. (A.5), $E(\Delta_k \mid \underline{m}, \underline{m}', \tau)$ has a sensible form: it is a weighted average of the sample proportion from cluster k and $\hat{\eta}_1$, the posterior expected value of $\eta_1 = E(\Delta_k \mid \tau)$. Since $\hat{\eta}_1$ is a function of $\underline{m}'$, there is a gaining of strength. However, if $m_k = M_k$, $\lambda_k = 1$ and no pooling is required since all information is obtained about Δ_k. Conversely, if $m_k = 0$, $\lambda_k = 0$ and $E(\Delta_k \mid \underline{m}, \underline{m}', \tau) = \hat{\eta}_1$ with cluster k contributing nothing to $\hat{\eta}_1$.

In Eq. (A.5), since λ_k is a monotone decreasing function of τ, there is more pooling as τ increases. This is to be expected because large τ reflects a belief that the proportions of individuals possessing the characteristic of interest are similar over the clusters.

Defining $\phi = 1 - \sum_{k \in s} \rho_k \lambda_k$, the posterior mean of P is

$$
E(P \mid \underline{m}, \underline{m}', \tau) = \phi \hat{\eta}_1 + (1 - \phi)\hat{P} \tag{A.7}
$$

where

$$
\hat{P} = \left(\sum_{k \in s} \rho_k \lambda_k\right)^{-1} \sum_{k \in s} \rho_k \lambda_k \left(\frac{m_k'}{m_k}\right)
$$

Thus the posterior mean of P is a weighted average of $\hat{\eta}_1$, the posterior expected value of the prior mean of P, and $\hat{P}$, a weighted average of the sample proportions within each cluster.

Finally, the posterior variance of P can be shown to be

$$
\text{var}(P \mid \underline{m}, \underline{m}', \tau)
$$
$$
= \hat{\eta}_2 \sum_{k=1}^{N} \sum_{l=1}^{N} \rho_k \rho_l (1 - \lambda_l)(1 - \lambda_k) + \sum_{k=1}^{N} \rho_k^2 (1 - \lambda_k)(\tau^{-1} + M_k^{-1})\hat{v}_k^2 \tag{A.8}
$$

Note that Eq. (A.8) is a correction of formula (2.12) in Nandram and Sedransk (1993).

A sampling-based method is needed to obtain the posterior distribution of P. It is also needed if one wishes to express uncertainty about τ or to make inferences for a subpopulation. For this application, a straightforward simulation is easy to implement.

Using i to denote the ith iteration:

1. Draw β^i from

$$\Pr(\beta = a_r \mid \underline{m}, \underline{m}', \tau) \text{ as given by Eq. (A.2)}.$$

2. Make independent selections for the N members of $\underline{\theta} = (\theta_1, \ldots, \theta_N)'$, and draw $\underline{\theta}^i$ using

$$p(\theta_k \mid \beta^i, \tau), \qquad\qquad k \notin s$$

$$p(\theta_k \mid \beta^i + m'_k, \tau + m_k), \qquad k \in s$$

where $p(\theta \mid a, b)$ is the beta distribution defined in Eq. (3). Thus, θ^i is an observation from Eq. (A.1). Finally,

3. Select independently

$$(M'_k)^i \sim \mathrm{bin}(M_k, \theta^i_k), \qquad\qquad k \notin s$$

$$(M'_k)^i - m'_k \sim \mathrm{bin}(M_k - m_k, \theta^i_k), \qquad k \in s$$

and evaluate

$$P^i = \frac{\displaystyle\sum_{k=1}^{N} (M'_k)^i}{\displaystyle\sum_{k=1}^{N} M_k} \tag{A.9}$$

To simplify the calculations, normal approximations to the binomial distributions in Eq. (A.9) were used in this application.

ACKNOWLEDGMENT

The authors are grateful to Dr. Jean Owen and Alex Hanlon of the American College of Radiology for providing the data.

REFERENCES

Box, G. E. P. (1980). Sampling and Bayes' inference in scientific modelling and robustness. *Journal of the Royal Statistical Society*, A 143: 383–430.

Box, G. E. P. and Tiao, G. (1973). *Bayesian Inference in Statistical Analysis*. Addison-Wesley, Reading, MA.

Calvin, J. and Sedransk, J. (1991). Bayesian and frequentist predictive inference for the Patterns of Care Studies. *Journal of the American Statistical Association* 86: 36–48.

Cochran, W. G. (1977). *Sampling Techniques*, 3rd ed. Wiley, New York.

Coia, L., Wizenberg, M., Hanlon, A., Gunderson, L., Haller, D., Hoffman, J., Kline, R., Mohiuddin, M., Russell, A., Tepper, J., Owen, J., and Hanks, G. (1994). Evaluation and treatment of patients receiving radiation for cancer of the rectum or sigmoid colon in the United States: Results of the 1988–1989 Patterns of Care Study process survey. *Journal of Clinical Oncology* 12: 954–959.

Cooper, J., Davis, L., Diamond, J., Sedransk, J., and Curley, R. (1979). Evaluation of carcinoma of the uterine cervix before radiotherapy. *Journal of the American Medical Association* 242: 1996–1997.

Kramer, S. and Herring, D. (1976). The Patterns of Care Study: A nationwide evaluation of the practice of radiation therapy in cancer management. *International Journal of Radiation Oncology, Biology and Physics* 1: 1231–1236.

MacLean, C., Davis, L., Herring, D., Powers, W., and Kramer, S. (1981). Variation in work-up and treatment procedures among types of radiation therapy facilities: The Patterns of Care process survey for three head and neck sites. *Cancer* 48: 1346–1352.

Malec, D. and Sedransk, J. (1985). Bayesian methodology for predictive inference for finite population parameters in multistage cluster sampling. *Journal of the American Statistical Association* 80: 897–902.

Malec, D. and Sedransk, J. (1994). Small Area Inference for Binary Variables in the National Health Interview Survey. Technical Report Department of Statistics, Case Western Reserve University, Cleveland, Onio.

Nandram, B. and Sedransk, J. (1993). Bayesian predictive inference for a finite population proportion: Two-stage cluster sampling. *Journal of the Royal Statistical Society, B* 55: 399–408.

Scott, A. and Smith, T. M. F. (1973). Survey design, symmetry and posterior distributions. *Journal of the Royal Statistical Society, B* 35: 57–60.

Sugden, R. A. and Smith, T. M. F. (1984). Ignorable and informative designs. *Biometrika* 71: 495–506.

Index

ISBN 0-8247-9334-X

9 780824 793340 >